Plasmas and Controlled Fusion

Plasmas and Controlled Fusion

DAVID J. ROSE

MELVILLE CLARK, JR.

The M.I.T. Press
Massachusetts Institute of Technology
Cambridge, Massachusetts

Second printing, March 1965
Third printing, November 1973

ISBN 0 262 18006 5
Library of Congress catalog card number: 61-8814
Printed in the United States of America

Preface

We have attempted primarily to write a textbook suitable for graduate study in either the first or intermediate years. To this end, we start with first principles, so that the book is useful to those engineers and scientists who have had no previous experience in the field. But it is more than an elementary tract: we introduce the reader to many of the advanced concepts that are still subjects for lively discussion in the journals.

We also expect that the book will prove useful to persons not engaged in a formal course but who have become interested in the plasma field in general and controlled thermonuclear processes in particular. To those already heavily engaged in plasma research, we recommend the book as a reference on the basic atomic and nuclear processes, the physical principles, energy recovery concepts, and confinement schemes.

The first twelve chapters provide a course in plasma physics, hydromagnetics, and elementary gaseous electronics, in association with transport and electromagnetic theories. The last four chapters summarize recent approaches to controlled fusion and energy recovery from hypothetical systems. Here, we illustrate principles and modern practice, rather than dwell upon a welter of ephemeral experiments. Except for the recording of cross-section data and a few other operations, we use rationalized mks units throughout and wish all other systems a speedy trip to oblivion; a service is provided in Appendix B for translating from other systems of units and expediting this trip.

A moderate knowledge of atomic physics, differential equations, electricity and magnetism, and thermodynamics is required of the prospective reader. He should have completed a substantial undergraduate curriculum in physics, electrical engineering, mechanical engineering, or some similar trade. We have taught first and second courses in plasmas and controlled fusion to such mixed assemblies in our department.

No claim is made for completeness in our coverage. The choice of material is limited both by our disinclination to double the size of the book and by our intention to keep the analytic complexity within bounds (whether we succeeded in this intent may be disputable). For example, important subjects that we do *not* cover at all are the Fokker-Planck transport equation, modern kinetic plasma formulations and the related particle correlation theories, and plasma confinement by high-frequency electromagnetic fields (radiation pressure). The subject of plasma diagnostics is not included by title but is interrelated with other material, chiefly in the last three chapters. The advanced reader will no doubt discover further omissions. We expect to prepare the reader so that he will be able to pursue these and other topics in the plasma field for himself. A majority of the references cited pertain to work in the United States. This selection is largely one of convenience.

In general, the pace of the book quickens through successive chapters, at least through Chap. 12. This accelerando is meant to match the increasing maturity of a graduate student through one or two terms of graduate study. The problems appended to each chapter serve three useful purposes: providing examples for the text material, extending or supplementing the principles set forth, or anticipating developments to follow later.

Parallel references are useful for general reading. We list here a number of them pertaining to the book as a whole:

Proc. 2nd U.N. Conference on Peaceful Uses of Atomic Energy, Vols. 31 and 32, United Nations, Geneva (1958) (from which many of our specific references are taken).

Nuovo Cimento (Suppl.) **13**, No. 1 (about 300 pages of review articles in English).

W. P. Allis (Ed.), *Nuclear Fusion*, D. Van Nostrand Co., Princeton, N.J. (1960) (selected U.N. Conference papers plus discussion).

L. A. Artsimovich, *Plasma Physics and the Problem of Controlled Thermonuclear Power*, 5 vols., Pergamon Press, London.

J. L. Delcroix, *Introduction to the Theory of Ionized Gases* (translated by M. Clark, Jr., D. J. BenDaniel, and J. M. BenDaniel), Interscience Publishers, Inc., New York (1960).

S. Glasstone and R. H. Lovberg, *Controlled Thermonuclear Reactions*, D. Van Nostrand Co., Princeton, N.J. (1960).

R. K. M. Landshoff (Ed.), *The Plasma in a Magnetic Field*, Stanford University Press, Palo Alto, Calif. (1958).

J. G. Linhart, *Plasma Physics*, North-Holland Publishing Co., and Interscience Publishers, Inc., New York (1960).

C. L. Longmire, J. L. Tuck, and W. B. Thompson (Eds.), *Progress in Nuclear Energy, Plasma Physics, and Thermonuclear Research*, Pergamon Press, London (1959) (selected U.N. Conference papers plus others).

A. Simon, *An Introduction to Thermonuclear Research*, Pergamon Press, New York (1959).

L. Spitzer, Jr., *Physics of Fully Ionized Gases*, Interscience Publishers, Inc., New York (1956).

At a much more elementary level is A. S. Bishop, *Project Sherwood, The U.S. Program in Controlled Fusion*, Addison-Wesley Publishing Co., Inc., Reading, Mass. (1958).

If the book is to be used for a one-semester course, we recommend that some of the more difficult sections concerning stress tensors, cyclotron radiation, stability, and energy balance be omitted or highly attenuated. These sections are starred in the table of contents. Further sections may be abstracted. For a two-semester course, we supplement the book by treating the following subjects: Fokker-Planck and plasma kinetic formulations, electromagnetic propagation theory in magnetized plasmas, plasma oscillations, magnetic-field analysis, and other special topics that vary from year to year.

We are moved to make some comments about the plasma field in general and thermonuclear research in particular. Although the principles stem chiefly from Newton's laws and Maxwell's equations, plasmas are yet poorly understood, because of the long-range interactions between particles that make the problem a many-body one. Nevertheless, considerable technological and social pressure is put upon this imperfect science to solve problems related to outer space, hydromagnetic energy conversion, controlled thermonuclear power, and the like.

As a result of these pressures and the present discontentful state of the art, we detect a certain alchemical flavor in thermonuclear plasma research. In spite of many fine words,† this quintessence of materials is not yet safely in our hands. We have detected here and there carefully nurtured airs of mystery‡ and proprietorship.§ In truth, we are hopeful about the eventual outcome of these plasma aspirations; some of them—controlled fusion in particular—have great potentiality for good. Were we not so encouraged, we would not have written our text. On the other hand, let us proceed carefully—hope is a necessary but not sufficient condition for its own fulfillment.

†". . . This night I'll change
All, that is metal, in my house, to gold."
—BEN JONSON, *The Alchemist*.

‡". . . Can you sublime, and dulcifie? calcine?
Know you the sapor pontick? sapor stiptick?
Or, what is homogene, or heterogene?"
"I understand no heathen language, truely.
. . . .
What a brave language here is? next to canting?
—*Ibid.*

§"All arts have still had, sir, their adversaries,
But ours the most ignorant."
—*Ibid.*

> Mental things which have not passed through the understanding are vain and give birth to no truth other than what is harmful. And because such discourses spring from poverty of intellect those who make them are always poor, and if they have been born rich they shall die poor in their old age . . . . And those who wish to grow rich in a day shall live a long time in great poverty, as happens and will to all eternity happen to the alchemists, the would-be creators of gold and silver . . .
>
> LEONARDO DA VINCI, *Quaderni d'Anatomia* I 13 V.

In the preparation of this text we have had much help. Discussions with our students, particularly with L. de Sobrino and S. Yoshikawa, have been most stimulating. Our colleagues, Dr. S. J. Buchsbaum and Professors W. P. Allis, M. Benedict, and S. C. Brown, have made valuable suggestions regarding the organization and content of the text and have encouraged us in many ways at our task. We are grateful to The M.I.T. Press—particularly to Miss C. D. Boyd—for their patience and forbearance toward us in the editorial stages. Our thanks go also to Miss A. Carbone, Mrs. R. Kugelman, and Miss N. Milner, who typed the manuscript and related material. Finally, we acknowledge with pleasure the support given to M.I.T.—especially to the Research Laboratory of Electronics—by many U.S. Government Agencies for research in the plasma field; for without the stimulation of that research, we would never have been motivated to write the book.

DAVID J. ROSE
MELVILLE CLARK, JR.

Cambridge, Massachusetts
February, 1961

Contents

★ Starred sections may be omitted or abbreviated for a short course.

1

Energy needs and resources

Man's search for energy to meet his needs has proceeded since the dawn of civilization. At least some of our international friction is directly traceable to disputes over possession of the areas favored with abundant energy resources. Classical conquest and exploitation rarely are undertaken for barren areas alone. The acquisition of land endowed with energy and material resources motivates much of our present strife. It is to be hoped that the nuclear processes of fission and fusion may permanently solve at least the need for energy and thus contribute substantially to the peaceful relations among mankind.

The demands for energy are now increasing so rapidly that the present reserves of nonnuclear fuels of all types, particularly coal and oil, seem very small indeed when measured against the use anticipated over even the next few decades. It would therefore be prudent to develop the use of nuclear fuels, which represent stores of energy vastly greater than those of all nonnuclear fuels.

In this first chapter, we survey the needs and the energy reserves of every kind. It appears that fissionable materials might satisfy all energy requirements that might be validly presented for thousands of centuries. Unfortunately, there are certain disadvantages in their use. Even in the absence of these drawbacks, the importance of the problem justifies a multiple attack. The process of controlled fusion appears attractive in principle, but its achievement involves the solution of many difficult problems. Creating stellar temperatures is one of these problems, and controlling material at these temperatures is another. It is our purpose in this book to present some of the current attacks on these problems and the principles underlying these attacks. In this chapter we survey briefly the problems before us.

1.1. Energy Needs

A brief review of the projected energy requirements of the world and of the expected reserves will document our claim for the great interest in thermonuclear energy. Neither the energy needs nor the reserves are known with any accuracy. The unit used in discussing such large amounts of energy is the Q = 10^{18} Btu (about 1.05×10^{21} joules). We may obtain some feeling for its size by noting that 1 Q will raise 2400 km^3 of water from 0° C to 100° C. This volume is larger than that of Lake Ontario and smaller than that of Lake Michigan.

Data on the energy consumption in the United States and the world are shown in Table 1.1. The consumption of fuels in the United States

Table 1.1. Energy consumption of the United States and the world

	United States	World
A.D. 1–1850	—	0.3–0.5 Q/century
A.D. 1850–1950	—	4 Q/century
A.D. 1950–2050 (projected)	—	80–500 Q/century
1860	0.004 Q/yr	0.017 Q/yr
1946	0.032 Q/yr	0.093 Q/yr
1975 (projected)	0.067 Q/yr	0.34 Q/yr
Annual growth 1897–1950	3%	3%
Annual growth 1900–1950 of electric energy only	7%	—
Projected annual growth (total energy)	3%	5%

is shown in Fig. 1.1, and an estimate for the whole world is shown in Fig. 1.2. The rapid increase in the use of gas and oil is particularly to be noted. The projected growth of energy needs is based on both the projected population growth and the increase of energy input per capita. According to Putnam,[1] the input per capita has been relatively constant up to now, although the output has been markedly increasing because of increasing efficiency; this situation cannot persist much longer. As a result of these various factors, it appears that the energy use rate is increasing at 3 to 5% per year throughout the world, so that with present consumption of about 0.1 Q/yr, very large numbers are predicted for the next few centuries.

There is an upper limit to the permissible rate. Solar radiation is absorbed (and reradiated) by the earth at about 4 Q/day. If climate is not to be altered drastically, the additional energy input must be limited to about 1% of this value, perhaps 15 Q/yr. At an average increase of 4% per year, this rate will be achieved in about 130 years. To be sure, we might circumvent this limit to some extent by using solar radiation. Even at this rate, a considerable fraction of the earth's surface would have to be covered with energy conversion devices—thermoelectric elements, for instance. Such a solution is not appealing.

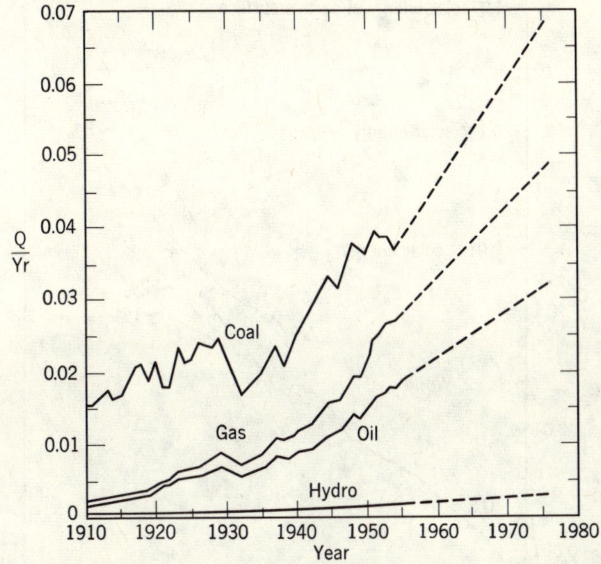

Fig. 1.1. Energy consumption in the United States. After L. E. Link and W. H. Zinn, "Economics of Nuclear Power," in H. Etherington, *Nuclear Engineering Handbook*, McGraw-Hill Book Co., New York (1958), p. **12**–86.

1.2. Reserves of Nonnuclear Fuels

The estimated reserves of nonnuclear fuels are listed in Table 1.2. Coal is the principal fuel in use throughout the world and is in greatest supply by far. The very modest reserves of liquid and gaseous fuels should be particularly noted, in view of the large expected energy demands. Their convenience has led to rapid exploitation; the cost of new oil-field development is now rising rapidly, and estimated reserves should be well depleted in about 50 years if current trends continue. With the 4% projected rate increase, all reserves (about 150 Q) listed in Table 1.2 would be exhausted in about 100 years.

It is sometimes stated that this picture is unduly pessimistic, for great reserves remain to be discovered. So they do, but the estimates given above have taken such factors into account and do not represent proved reserves. There is a historical example of the virtual exhaustion of a principal energy resource: during and shortly following the time of the Greek and Roman Empires, much of Mediterranean Europe was largely deforested, and so it remains to the present day.

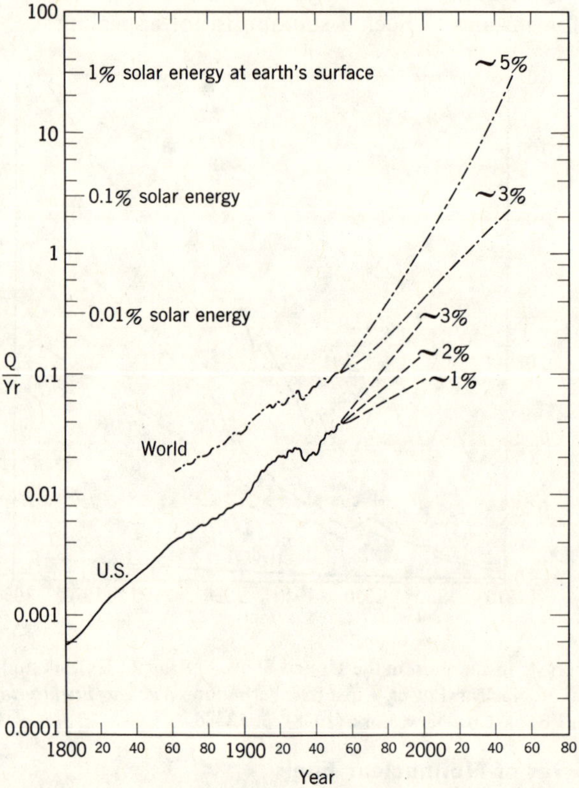

Fig. 1.2. Input of energy per year as a function of time. After P. C. Putnam, *Energy in the Future*, D. Van Nostrand Co., New York (1953).

Equally significant in these considerations is the fact that coal, oil, and the like are very useful raw materials, and their use as simple energy sources is extremely wasteful.

The use of other nonnuclear energy sources does not at present look hopeful. Hydroelectric power has already largely been exploited in the United States; it will be important in specific localities, but the total

Table 1.2. World reserves of nonnuclear fuels

Heat value (Q) of fuel listed

After L. E. Link and W. H. Zinn, Reference 2, page 12–88

Area	Coal	Petroleum	Oil Shale	Tar Sands	Natural Gas
North America					
United States	50.3 (6)	0.6	2.1	0.6	0.4
Canada	2.5 (2)	0.2	—	2.9	—
Remainder (incl. Alaska)	2.7	—	—	—	—
Asia					
U.S.S.R. (both Europe and Asia)	33.1 (10)	0.9	—	—	0.04
China	27.9 (6)	—	—	—	—
Middle East (incl. Egypt)	—	0.9	—	0.3	—
Remainder	2.4	0.3	—	—	—
Europe (excl. U.S.S.R.)		0.1	0.02	—	—
Germany	9.3	—	—	—	—
United Kingdom	4.7 (1)	—	—	—	—
Remainder	4.0 (7)	—	—	—	—
Africa (excl. Egypt)		0.05	—		—
Union of South Africa	5.6	—	—	—	—
Remainder	0.05	—	—	—	—
Australia	1.5	0.05	—	—	—
South America	0.1				—
Brazil	—	—	1.7	—	—
Venezuela	—	0.5	—	0.8	—
World Total: 158 Q	144.2 (32)*	3.6 (5.6)†	3.8 (2)	4.6 (1.7)	0.44 —

* Figures in parentheses are Putnam's.[1] His reserves are based on costs no higher than twice present costs.

† Includes gas.

amount available is small compared with the amounts considered here. Wind and tides have similarly been proposed to take advantage of peculiar local situations. Temperature gradients in the ocean and the earth have been considered. The heat of the earth's interior is great, but it is not readily accessible, except in a few places. The drilling of deep holes does not appear economical.

Solar energy was discussed briefly in Sec. 1.1. Its principal disadvantage is its dilution; the incident noon flux is about 1.3 kw/m^2 at the surface. Photosynthesis by algae is promising as a food source, but not as a low-cost fuel.

1.3. Reserves of Fissionable and Fertile Fuels

There are two diametrically opposed nuclear processes that yield energy: fission and fusion. In fission a heavy nucleus is split by a neutron into two (or more) parts of almost always different mass. Energy is released in the process along with some neutrons. These neutrons may in turn be used to cause other heavy nuclei to fission. In fusion two very light nuclei combine if they have sufficient energy and create various light reaction products that have much more kinetic energy than the reactants. The energy in both cases is of nuclear origin. In this book we do not inquire into the origin of this energy or the details by which it is created, except to state that mass is transformed in both processes into energy. Nuclides of intermediate mass suffer neither fission nor fusion.

Among the nuclides that occur in nature, only the U^{235} nuclide is fissionable by slow neutrons and suitable for use in most reactors as a fuel. (Reactors that can produce energy from fast neutrons and that can use other fuels tend to be quite small and therefore unsuited to the generation and removal of large quantities of heat. They are still very much in the developmental stage.) However, Th^{232} and U^{238}, both of which occur naturally, can be made into U^{233} and Pu^{239}, respectively, by the absorption of a neutron, including the neutrons emitted in fission. The U^{233} and Pu^{239} nuclides are suitable as a fuel for a reactor. For this reason, Th^{232} and U^{238}, which are by far the most common naturally occurring isotopes of thorium and uranium, are called fertile fuels. The nuclide U^{235} occurs only to the extent of 0.71% in natural uranium. Uranium is estimated to occur at a concentration of 4 parts per million and thorium at 12 parts per million in the earth's crust. Their ores are widely distributed. Fewer specific sources of thorium are known because there has been less economic incentive to find it.

Recent estimates indicate that the Western nations have a reserve of 1 to 2 million tons of uranium available at rather low cost (about \$10/lb of U_3O_8). There are several more million tons available at higher cost ($10–$30/lb), and perhaps 25 million tons at a cost of approximately $50/lb. Such amounts must be extrapolated to cover the rest of the earth and to include thorium also. It is therefore not unreasonable that about 10^8 tons of fissionable or fertile material are available. In addition there is a virtually inexhaustible reserve at very low concentration in granitic rocks. Weinberg[3] discusses the problems of extraction of fertile materials from granitic rocks and concludes that it is probably feasible. Since we are concerned with thermonuclear power, we avoid speculation about such extraction problems and fix on the number 10^8 tons of reasonably available fissionable or fertile material. From this figure we find about 5000 Q available if it all were bred, reprocessed, and used completely. If only a few percent of the U^{235} is fissioned, as in some nonbreeding reactors, the reserve is not very great. But if it all would be completely fissioned, the reserves would last for several hundred years at the greatest permissible rate of energy release.

Beside the fact that several hundred years is a short historical time, the problem of radioactive waste disposal is severe with fission processes. At the rates considered, approximately 10^{13} curies/yr of long-lived fission products would be produced.

1.4. Reserves of Fusionable Fuel

The energy resources available for the fusion process, on the other hand, are enormous and available at low cost. Deuterium (D), the most useful of the naturally occurring nuclides available for fusion, exists as 0.0153 atm% of the hydrogen in sea water. This amount of deuterium, even at the rate of 15 Q/yr, could supply enough energy to last 10^9 years.

At present D_2O is sold by the U.S. Atomic Energy Commission for $28/lb. The real costs of nonnuclear and fissionable fuels may be expected to increase with time, whereas that of deuterium (already low) should decrease.

The fusion process itself does not produce very dangerous or long-lived radioactivity. Thus, any such radioactivity will arise only on account of surrounding materials and presumably would be less than that from a fission device.

Ordinary hydrogen (${}_1H^1$) is of course the raw material in stellar thermonuclear or fusion processes. Its very slow reaction rate (see Chap. 2) makcs it unsuitable for man's technological purposes.

Granted the fact that an immense nuclear energy reserve exists in the form of deuterium, the question arises of how it may be exploited. As a preamble to our main development, we survey these problems in the next section.

1.5. Outline of the Controlled Fusion Problem

We present here in outline the principles upon which we believe a controlled fusion device must be based. In subsequent chapters, where the principles are developed in detail, we shall see that many of the basic problems suggested here are not yet solved.

Our discussion is based on the block diagram, Fig. 1.3, of almost any

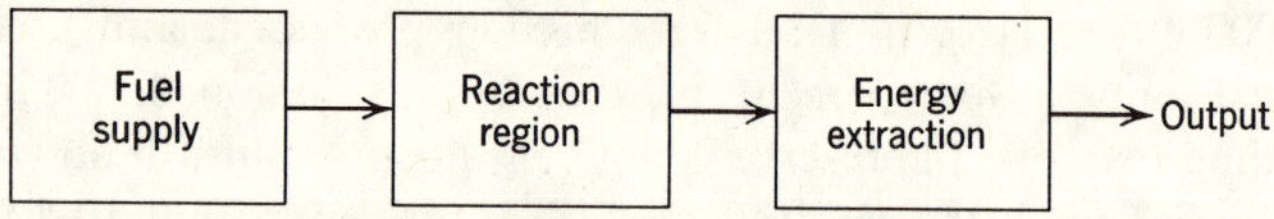

Fig. 1.3. Simplified block diagram of an energy conversion system.

energy conversion system. First, consider the fuel to be burned. While a pure deuterium reaction is possible in principle, a deuterium-tritium mixture (D-T) reacts about 100 times faster, so this reaction is much to be preferred. However, tritium does not occur naturally; it decays (into He^3) with a half-life of about 12 years. Thus, if a deuterium-tritium reaction is envisaged, the tritium must be somehow regenerated in a side reaction. Here we find our first problem, to which we shall return shortly.

The nature of the reaction itself is the source of the greatest difficulty. In order for two nuclides to fuse, they must approach each other very closely, in spite of the electrostatic repulsion between them. To achieve these close encounters, we must either make the atoms much smaller or give the nuclides high energy so that they can approach closely on chance encounters. The first of these two choices is not entirely absurd, but it will not work. In this connection, a particle called the mu-meson, which is much heavier than an electron, is useful. A negative mu-meson may replace the orbital electron in a deuterium atom, and because of its great mass its orbit lies very close to the nucleus. The atom is effectively much smaller. Similarly, the mu-meson may bind together two deuterons in a so-called mesic-deuterium ion; the deuterons are close enough to have appreciable chance of undergoing fusion. The mu-meson leaves unchanged and goes on to catalyze further fusions. The scheme will not work because the mu-meson has a short natural lifetime and decays into an electron and two neutrinos. In its short lifetime, it cannot catalyze enough fusions to release energy sufficient to create another meson.

From the known fact that large volumes of liquid deuterium do not mysteriously start fusing, we conclude also that no suitable catalyzing particle remains to be discovered.

We are thus forced to the second choice—heating the fuel. The temperature required is higher than that in the interior of stars, and is estimated to be in the range 10^8 to 10^9 °K. At this temperature, all light atoms are completely stripped of their electrons, and the gas consists of bare nuclei and free electrons. This assembly of charged particles is called a plasma. (The name is also properly applied to ionized gases at lower temperature, where a considerable fraction of the original neutral atoms or molecules may still remain.)

The reaction region in Fig. 1.3 therefore must contain a deuterium (or deuterium-tritium) gas at stellar temperatures; the proper confinement of this gas presents the greatest difficulty. It is evident that material walls alone will not suffice. We must take advantage of the fact that all the particles are charged, and use electromagnetic forces. These electromagnetic fields can be thought of as exerting a pressure on the plasma and keeping it confined in one place (remember that the plasma is basically a very hot gas). In later chapters, we shall find a number of confinement configurations, of the general sort shown in Fig. 1.4, where an electromagnetic field isolates the plasma from any part of the surrounding

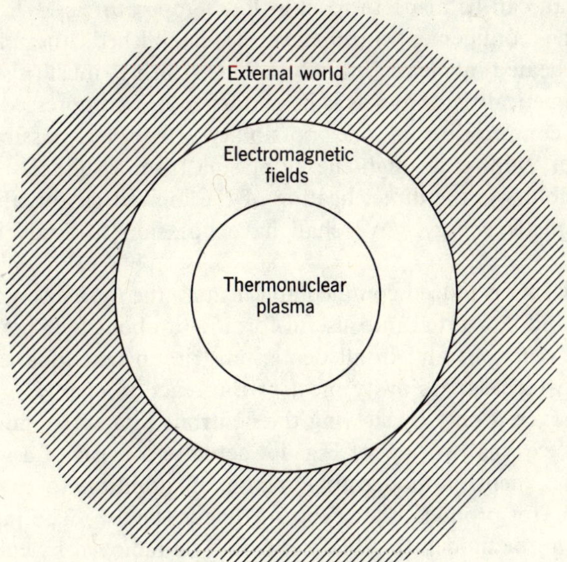

Fig. 1.4. Schematic confinement of a plasma by electromagnetic fields.

material world. Unfortunately, many of the confinement schemes are very unstable, in that the plasma and field regions tend to exchange positions rapidly. Whether any confinement scheme is stable long enough to give a practical fusion device is still an open question.

Not only must the plasma be kept from leaking out, but also foreign material must be kept from leaking in. There are several reasons for this. First, when the charged particles "collide," they scatter each other more often than they fuse. Scattering leads to diffusion, which in turn leads to loss of the plasma. Impurity nuclei increase the scattering and hence the gradual escape of the plasma, without aiding the fusion process. Second, it will prove difficult enough to confine the wanted material without confining useless material as well. Finally, and most important, the plasma is very hot and loses energy by radiation. Multiply charged impurity nuclei vastly increase this radiation; a few percent admixture of such impurities would increase it to an intolerable level. Thus the thermonuclear plasma with its confining field must be surrounded by a vacuum wall. Since the influx of impurities cannot be suppressed altogether, we must study carefully the various gas-kinetic and surface reactions that are to be minimized.

In view of these comments, we see that the achievement of controlled nuclear fusion is much more difficult than that of nuclear fission. The difference is particularly striking if we note that a simple solution of uranium salt can be made to react at low temperature.

Once the confinement mechanism is established, the plasma itself must be heated at least initially. The heating must also be done electromagnetically. Any gas is heated when it is compressed, so electromagnetic compression is one possibility. Another possibility is the injection of "blobs" of high-energy plasma from a suitable ion gun. A third possibility is the direct heating of the ionized gas by absorption of electromagnetic energy. We shall have occasion to study all of these methods.

If the plasma is indeed confined and heated, the nuclear energy must be extracted and converted into useful form. In about half the deuterium-deuterium reactions and in all deuterium-tritium reactions, a neutron is produced which carries away most of the reaction energy. This energy can be recovered only by slowing the neutrons in a surrounding blanket. Thus the "external world" of Fig. 1.4 actually consists of a vacuum wall plus a thick neutron-moderating region, in addition to a structure of conductors to produce the electromagnetic fields. The nuclear heat deposited in the moderating blanket would be removed by cooling fluids; its further conversion is not within the scope of our book.

Beside the energy of the neutrons, some of the fusion energy appears in charged nuclei formed in the reaction. It is sometimes proposed that this energy can be extracted directly as electricity, by appropriate pulsing of the confining electromagnetic fields. For reasons that we shall develop at some length in later chapters, we do not feel optimistic about such schemes. However, all fusion energy should be recoverable at least as heat.

It is important to know whether a thermonuclear system, once achieved, will be large or small. The surrounding blanket must be of the order of 1 m thick to accomplish its various purposes. For that reason alone, the system must be at least as large as a conventional fission reactor. We shall in fact conclude that the system will probably be considerably larger. The reason is that the thermonuclear power will be proportional to the volume of the plasma, while much of the escaping radiation is proportional to the surface area of the plasma. These losses are so severe that they can be overcome only if the system is large. By this qualitative term, we envisage a long cylinder of confined plasma, for example, whose diameter is 1 m or more. Outside this cylinder must be a vacuum region, the vacuum wall, the blanket, and the field-producing structure. Power output from such a device might be hundreds or thousands of megawatts.

Finally, we return to the fact that a deuterium-tritium mixture reacts much faster than pure deuterium, and we inquire about the regeneration of tritium. A neutron appears at each fusion. Fortunately, Li^6 fissions into He^4 and tritium on capture of a slow neutron. Therefore, it is proposed that the fusion neutrons be ultimately captured in the moderating blanket by Li^6 and the resulting tritium reinjected. Under those conditions, the raw materials are actually deuterium and Li^6, and the energy extraction and fuel supply regions of Fig. 1.3 are partly combined. Lithium (7.5% Li^6) is abundant.

This brief summary has covered only the principal considerations of controlled nuclear fusion in a qualitative way. We shall develop each of the physical concepts in detail in the following chapters.

PROBLEMS

1. Present energy consumption is about 0.1 Q/yr. Assume a 5% annual increase and a limit given by 1% of the solar absorption. Compute:

(*a*) The power and total energy used after 25, 50, and 100 years.
(*b*) When the limit of energy consumption will be reached.
(*c*) How long 7000 Q of fissionable material will last.

2. The volume of the oceans is 1.37×10^9 km^3. At the rate of 7.2 Mev per deuteron for complete fusion to He^4, compute:

(*a*) The energy available.
(*b*) How long the source will last at the rate of 15 Q/yr.

3. Let fossil fuel cost be 0.4 cent/kw-hr. At the standard rate of 1 megawatt-day/g of uranium fissioned, how much can be spent per kilogram to find, process, and reprocess uranium if fossil fuel costs are not to be exceeded?

4. It is proposed to explode a thermonuclear device underground and tap the heat. Let the surrounding rock have a heat capacity of 0.5 cal/cm^3-°C, and conductivity of 0.005 cal/sec-cm. Discuss the heat transfer to a 1-m diam sphere at 200° C with the simplifying assumption that a 500° C temperature is maintained at all times at a large radius R for (*a*) R = 10 m, (*b*) R = 100 m.

Under what circumstances might the heat transfer be brought up to a technologically more interesting level?

REFERENCES

1. P. C. Putnam, *Energy in the Future*, D. Van Nostrand Co., New York (1953).
2. L. E. Link and W. H. Zinn, "Economics of Nuclear Power," in H. Etherington, *Nuclear Engineering Handbook*, McGraw-Hill Book Co., New York (1958), pp. **12**–78 ff.
3. A. M. Weinberg, *Physics Today*, **12**, 11, 19–25 (1959).

2

Collision phenomena: basic theory and fusion cross sections

A thermonuclear reaction, or for that matter any associative reaction, proceeds via interactions at the microscopic level. We shall need to know the rates at which these reactions proceed in order to compute, among other things, the power output from a fusion device, the rate at which fuel must be added and products abstracted, and the optimum operating conditions. These rates will, for example, give us information about the time a reaction must be contained in order for some significant energy to appear.

It will be necessary to compute the rate at which reactions take place between various constituents. As might be surmised, and as we shall shortly see, the temperatures of interest are such that the components are gaseous and, in simple idealizations, also completely ionized. Many of the reactions among the reactants therefore involve simple two-particle collisions. The concepts involved and associated experimental data are described in this chapter.

It is not at all sufficient to consider a thermonuclear gas plasma made up merely of deuterons (for example) and electrons, scattering each other by Coulomb forces, and fusing by D-D collisions. In any practical situation, problems arise involving the influx of impurity gas from external regions, initial ionization of the gas, and so on. If any useful plasma is to be produced, strict attention must be paid to these matters. Thus a large number of such processes will be considered in Chap. 3, after the basic development presented here.

2.1. Cross Section, Mean Free Path, and Collision Frequency

We start our discussion by first considering the concept of a cross section. In order to define a microscopic cross section, a statistically uniform density N of free target particles in space is considered; for simplicity, let the target particles have no motion. Now imagine a beam of test particles of density n and uniform velocity $\mathbf{v}$ passing through the target ensemble. Let us fix attention on a slab of unit area normal to $\mathbf{v}$ and thickness dx, and inquire about an interaction of some sort between the two types. The number of particles dn undergoing the collision is evidently proportional to the density of target particles, the density of test particles, and the thickness dx. Thus

$$dn = \sigma n N \, dx, \tag{2.1}$$

where σ is the constant of proportionality, defined as the cross section for the interaction considered.

The choice of slab orientation normal to $\mathbf{v}$ was not necessary. If the slab is tilted at angle θ, the number of target particles passed by a test particle in penetrating the slab is increased by $\sec\theta$. The distance $dx' = dx \sec\theta$ traveled by the particles is similarly increased. Therefore, the number of interactions per unit distance is independent of θ. The cross section σ has the dimension of area, and, in a very simple case, a geometrical interpretation may be given it. If the target and test particles are hard spheres of radii r_1 and r_2, respectively, any test particle whose center passes within $r_1 + r_2$ of a target center will suffer a collision. Then $\sigma = \pi(r_1 + r_2)^2$. Such interpretations are not generally useful. The quantity σ is very often not comparable with the "size" of the particles; it usually depends on the relative speed, and we can define various σ's for competing processes between the particles.

Equation 2.1 yields

$$n = n_0 \exp(-N\sigma x), \tag{2.2}$$

where n_0 is the density of test particles at $x = 0$. It is assumed that the density N is not disturbed by the density n, which is true if $n \ll N$. There are a number of useful allied concepts. By Eq. 2.2, the probability that a particle goes a distance x without a collision is $\exp(-N\sigma x)$. The mean free path l is defined as the total distance traveled between collisions by the test ensemble divided by the number of test particles. To find this number, consider particles that travel a distance x without collisions and that collide between x and $x + dx$. The total path length is $n_0 x \exp(-N\sigma x)\,dx$, and the number of particles interacting there is $n_0 \exp(-N\sigma x)\,dx$. Thus

$$l = \frac{\int_0^\infty dx\, x \exp(-N\sigma x)}{\int_0^\infty dx \exp(-N\sigma x)} = \frac{1}{N\sigma}. \tag{2.3}$$

This general method of averaging and normalizing will be used extensively in later chapters. Sometimes the quantity $N\sigma = 1/l$ is used. It is called the macroscopic cross section.

Often it is more convenient to consider the number of collisions per unit time rather than per unit path length. By following a test particle of speed v, we may construct the analogues of Eqs. 2.1 and 2.2 and find the collision frequency ν, given by

$$\nu = vN\sigma = v/l. \tag{2.4}$$

A particle has a chance $\exp(-\nu t)$ of surviving a time t without a collision.

If a number of different processes may happen at a collision, we may define a collision frequency or cross section for each. Thus for k different possibilities, for example,

$$\nu = \sum_{\zeta=1}^{k} \nu_\zeta. \tag{2.5}$$

The mean free paths add reciprocally, of course.

There are a number of methods in practical use for specifying the magnitudes of cross sections, and we must be able to convert readily. For nuclear processes, the barn $= 10^{-24}\ \text{cm}^2$ is used and will be used here. For atomic processes, the unit $10^{-16}\ \text{cm}^2$ ($= 1$ square angstrom) is convenient, being comparable with the geometric size of atoms. We shall adopt this unit with the symbol A^2. Cross sections are sometimes quoted in units of $\pi a_0^2 \approx 0.88\ \text{A}^2$, where a_0 is the Bohr radius. In the literature on gaseous electronics, the probability P of collision is used. The probability is defined as the reciprocal of the mean free path at a target gas pressure of 1 mm Hg. The probability P is then related to the cross section σ by

$$\sigma = pP/N, \tag{2.6}$$

where p is the pressure in mm Hg. We must often refer to the experiment by which P was measured to determine just what was meant by 1 mm Hg: whether at 0° C, 20° C, gas density referred to 0° C but with the experiment done at 20° C, and so forth. Usually the gas density at

20° C is meant. In this case, the density N is

$$N = \frac{A}{22{,}400}\left(\frac{p}{760}\right)\left(\frac{273}{293}\right) = 3.3 \times 10^{16}\, p/\text{cm}^3, \tag{2.7}$$

where A = 6.02×10^{23} particles/mole; thus

$$\sigma = 3.03 \times 10^{-17} P \text{ cm}^2. \tag{2.8}$$

In addition, for a given mass density ρ_m of particles with molecular weight M, the particle density is

$$N = \rho_m A/M. \tag{2.9}$$

2.2. Speed, Energy, and Temperature

The energy U of any particle of mass m and speed v is

$$U = \tfrac{1}{2}mv^2. \tag{2.10}$$

For an assembly of n particles with different energies, the average energy $\bar{U}$ per particle is

$$\bar{U} = \frac{1}{2n}\sum_{\zeta=1}^{n} m_\zeta v_\zeta^2. \tag{2.11}$$

There are other ways of measuring energy. It will be derived later that, for any gas in thermodynamic equilibrium at temperature T, the average energy per particle is $3kT/2$, where $k = 1.38 \times 10^{-23}$ joule/° K is Boltzmann's constant and T is the absolute temperature. Thus a gas at 1° K corresponds to an average energy of 2.07×10^{-23} joule per particle. It is also convenient to measure energy in terms of electron volts, as follows. Let a particle have an electric charge (positive or negative) equal in magnitude to an electronic charge, 1.602×10^{-19} coulomb. Let the charge be accelerated through a potential difference of 1 volt. Then it has energy 1.602×10^{-19} joule. This unit of energy is 1 electron volt (ev). Thus to measure energy in terms of electron volts, we must divide the kinetic energy by $e = 1.602 \times 10^{-19}$ coulomb. Hence Eq. 2.10 becomes

$$u = \frac{mv^2}{2e} \text{ ev}, \tag{2.12}$$

where now u has the dimension of volts. The unit is particularly useful in dealing with charged particles, for it indicates directly the potential necessary to produce a singly charged particle of some particular energy. Multiples of this unit will often be used: 1 kev = 10^3 ev, 1 Mev = 10^6 ev.

For a gas in thermodynamic equilibrium, we may readily check that the conversion is

$$\text{Average energy (ev)} = 1.292 \times 10^{-4} \text{ ev/° K}. \tag{2.13}$$

Thus 300° K corresponds to 0.0388 ev, and 10-kev average energy is 7.75×10^{7}° K. Very often a plasma energy kT is referred to because the quantity $3kT/2$ appears infrequently in computations. In addition, we find it useful to use the symbol T alone to represent the quantity kT/e measured in volts or kilovolts. If the unit ev is used, a conversion factor of 11,600° K/ev is implied. The context will make the meaning clear in each case. It should be noted in passing that temperature is an equilibrium (thermodynamic) concept, and we shall not always deal with equilibrium situations. Thus a true temperature cannot always be assigned, although we shall sometimes use the term in the sense of average energy. In addition, by temperature we mean the quantity sometimes called kinetic temperature, that is, the state of energy of the particles alone. Any other use of the concept, for example, consideration of a radiation field, will be specifically pointed out.

The cross sections σ to be discussed in this chapter are functions of the relative speed of the two interacting particles. Usually, however, σ is expressed as a function of the energy of one of the particles, the other being assumed to be at rest. Arguments of functions involving the energy must be chosen very carefully because the difference of the energies is not invariant under transformation between moving coordinate systems. For example, the fusion cross section for deuterons striking stationary tritons at energy u, $\sigma_{\mathrm{DT}}(u)$, is unequal to $\sigma_{\mathrm{TD}}(u)$ for tritons having the same energy. In the latter case, the relative speed is just $\sqrt{2/3}$ as great.

2.3. Experimental Fusion Cross Sections

In Sec. 2.1 we introduced the concepts of cross sections and related quantities. Let us consider now the particular properties of fusion cross sections. We shall find that only nuclides of low mass are of interest.

From the equivalence of mass and energy (931 Mev/atomic mass unit), we see that energy can be extracted from any source only by decreasing the mass of the system. Reactions leading to a decrease of mass can be quickly found by inspection of the energy available per nucleon. This quantity is conveniently measured by the packing fraction P, defined by

$$P = \frac{M - A}{A}, \tag{2.14}$$

where M is the actual mass in amu and A is the nuclear mass number. Figure 2.1 displays the packing fraction of the nuclides. Zero on the ordinate is derived from the fact that the oxygen mass is 16 amu. The most stable nuclei are found in the region $40 < A < 80$. Clearly, energy can be achieved either by splitting very heavy nuclides or fusing light ones.

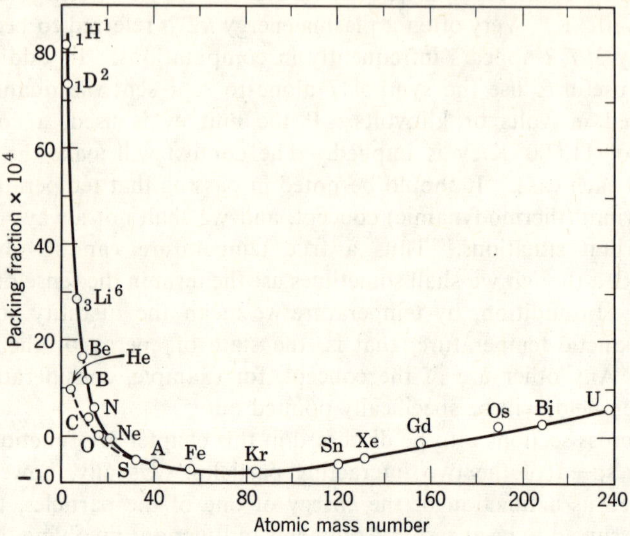

Fig. 2.1. Packing fractions.

The fusion cross sections of likely interest are shown in Fig. 2.2 as functions of energy. In this notation $a(b, c)d$ means $a + b \rightarrow c + d$, with particle a initially at rest. The number associated with each curve is the energy produced in Mev. A number of additional reactions may take place with small probability, as summarized in Table 2.1. We see that no reactions involving protons as fuel have reasonably large cross sections. Lithium reactions have smaller cross sections than D, T, or He^3 reactions, because of the higher Coulomb barrier between the particles. Furthermore, we shall find later that the radiation loss from a thermonuclear plasma is more severe if heavy elements are present. Consequently, lithium reactions are unlikely, and we must consider deuterium, tritium, or helium.

For all these reactions of interest, we see first that the energy output is far greater than the initial energy of the interacting particles. All these reactions are highly exothermic. Then to a good approximation, we may compute the division of energy between the reaction products on the basis of stationary initial reactants. Conservation of linear momentum therefore yields, for example,

$$\mathrm{D} + \mathrm{D} \begin{cases} \nearrow (\mathrm{T} + 1.01\ \mathrm{Mev}) + (p + 3.03\ \mathrm{Mev}), \\ \searrow (\mathrm{He}^3 + 0.82\ \mathrm{Mev}) + (n + 2.45\ \mathrm{Mev}), \end{cases} \quad \text{about equal probability}$$

$$\mathrm{D} + \mathrm{T} \rightarrow (\mathrm{He}^4 + 3.52\ \mathrm{Mev}) + (n + 14.06\ \mathrm{Mev}),$$

$$\mathrm{D} + \mathrm{He}^3 \rightarrow (\mathrm{He}^4 + 3.67\ \mathrm{Mev}) + (p + 14.67\ \mathrm{Mev}).$$

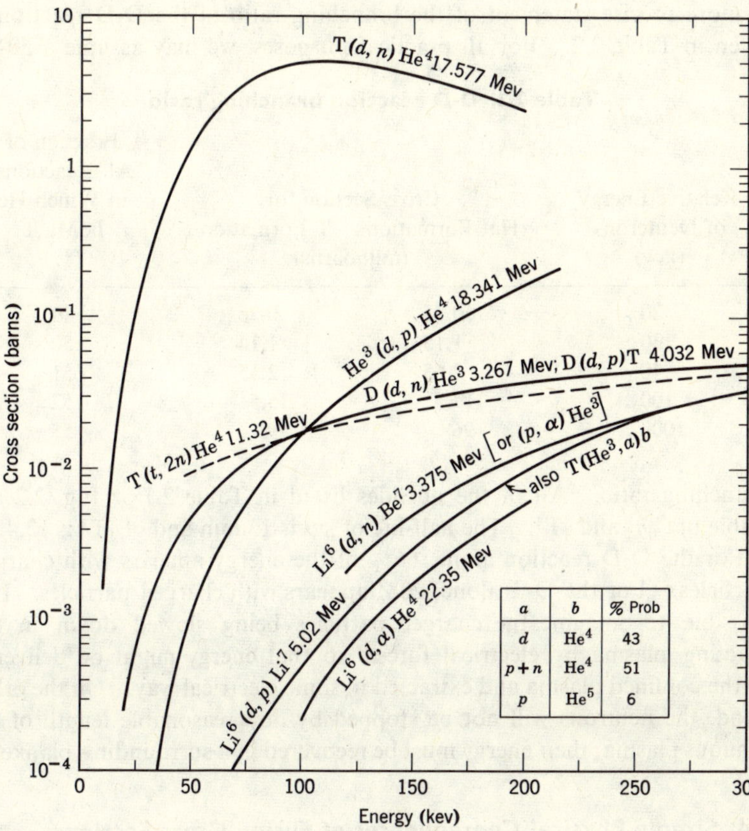

Fig. 2.2. Fusion cross sections.

Table 2.1. Fusion reactions of low probability

Reaction	Cross Section
$D(p, \gamma)He^3$	10^{-6} barn
$T(p, \gamma)He^4$	10^{-6} barn
$T(p, n)He^3$	zero below 1 Mev (not exothermic)
$Li^6(p, He^3)He^4$	$\approx 10^{-4}$ barn at 300 kev
$Li^6(d, T)Li^5$	0.05 barn at 400 kev (no data at lower energy)
$Li^7(p, \gamma)Be^8$	$\approx 10^{-5}$ barn at 250 kev
$Li^7(p, \alpha)He^4$	$\approx 10^{-3}$ barn at 250 kev
$Li^7(d, n)Be^8$	$\approx 10^{-3}$ barn at 200 kev

A more precise statement of the branching ratio of the D-D reaction is given in Table 2.2. For all practical purposes, we may assume a 50–50

Table 2.2. D-D reaction branching ratio

Relative Energy of Deuterons (kev)	Cross Section for He^3 Formation (millibarns)	Cross Section for T Formation (millibarns)	Fraction of All Reactions in Which He^3 Is Made (%)
20	0.25	0.26	49
30	1.15	1.14	50
40	2.65	2.55	51
100	17	15.5	52
1000	96	79	55

branching ratio. All of the nuclides listed in Table 2.1 or Fig. 2.2 are stable but ${}_0n^1$ and ${}_1T^3$. The half-life of ${}_0n^1$ is 13 min and of ${}_1T^3$ is 12.4 yr.

For the D-D reaction alone, 66% of the energy appears with charged particles. For the D-T alone, 20% appears with charged particles. It is possible to imagine the charged particles being slowed down in the reacting plasma by electrical forces, so that energy might be delivered to the confined plasma and extracted in some electrical way. On the other hand, the neutrons will not be stopped by any reasonable length of the tenuous plasma; their energy must be recovered in a surrounding blanket.

2.4. Simple Physical Consequences of Fusion Cross Sections

A number of important conclusions can be drawn from inspection of the cross sections, and the nature of the problem of controlled fusion can be demonstrated. Because the particles are charged, their Coulomb repulsion precludes any interaction at low energies. In contradistinction, a fission reactor may operate at low temperature because the neutrons, which induce fission, are able to penetrate to the nucleus with no repulsion. The fusion cross sections are significant only at energies greater than about 5 kev, and have maxima at 100 kev or above. The cross sections themselves are 5 barns or less. It will be seen in later sections that the cross sections for Coulomb scattering are much larger, and cross sections for processes involving incompletely stripped atoms are very much larger. The gas must be kept fully ionized, and confined so that the unavoidable Coulomb scattering does not disperse it. Thus colliding beams of deuterons, for example, cannot undergo significant fusion and still remain beams.

From such an argument we conclude that the reacting medium must be looked upon as a gas at a temperature corresponding to the energy required for fusion, that is, at 5×10^7 to 10^9 °K. The gas is completely ionized, since the ionization potential of hydrogen is 13.6 ev, and the first and second ionization potentials of helium are 24.5 ev and 54.1 ev, respectively. Such a gas, in which electrical interactions between the charged particles play a dominant role, is called a plasma.

The plasma will be quasi-neutral, consisting both of the bare nuclei (positive ions) and negative charges (principally electrons). That the plasma is nearly neutral can be seen from Poisson's equation

$$\nabla \cdot \mathbf{E} = \rho/\epsilon_0, \tag{2.15}$$

where **E** is the macroscopic electric field, ρ is the net charge density, and $\epsilon_0 = 8.85 \times 10^{-12}$ farad/m is the permittivity of free space. If a one-dimensional system is considered with a density of 10^{18} ions/m^3, then in a distance of 1 m, the potential created by the space charge is 5×10^{10} volts. At very much lower electric potentials, excess ions would be expelled and electrons retained, making the positive and negative charge densities very nearly equal.

A plasma at temperatures of thermonuclear interest cannot be confined by material walls. The difficulty is not just that the walls would melt but that they would cool the plasma. In view of the weakness of gravitational fields and the short-range character of nuclear fields, only combinations of electric and magnetic fields may be used to confine the plasma.

Turning again to the cross sections of Fig. 2.2, we note that the D-T reaction has the largest cross section. If this reaction is used, the tritium must be regenerated in a blanket surrounding the thermonuclear machine. The reaction $\mathrm{Li}^6(n, \alpha)\mathrm{T}$ looks promising for this purpose, using the neutrons produced in the D-T reaction. Since 100% efficiency cannot be realized in the process, it will be necessary to use $(n, 2n)$ reactions to make up for the losses. Fortunately, cross sections are large for the 14-Mev neutrons produced in the D-T reaction. We consider the problems of energy recovery and tritium regeneration in Chap. 13.

If only naturally occurring fuel is used, the D-D reaction must be used, and much higher energies are required. If the reaction can be made to go, the tritium formed will undoubtedly also burn up in a D-T reaction, increasing the energy output severalfold. If the He^3 formed can be confined for a reasonable length of time and its energy maintained, the $\mathrm{He}^3(d, p)\mathrm{He}^4$ reaction will also take place. Thus we get complete burnup: $3\mathrm{D} \rightarrow \mathrm{He}^4 + n + p + 21.6$ Mev by either process. The protons have 8.8-Mev energy total, the neutrons have a total energy of 8.3 Mev, the He^4 of 3.6 Mev. The difference 0.9 Mev between 21.6 Mev and the sum of these figures is the kinetic energy of the intermediate T and He^3 nuclei.

The secondary reactions need not be considered. The T and He^3 will not interact among themselves because the D-T and D-He^3 reactions, respectively, are more probable.

In fission technology, energy calculations are facilitated by using the fact that 1 g of U^{235} completely fissioned yields about 1 megawatt-day of energy. For fusion of deuterium, complete burnup as described yields 4 megawatt-days per gram.

2.5. Classical Theory of Elastic Collisions

Collisions between particles may be roughly divided into two major categories—those in which the nature of one or both of the particles is altered, and those in which the nature of both particles remains strictly unaltered. Collisions in the first category must be specifically identified in terms of the changes experienced by the particles, for example, fusion, ionization, excitation, dissociation, and so forth. We have already discussed the fusion cross sections and shall treat other such collision phenomena in Chap. 3.

Collisions in the second major category are called elastic collisions. It is to these interactions that we now turn our attention. For any collision, whether elastic or inelastic, the cross section σ is determined by the nature of the interparticle forces and the relative velocity of the particles.

We assume that the reader is familiar with the elementary concepts of quantum mechanics. As a further parenthetical preamble, we first justify our use of a classical model. Strictly speaking, we should calculate the interaction quantum-mechanically, because classical concepts tend to fail at the atomic level. The desirability of such a procedure becomes a necessity in cases of inelastic collisions, where changes occur in the quantized energy levels occupied by the particles. On the other hand, elastic collisions may often be calculated accurately enough for our purposes by a simple classical model. The reason for this fact is that many elastic collisions of interest to us are small-angle scatterings in which the two particles never approach each other very closely. In other words, the forces are weak at large distances, and the energy of interaction is small. Under these circumstances, the quantized energy levels are closely spaced in energy, and may be replaced by the classical continuum.

Very roughly speaking, the classical calculation will be expected to fail if the square of the wavelength of the impinging particle is comparable with or greater than the calculated cross section. If the converse of that criterion is well satisfied, classical theory often suffices. Thus we must have

$$\left(\frac{h}{mv}\right)^2 \ll \sigma, \tag{2.16}$$

if classical theory is to be applied, where h is Planck's constant. Note that Inequality 2.16 implies that classical theory is a better approximation at high energy than at low energy, provided that σ is found to vary less rapidly than v^{-2}.

With these preliminary remarks, we now calculate the classical elastic-scattering cross section.

It is assumed that the forces are central, that is, that the forces $\mathbf{F}_1$ and $\mathbf{F}_2$ on two particles of masses m_1 and m_2 at $\mathbf{r}_1$ and $\mathbf{r}_2$ obey

$$\mathbf{F}_1 = m_1\ddot{\mathbf{r}}_1, \tag{2.17}$$

$$\mathbf{F}_2 = m_2\ddot{\mathbf{r}}_2 = -\mathbf{F}_1. \tag{2.18}$$

We thus exclude certain classes of collisions, as between fixed dipoles, and so forth, and further assume that the force field is conservative, so that a potential function $\phi(\mathbf{r})$ alone describes the nature of the interaction.

By multiplying Eqs. 2.17 and 2.18 by m_2 and m_1, respectively, and subtracting, we obtain

$$\mathbf{F}_1 = m_r \frac{d^2}{dt^2}(\mathbf{r}_1 - \mathbf{r}_2), \tag{2.19}$$

where

$$m_r = m_1m_2/(m_1 + m_2) \tag{2.20}$$

is the reduced mass. Equation 2.19 states that the scattering process is identical to the motion of a particle of reduced mass m_r about a fixed point at the distance $\mathbf{r} = \mathbf{r}_1 - \mathbf{r}_2$ and acted upon by one of the forces.

The cross product $\mathbf{F}_1 \times \mathbf{r} = 0$, by virtue of the central-field assumption. By taking this product and rewriting the derivatives, we obtain

$$\frac{d}{dt}(\mathbf{r} \times \dot{\mathbf{r}}) = 0, \tag{2.21}$$

whence

$$\mathbf{r} \times \mathbf{v} = \mathbf{K} = \text{const}, \tag{2.22}$$

where

$$\mathbf{v} = \mathbf{v}_1 - \mathbf{v}_2 = \dot{\mathbf{r}}_1 - \dot{\mathbf{r}}_2 \tag{2.23}$$

is the relative velocity. Equation 2.22 states that the two particles move always in a plane perpendicular to $\mathbf{K}$ and only a two-dimensional problem need be considered.

Figure 2.3 shows the geometry of the scattering process in the plane normal to **K**. Point 0 is the center of gravity, which generally will move in the laboratory system. Particles 1 and 2 are separated by the total distance r, and the lengths of the segments 01 and 02 are in ratio m_2/m_1. The angle between the line joining the two particles when closest together and the line between the two particles initially is called θ_m. The total deflection of particle 1 in the center of mass system is denoted by χ.

$$\chi = \pi - 2\theta_m. \tag{2.24}$$

The distance b is the impact parameter, defined as the perpendicular distance between the trajectories when the particles are remote.

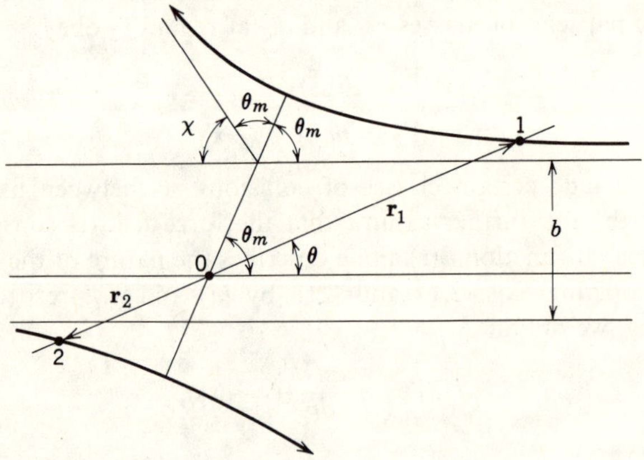

Fig. 2.3. Trajectories of particles in a central-field collision.

The total kinetic energy U of the particles is

$$U = \tfrac{1}{2}m_r(\dot{r}^2 + r^2\dot{\theta}^2), \tag{2.25}$$

a result which can be readily derived. The total angular momentum $\mathfrak{p}$ about point 0 is also

$$\mathfrak{p} = m_r r^2 \dot{\theta} \tag{2.26}$$

and is an invariant of the motion.

At some time before the collision takes place, when the interaction between the two particles is negligibly small, the angle $\theta \to b/r$, $\dot{\theta} \to -b\,\dot{r}/r^2$, and we have

$$U_0 = \text{total kinetic energy} = \tfrac{1}{2}m_r v_0^2 \tag{2.27}$$

$$\mathfrak{p}_\theta = m_r b v_0 = m_r r^2 \dot{\theta}, \tag{2.28}$$

where v_0 is the relative speed when the particles are remote. The sum of kinetic and potential energies is constant:

$$U_0 = \tfrac{1}{2}m_r(\dot{r}^2 + r^2\dot{\theta}^2) + \phi(r). \tag{2.29}$$

Equations 2.28 and 2.29 describe the orbit. We may eliminate the time, and obtain the trajectory in polar coordinates:

$$\frac{dr}{d\theta} = \pm \frac{r^2}{b}\sqrt{1 - \frac{b^2}{r^2} - \frac{\phi(r)}{U_0}}. \tag{2.30}$$

The trajectories are symmetric about the configuration of closest approach; the minus sign applies to the particles when approaching, and the plus sign when they are receding.

At the angle θ_m of closest approach, we have

$$1 - \frac{b^2}{r_m^2} - \frac{\phi(\mathbf{r})}{U_0} = 0 \tag{2.31}$$

determining the distance r_m of closest approach.

It is worth noting in passing that a solution of Eq. 2.31 does not always exist. We have chosen, by the sign of ϕ in Eq. 2.29, a repulsive potential, for which a solution of Eq. 2.31 always exists. But suppose that $\phi(r) = -C/r^n$, an attractive potential. If $n \geqslant 2$, there are values of initial energy and impact parameter for which no solution exists. Physically, this condition means that the particle spirals in until the nature of the interaction changes to an attractive force with $n < 2$, or to a repulsive force; then it spirals out again. For Coulomb attractive scattering, $n = 1$, and the problem does not arise.

If a solution exists,

$$\theta_m = \int_{r_m}^{\infty} dr\,\frac{d\theta}{dr}, \tag{2.32}$$

and the total deflection is

$$\chi(b, U_0) = \pi - 2b\int_{r=r_m}^{\infty} \frac{dr/r^2}{\{1 - [\phi(r)/U_0] - [b^2/r^2]\}^{1/2}}. \tag{2.33}$$

We now calculate the differential cross section $\sigma(\chi)$ for scattering into the angle χ. A particle incident in an annular ring of area $2\pi b\,db$ centered on an atom scatters into the solid angle $d\Omega = 2\pi \sin\chi\,d\chi$. Conservation of particles then requires that

$$2\pi b\,db\,Nnv = 2\pi \sin\chi\,d\chi\,N\sigma nv, \tag{2.34}$$

where N is the atomic density and n is the density of scattered particles. Since both sides of Eq. 2.34 are merely different representations of the number of particles scattered through the angle χ,

$$\sigma(\chi) = \frac{b}{\sin \chi} \frac{db}{d\chi}. \tag{2.35}$$

The quantity $\sigma(\chi)\, d\Omega$ has the dimensions of area, and is the number scattered into a solid angle $d\Omega$ at χ.

The total cross section σ is

$$\sigma = \int_{\chi=0}^{\pi} d\Omega\, \sigma(\chi). \tag{2.36}$$

This cross section is infinite for some scattering potentials, because of a pole in $\sigma(\chi)$ at $\chi = 0$ (large cross section for small-angle scattering).

Another cross section of practical interest is that for momentum transfer. In the center of mass system, the momentum change of the particle in the forward direction is $m_r v_0(1 - \cos \chi)$; that is, scattering nearly forward produces little momentum change. Thus we introduce the cross section σ_m for momentum transfer:

$$\sigma_m = \int_{\chi=0}^{\pi} d\Omega\, \sigma(\chi)(1 - \cos \chi); \tag{2.37}$$

σ_m is often finite when σ is not.

We shall now show by dimensional arguments that for a scattering potential $\phi \propto r^{-n}$, the differential scattering cross section $\sigma(\chi) \propto v^{-4/n}$. Let us outline a proof, starting with Eq. 2.30, for instance. In this case,

$$\frac{dr}{d\theta} = \pm \frac{r^2}{b} \sqrt{1 - \frac{b^2}{r^2} - \frac{C'}{{v_0}^2 r^n}}, \tag{2.38}$$

where the constant C' combines the various fixed parameters. Note that, once C', b, and v_0 are determined, the trajectory—in particular, the asymptotic angle—is determined. Thus let us scale v_0 by a factor k, and rescale the distances in such a way as to recover a formally identical equation. Then, to preserve the last term under the radical, we must set a new distance scale $r' = (k^{-2/n})r$. The remainder of the equation is preserved if we scale the impact parameter b the same way so that

$$\frac{dr'}{d\theta} = \pm \frac{r'^2}{(b')} \sqrt{1 - \frac{(b')^2}{r'^2} - \frac{C'}{(v_0')^2 r'^n}}, \tag{2.39}$$

where

$$v_0' = k v_0, \tag{2.40}$$

$$\begin{Bmatrix} r' \\ b' \end{Bmatrix} = (k^{-2/n}) \begin{Bmatrix} r \\ b \end{Bmatrix}. \tag{2.41}$$

Since Eq. 2.39 is formally identical to Eq. 2.38, χ is the same: only the scale of the interaction region of Fig. 2.3 is changed. Since $\sigma(\chi)$ is proportional to b^2 by Eq. 2.35, we find immediately that

$$\sigma(\chi, kv_0) = k^{-4/n}\sigma(\chi, v_0), \tag{2.42}$$

which is the desired relation.

This result has many useful applications. The Coulomb scattering cross section is inversely proportional to $v_0{}^4$, or the square of the incident energy. This result (Rutherford scattering) leads to an infinite cross section both for σ and the momentum transfer value σ_m. This circumstance has led to great difficulty in formulating precise plasma theories, and the matter is not yet satisfactorily resolved. Further discussion of this point will be given in Chap. 8.

Scattering of charges by atoms or molecules often proceeds through polarization forces, in which the incident charge induces an electric dipole moment in the atom or molecule. The induced dipole potential $\phi(r)$ varies as r^{-4}, and scattering proceeds in the resulting attractive field. Thus we expect in this case that $\sigma \propto 1/v_0$. Elastic-scattering data for electrons in many gases (H_2 and He are good examples) show this effect: at higher electron energies, typical cross sections fall off about as $1/v_0$; that is, the collision frequency $\nu_c \to$ constant. Examples may be seen in the next chapter.

It is necessary to remember that these elastic collisions, while leading to no energy exchange in the center of mass system, may do so in the laboratory system. If the center of mass is not stationary, some energy exchange takes place. Thus high-energy electrons will heat up slow ions, for example.

PROBLEMS

1. Compute the mean time between collisions in terms of the pressure and probability of collision.

2. A particle X moves with speed v in a mixture of gases A, B, and C, with partial pressures p_A, p_B, and p_C measured at 20° C. The collision cross sections for X with each type of molecule is σ_A, σ_B, and σ_C. Find the mean free path of X at 20° C.

3. The atmospheric density n above the earth varies with altitude h approximately as $n = n_0 \exp(-h/H)$, where H is called the scale height. For a collision cross section σ, calculate the height h_0 at which a particle shot upward with very high speed has probability 0.5 of escaping without a collision.

4. Calculate the speed of a 10-kev triton. What is the energy of an electron with this same speed?

5. The force on a particle of mass m, charge q, and velocity $\mathbf{v}$ is $\mathbf{F} = q(\mathbf{v} \times \mathbf{B})$. Prove that the particle moves in a circle with angular frequency $\omega_b = qB/m$ (ω_b is called the cyclotron frequency).

6. From the results of Problem 5, find the radius of the circular motion of a 10-kev triton in a magnetic induction of 1 weber/m^2 (10,000 gauss).

7. Compute the division of energy among the products of the $Li^6(d, n)Be^7$ and the $Li^6(d, \alpha)He^4$ reactions.

8. Compute the amount of gasoline equivalent to 1 liter of natural water if deuterium be burned to completion. The energy released in the burning gasoline is about 4.6×10^4 joules/g.

9. An infinite slab of plasma 0.1 m thick is made up of deuterons of uniform density $n_i = 10^{20}/\text{m}^3$ and electrons of uniform density n_e. Both surfaces are at the same potential, and the potential difference between the midplane and surfaces is 5000 volts. Compute n_e.

10. A 0.5-amp beam of 600-kev deuterons 1 cm in diameter and uniform density is produced for injection into the Oak Ridge DCX thermonuclear device. Calculate the electric field on the surface of the beam and the voltage at the surface relative to that at the center.

11. What is the wavelength of a 5-ev photon? Of a 10-kev photon?

12. Let α be the polarizability of a molecule (dipole moment $M = \alpha E$). Compute the potential arising from the molecule at a charge q a distance r away.

GENERAL REFERENCES

S. C. Brown, *Basic Data of Plasma Physics*, The Technology Press and John Wiley & Sons, New York (1959).

R. B. Leighton, *Principles of Modern Physics*, McGraw-Hill Book Co., New York (1959).

D. C. Peaslee and H. Mueller, *Elements of Atomic Physics*, Prentice-Hall, New York (1955).

R. K. Richtmyer, E. H. Kennard, and T. Lauritsen, *Introduction to Modern Physics*, McGraw-Hill Book Co., New York (1955).

3

Collision phenomena: atomic and surface effects

A particle can experience any one of a large number of different types of interactions with other atoms, molecules, ions, electrons, photons, or surfaces. Hot terrestrial plasmas do not occur spontaneously in nature. They must be created from initially neutral, un-ionized molecules by heating and by collisions. Many of these collisions take place between the particles in the plasma. Other collisions will take place between the plasma particles and any nearby surfaces, because we cannot hope to isolate a plasma ideally. These collisions may knock off material from the surface and so introduce foreign matter into the plasma.

In the last three chapters of this book, we survey current experimental attempts to overcome some of these undesirable side effects in producing confined plasmas. If the attempts are to be successful, a good understanding of the various interactions is necessary. Following the general cross-section formulations of the previous chapter, we propose to describe the important interactions here.

3.1. Elastic Scattering of Electrons and Ions

Among the more probable events for an electron is an elastic collision with a molecule or atom. Figure 3.1 shows the cross section for elastic scattering of an electron by H, H_2, D_2, and He. The abscissa is marked in units of the square root of the electron energy in ev. This scale is often convenient, being proportional to speed of the test particle. Figure 3.2

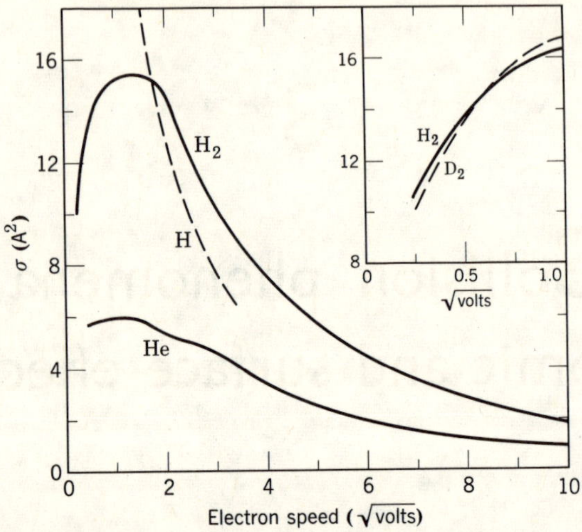

Fig. 3.1. Elastic-collision cross section of electrons in H, H_2, D_2, and He. After R. T. Brackmann, W. L. Fite, and R. H. Neynaber, *Phys. Rev.*, **112**, 1157 (1958); R. B. Brode, *Revs. Modern Phys.*, **5**, 257 (1933); L. Gould and S. C. Brown, *Phys. Rev.*, **95**, 897 (1954).

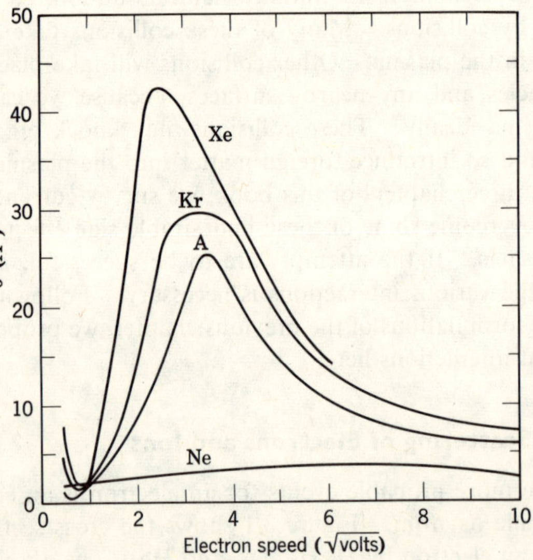

Fig. 3.2. Elastic-collision cross section of electrons in Ne, A, Kr, and Xe. After R. B. Brode, *Revs. Modern Phys.*, **5**, 257 (1933).

shows the elastic-collision cross section for electrons in the remainder of the rare gases; Fig. 3.3 illustrates the elastic scattering in O_2, N_2, and CO, which are gases commonly found as contaminants.

Several remarks may be made about these figures:

1. Above a few electron volts, the elastic cross sections decrease with increasing speed; for H_2, H, and He, the dependence is v^{-1} very closely, and hence indicates polarization forces.

2. In the monatomic gases, the cross section increases with increasing gas polarizability.

3. At low energies, the scattering is complicated by quantum-mechanical effects. Most striking is the virtual transparency of A, Kr, and Xe near 1 ev. In these cases, the electron is diffracted around the atom (the Ramsauer effect).

4. In general, atoms or molecules having similar electronic structure have similar cross sections, as is the case with CO and N_2 (which are very similar) and with the rare gases. (Ne and He are different from A, Kr, and Xe in that their scattering potentials are too weak to produce electron-wave phase shifts sufficient for the diffraction effect.) The similarity is also apparent in the hydrocarbons and alkali metals.

5. At electron energies so high that the electron wavelength is shorter than the internuclear separation, a diatomic molecule looks like two atoms. Note the cases of H vs. H_2.

6. A discrepancy will be noted in the data for H_2 if the main curve and inset of Fig. 3.1 are compared. This discrepancy is purposely included

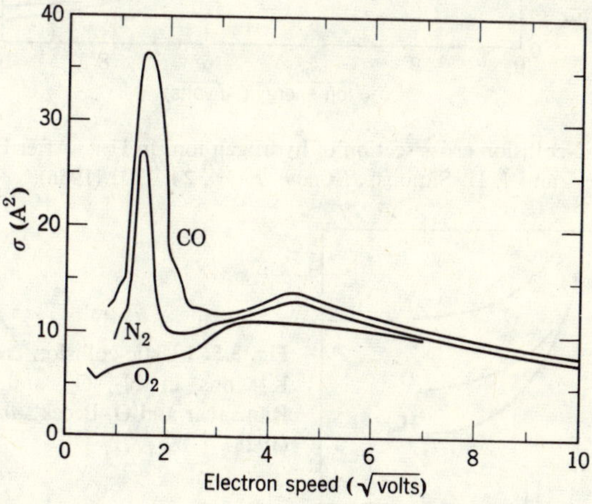

Fig. 3.3. Elastic-collision cross section of electrons in O_2, N_2, and CO. After R. B. Brode, *Revs. Modern Phys.*, **5**, 257 (1933).

to illustrate the fact that data derived from different sources and by different methods often disagree. The situation in H is particularly bad; the single curve shown is believed to be the best available data. Others exist differing by a factor of 2. In all these curves and others to follow, we have exercised some personal judgment in selection of data. The reader who studies the literature carefully must be prepared to do the same.

Angular distributions of the scattered electrons have been measured; in general, they are not isotropic.

A great deal of experimental data concerning collision cross sections of all kinds may be found in the standard literature.[1,2]

Ion-atom elastic-collision cross sections have also been measured. They tend to fall off at higher energies, often leading to an approximate $1/v$ dependence. Hydrogenic ions in H_2 are shown in Fig. 3.4; and K^+ in N_2, O_2, and H_2 are shown in Fig. 3.5. Elastic scattering of He^+ ions in He is shown in Fig. 3.9.

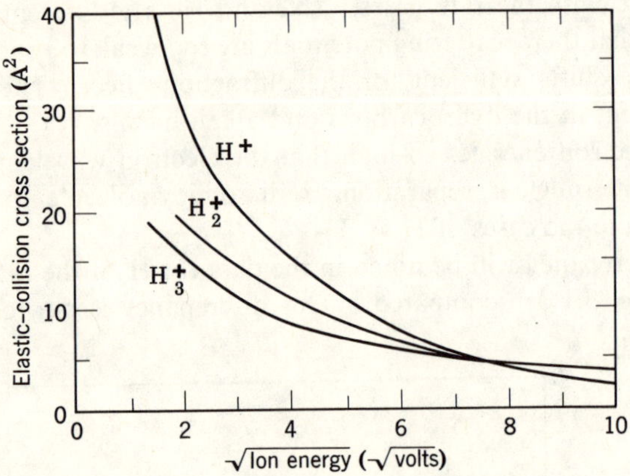

Fig. 3.4. Elastic-collision cross section of hydrogen ions in H_2. After E. E. Muschlitz, Jr., T. L. Bailey, and J. H. Simons, *J. Chem. Phys.*, **24**, 1202 (1956).

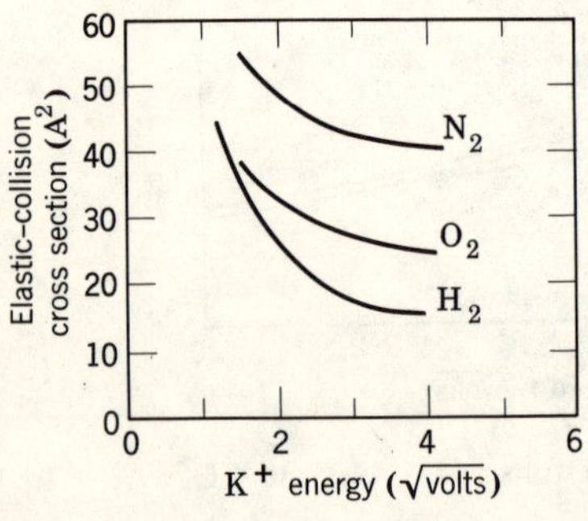

Fig. 3.5. Elastic-collision cross section for K^+ ions in N_2, O_2, and H_2. After C. Ramsauer and O. Beeck, *Ann. Physik*, **87**, 1 (1928).

3.2. Charge Transfer

Another kind of interaction between particles is the transfer of charge from a fast ion to a slow atom. The process is quite important in thermonuclear devices because an energetic ion may escape from the system as a fast neutral, leaving a low-energy ion behind. The cross sections σ_t are appreciable at both high and low energies.

Charge-transfer cross sections for high-energy protons in H_2, He, N_2, and A are shown in Fig. 3.6; data for He^+ ions in the same gases are shown

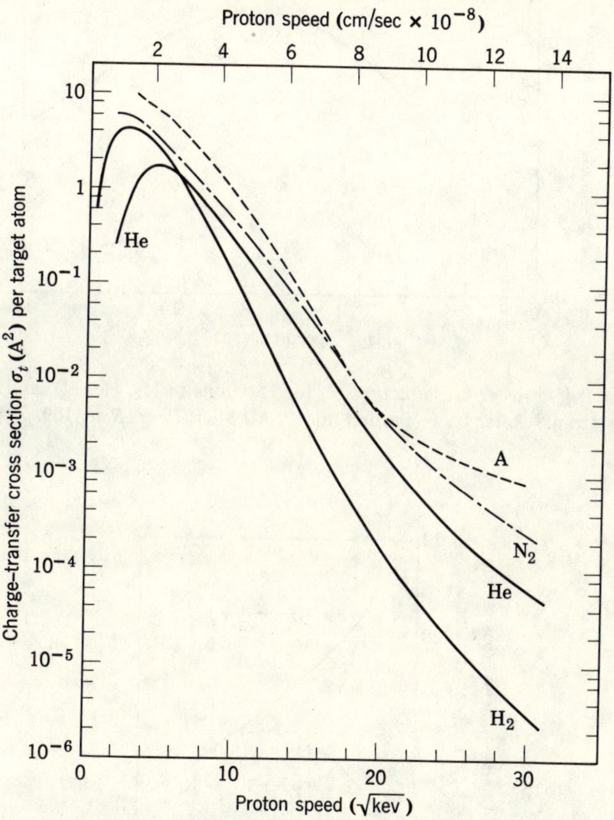

Fig. 3.6. Charge-transfer cross sections σ_t for H^+ ions in H_2, He, N_2, and A; note that data are plotted per atom, not per molecule. After P. M. Stier and C. F. Barnett, *Phys. Rev.*, **103**, 896 (1956); C. F. Barnett and H. K. Reynolds, *ibid.*, **109**, 355 (1958).

in Fig. 3.7. Note that the results are plotted per atom, as if the H_2 and N_2 were dissociated. An H_2 molecule will have twice the indicated cross section, and so on. In actuality, an H_2 molecule will not appear as two

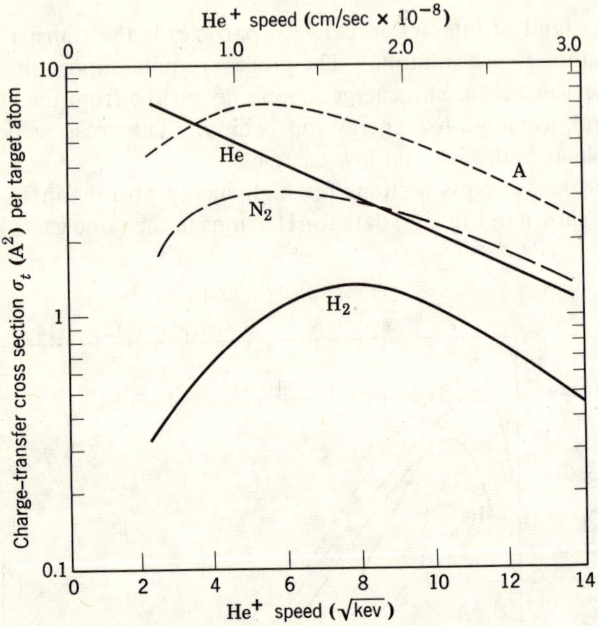

Fig. 3.7. Charge-transfer cross sections σ_t for He$^+$ ions in H_2, He, N_2, and A; results plotted per atom. After C. F. Barnett and P. M. Stier, *Phys. Rev.*, **109**, 385 (1958).

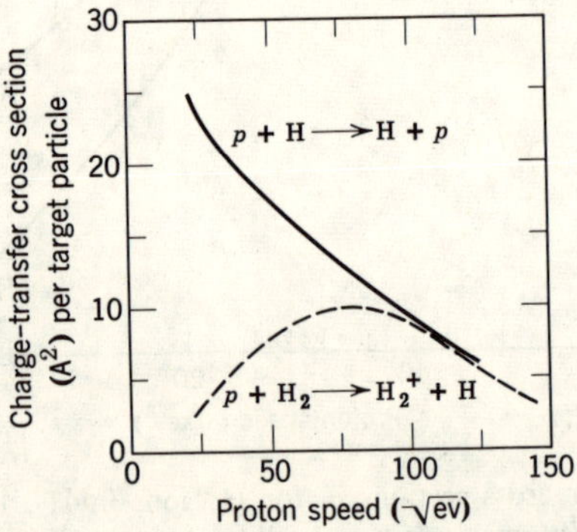

Fig. 3.8. Charge-transfer cross sections for protons in atomic and molecular hydrogen gas. Atomic data are plotted per atom and molecular data, per molecule. After W. L. Fite, R. T. Brackmann, and W. R. Snow, *Phys. Rev.*, **112**, 1161 (1958).

protons, because the reaction $p + H \rightarrow H + p$ is a resonance phenomenon, whereas $p + H_2 \rightarrow H + H_2^+$ (or $2H + p$) is not. This point is illustrated in Fig. 3.8, which shows the charge-transfer cross sections by protons in atomic hydrogen and molecular hydrogen, the latter being plotted per molecule. Note that above about 10 kev the cross sections in H or H_2 are about the same; this equality would not be the case if H_2 looked like two atoms at this energy. The cross section σ_t for $d + D \rightarrow D + d$ is shown by Fite, Brackman, and Snow to be almost identical to σ_t for $p + H \rightarrow H + p$. The resonance effect also shows up in $He^+ + He \rightarrow He + He^+$, as shown in Fig. 3.9. Figure 3.10 illustrates the case of protons and deuterons in D_2 gas.

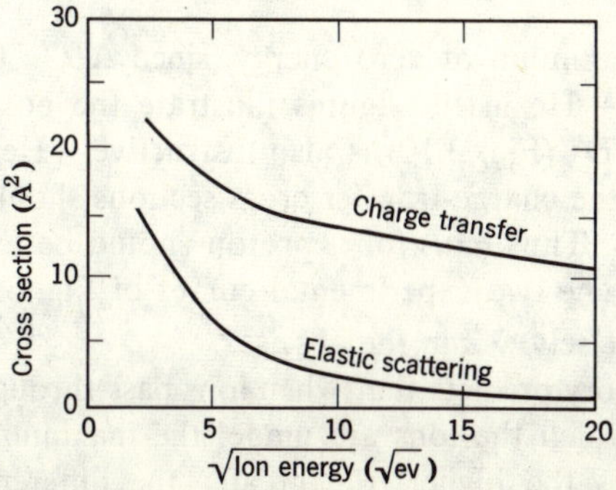

Fig. 3.9. Interactions of He^+ in helium. After W. H. Cramer and J. H. Simons, *J. Chem. Phys.*, **26**, 1272 (1957).

We observe in these data that the charge-transfer cross section σ_t is large at high energies and has a maximum generally in the kilovolt region, except for $p + H$ and $He^+ + He$. An approximate condition for this maximum is readily derived.

Fig. 3.10. Charge-transfer cross sections per molecule for protons and deuterons in D_2 gas. After W. L. Fite, R. T. Brackmann, and W. R. Snow, *Phys. Rev.*, **112**, 1161 (1958).

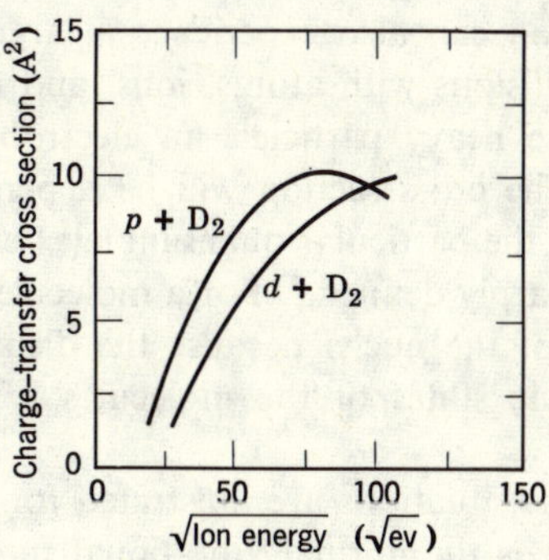

The approaching particle passes by the target in some characteristic time τ, and the principal component of the frequency spectrum of the interaction is $\nu = 1/\tau$. Since the interaction is governed by quantum mechanics, we expect that the maximum cross section will be found when the energy $h\nu = h/\tau$ is equal to the energy difference ΔU involved in the interaction. Thus

$$a\,\Delta U/hv = 1, \tag{3.1}$$

where a is the range of interaction, v the relative speed, and ΔU the difference in ionization potentials.

According to this view, resonant charge transfer

$$M^+ + M \rightarrow M + M^+ \tag{3.2}$$

should have a maximum at zero energy, since $\Delta U = 0$. The cases of $p + H$ and $He^+ + He$ in the figures illustrate the point. The case of $p + D_2$ and $d + D_2$ (Fig. 3.10) is also instructive. Here a and ΔU are the same; hence, the charge-transfer cross sections should be the same at equal ion speeds. Thus $\sigma_t(u)$ for a proton should be the same as $\sigma_t(2u)$ for a deuteron. The two experimental curves of Fig. 3.10 are separated by a factor of precisely $\sqrt{2}$ in the abscissa.

This simple theory predicts that when ions pass through a gas different from that from which the ions are made, the maximum of the transfer cross section σ_t is the higher, the greater the difference in ionization potentials. In some molecular gases there are a large number of vibrational and rotational states of the ions and neutrals, and for some combinations ΔU may be small. The maxima in these cases may occur at very low energies.

3.3. Excitation of Atoms, Ions, and Molecules by Electrons

An electron can also experience a large number of different types of inelastic collisions with atoms, ions, and molecules. In a collision with one of these heavy particles, an electron may merely excite the heavy particle. The cross section will be dependent on the energy of relative motion and the particular quantum level excited. For an atom the levels are quite sharply defined. For a molecule the energy is a function of the separation of the nuclei because the distance between the atoms affects the electronic states of the molecules. The effect is important in ion sources.

Both these situations are illustrated for H_2 in Figs. 3.11*a* and *b*. The shaded area is the normal vibrational range of the nuclei in the ground state (labeled $^1\Sigma_g$). To pull the atom apart requires about 4.4 ev, as shown by two H atoms at large nuclear separation. Various excited states of the molecule are shown at high energies. The first of these can be excited with a minimum energy of about 8.9 ev ($^3\Sigma_u$). This state is repulsive; the molecule dissociates into two H atoms, each with about 2.25 ev. Other excited states are shown, and there are more beside.

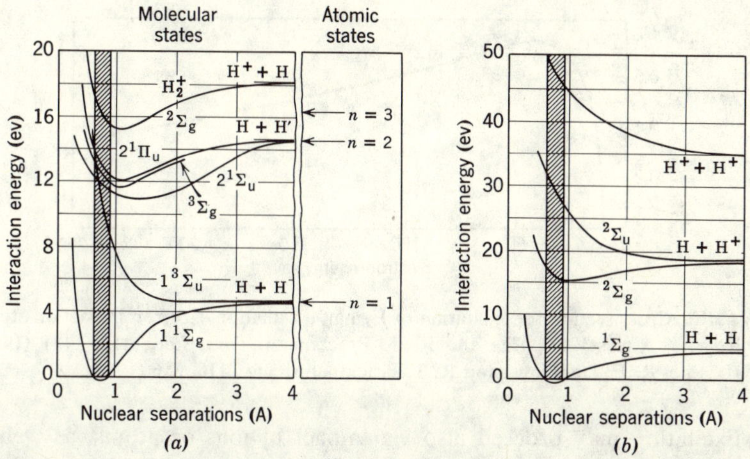

Fig. 3.11. Potential-energy curves for electronic states of H_2 and $H_2{}^+$. (*a*) States within 20 ev of the ground state. (*b*) Higher energy states. After H. S. W. Massey and E. H. S. Burhop, *Electronic and Ionic Impact Phenomena*, Clarendon Press, Oxford (1952), pp. 230, 231.

If the excited molecule (for example, $^1\Sigma_u$) is pulled apart, we find a H atom in the ground state and an excited H atom. In this particular case, the H is in its lowest excited state (principal quantum number $n = 2$). Thus, we associate the asymptotic values at the right of Fig. 3.11*a* with the energy of the excited states of one H atom if the zero of the energy scale is shifted up to the H + H level.

Once an atom, ion, or molecule is excited, it can decay to some lower excited state or to the ground state with the emission of a photon, unless some selection rule of quantum mechanics inhibits the transition. Selection rules result from incommensurate symmetry between the two states. In such a case, the state is said to be metastable, since the lifetime is relatively very long (greater than 10^{-6} sec). In general, radiating states decay in about 10^{-7} sec.

Data on excitation cross sections σ_x by electron impact are meager.

The cross section for excitations in atomic hydrogen which lead to the transition $n = 2 \rightarrow n = 1$ (10.2-ev Lyman α radiation) in Fig. 3.11*a* is shown in Fig. 3.12. The shape of the curve is typical: a sharp rise above the excitation energy u_x and a slow decline at high energy.

The sum of all the excitation cross sections to the various states would be a very useful number, but it has been measured only at electron impact energies up to approximately the ionization potential.

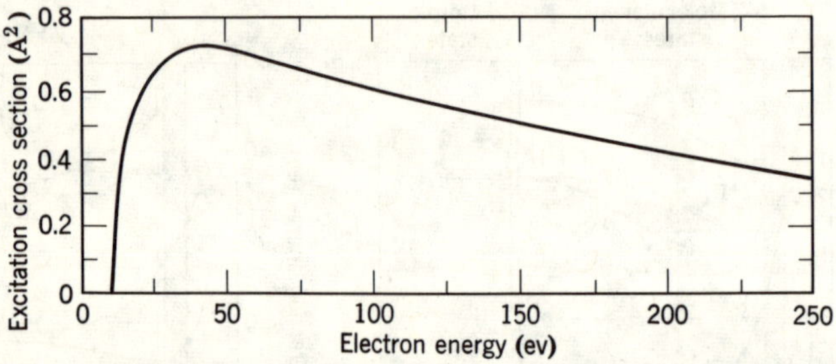

Fig. 3.12. Cross section for excitation of Lyman α radiation–electron impact on atomic hydrogen. After W. L. Fite and R. T. Brackmann, *Phys. Rev.*, **112**, 1151 (1958); W. L. Fite, R. F. Stebbings, and R. T. Brackmann, *ibid.*, **116**, 356 (1959).

Excitation may proceed also via impact of ions or atoms, as well as electrons. Few data exist. In general, the cross sections become large at much higher impact energy than for the case of electron impact. Even if the energy is quite sufficient, the ion must in addition be fast enough for a resonance of the sort described in Sec. 3.2 to occur, if the cross section is to be appreciable.

3.4. Ionization by Electron, Ion, and Atom Impact

An electron may ionize an atom or molecule, or further ionize an incompletely stripped atom. This is the way ions are made in ion sources.

In general, the ionization cross sections σ_i for electron impact are well known at least up to several hundred volts. Above such energies, theoretical extrapolations usually suffice. The case for H_2 and H is shown in Fig. 3.13; the cross sections for D_2 and D or T_2 and T will be virtually identical. Note that H_2 appears as two H atoms to a fast electron. At high energy, theory predicts $\sigma_i \propto (1/\text{energy})$; this dependence is borne out by the data.

Figures 3.11*a* and *b* show more of the events in H_2. The minimum ionization potential is about 15.8 ev, and the onset for formation of

$H^+ + H$ is 18.1 ev. The reaction $H_2^+ + e \rightarrow 2H^+ + 3e$ is unlikely, and has a threshold at about 46 ev. Thus, with a distribution of electron energies between 16 and 50 ev, for example, only about half the hydrogen atoms can be transformed into H ions by single impacts on H_2. The rest appear as H atoms of several ev energy, or as the H in H_2^+.

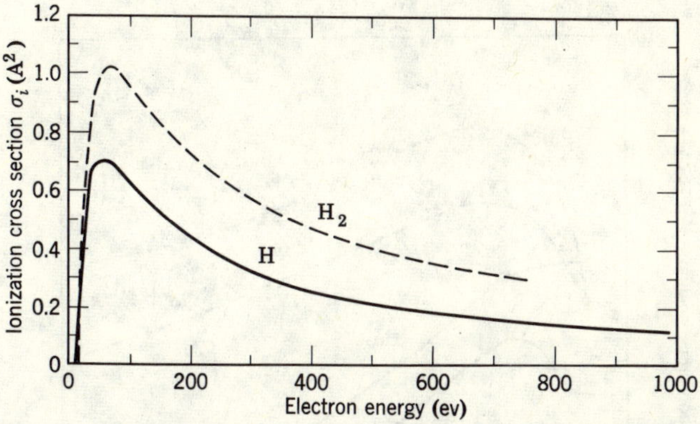

Fig. 3.13. Ionization cross sections σ_i for atomic and molecular hydrogen by electron impact. After W. L. Fite and R. T. Brackmann, *Phys. Rev.*, **112**, 1141 (1958).

The ionization cross sections σ_i for electrons in helium and neon are shown in Fig. 3.14. Note that the cross section for direct formation of Ne^{++} is much less than that for formation of Ne^+; $\sigma_i(Ne^{+++})$ is about twenty times lower still.

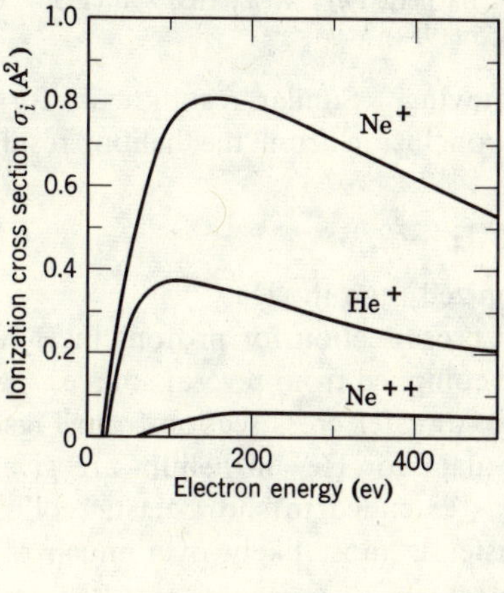

Fig. 3.14. Cross sections σ_i for formation of He^+, Ne^+, and Ne^{++} from the neutral atom by single-electron impact.[1]

Some data are available for impact by fast atoms or ions, particularly for H atoms and protons. Barnett and co-workers[3,4] have obtained data using fast neutral H beams, which may be interpreted as the cross section for ionization of H on impact with various target atoms. Whether the target atom is itself ionized was not discovered. The data are shown in Fig. 3.15. Only the relative speed is shown; we may consider either

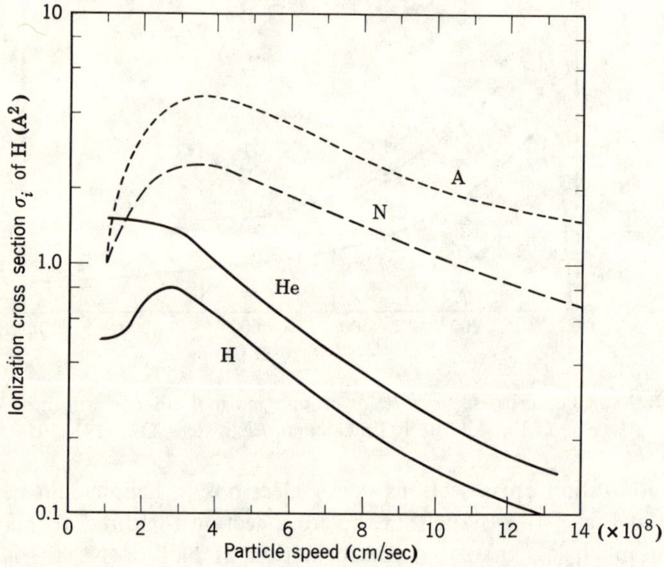

Fig. 3.15. Cross sections σ_i for ionization of H atoms on impact with H, He, N, or A atoms. Curves are plotted per atom, but H and N data are derived from molecular gas targets. After P. M. Stier and C. F. Barnett, *Phys. Rev.*, **103**, 896 (1956); C. F. Barnett and H. K. Reynolds, *ibid.*, **109**, 355 (1958).

the H atom or the target to be moving. Similar results exist for fast helium neutrals.[3] An interesting conclusion from the helium results is that in a reaction of the sort

$$\text{He} + \text{H} \rightarrow ? \tag{3.3}$$

the H is rather more likely to be ionized than the He.

Figure 3.16 shows the ionization cross section for protons in H_2 on a per molecule basis. The data were compiled from several sources. Also plotted for comparison is the charge-transfer cross section from Figs. 3.6 and 3.8. Similar and less reliable data for He^+ in helium are given in Fig. 3.17, where the energy range is extended to show elastic collisions also. One notes that charge transfer is most likely over much of the energy range of thermonuclear interest.

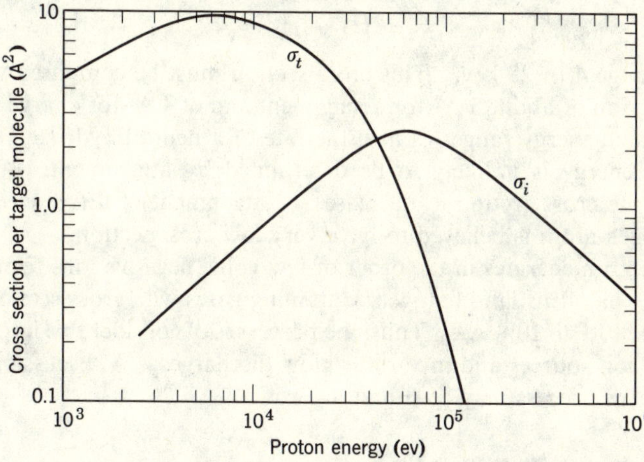

Fig. 3.16. Ionization cross section σ_i per target molecule for protons in H_2. After J. P. Keene, *Phil. Mag.*, **40**, 369 (1949); V. V. Afrosimov, R. N. Ilin, and N. V. Fedorenko, *Sov. Phys. JETP*, **34**, 968 (1958); E. W. McDaniel, D. W. Martin, J. W. Hooper, and D. S. Harmer, to be published. Plotted for comparison is the charge-transfer cross section σ_t from Figs. 3.6 and 3.8, converted to a per molecule basis.

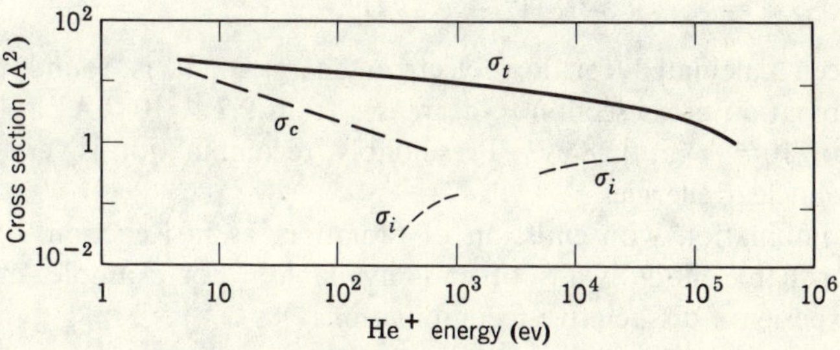

Fig. 3.17. Cross sections per target atom for He^+ in He. After A. Rostagni, *Nuovo Cimento*, **13**, 389 (1936); J. P. Keene, *Phil. Mag.*, **40**, 369 (1949); see Figs. 3.7 and 3.9 for other references.

3.5. Electron Attachment

An electron may attach itself to a neutral atom to form a negative ion. The binding energy ranges from a few tenths of a volt to a few volts. The attachment energy is greatest for the halogens, since the outer electron shell lacks but one electron of being filled.

In general, electron attachment is of little significance in a plasma of thermonuclear interest. To be sure, Stier and Barnett[4] measure an attachment cross section σ_a of about 0.1 A^2 per target atom for the reaction

$$H + H_2 \rightarrow H^- + ? \tag{3.4}$$

in the range 4 to 25 kev. This cross section must be compared with the cross section of about 5 A^2 for detachment and of 4 A^2 for charge transfer in the same energy range. Thus, the fate of a neutral hydrogen atom at kilovolt energy is unlikely to be determined by attachment. At lower energy, the cross section σ_a decreases. Attachment of a free electron to a hydrogen atom is believed to have very low cross section.

At electron energies in the order of few volts, negative ions form in O_2, H_2O, the halogens, and halogen containing gases with cross sections σ_a in the range 10^{-2} to 10^{-1} A^2. Thus, the process is of considerable importance in some ion sources and in various glow discharges. A discussion of the attachment process may be found in Brown.[1]

3.6. Recombination

Electrons and positive ions, or ions of opposite sign, may recombine with or without the emission of radiation. Here again, the process is not thought to be of much importance in a thermonuclear plasma. For example, the reaction

$$H^+ + e \rightarrow H^0 + h\nu \tag{3.5}$$

has been calculated for low electron energies. It is found that the recombination cross section σ_r decreases from 2.7×10^{-3} A^2 at 0.034 ev to 2.3×10^{-4} A^2 at 0.28 ev. Presumably, recombination is very small at thermonuclear energies.

Recombination with emission of radiation, as in Reaction 3.5, is the least probable mode if any other is available. For example, molecular ions experience dissociative recombination:

$$XY^+ + e \rightarrow X + Y, \tag{3.6}$$

with the energy appearing as kinetic energy of the fragments. Here, we find $\sigma_r \approx 10^2$ A^2 at an energy of 0.03 ev. It is expected that the cross section σ_r will decrease rapidly with increasing energy. In addition, molecular ions will be very rare in a thermonuclear plasma. Negative and positive ions may similarly recombine, with or without dissociation.

While the actual recombination illustrated in Reaction 3.5 is itself not important, a process somewhat related to it is very important: the radiation of a fast electron not actually captured. Its acceleration in the vicinity of a proton may induce radiation, which is an important energy loss in a thermonuclear plasma, and will be discussed in more detail later.

Further details regarding recombination can be found in Brown[1] and in Massey and Burhop.[2]

3.7. Collision of Particles with Surfaces. Nature of the Surface

Very small traces of impurity gases can have rather drastic effects on high-energy plasmas. The impurities may come from the initial gases themselves, but these gases can generally be made very pure. It is much more likely that the contamination arises from the surfaces in the vicinity.

Reactions with surfaces are quite complex. It is necessary to understand first the composition of the surface. An atomically clean surface consists, of course, of atoms of only the underlying material, and for the usual refractory metals the binding energy is a few volts (2 to 8 ev). For the alkali metals it is very low. We rarely encounter an atomically clean surface. Heating at 2000 to 3000° K will produce such a surface in a short time ($\leqslant 1$ sec) in a good vacuum.

Gases are adsorbed on any clean surface. In general, gas atoms striking the surface stick with good probability (0.3 to 0.5) until one foreign atom lies on the surface for about each four metal atoms. The binding energy of each foreign atom to the surface is maximized by such a disposition. The energy of adsorption is usually 2 to 4 ev per adsorbed atom. Somewhat arbitrarily, such coverage is called a monolayer; a monolayer is shown idealized in Fig. 3.18 for a simple lattice. We should expect the

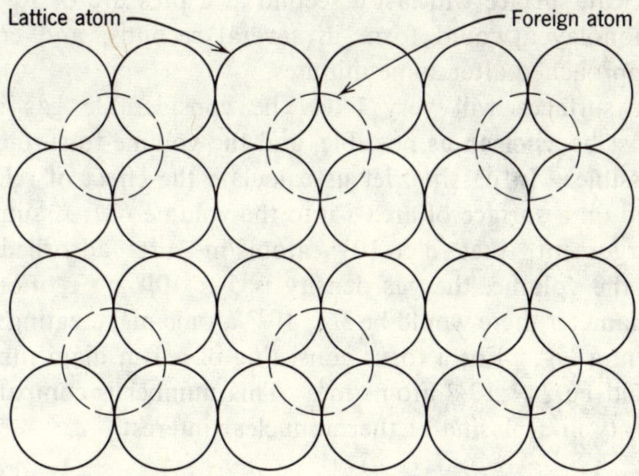

Fig. 3.18. One monolayer viewed from above.

adsorption to be modified by the chemical and physical nature of the gas. Rare gases are only weakly bound to a surface. Since the hydrogen atom is small, it may penetrate deeply into the lattice. Hydrogen is strongly adsorbed.

Once a monolayer is formed, further atoms find fewer free valence bonds. The sticking probability drops by a factor of 10^{-1} to 10^{-2}, and the binding energy is lower for a second such monolayer. Continued adsorption, if allowed, builds up a layer of adsorbed gas several monolayers deep. The sticking probability becomes 10^{-3} or less, and the energy decreases as monolayers are added. If the surface is in equilibrium with the surrounding gas, eventually atoms leave by thermal agitation as fast as they arrive. At room temperature, the binding energy of the last layers is a few tenths of 1 ev, typical of van der Waals' polarization forces. The heavier rare gases are quite polarizable and will be adsorbed on room-temperature surfaces to the extent of a monolayer or so.

The gas monolayers may be successively removed by heating. In a good vacuum, heating at 700° K will remove in a short time all but the last few monolayers. Heating at 1300° K will remove all but the last monolayer, and flashing at 2500° K removes the last one, provided the surface is not destroyed.

The monolayers form very rapidly. At a pressure of 1 mm Hg at room temperature, the gas density is about $3.5 \times 10^{22}/m^3$. Typical thermal gas speeds are about 5×10^2 m/sec. An average over directions of arrival then yields an arrival rate of about $4 \times 10^{24}/m^2$-sec at the surface. A monolayer has about 4×10^{18} atoms/m^2; with unity sticking probability, the first monolayer forms in about 10^{-6} sec. In other words, an atomically clean surface will last a second at a pressure of 10^{-6} mm Hg, the next monolayer would form in several seconds, and equilibrium would be approached after some minutes.

Quiescent surfaces will very likely be considerable gas reservoirs. Surfaces must be as clean as possible, and the volume-to-surface ratio as large as possible. In passing, let us calculate the effect of releasing the gas adsorbed on a surface of area s into the volume $\mathscr{V}$. Assume, for the sake of argument, that 5×10^{19} atoms/m^2 are adsorbed. Spread throughout the volume, the gas density is $5 \times 10^{19}\ s/\mathscr{V}$. For a sphere 0.01 m in diameter there would be 3×10^{22} atoms/m^3, creating a pressure of nearly 1 mm Hg. For a torus consisting of a 1-m diam tube, the gas density would be 2×10^{20} atoms/m^3. This number is comparable with the ion density in a plasma of thermonuclear interest.

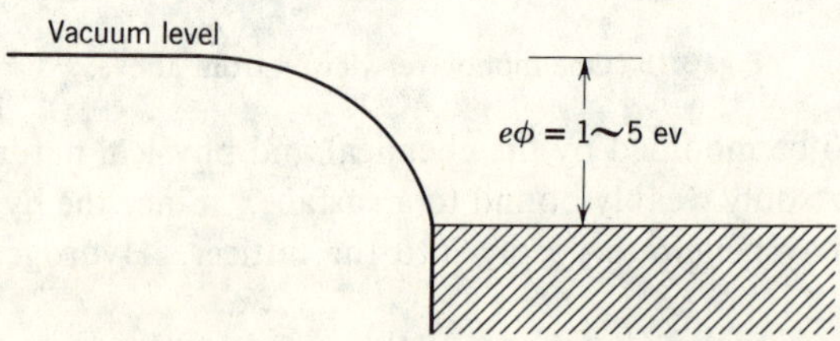

Fig. 3.19. Energy near the surface of a metal.

We may draw an energy-level diagram for the clean surface, as in Fig. 3.19. The electrons in the conduction band lie below the vacuum level by the energy $e\phi$, which is the work function. The magnitude of this energy is generally increased by adsorption of atoms with high ionization potential and decreased by adsorption of atoms with low ionization potential.

3.8. Secondary Emission by Electrons

Electrons when they hit a surface may knock out other electrons. This phenomenon, called secondary emission, is somewhat dependent upon surface conditions. Generally, secondary emission is of little importance in thermonuclear plasmas; it is often important in ion sources and the like, where the phenomenon must be considered in arriving at current balances.

Let δ denote the number of electrons emitted for each electron incident. The secondary-emission coefficient for typical metals is shown in Fig. 3.20.

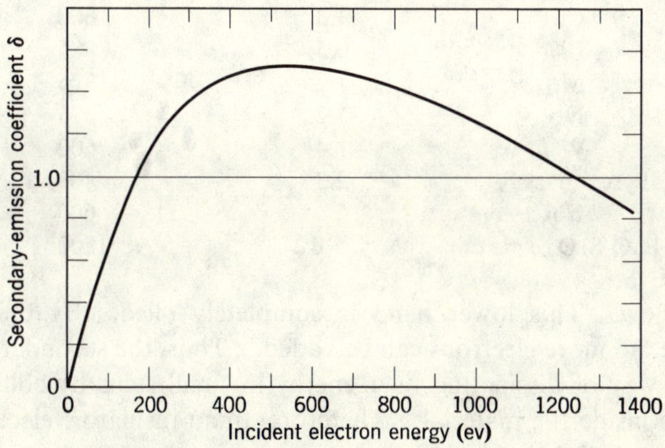

Fig. 3.20. Electron secondary emission for normal incidence on a typical metal.

The shape of the plot of the secondary-emission coefficient versus the energy of the primary electron is rather similar for all materials. Data concerning the maximum value of this coefficient and the energy of the primary electron at which this maximum occurs are given for several materials in Table 3.1. At very low and very high energies, few secondary electrons are emitted, but at intermediate energies more than one secondary electron may be emitted for each incident electron. At low energies, the secondary electrons have energies less than the work function and cannot escape. At high energies, the primary electron penetrates so deeply into the base material that the secondaries formed cannot reach the surface before losing their energy by collision with other electrons. They too cannot get over the potential barrier at the surface.

For an insulator, δ can be high, for example, 8. The reason δ is high is that there are no electrons in the conduction band apart from the few secondaries, and that there is a quantum mechanically forbidden energy gap between the bottom of the conduction band and the next allowed

Table 3.1. Maximum values of the secondary-emission coefficient δ and the energies at which the maxima occur

Material	Maximum Value of Secondary-Emission Coefficient δ_{max}	Energy at Which Maximum Occurs (ev)
Al	1.0	300
Au	1.46	800
Ba	0.83	400
C	1.0	300
Cu	1.3	600
Li	0.5	85
Mo	1.25	375
Nb	1.2	375
W	1.4	600
Pyrex glass	2.3	400
NaCl	6	600
BaO-SrO oxide mix	5–12	$\approx$ 1200

band below. This lower band is completely filled. By the exclusion principle, no more electrons can be added. Thus, the secondary electrons have no way of dissipating their energy in small steps by collisions with electrons inside the material. Therefore, in an insulator, electrons reach the surface with less energy loss. The surface of an insulator will charge positively or negatively depending upon whether the secondary-emission coefficient is larger or smaller than 1, respectively.

Cleanliness, roughness, and the crystalline structure of the surface can affect secondary emission. The secondary electrons have predominantly energies in the range 0 to 10 ev. A magnetic induction parallel to the surface will return the secondaries to the surface.

Because of the inefficient energy transfer resulting from the unfavorable mass ratio, electrons have low probability of ejecting atoms.

3.9. Secondary Electron Emission by Ions

Electrons may be emitted from a surface as a result of ion impact. Here again, the phenomenon is not expected to be of great importance in a thermonuclear plasma, but it is generally important in the operation of arc cathodes, dc ion sources, and so forth.

The number of secondary electrons emitted per positive ion is denoted by γ_i. It generally proceeds by either of two methods: (1) potential ejection, (2) kinetic ejection.

In the first and only important case, the ion with ionization potential V_i approaches the surface as in Fig. 3.21. An electron from the conduction

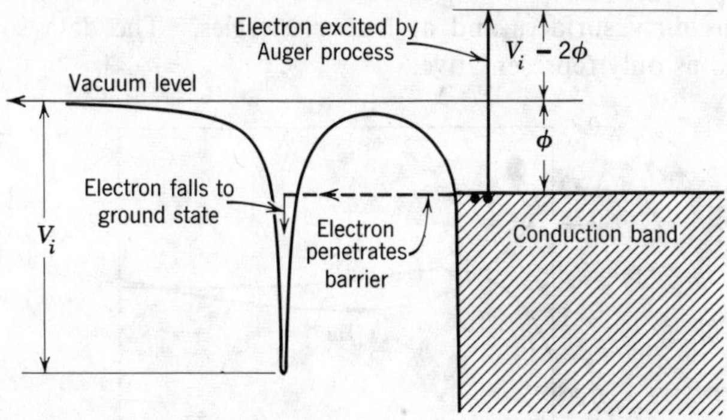

Fig. 3.21. Auger ejection of an electron by a positive ion.

band either tunnels through the reduced potential barrier to an excited state of the atom and then falls to the ground state, or tunnels directly from the metal to the ground state. The energy available, which is a maximum of $e(V_i - \phi)$, goes into Auger excitation of a second conduction-band electron, which can have at most an energy $e(V_i - 2\phi)$ above the vacuum level. The electron then will escape from the metal if it is headed in the right direction and is not reflected by the potential change at the surface. No photon is involved in the interaction.

The probability of secondary emission γ_i should then be rather independent of ion speed, because the electronic transitions are rapid. It should also be more or less an increasing function of $V_i - 2\phi$. Figure 3.22 shows the situation for the rare gases on tungsten. For the most part, γ_i increases with V_i. The inversion of He^+ and Ne^+ in the energy range 200 to 900 ev is explained partly on the basis of the atomic excitation levels in the gas atoms.

Few reliable data exist for γ_i by hydrogen-like or deuterium-like ions. Experiments by Rose[5] indicate that γ_i should be about 0.1 for H_2^+ or D_2^+ on clean molybdenum. For H^+ or D^+, we should guess that $\gamma_i \approx 0.05$ because V_i is lower for these ions.

Multiply charged ions and excited singly charged ions have a higher secondary-emission coefficient, corresponding to the possibility of two

Auger processes taking place in succession. For example, $\gamma_i \approx 0.7$ for He^{++} on clean refractory metals.

For surfaces covered with adsorbed gases, in general, the secondary-emission coefficient is reduced at low ion energy, corresponding to the inability of an ion to penetrate to the metal surface. The effect has been discussed by Hagstrum[6] and by Rose.[5] In general, surfaces of experimental tubes are covered in an uncontrolled way, and there is a great mélange of effects. The curves of Fig. 3.23 are typical of results obtained on various dirty surfaces and at higher energies. The data should be considered as only representative.

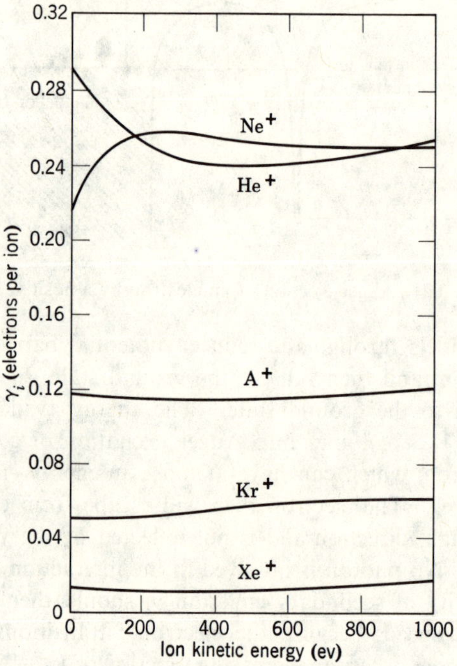

Fig. 3.22. Total electron yield for singly charged ground-state ions on atomically clean tungsten. After H. D. Hagstrum, *Phys. Rev.*, **104**, 317 (1956).

The secondary electrons have energies in the approximate range 0 to 15 ev if produced by singly charged ions of energy less than 1 kev. Faster and multiply charged ions eject more energetic secondaries.

3.10. Sputtering

Sputtering is a process that blackens vacuum tubes and lamp bulbs. Ions of the gas strike surfaces and eject atoms of the surface material, which deposit elsewhere. In a thermonuclear device, these atoms may become ionized and contribute to the impurities in the plasma. Therefore, sputtering is an important consequence of ion-surface impacts.

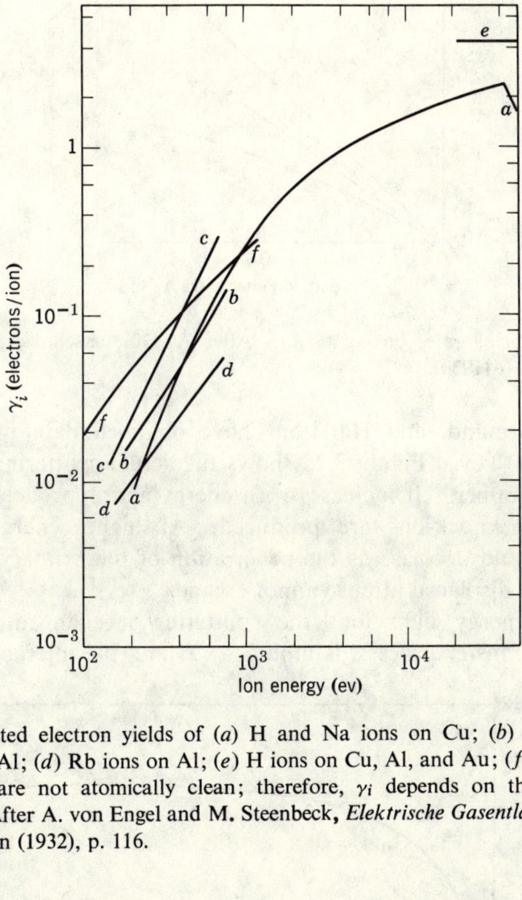

Fig. 3.23. Ejected electron yields of (*a*) H and Na ions on Cu; (*b*) K ions on Al; (*c*) Li ions on Al; (*d*) Rb ions on Al; (*e*) H ions on Cu, Al, and Au; (*f*) H ions on Ni. The surfaces are not atomically clean; therefore, γ_i depends on the experimental conditions. After A. von Engel and M. Steenbeck, *Elektrische Gasentladungen*, Vol. 1, Springer, Berlin (1932), p. 116.

Sputtering is largely a momentum transfer mechanism, as shown by Wehner[7] and others. Very roughly, sputtering is expected to be most prolific if the incident ion mass is equal to that of a surface atom. Figure 3.24 illustrates this fact: the sputtering rate increases as the masses approach equality. The scale on the right is the quantity observed in the particular experiment. If the unmeasured secondary-emission coefficient $\gamma_i \neq 0$, we must correct the target current for electrons that leave. Since $\gamma_i \leqslant 0.3$ in most cases, the correction is not large. Other data involving Hg^+ ions are given by Wehner.[7]

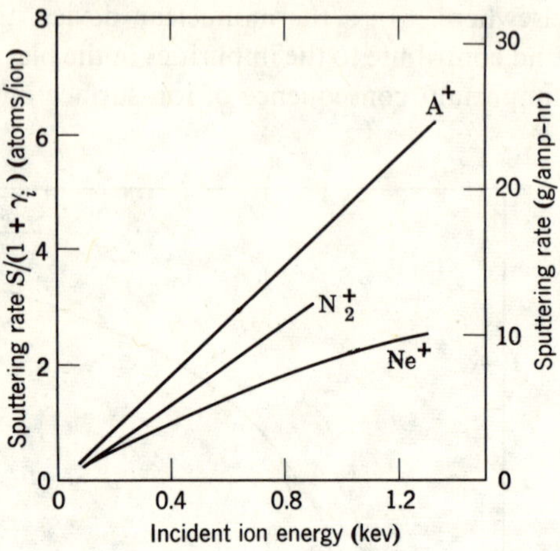

Fig. 3.24. Sputtering rates on Ag target. After A. Güntherschulze and S. Meyer, *Z. Physik*, **62**, 607 (1930).

Yonts, Normand, and Harrison[8] have obtained data in the energy region 5 to 50 kev. Figure 3.25 shows the actual sputtering ratio S for A^+ ions on copper. It increases with energy and approaches a limit as higher-energy knock-ons are produced. At higher energy than that shown, S should decrease as the penetration of the primary becomes so deep that the displaced atoms cannot escape.

For high-energy light ions the sputtering mechanism, which is a momentum transfer process, is modified. Here, the interaction is not a

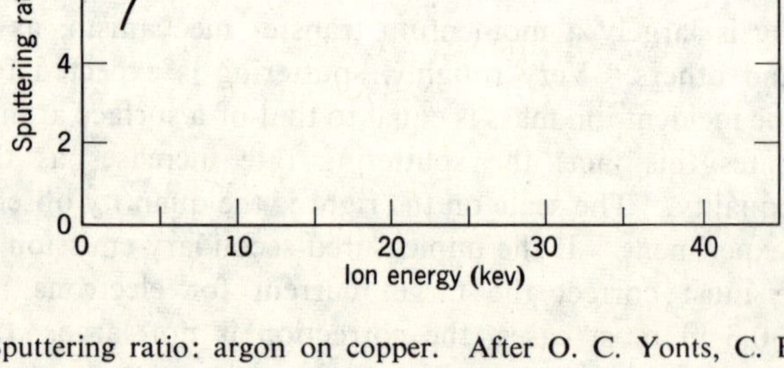

Fig. 3.25. Sputtering ratio: argon on copper. After O. C. Yonts, C. E. Normand, and D. E. Harrison, Jr., *J. Appl. Phys.*, **31**, 447 (1960).

hard-sphere type but is rather a Coulomb interaction of the ionic charge with the weakly screened target nucleus. Then, because the Coulomb cross section decreases with (*a*) increasing energy or (*b*) decreasing target atomic number, the sputtering ratio should show a corresponding behavior. Figure 3.26 shows S for D^+ and He^+ on copper, and the decrease is evident. Pease[9] has calculated the sputtering ratio to be expected for unscreened collisions. Table 3.2 shows his calculation for

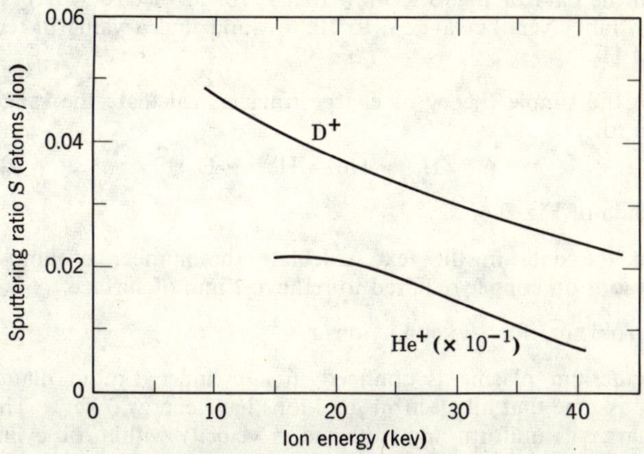

Fig. 3.26. Sputtering ratio: D^+ and He^+ on copper. After O. C. Yonts, C. E. Normand, and D. E. Harrison, Jr., *J. Appl. Phys.*, **31**, 447 (1960).

Table 3.2. Sputtering by deuterons (theoretical)

Material	Sputtering Rate (atoms/ion)		
	10 kev	50 kev	100 kev
Be	0.005	0.0013	0.0007
Al	0.025	0.0076	0.0037
Ti	0.039	0.010	0.0058
Fe	0.043	0.012	0.0065
Cu	0.048	0.013	0.0072

sputtering by deuterons; we observe an approximate check with the experimental data of Fig. 3.26. We expect that S for protons would be several times lower. Wehner's theory predicts that the threshold for sputtering by D^+ on heavy metals should be 500 to 1000 ev.

For thermonuclear devices, a sputtering ratio of even 10^{-3} atom/ion can be a serious source of contamination; in addition, surface monolayers of impurity gas can be readily dislodged by sputtering because of the low binding energy. Either the ions must be kept rigorously away from the surface, or the sputtered atoms must be ionized near the wall and somehow removed.

PROBLEMS

1. From the data of Fig. 3.1, show that ν_c for an electron in H_2 is approximately constant over the range 6 to 100 ev, and find a value in terms of the pressure of H_2.

2. From the simple theory of charge transfer, calculate the range a of the interaction for

$$H^+ + He \rightarrow He^+ + H$$

from the data of Fig. 3.6.

3. From the data in the text, calculate the number of amp-hr/cm^2 of 10-kev A^+ ions on copper required to remove 1 mm of surface.

4. Do Problem 3 for 10-kev D^+ ions.

5. A deuterium plasma is confined in a cylinder 1 m in diameter. For simplicity, assume that all electrons and ions have energy 3 kev. The distribution of charges is uniform and isotropic in velocity within the cylinder. The deuterium charge density is $10^{20}/m^3$. Deuterium gas is being slowly desorbed from the walls, and the molecules enter the plasma. Assume that they do so radially and all have thermal speed $= 2 \times 10^3$ m/sec. The deuterium molecules are destroyed only by:

(*a*) Ionization by electrons, for which the Born approximation gives $\sigma_i = 0.2/u$ (kev) A^2.
(*b*) Charge transfer, $\sigma_t = 6$ A^2 at 3 kev for $D_2 + D^+ \rightarrow D_2^+ + D$.

Neglecting second-order effects, calculate:

(*a*) The relative loss rate of D_2 through ionization by electrons versus charge exchange with D^+.
(*b*) The D_2 density variation as a function of distance into the plasma, ignoring the curvature.

REFERENCES

1. S. C. Brown, *Basic Data of Plasma Physics*, The Technology Press and John Wiley & Sons, New York (1959).
2. H. S. W. Massey and E. H. S. Burhop, *Electronic and Ionic Impact Phenomena*, Clarendon Press, Oxford (1952).
3. C. F. Barnett and P. M. Stier, *Phys. Rev.*, **109**, 385 (1958).
4. P. M. Stier and C. F. Barnett, *Phys. Rev.*, **103**, 896 (1956).
5. D. J. Rose, *Proc. 3rd International Conference on Ionization Phenomena in Gases*, Venice (1957), p. 888.

6. H. D. Hagstrum, *Phys. Rev.*, **104**, 1516 (1956).
7. G. K. Wehner, *Phys. Rev.*, **102**, 690 (1956); **108**, 35 (1957).
8. O. C. Yonts, C. E. Normand, and D. E. Harrison, Jr., *J. Appl. Phys.*, **31**, 447 (1960).
9. See J. L. Craston and co-workers, *Proc. 2nd U.N. Conf. on Peaceful Uses of Atomic Energy*, Vol. 32, United Nations, Geneva (1958), p. 414.

GENERAL REFERENCES

S. C. Brown, *Basic Data of Plasma Physics*, The Technology Press and John Wiley & Sons, New York (1959).

F. H. Field and J. L. Franklin, *Electronic Impact Phenomena*, Academic Press, Inc., New York (1957).

H. S. W. Massey and E. H. S. Burhop, *Electronic and Ionic Impact Phenomena*, Clarendon Press, Oxford (1952).

4

Velocity distributions and averages

Up to now, we have considered the interaction of two particles of known speed or energy. In an ensemble of particles, the velocities are generally not all the same. A particle may change its velocity or even its nature by interacting with its neighbors. In fact, the particle itself is not normally observed at all: it is some average property of the plasma that is observed. We must therefore develop the concept of average coefficients in order to describe the properties of a plasma. To this end, we first establish the basic concepts of velocity space and derive the Boltzmann equation. It is then possible to consider a number of average coefficients. Their use will be illustrated by kinetic equations describing the rates at which reactions take place.

4.1. Velocity Space and Distribution Functions

Abstract spaces are frequently used by mathematicians to systemize their thinking. There are spaces of many types. One of these is particularly useful in the present connection, namely, the velocity space. A velocity space is simply a space in which one point describes the velocity of a particle. A Cartesian coordinate system in this space specifies the three components v_x, v_y, and v_z of the velocity of the particle. A particle may move continuously in this space if its velocity changes continuously. For example, the rotation of a charged particle in a magnetic induction would be a circle centered on the origin in this space. As another example, uniform acceleration of a particle in one direction would be represented by a uniform motion of the particle in velocity space. A particle may also move discontinuously in this space. For example, the forces between two molecules are of very short range. Consequently, the scattering between molecules takes place in times short compared with any of interest. In this approximation, the direction of travel of a molecule in a scattering process may suddenly change. Then the point describing the particle velocity in this space will disappear from one point and reappear at another.

Generally, a large collection of particles must be discussed. Such a collection is called an ensemble. The velocities of all of these particles will form a cloud in the velocity space. Figure 4.1 illustrates this concept: particles are spotted in the velocity space according to their instantaneous values of v_x, v_y, and v_z.

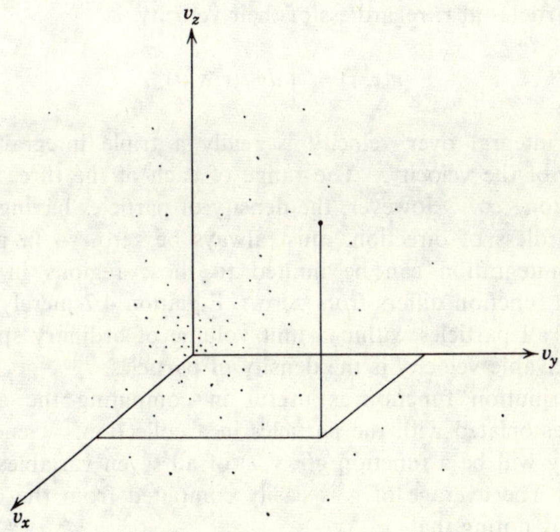

Fig. 4.1. Particles in velocity space.

A particle may be completely described by its location (x, y, z) in ordinary space, by its velocity (v_x, v_y, v_z), and by the time t at which it is observed. Because so many variables are involved, we shall let the symbol $\mathbf{r}$ stand for the variables x, y, and z and the symbol $\mathbf{v}$ stand for the variables v_x, v_y, and v_z. Thus, three symbols represent the seven variables. In addition, elements of volume must be considered. The element $dx\,dy\,dz$ of volume in ordinary space will be denoted by $d\mathbf{r}$; the element $dv_x\, dv_y\, dv_z$ of volume of velocity space will be denoted by dv. Note very well that dr and dv are not to be interpreted as vectors.

The distribution function f of a collection of particles all of one kind (and in this section we consider only one kind) may be defined in the following manner. Consider the number dn of particles in the volume element dr of ordinary space and also in the volume element dv of velocity space at time t. This number is a function of the seven variables $\mathbf{r}$, $\mathbf{v}$, and t, namely:

$$dn = f(\mathbf{r}, \mathbf{v}, t)\, dr\, dv. \tag{4.1}$$

The distribution function is then just the number of particles per unit volume of ordinary space and per unit volume of velocity space which at time t is located at $\mathbf{r}$ and which has the velocity $\mathbf{v}$. The distribution function has the dimensions of $t^3 l^{-6}$. It is a density in two spaces at once. Clearly, the least value of the distribution f is zero.

Useful relations may be deduced from Eq. 4.1. The total density $n(\mathbf{r}, t)$ of particles at $\mathbf{r}$, regardless of their velocity, is

$$n(\mathbf{r}, t) = \int dv\, f(\mathbf{r}, \mathbf{v}, t), \tag{4.2}$$

where the integral over velocity is really a triple integral over each component of the velocity. The range of each of the three integrals is from $-\infty$ to $+\infty$. However, the density of particles having an infinite speed, regardless of direction, must always be zero, so in practice the ranges of integration can be limited to those regions in which the distribution function differs from zero. Equation 4.2 merely states that the sum of all particles within a unit volume of ordinary space and of every conceivable velocity is the density of particles.

The distribution function is useful in computing the averages of quantities associated with the particles in a collection. Generally, such a quantity g will be a function $g(\mathbf{r}, \mathbf{v}, t)$ of all seven variables describing a particle. The average of g is easily computed from the distribution function f by noting that

$$\frac{f(\mathbf{r}, \mathbf{v}, t)}{\int dv\, f(\mathbf{r}, \mathbf{v}, t)} \tag{4.3}$$

is the probability of finding one particle in a unit volume of ordinary space centered at $\mathbf{r}$ and in a unit volume of velocity space centered at $\mathbf{v}$ at time t. The average $\overline{g(\mathbf{r}, t)}$ of the quantity g is then the value of this quantity at $\mathbf{r}$, $\mathbf{v}$, t times the probability that a particle will be found with such coordinates:

$$\overline{g(\mathbf{r}, t)} = \int \frac{dv\, g(\mathbf{r}, \mathbf{v}, t) f(\mathbf{r}, \mathbf{v}, t)}{\int dv\, f(\mathbf{r}, \mathbf{v}, t)}$$

$$= \frac{\int dv\, g(\mathbf{r}, \mathbf{v}, t) f(\mathbf{r}, \mathbf{v}, t)}{\int dv\, f(\mathbf{r}, \mathbf{v}, t)}. \tag{4.4}$$

Again the integrals are to be taken over all velocities for which $f(\mathbf{r}, \mathbf{v}, t)$ differs from zero. The last step of Eq. 4.4 follows because the denominator is independent of $\mathbf{v}$. More conveniently,

$$n(\mathbf{r}, t)\overline{g(\mathbf{r}, t)} = \int dv\, g(\mathbf{r}, \mathbf{v}, t) f(\mathbf{r}, \mathbf{v}, t). \tag{4.5}$$

Here $g(\mathbf{r}, \mathbf{v}, t)$ may be a scalar, a vector, or any other type of quantity. As an important example, the average energy $\bar{U}$ of the particles in an ensemble is given by

$$n(\mathbf{r}, t)\overline{U(\mathbf{r}, t)} = \int dv \left(\frac{mv^2}{2}\right) f(\mathbf{r}, \mathbf{v}, t). \tag{4.6}$$

The average velocity $\bar{\mathbf{v}}$ is given by

$$n(\mathbf{r}, t)\overline{\mathbf{v}(\mathbf{r}, t)} = \int dv\, \mathbf{v} f(\mathbf{r}, \mathbf{v}, t). \tag{4.7}$$

Equation 4.7 means that each component of $\overline{\mathbf{v}(\mathbf{r}, t)}$ is equated to the integral over the corresponding component of $\mathbf{v}f(\mathbf{r}, \mathbf{v}, t)$. It is to be noted that the average velocity will generally be a function of $\mathbf{r}$ and t, even though $\mathbf{v}$ itself is not, because the distribution function may vary with position and time. Further, if the distribution is independent of the direction in which $\mathbf{v}$ points, then the average velocity will be zero.

For any isotropic distribution, the remark just made may be extended. Let

$$g_{\zeta_1\zeta_2\cdots\zeta_k} = v_{\zeta_1} v_{\zeta_2} \cdots v_{\zeta_k}, \tag{4.8}$$

where the ζ's represent any coordinate directions and where k is an odd integer. Then

$$\overline{g_{\zeta_1\zeta_2\cdots\zeta_k}} = 0. \tag{4.9}$$

The result is proved by noting that, if the distribution is isotropic, the distribution function is independent of the direction of the velocity by definition. Therefore, in this case $f(\mathbf{r}, \mathbf{v}, t)$ is an even function of v_x, v_y, and v_z. Then

$$n(\mathbf{r}, t)\bar{g} = \int dv_x \int dv_y \int dv_z \, v_{\zeta_1} v_{\zeta_2} \cdots v_{\zeta_k} f(\mathbf{r}, \mathbf{v}, t); \qquad (4.10)$$

whereupon Eq. 4.9 follows if k is an odd integer, for in one or more of the integrals the contribution from $-\infty$ to 0 will be equal and opposite to that from 0 to $+\infty$.

The difference of the average velocity $\bar{\mathbf{v}}$ and the average speed $\bar{v}$ is to be noted. The former is zero for isotropic distributions, but the latter is not, being an integral over two everywhere-positive factors.

Spherical coordinates are frequently used in velocity space because integrations over the angles can be easily carried out if the product gf is independent of the direction of the velocity. Usually this requirement demands that both g and f be independent of direction of the velocity vector. If gf is independent of direction, then

$$n(\mathbf{r}, t)\overline{g(\mathbf{r}, t)} = 4\pi \int_0^\infty dv \, v^2 g(\mathbf{r}, v, t) f(\mathbf{r}, v, t). \qquad (4.11)$$

4.2. Formulation of the Boltzmann Equation

We have seen that the density of particles may be a function of the three variables describing the position $\mathbf{r}$ of a point in space, the time t, and three more variables specifying the velocity $\mathbf{v}$ of the particles. The Boltzmann equation is merely a conservation relation for the particles. It states that the fate of all particles must be accounted for. Thus we search for an equation for the time derivative of f.

The distribution function changes with time because of particles flowing across the six surfaces bounding an element in phase space. Three of these surfaces are normal to the $\mathbf{v}$-axes, and three are normal to the $\mathbf{r}$-axes. This flow may be continuous or discontinuous. The discontinuous processes arise because of collisions, in which a particle changes its velocity in a time less than the shortest time of interest. Thus we have a contribution

$$\left(\frac{\partial f}{\partial t}\right)_{\text{coll}}$$

to the rate of change $\partial f/\partial t$ of the distribution function.

As regards the continuous flow, we use the divergence theorem applied to six dimensions rather than to the more familiar three dimensions. Let us define a vector $\mathscr{V}$ with six components $(\mathbf{v}, \mathbf{a})$, where $\mathbf{a}$ is the acceleration. This vector $\mathscr{V}$ is a generalized velocity vector for a hypothetical, mathematical space composed of the six coordinates (**r**, **v**) all taken to be mutually orthogonal. Such a space is called a phase space. We now see that the distribution function is merely the density of points in phase space. The current of particles is then $\mathscr{V} f(\mathbf{r}, \mathbf{v}, t)$. Conservation of particles now requires that the rate at which particles flow in over the surface of an element of volume plus those made inside the volume per unit of time be equal to the rate at which the density of particles changes with time. The rate of flow over the surface into the volume element is given by

$$-\int d\mathbf{s}\cdot\mathscr{V} f(\mathbf{r}, \mathbf{v}, t), \tag{4.12}$$

where $d\mathbf{s}$ is a vector in the six-dimensional space normal to an element of the surface bounding a volume element in this space, pointing outward, and of magnitude equal to ds. By the divergence theorem, the flow of particles into the volume element of interest is given by

$$\int dr\, dv\, [\nabla_r\cdot\mathbf{v}f(\mathbf{r}, \mathbf{v}, t) + \nabla_v\cdot\mathbf{a}f(\mathbf{r}, \mathbf{v}, t)], \tag{4.13}$$

where $\nabla_r\cdot$ is the divergence with respect to the coordinates **r**, and $\nabla_v\cdot$ is the divergence with respect to the coordinates **v**. Since the volume element is perfectly arbitrary, the leakage into a very small volume element $dr\, dv$ is given by

$$-[\nabla_r\cdot\mathbf{v}f(\mathbf{r}, \mathbf{v}, t) + \nabla_v\cdot\mathbf{a}f(\mathbf{r}, \mathbf{v}, t)]\, dr\, dv, \tag{4.14}$$

since over the volume element $\mathscr{V} f(\mathbf{r}, \mathbf{v}, t)$ will change negligibly.

The addition of the two contributions yields

$$\frac{\partial f(\mathbf{r}, \mathbf{v}, t)}{\partial t} = -\nabla_r\cdot\mathbf{v}f(\mathbf{r}, \mathbf{v}, t) - \nabla_v\cdot\mathbf{a}f(\mathbf{r}, \mathbf{v}, t) + \left(\frac{\partial f(\mathbf{r}, \mathbf{v}, t)}{\partial t}\right)_{\text{coll}}. \tag{4.15}$$

This equation is known as the Boltzmann equation. It is relativistically correct up to this point. The acceleration is usually expressed in terms of the force **F** acting:

$$\mathbf{a} = \mathbf{F}/m, \tag{4.16}$$

where m is the mass of a particle. We have made the nonrelativistic approximation at this point.

A simplification can be made in our case, because only electric, magnetic, and gravitational forces are of interest:

$$\mathbf{F} = q(\mathbf{E} + \mathbf{v}\times\mathbf{B}) - m\nabla\phi, \tag{4.17}$$

where **E** is the electric field, **B** the magnetic induction, ϕ the gravitational potential, q the charge, **v** the velocity, and m the mass. Since

$$\nabla_v\cdot(\mathbf{v}\times\mathbf{B})f(\mathbf{r}, \mathbf{v}, t) = (\mathbf{v}\times\mathbf{B})\cdot\nabla_v f(\mathbf{r}, \mathbf{v}, t), \tag{4.18}$$

and since **r** and **v** are independent variables, the Boltzmann equation becomes

$$\frac{\partial f}{\partial t} + \mathbf{v}\cdot\nabla_r f + \frac{\mathbf{F}}{m}\cdot\nabla_v f = \left(\frac{\partial f}{\partial t}\right)_{\text{coll}} \tag{4.19}$$

in nonrelativistic form.

Some approximations have been implicitly made in this formulation, and we now bring them to light. The Boltzmann equation requires that we can distinguish between long-range forces appearing in **F** that affect many neighboring particles in the same way and short-range forces that lead to events describable as collisions. This concept is true for neutrons, widely separated atoms, and so on, which interact in this way; the collisions are well represented as binary events. The formulation is also valid for a weakly ionized plasma, in which the charges interact principally with neutrals by virtue of short range, for example, polarization forces. For a highly ionized plasma, the matter is not so clear, because the particle interaction proceeds via long-range Coulomb fields. Thus the division between **F** and the collision term is somewhat vague. This matter cannot be resolved in the simple Boltzmann formulation because, if we include the neighboring interparticle forces in **F**, then f itself must be chosen to contain so few particles in the volume $d\mathbf{r}\, d\mathbf{v}$ that f cannot be approximated by a continuous function. There are, in fact, other possible formulations, for example, the Fokker-Planck equation, that attempt to remove this difficulty, but at the expense of greater mathematical complexity. Because of this complexity involved in more rigorous approaches, we shall use the Boltzmann equation.

In summary, then, the electric and magnetic fields are treated as continuous macroscopic quantities rather than as functions singular at each charge. The electric and magnetic fields may be averaged over volumes containing many particles, yet very small compared with other characteristic dimensions. A typical particle density of interest to us is $10^{21}/\text{m}^3$ ($10^{15}/\text{cm}^3$). If several kinds of particles are present, then a distribution function and a Boltzmann equation for each kind of particle may be used.

4.3. The Maxwell-Boltzmann Distribution

Knowledge of the distribution function f is important in calculation of averaged quantities of the plasma. A very useful distribution is that pertaining to a collection of particles in thermal equilibrium. That is, if a large collection of particles is allowed to interact with each other for a long period of time in some confined, insulated, and force-free space, then an equilibrium distribution will be set up in which as many particles enter an element of velocity space as leave. As a result of collisions, some particles will be speeded up and others will be slowed down. The distribution achieved is called the Maxwell-Boltzmann distribution, and the ensemble is called Maxwellian for short.

There are a number of derivations of this distribution, some of which are aesthetically more satisfying than others. One direct approach stems from the Boltzmann equation (Eq. 4.19), by demanding no change with time or position, and no external forces. Then $(\partial f/\partial t)_{\text{coll}} = 0$. If the mechanics of any two-particle collisions in which total energy is conserved is then analyzed in detail, the Maxwellian distribution results. We shall not reproduce this derivation, which may be found in the standard literature.[1] The trouble with this derivation is that it is too restrictive: in actuality, f is Maxwellian (if quantum effects are neglected) even if the interactions are of the many-body type; the derivation does not bring out this point.

A more general view consists in noting that the assembly will tend toward the most probable distribution as the equilibrium one, whatever that might be. We therefore look for the most probable distribution of the points in Fig. 4.1 such that their center of gravity is at the origin, but all the points are not at the origin. We now have recourse to elementary error theory, which states that if we try to spot the points at the origin, but subject to some randomizing process, the distribution of points will be closely approximated by a Gaussian function. Thus,

$$f(\mathbf{r}, v, t) = A(\mathbf{r}, t) \exp(-\beta^2 v^2). \tag{4.20}$$

Here, β is an inverse measure of the amount of randomizing, and A is a normalization factor independent of v. The advantage of this derivation is that the true nature of a statistical distribution is made more clear: the one given by Eq. 4.20 is more probable than any other; in the limit of a large number of particles, departures from it will be small almost all of the time. These statements are equivalent to requiring that the entropy of the system be a maximum. The quantities A and β are determined by the requirements that the total number of particles and the average energy of the particles in the distribution be correctly predicted. The distribution function f is normalized to the correct particle density by integrating Eq. 4.20 over v_x from $-\infty$ to $+\infty$, and v_y and v_z over the same range. It is immediately found that

$$A = n\beta^3/\pi^{3/2}. \tag{4.21}$$

The average energy may be found by an entirely similar calculation:

$$\bar{U} = m\overline{v^2}/2 = 3m/4\beta^2. \tag{4.22}$$

The perfect gas law relates the pressure p to the density n of particles in ordinary space and the absolute temperature T:

$$p = nkT. \tag{4.23}$$

It will be shown in Chap. 6 that

$$p = nm\overline{v^2}/3. \tag{4.24}$$

Elimination of p and β yields the distribution function

$$f(v) = n(m/2\pi kT)^{3/2} \exp(-mv^2/2kT). \tag{4.25}$$

This distribution function is the density of particles in ordinary space per unit volume of velocity space. The density of particles in the interval $dv_x\, dv_y\, dv_z$ centered at $\mathbf{v}$ is

$$f(v)\, dv_x\, dv_y\, dv_z. \tag{4.26}$$

This density of particles must be carefully distinguished from the density of particles per unit *speed*, which is

$$4\pi v^2 f(v), \tag{4.27}$$

in view of the isotropy of the distribution (Eq. 4.25). The distribution function given by Eq. 4.27 is the total density of particles per unit speed going in any direction whatever that have a speed v. The distribution function given by Eq. 4.25 is the density of particles per unit velocity going in the direction of the velocity vector $\mathbf{v}$. Again, the density $f(U)$ of particles in ordinary space per unit energy is given by

$$f(U) = [4\pi v^2 f(v)]\, |dv/dU| = 4\pi v^2 f(v)\, (2mU)^{-1/2}. \tag{4.28}$$

Thus we find

$$f(U) = n(4/\pi)^{1/2}(kT)^{-3/2}U^{1/2} \exp(-U/kT). \tag{4.29}$$

This distribution function is the total density of all particles per unit energy going in any direction whatever that have an energy U. The distribution in speed is different from that in energy because a unit speed interval corresponds to an energy interval that increases monotonically with energy or speed, as may be seen from Eq. 4.28. The speed and energy distribution functions are compared in Fig. 4.2.

Distributions that depart relatively little from a Maxwell-Boltzmann distribution are often represented by such a distribution plus a small perturbation.

A number of averages may be found from the distributions given by Eqs. 4.25, 4.27 and 4.29 and by integration of Eq. 4.11. Thus, the average speed is

$$\bar{v} = (8kT/\pi m)^{1/2} \tag{4.30}$$

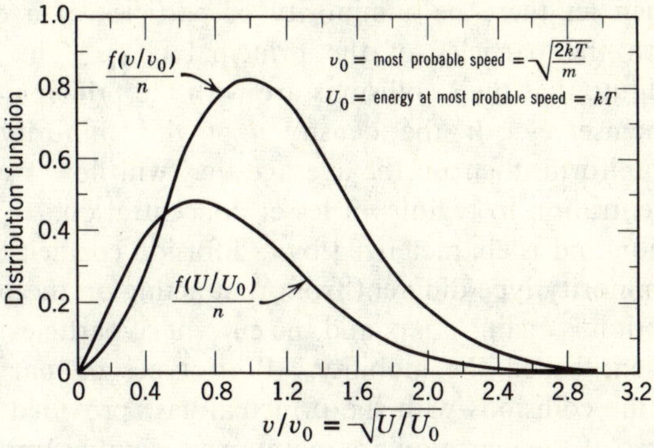

Fig. 4.2. Maxwell-Boltzmann distributions in speed and in energy.

by Eq. 4.27, and the energy at this speed is

$$U_1 = 4kT/\pi. \tag{4.31}$$

The average energy is

$$\bar{U} = 3kT/2 \tag{4.32}$$

by Eq. 4.29, and the corresponding speed is $(3kT/m)^{1/2}$. By finding the maximum of $4\pi v^2 f(v)$, Eq. 4.27, the most probable speed is seen to be

$$v_2 = (2kT/m)^{1/2}, \tag{4.33}$$

and the corresponding energy U_2 is

$$U_2 = kT. \tag{4.34}$$

The most probable energy U_3 is that for which Eq. 4.29 is a maximum:

$$U_3 = kT/2; \tag{4.35}$$

and the corresponding speed is

$$v_3 = (kT/m)^{1/2}. \tag{4.36}$$

4.4. Diffusion and Mobility

The concept of distribution functions, as developed in the previous sections, is useful in the definition of many macroscopic properties of a gas. In particular, we discuss in this section two processes, diffusion and mobility, that can lead to loss of material from the plasma.

Consider a gas or other medium consisting principally of particles, the density of which is statistically uniform and which has no macroscopic motion. Then let there be a minority of particles of a different kind admixed with the particles of the principal class. The minority are sufficiently dilute that their collisions are with the principal particles and not with themselves. If the density n of this minority type is not everywhere uniform, then on the average they will flow from regions of greater concentration to regions of lesser concentration. This process is called diffusion and is characterized by a diffusion coefficient. If a force acts on the minority type different from that acting on the principal type, a relative motion is again set up, and the current of particles is determined by a coefficient called the mobility. Both types of particle flow are impeded by the collisions with the principal gas; provided the gradients and/or sources of the density n are maintained, equilibrium currents may be derived. We shall now derive these currents and corresponding coefficients in the linear limit, where the perturbing gradients and/or forces modify the distribution function $f(\mathbf{v})$ of the minor class only slightly.

Our derivation will proceed via a momentum equation for the minority particles. They lose momentum by colliding with the principal gas, and we assume that it is dense enough to be unaffected. Thus the scattering centers are at rest. Such an ensemble is called a Lorentzian gas; it serves for our introductory discussion of diffusion and mobility. The model applies well to a weakly ionized gas, where neutral particles are available in abundance. The model will *not* apply to a highly ionized gas: there, no passive scattering centers exist, and we must consider the motions of both ions and electrons together. These complications of the fully ionized plasma are considered in Chaps. 7 and 8. Also, we shall restrict our derivation here to cases where there is no magnetic induction. If an induction exists, the particle flows are not collinear with the driving forces. Again, we postpone extended discussion of such effects until Chap. 7.

To find the particle current, let the distribution $f(\mathbf{r}, \mathbf{v})$ be expressed as

$$f(\mathbf{r}, \mathbf{v}) = f_0(\mathbf{r}, v) + f_1(\mathbf{r}, \mathbf{v}). \tag{4.37}$$

Here $f_0(\mathbf{r}, v)$ is isotropic, and

$$n(\mathbf{r}) = \int d\mathbf{v}\, f_0(\mathbf{r}, v) \tag{4.38}$$

represents all the particles of the minor class. The term $f_1(\mathbf{r}, \mathbf{v})$ represents the anisotropy of the distribution, arising from the presence of the density gradients and forces. Incidentally,

$$\int d\mathbf{v}\, f_1(\mathbf{r}, \mathbf{v}) = 0; \tag{4.39}$$

that is, f_1 is positive and negative in different parts of velocity space. Note further that $|f_1| \ll f_0$ according to our assumptions, so that f itself is everywhere positive.

We now substitute Eq. 4.37 into Eq. 4.19 and look for a steady-state solution. It is clear that $\partial f_0/\partial t = \partial f_1/\partial t = 0$ in the steady state, but for the nonce we shall retain $(\partial f/\partial t)_{\text{coll}}$. Thus

$$\mathbf{v}\cdot\nabla_r(f_0 + f_1) + \frac{\mathbf{F}}{m}\cdot\nabla_v(f_0 + f_1) = \left(\frac{\partial f_0}{\partial t}\right)_{\text{coll}} + \left(\frac{\partial f_1}{\partial t}\right)_{\text{coll}}. \tag{4.40}$$

The terms $\nabla_r f_1$ and $\nabla_v f_1$ are higher-order quantities than $\nabla_r f_0$ and $\nabla_v f_0$, and are now neglected. In addition, $(\partial f_0/\partial t)_{\text{coll}} = 0$, because f_0 is by definition the equilibrium solution if the forcing terms should be removed. Equation 4.40 then becomes

$$\mathbf{v}\cdot\nabla_r f_0 + \frac{\mathbf{F}}{m}\cdot\nabla_v f_0 = \left(\frac{\partial f_1}{\partial t}\right)_{\text{coll}}. \tag{4.41}$$

The term $(\partial f_1/\partial t)_{\text{coll}} \neq 0$. The component f_1 is in fact generated by the forcing terms, consisting of the gradient with respect to $\mathbf{r}$ and the force $\mathbf{F}$. It tends to be destroyed by the randomizing collisions. If we assume that a momentum-transfer collision randomizes the velocity of a particle (that is, removes a particle from f_1), then

$$\left(\frac{\partial f_1}{\partial t}\right)_{\text{coll}} = -\nu_m f_1. \tag{4.42}$$

Next we define the particle current $\mathbf{\Gamma}$ as

$$\mathbf{\Gamma} = n\bar{\mathbf{v}} = \int dv\, \mathbf{v}f = \int dv\, \mathbf{v}(f_0 + f_1) = \int dv\, \mathbf{v}f_1, \tag{4.43}$$

the last equality resulting from the fact that f_0 is isotropic. Substitution of Eqs. 4.41 and 4.42 into Eq. 4.43 yields

$$\mathbf{\Gamma} = -\int dv\, \mathbf{v}\left(\frac{\mathbf{v}}{\nu_m}\cdot\nabla_r f_0 + \frac{\mathbf{F}}{m\nu_m}\cdot\nabla_v f_0\right). \tag{4.44}$$

The first term on the right represents diffusion, and the second term represents mobility. At this point we make the simplifying assumption that $\mathbf{F}$ does not depend on $\mathbf{v}$ in any way. Let us now discuss the diffusion term.

Suppose that the gradient is in the direction $\mathbf{k}$. Then $\mathbf{v}\cdot\nabla_k f_0 = v_k\, \partial f_0/\partial x_k$. The flow will also be in the $\mathbf{k}$ direction; therefore in Eq. 4.44 we are bid to discover the value of

$$\Gamma_k = -\int dv\, v_k\left(\frac{v_k}{\nu_m}\frac{\partial f_0}{\partial x_k}\right). \tag{4.45}$$

Because

$$v_x^2 + v_y^2 + v_z^2 = v^2, \tag{4.46}$$

then in any direction and close to equilibrium it follows that

$$\int d\mathbf{\Omega}\, v_k^2 = \int d\mathbf{\Omega}\, v^2/3. \tag{4.47}$$

The diffusion current therefore becomes, after generalizing to any arbitrary direction,

$$\mathbf{\Gamma}_{\text{diff}} = -\int d\mathbf{v}\, \frac{v^2}{3\nu_m} \nabla_r f_0. \tag{4.48}$$

Since **r** and **v** are independent, we obtain finally

$$\mathbf{\Gamma}_{\text{diff}} = -\nabla \int d\mathbf{v}\, \frac{v^2}{3\nu_m} f_0 = -\nabla(\mathsf{D}n). \tag{4.49}$$

Here,

$$\mathsf{D} = \overline{v^2/3\nu_m} \tag{4.50}$$

is the diffusion coefficient. Note that D is inversely proportional to the background gas density and is directly proportional to the average energy if ν_m = constant. Note also that currents flow if the density n is uniform, but the energy is different at different places. In retrospect, the reason is obvious: hotter particles travel faster and diffuse out of a region more quickly than an equal density of cold particles diffuses in.

In deriving the mobility, let

$$\mathbf{F} = q\mathbf{E}, \tag{4.51}$$

and let $\mathbf{E} = E_k\mathbf{k}$ in the same sense as before. Then we must evaluate $\partial f_0/\partial v_k$. By differentiation of Eq. 4.46 we find that

$$\frac{\partial f_0}{\partial v_k} = \frac{v_k}{v}\frac{\partial f_0}{\partial v}. \tag{4.52}$$

Since again **v** is in the **k** direction, the mobility current is given by

$$\Gamma_k = -E_k \int d\mathbf{v}\, \frac{qv_k^2}{mv\nu_m}\frac{\partial f_0}{\partial v}. \tag{4.53}$$

The equation can be integrated by parts, by using Eqs. 4.47, 4.11, and the fact that $v^3 f_0$ is zero at $v = 0$ and $v = \infty$. If all averages of v^n are to exist, f_0 must decrease exponentially as v approaches ∞. Finally, we obtain

$$\mathbf{\Gamma}_{\text{mob}} = n\bar{\mathbf{v}}_{\text{mob}} = \mathbf{E}\int_0^\infty dv\, f_0 \frac{d}{dv}\left(\frac{4\pi q v^3}{3m\nu_m}\right) = \mathbf{E}(\mu n). \tag{4.54}$$

The coefficient μ is the mobility and is defined here as positive or negative, depending on the sign of q. Conventionally, however, μ is considered a positive quantity, with an appropriate sign inserted in equations where it is used.

If ν_m = constant, we find by direct integration of Eq. 4.54 that

$$\mu = q/m\nu_m, \tag{4.55}$$

and further that

$$\mathsf{D}/\mu = 2\bar{U}/3q, \tag{4.56}$$

where $\bar{U}$ is the average energy. In addition, if f_0 is Maxwellian, we may also find by direct integration that

$$\mathsf{D}/\mu = kT/q \tag{4.57}$$

independent of the form of ν_m; this result is known as the Einstein relation.

Expressed in terms of these averaged coefficients, Eq. 4.44 becomes

$$\mathbf{\Gamma} = -\nabla(\mathsf{D}n) + \mu\mathbf{E}n \tag{4.58}$$

for the total current, for each type of particle of interest.

4.5. The Parameter E/p in Weakly Ionized Gases

Although highly ionized gases are of principal interest, all such media must pass through a stage of weak ionization. Several concepts for such media are useful. In a weakly ionized medium the density of charges is so low that all interactions between the charges can be neglected.

Let us inquire into the experimental variables upon which the average energy $\bar{u}$ (ev) of a collection of charges in a gas depends. If the boundaries of the system are remote, $\bar{u}$ must depend on local properties, and we must find local parameters having this dimension. There are only two: the gas temperature u_g, which may as well be expressed in ev, and the product El, or potential per mean free path. If the gas energy u_g is either constant (T_g = constant) or so low compared with u that the atomic motion of the gas may be neglected, we then find

$$\bar{u} = Ely_1(El/u_c). \tag{4.59}$$

Here y represents a functional dependence, and u_c represents all the characteristic energies (u_g, u_x, u_i, etc.). Since all these are fixed, then

$$\bar{u} = y_2(E/p). \tag{4.60}$$

The second form follows from the fact that any mean free path is inversely proportional to pressure.

Moments other than $\bar{u}$ of the electron velocity distribution will also be some function of E/p by the arguments just presented. Since all

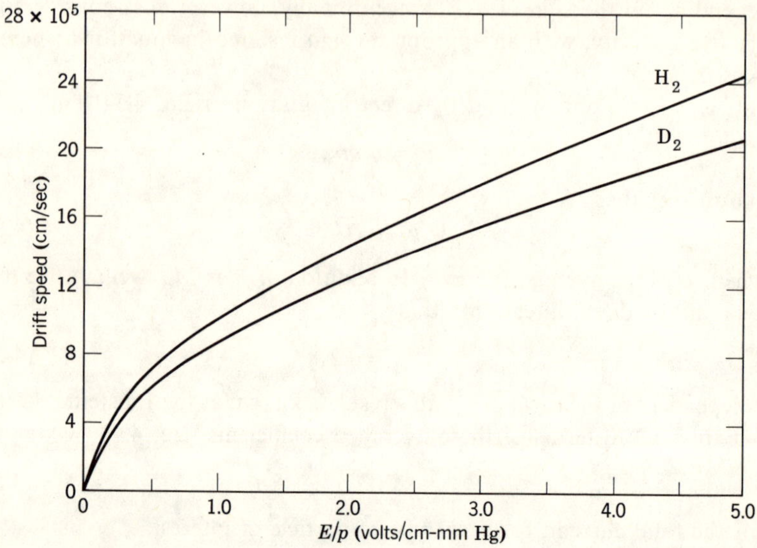

Fig. 4.3. Drift velocity of electrons in hydrogen and deuterium. After B. I. H. Hall, *Australian J. Phys.*, **8**, 468 (1955).

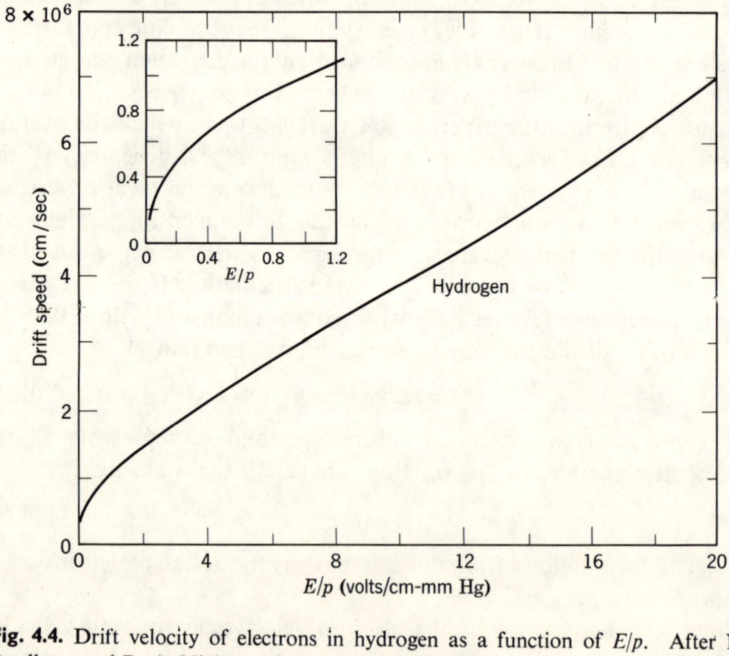

Fig. 4.4. Drift velocity of electrons in hydrogen as a function of E/p. After N. E. Bradbury and R. A. Nielsen, *Phys. Rev.*, **49**, 388 (1936).

moments of the distribution function $f(v)$ are functions of E/p, the distribution function must itself be a function of E/p, other parameters (u_g, for example) being kept constant.

All integrals of the form

$$\int dv\, y(v, E/p) f(v, E/p) = \bar{y} \tag{4.61}$$

must therefore be functions of E/p.

Equation 4.54 is a special form of Eq. 4.61. There, ν_m is some function of v and is proportional to pressure. Therefore, the drift velocity of charges is a function of E/p. Figure 4.3 shows the drift velocity $\bar{\mathbf{v}}$ of electrons in H_2 and D_2 as functions of E/p at low E/p. Figure 4.4 shows $\bar{\mathbf{v}}$ for electrons in H_2 at higher E/p. Note that in this higher range of E/p the electrons have several volts energy, and in this range $\nu_m \approx$ constant. Thus $|\bar{\mathbf{v}}| \propto E/p$ approximately. The drift velocities of hydrogen and deuterium ions in their parent gases are shown in Fig. 4.5.

The average ionization rate per unit volume per electron

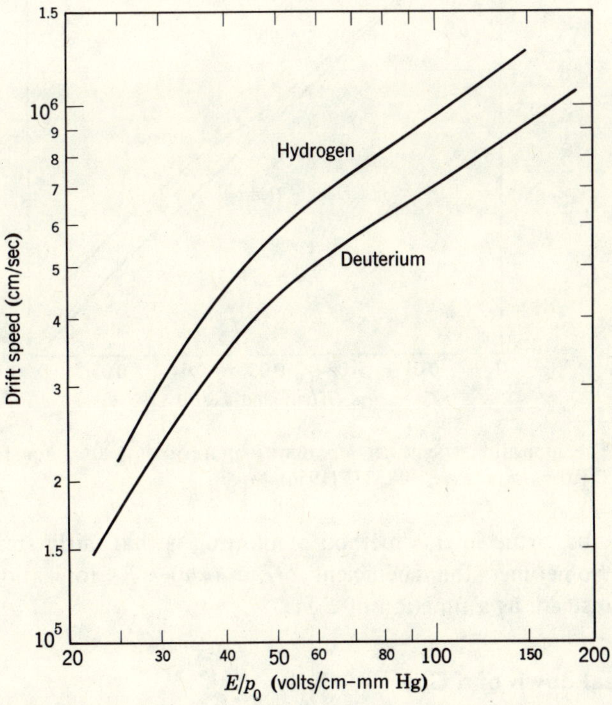

Fig. 4.5. Drift velocity of ions in parent gases H_2 and D_2 as functions of E/p_0 (density corrected to 0° C). After D. J. Rose, *J. Appl. Phys.*, **31**, 643 (1960).

$$\bar{\nu}_i = \int dv\, n_g \sigma_i(v) v f(v)/n_e \tag{4.62}$$

is not a single-valued function of E/p because it contains the gas density n_g and, hence, pressure p explicitly. However, the quantity

$$\bar{\nu}_i/p = y_3(E/p) \tag{4.63}$$

is single-valued. Since also $\bar{\mathbf{v}}$ is a function of E/p, we find that

$$\bar{\nu}_i/(|\bar{\mathbf{v}}|p) = \alpha/p = y_4(E/p). \tag{4.64}$$

The quantity $\alpha = \nu_i/|\bar{\mathbf{v}}|$, called the Townsend ionization coefficient, is a readily observable parameter that represents the number of ionizations produced by one electron as it drifts unit distance in the field direction. The coefficient α/p for hydrogen and deuterium is shown in Fig. 4.6

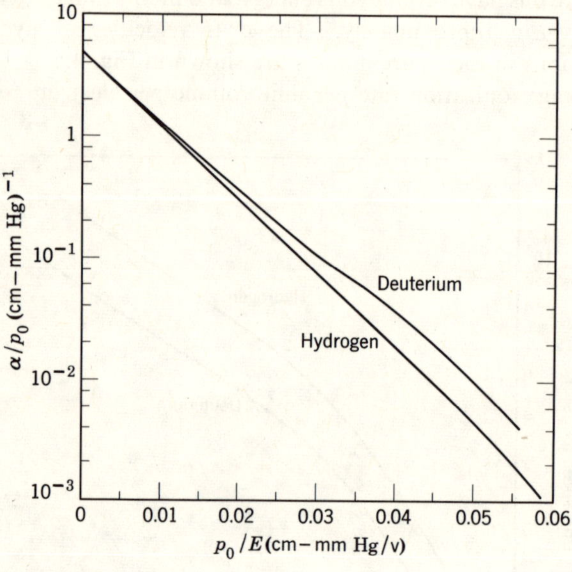

Fig. 4.6. The ionization coefficient α/p_0 (cm × mm Hg)$^{-1}$ at 0° C for H_2 and D_2. After D. J. Rose, *Phys. Rev.*, **104**, 273 (1956).

vs. p/E; the virtue in this method of plotting is that fairly straight lines result. Sometimes the coefficient $\alpha/E \equiv (\alpha/p)(p/E)$ ionizations/volt is plotted instead, as a function of E/p.

4.6. Breakdown of a Gas

Electrical breakdown of a gas must often be considered, not only in ion sources and so forth, where we may desire it, but also in high-voltage circuitry, where we do not. The breakdown generally proceeds via multiplication of the electrons in a gap; their rate of change is

$$\frac{\partial n_e}{\partial t} = (\bar{\nu}_i - \bar{\nu}_a)n_e - \nabla \cdot \mathbf{\Gamma}_e. \tag{4.65}$$

In words, the rate at which the density of electrons changes equals their rate of creation minus their rate of loss. Here $\bar{\nu}_i$ and $\bar{\nu}_a$ are the average ionization and attachment rates, respectively. The discussion is here restricted to the case of a uniform electric field across a parallel-plane gap d. Then

$$\mathbf{\Gamma}_e = \bar{\mathbf{v}}_e n_e, \tag{4.66}$$

and $\bar{\mathbf{v}}_e$ is a function of E/p alone and not of position. Thus

$$\frac{\partial n_e}{\partial t} = (\bar{\nu}_i - \bar{\nu}_a)n_e - \bar{v}_e \frac{\partial n_e}{\partial x}. \tag{4.67}$$

For simplicity, we now ignore $\bar{\nu}_a$, noting that $\bar{\nu}_i/p$ and $\bar{\nu}_a/p$ are functions of E/p, and both may be combined in one adjusted parameter. Then the steady-state solution of Eq. 4.67 is

$$n = n_0 \exp(\alpha x), \tag{4.68}$$

where α is defined in Eq. 4.64. In order that a steady-state solution exist, a current $I_0 = en_0\bar{v}_e$ must be supplied at the negative plate located at $x = 0$.

Equation 4.68 leads to no instability, and some feedback mechanism is required to achieve breakdown. One of the most common is secondary emission of electrons by positive-ion arrival at the cathode. The total ion production per unit time in the gap is

$$\int_0^d dx\, n(x)\bar{\nu}_i = \bar{\nu}_i \frac{n_0}{\alpha}[\exp(\alpha d) - 1]. \tag{4.69}$$

Since ions can go only to the cathode, there is an ion current of this amount at the cathode. Each ion releases γ_i electrons. This process creates an electron density at the cathode, given by Eq. 4.69 multiplied by γ_i and divided by $\bar{v}_e$. This density is then $\gamma_i n_0[\exp(\alpha d) - 1]$ at the cathode. The total electron density n_0 there is then equal to the sum of this term and a contribution n_{0o} from any outside stimulus. Reduction of algebra gives

$$n(x) = \frac{n_{0o}\exp(\alpha x)}{1 - \gamma_i[\exp(\alpha d) - 1]}. \tag{4.70}$$

As the gap potential is increased, α increases rapidly, and $\gamma_i[\exp(\alpha d) - 1]$

approaches unity. If

$$\gamma_i[\exp(\alpha d) - 1] = 1, \tag{4.71}$$

no external stimulus is needed except for a single event to incite the multiplicative process. Any slight increase in potential causes the current to increase indefinitely until some nonlinear mechanism, for example, power-source limitation, becomes important.

Photoemission at the cathode by photons from the excited gas is another important secondary mechanism, especially in H_2 and D_2. In that case, breakdown proceeds much more rapidly, because the photons return to the cathode quickly compared with the speed with which the slow ions return.

Electrons are not induced to leave a plasma by a high-frequency electric field but merely oscillate. In the previous example, the dominant loss process was mobility, and diffusion was negligible. If only a high-frequency field is present, diffusion becomes the dominant process, and Eq. 4.49 is applicable. Under the assumption that D_e is independent of position, Eq. 4.65 becomes

$$\mathsf{D}_e \nabla^2 n_e + \bar{\nu}_i n_e = \frac{\partial n_e}{\partial t}. \tag{4.72}$$

If n_e is to be just on the verge of increasing, that is, if the gas is to be on the verge of breakdown, the right side must be zero. Thus the breakdown problem reduces to a solution of the spatial equation

$$\nabla^2 n_e + n_e/\Lambda^2 = 0, \tag{4.73}$$

where

$$\Lambda^2 = \mathsf{D}_e/\bar{\nu}_i \tag{4.74}$$

is the diffusion length of the cavity, and is determined as the characteristic value of Eq. 4.73 solved in the specific geometry. The density of electrons is approximately zero, $n_e \approx 0$ at the boundaries, because electrons are assumed to be absorbed there. It may readily be demonstrated that $\bar{\nu}_i/\mathsf{D}_e p^2$ is a function of E/p, so that, from Eq. 4.74, E/p at breakdown is a function of $p\Lambda$. In the case of high-frequency breakdown, the frequency ω of the field is a new variable; it has been assumed to be fixed in this derivation, and the frequency dependence will not be discussed.

4.7. Breakdown and Initial Charge Build-Up in a Long Torus

Diffusion theory may readily be applied to investigate breakdown and initial charge build-up in a gas-filled torus. The torus forms a one-turn secondary of the transformer, and a pulse is applied to the primary, giving a field E_z around the torus, as shown in Fig. 4.7. The scheme is commonly applied to make a plasma in a torus.

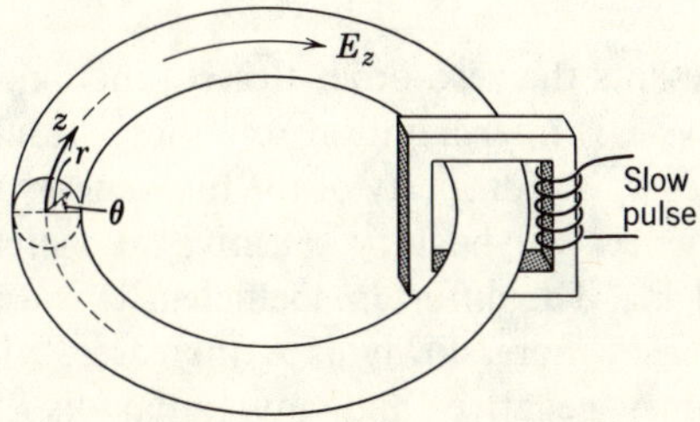

Fig. 4.7. Torus as transformer secondary.

For convenience in analysis, we shall assume:

1. The pulse is so shaped that E_z is constant over the time of interest. This situation cannot continue indefinitely, for the primary current and transformer flux must eventually saturate. But it is simple enough to keep **E** substantially constant for times long compared with the transit time of any charge around the torus.

2. There is sufficient gas in the torus that the charges move by diffusion and mobility rather than by free fall along the potential lines.

3. The charge build-up takes place slowly compared with the time for a charge to encircle the torus. Conditions around the torus will then be substantially independent of z.

Under these conditions, the electric field E_z causes ionization but no loss of charge. The particles are lost by diffusion in the radial direction. (Note that there is no radial field E_r.) Under the present circumstances, the conservation equation for electrons is merely Eq. 4.72, where the ionization frequency $\bar{\nu}_i$ corresponds to E_z/p, which is here a constant. The boundary condition is $n_e = 0$ at the wall $r = R$.

The solution of Eq. 4.72 for a gas discharge in a torus is desired. The equation, of partial differential form, can be solved in this geometry by the method of separation of variables. The spatial factor for toroidal geometry becomes cumbersome. Cylindrical geometry furnishes an adequate approximation if the torus is long and thin. In view of the symmetry, we seek only a solution independent of angle and axial position. Thus, only derivatives of the solution with respect to r and t are not zero in Eq. 4.72. We find that the solution is then

$$n_e = CJ_0\left(\frac{2.405r}{R}\right) \exp{(t/\tau)}, \tag{4.75}$$

where C is a constant, and the separation constant τ^{-1} is given by

$$\frac{1}{\tau} = \bar{\nu}_i - \left(\frac{2.405}{R}\right)^2 \mathsf{D}, \tag{4.76}$$

and where J_0 represents the zero-order Bessel function.

We find a physical interpretation of these results. The average ionization frequency $\bar{\nu}_i$ is a property of the high-energy tail of the electron distribution and hence will be very sensitive to increasing the electron energy by the field E. The diffusion coefficient D is a property of all the electrons and increases more slowly as E increases. Thus if E becomes too small, τ becomes negative, and any initial distribution decays; at some critical value of E, the two terms of Eq. 4.76 are equal, and $\tau = \infty$. For larger E, ionization exceeds diffusion loss, and the density builds up. Stated in another way, which is entirely equivalent, an electron starting out at some point in the torus generates new ones as it makes its way around the torus. Some of the electrons are lost to the walls; the breakdown criterion is that the process followed once around the torus produces one net electron replacing the original one.

The criterion for breakdown (Eq. 4.76) can be rewritten

$$\frac{\alpha}{p} \equiv \frac{\bar{\nu}_i}{\bar{v}_e p} \geqslant \left(\frac{2.405}{R}\right)^2 \frac{\mathsf{D}}{\bar{v}_e p}. \tag{4.77}$$

If ν_m is constant, Eqs. 4.50, 4.54, and 4.55 may be used to eliminate D and $\bar{v}_e$. Thus

$$\frac{\alpha}{p} \geqslant \left(\frac{2.405}{pR}\right)^2 \frac{2\bar{u}}{3(E/p)}, \tag{4.78}$$

where $\bar{u}$, the average energy in ev, is itself a function of E/p. Note that the proper variables here are E/p and pR. The latter variable is a measure of the number of mean free paths across the tube.

Very frequently, a temporally constant magnetic induction B_z is applied to the torus by a simple solenoidal winding. This induction does not affect motions in the z-direction, but it cuts down sideways diffusion, which is the reason for its application. In this case, we state, but prove in Chap. 7, that the diffusion coefficient D_T transverse to B is reduced compared with $\mathsf{D}_\|$ parallel to B:

$$\mathsf{D}_\mathsf{T} = \frac{\overline{v^2}/3\nu_m}{1 + (\omega_b/\nu_m)^2},$$

$$= \frac{\mathsf{D}_\|}{1 + (\omega_b/\nu_m)^2}, \tag{4.79}$$

where

$$\omega_b = -eB/m \tag{4.80}$$

is the cyclotron frequency. Therefore, Eq. 4.78 is generalized to

$$\frac{\alpha}{p} \geqslant \left(\frac{2.405}{pR}\right)^2 \frac{2\bar{u}}{3(E/p)[1 + (\omega_b/\nu_m)^2]}. \tag{4.81}$$

The equation may be applied in any consistent units; in particular, if α/p represents ionizations/cm-mm Hg, then p represents mm Hg, E represents volts/cm, and so on.

Let us try a typical case of a small stellarator (model A-3), in which the minor radius is 2.5 cm, $p = 0.01$ mm Hg (of H_2), $E = 0.5$ volt/cm, $B = 0.10$ weber/m^2. Then $E/p = 50$ volts/cm-mm Hg and $\alpha/p = 0.3$ from Fig. 4.6. From data on $\bar{u}$ vs. E/p, we find $u \approx 4$ ev; also for electrons in H_2, $\nu_m \approx 5.9 \times 10^9 p$. Then $(\omega_b/\nu_m)^2 + 1 \approx 10^5$; the right side of Eq. 4.81 is approximately 5×10^{-3}. The field is well above breakdown; the rate of build-up could be calculated from Eq. 4.76 if desired. We could check from Fig. 4.6 and trial-and-error substitution in Eq. 4.81 that the field that just gives breakdown ($\tau = \infty$) is about 0.22 volt/cm.

If the magnetic induction had not been applied, a huge field E would have been required, as may readily be checked by insertion of the numbers.

The foregoing discussion is suitable if the charge density is low. As the charge density builds up, a complicating effect enters: the charge density gives rise to a radial space-charge electric field. The field term of Eq. 4.58 then has a radial component. This radial field can be determined only after considering the motion of the ions as well. The analysis is quite complicated and nonlinear, and it will not be attempted here. In the special case:

1. The electrons have much more energy than the ions.
2. The magnetic induction B_z is high, so that electron diffusion is cut down much more than ion diffusion: note that $|\omega_{bi}/\omega_{be}| = (m_e/m_i) \ll 1$.
3. There are no z or θ inhomogeneities in density.

The principal nonlinear effects coincidentally cancel, and the derivation for low charge density is approximately applicable.

4.8. Diffusion in the Presence of Space Charge

If the density of the charges in a plasma is very low, the particle currents arise from free diffusion generated by density gradients and from drift velocities generated by the applied electric fields. At breakdown of the gas, the charge density is low, and the particle flows have been so described in the previous sections. As the charge density rises, however, the space-charge electric field generated by the charges themselves will modify the flow.

To begin the discussion, we write again the basic particle current equations for electrons

$$\boldsymbol{\Gamma}_e = -\mathsf{D}_e \nabla n_e - \mu_e \mathbf{E} n_e, \tag{4.82}$$

and for ions

$$\boldsymbol{\Gamma}_i = -\mathsf{D}_i \nabla n_i + \mu_i \mathbf{E} n_i. \tag{4.83}$$

We assume here for simplicity that the diffusion coefficients and mobilities are independent of position. Since the electric field **E** will here be a function of position, the assumption may not be quite valid, for these constants may be functions of the average energy of the charges. However, within this limitation, Eqs. 4.82 and 4.83 are valid for any plasma in which the particle velocities are nearly isotropic. Even if a magnetic induction **B** is present, the coefficients appropriate to each spatial direction may be inserted. Poisson's equation

$$\nabla \cdot \mathbf{E} = e(n_i - n_e)/\epsilon_0 = \rho/\epsilon_0 \tag{4.84}$$

causes difficulties, however. The last terms of Eqs. 4.82 and 4.83 then contain nonlinear contributions, and the problem is very much more complicated.

The effect has already been referred to in a number of previous sections: if the charge density is high, the self-electrostatic field tends to expel charges of the type in excess, thus bringing the plasma to approximate charge neutrality. Neutrality is certainly closely approximated in all plasmas of thermonuclear interest, including those of high-density ion sources and the like. In actual fact, the plasma is not quite neutral, as we shall shortly see.

In discussing the equations, we make the further simplification that there is no externally applied dc field in the directions of interest. A plasma excited by radio frequency, direct injection of the charges, or the radial currents in a long cylindrical column sustained by an axial **E** are so describable. Then, in Eqs. 4.82 and 4.83, **E** arises entirely from space charge. Finally, we assume that the discharge parameters depend on just one coordinate direction, so that in the steady state $\boldsymbol{\Gamma}_i = \boldsymbol{\Gamma}_e$ everywhere. (In the more general case, $\nabla \times \boldsymbol{\Gamma}_i \neq \nabla \times \boldsymbol{\Gamma}_e$ near corners or if an induction exists; we shall not discuss this situation.)

In attacking nonlinear problems, we must use unconventional means. We first inquire into the limit of very small charge densities. In that case, the mobility terms are negligible; thus

$$n_i/n_e = \mathsf{D}_e/\mathsf{D}_i \tag{4.85}$$

from Eqs. 4.82 and 4.83 trivially. If $\mathbf{B} = 0$, then $\mathsf{D}_e \gg \mathsf{D}_i$; hence, $n_i \gg n_e$. Each type of charge independently obeys the diffusion equation;

the ions are more dense because, though they are assumed to be created in the space at the same rate as electrons, the ions diffuse more slowly.

Next we consider the case where the charge densities are very large. It is evident that all terms of Eqs. 4.82 and 4.83 will increase as n_i and n_e increase. If bounded solutions are to exist, it is also evident that the ratio of the various terms must remain finite at most points. Since ∇n and n are related by the fixed size of the plasma, the electric intensity must approach a limit as n_i and n_e increase indefinitely. Thus, by multiplying Eq. 4.82 by μ_i and Eq. 4.83 by μ_e, and with $\mathbf{\Gamma}_i = \mathbf{\Gamma}_e = \mathbf{\Gamma}$, we find

$$(\mu_i + \mu_e)\mathbf{\Gamma} = -(\mathsf{D}_i\mu_e\nabla n_i + \mathsf{D}_e\mu_i\nabla n_e) + \mu_i\mu_e\epsilon_0\mathbf{E}\nabla\cdot\mathbf{E}/e. \quad (4.86)$$

As n_i and n_e approach infinity, the last term becomes negligible, and by the saturation of Eq. 4.84 we may set $n_i \approx n_e = n$ in Eq. 4.86. Thus we find

$$\mathbf{\Gamma} = -\mathsf{D}_a\nabla n, \quad (4.87)$$

where

$$\mathsf{D}_a = (\mathsf{D}_i\mu_e + \mathsf{D}_e\mu_i)/(\mu_i + \mu_e) \quad (4.88)$$

is called the ambipolar diffusion coefficient.

The physical picture is made clearer if we return to Eq. 4.82, for example, and calculate $\mathbf{E}$, using Eq. 4.87 for $\mathbf{\Gamma}$. The result is

$$\mathbf{E} = -\left(\frac{\mathsf{D}_e - \mathsf{D}_i}{\mu_i + \mu_e}\right)\frac{\nabla n}{n}. \quad (4.89)$$

The actual net charge density ρ can then be calculated from the divergence of Eq. 4.89. Consider now Fig. 4.8, which illustrates the case, with the difference $n_i - n_e$ much exaggerated. In this figure, it is assumed merely for the sake of definiteness that $\nabla\cdot\mathbf{\Gamma} = \nu n_e$; that is, the electrons cause the ionization. Neglect for a moment the regions near the walls. For the example chosen, $\mathsf{D}_e > \mathsf{D}_i$; thus electrons diffuse faster, with the result that ions are in excess. Then the electric field $\mathbf{E}$, by Eq. 4.89, is directed outward. In Eq. 4.82, the two terms on the right are opposite in sign: the electric field created counteracts the diffusion of the faster electrons. On the other hand, the ion current is enhanced by the field. Thus with the two densities approximately equal, the two currents are equal. The resulting flow of either particle is given by Eq. 4.87; since it resembles a diffusion equation with an adjusted coefficient, the process is called ambipolar diffusion. It should be remembered, however, that the process is not one of simple diffusion.

Let us return to Fig. 4.8 and examine the regions near the bounding walls. Since n_e becomes small there, the assumption $n_i - n_e \ll n_i$ must

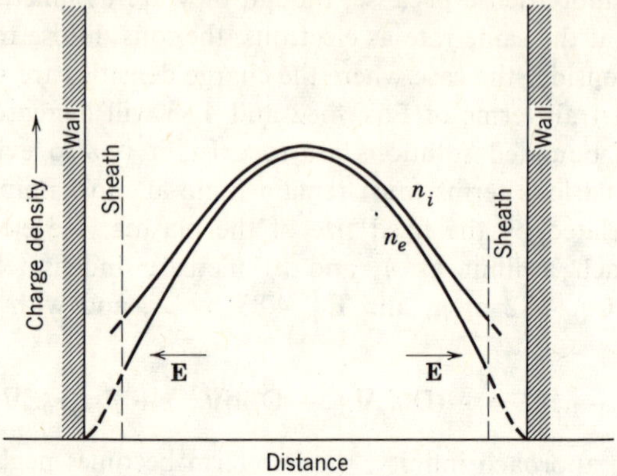

Fig. 4.8. Schematic representation of charge densities between parallel planes under near-ambipolar conditions.

eventually fail near the boundary. Indeed, Eq. 4.89 would indicate that **E** and ρ are very large there. The approximations fail, and this region where more sophisticated solutions are required is called the sheath.

Between the limits of free and ambipolar diffusion, the equations are highly nonlinear and have been discussed by Allis and Rose.[2] An approximate idea of the nature of the transition may be obtained by setting

$$n_i/n_e = C, \tag{4.90}$$

noting that the ratio C is independent of position in both the free and ambipolar limits, and assuming therefore that

$$\nabla C/C \ll \nabla n/n \tag{4.91}$$

always. Elimination of **E** in Eqs. 4.82 and 4.83 then gives

$$\begin{aligned}\mathbf{\Gamma} &= -\frac{C(\mathsf{D}_i\mu_e + \mathsf{D}_e\mu_i)}{(\mu_e + C\mu_i)}\nabla n_e \\ &= -\mathsf{D}_s\nabla n_e,\end{aligned} \tag{4.92}$$

where

$$\mathsf{D}_s = C(\mathsf{D}_i\mu_e + \mathsf{D}_e\mu_i)/(\mu_e + C\mu_i). \tag{4.93}$$

These manipulations show that the currents are approximately derivable from an effective diffusion coefficient, which reduces to D_e at the free limit ($C = \mathsf{D}_e/\mathsf{D}_i$) and to D_a at the ambipolar limit ($C = 1$). Note that for $C = 2$, or $n_i - n_e \approx n_e$, $\mathsf{D}_s \approx 2\mathsf{D}_a$. More precise calculation shows that D_s calculated in this way is smaller than the true value through the range of intermediate densities between free and ambipolar diffusion.

While the transition from free to ambipolar diffusion extends over many orders of magnitude in density, we may derive very approximately the charge density at which $\rho \approx n_e e$. Suppose that this density exists in the plasma at a distance L from the wall. Then $E \approx e n_e L/\epsilon_0$, $|\nabla n_e| \approx n_e/L$, and, if the electron flow is to be sharply reduced, we must have $\mathsf{D}_e|\nabla n_e| \approx \mu_e E n_e$. Then

$$L^2 \approx \left(\frac{\epsilon_0}{e n_e}\frac{\mathsf{D}_e}{\mu_e}\right) \equiv \lambda_e^2. \tag{4.94}$$

The quantity λ is called the Debye shielding length in the plasma; it is here calculated for the electrons. Further significance of the Debye shielding length will be brought out in Chap. 8. Suffice it to say at this time that the thickness of the sheath referred to in Fig. 4.8 is in the order of the Debye length of the more mobile particle. Then as the plasma density is increased, λ and the sheath thickness decrease; but the electric field there increases, so that the sheath potential changes rather slowly with changing density. This potential is in the order of a few times D/μ. As the density is decreased, the condition in Eq. 4.94 is reached. At this density, the sheath occupies all the space, and the sheath concept has no value at lower charge densities.

So far, the electrons have been assumed to be more mobile. If a magnetic induction **B** exists, $\mathsf{D}_{e\mathsf{T}}$ may be less than $\mathsf{D}_{i\mathsf{T}}$. In that case, $n_e > n_i$ in the plasma; the mathematical development is completely similar. If $\mathsf{D}_{e\mathsf{T}} = \mathsf{D}_{i\mathsf{T}}$ exactly, that is,

$$\frac{\mathsf{D}_{e\parallel}}{1 + (\omega_{be}/\nu_m)^2} = \frac{\mathsf{D}_{i\parallel}}{1 + (\omega_{bi}/\nu_m)^2}, \tag{4.95}$$

then no field **E** develops. A discussion of radial diffusion in a long cylindrical column with axial induction **B** is given by Lehnert[3] and by Simon.[4] If the cylinder is of finite length, the radial ambipolar diffusion may be enhanced by axial diffusion, as pointed out by Simon. Here, $\mathsf{D}_\mathsf{T} \ll \mathsf{D}_\parallel$; the charges can easily diffuse up and down the column and travel radially on the conducting end plates.

All the foregoing discussion applies quite well to a plasma in which the diffusion coefficients and mobilities can be recognized as such, as is the case in a weakly ionized plasma. In a fully ionized plasma, processes of the same general sort certainly take place, and space-charge fields develop. In fact, however, the theory of the fully ionized plasma confined in a magnetic field has not yet advanced to the state where space-charge problems can be answered with real certainty.

4.9. Reaction Rates

The average frequency $\bar{\nu}$ of an event occurring is by definition

$$\bar{\nu} = \int d\boldsymbol{v}\, \nu(v) f(\mathbf{v}), \tag{4.96}$$

where in this section $f(\mathbf{v})$ is normalized to unity for convenience, instead of to n. The equation appears deceptively simple; the calculation is sometimes complicated by the fact that the particles in the distribution $f(\mathbf{v})$ collide with other particles which also are moving. Thus $\nu(v)$ already contains hidden averages of a complicated nature over the distribution of target particles. We resolve the matter by considering in detail the reaction rate.

Consider Fig. 4.9 and two distributions of unlike particles in the velocity space. One distribution has the number density n_1 and the

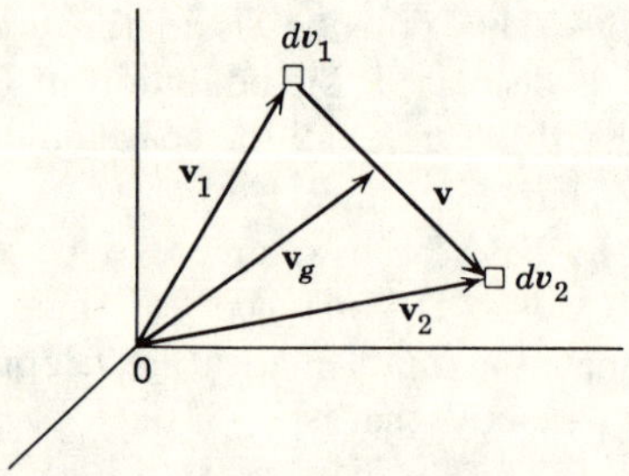

Fig. 4.9. Transformation from laboratory frame to center of mass frame.

velocity distribution $f_1(\mathbf{v}_1)$. We consider the number of particles $n_1 f_1(\mathbf{v}_1)\, d\boldsymbol{v}_1$ in a volume element $d\boldsymbol{v}_1$. These particles react with the number $n_2 f_2(\mathbf{v}_2)\, d\boldsymbol{v}_2$ of distribution 2 in $d\boldsymbol{v}_2$ at the rate

$$dR = n_1 f_1(\mathbf{v}_1) n_2 f_2(\mathbf{v}_2) |\mathbf{v}_1 - \mathbf{v}_2| \sigma(|\mathbf{v}_1 - \mathbf{v}_2|)\, d\boldsymbol{v}_1\, d\boldsymbol{v}_2. \tag{4.97}$$

The rate R is the sixfold integral over all space

$$R = \int_{\boldsymbol{v}_1, \boldsymbol{v}_2} d\boldsymbol{v}_1\, d\boldsymbol{v}_2\, [n_1 f_1(\mathbf{v}_1)][n_2 f_2(\mathbf{v}_2)] |\mathbf{v}| \sigma(|\mathbf{v}|), \tag{4.98}$$

where

$$|\mathbf{v}| = |\mathbf{v}_1 - \mathbf{v}_2|, \tag{4.99}$$

which counts all possible collisions. The averaged quantity $\overline{\sigma v}$ is useful and is given by

$$\overline{\sigma v} = \int_{v_1, v_2} d\boldsymbol{v}_1\, d\boldsymbol{v}_2\, f_1(\mathbf{v}_1) f_2(\mathbf{v}_2) |v| \sigma(|v|). \tag{4.100}$$

Clearly,

$$R = n_1 n_2 \overline{\sigma v} \tag{4.101}$$

if the particles are dissimilar.

If f_1 and f_2 are both Maxwellian at temperature T, it is possible to perform the integrals by changing variables to the center of mass system. Let the particles in distributions 1 and 2 have masses m_1 and m_2. The distributions for the two particles are given by Eq. 4.25 applied with the present normalization:

$$f_{1,2} = \left(\frac{m_{1,2}}{2\pi kT}\right)^{3/2} \exp\left(-\frac{m_{1,2}v^2}{2kT}\right), \tag{4.102}$$

the subscripts referring to distributions 1 and 2. Let the vector $\mathbf{v}_g$ be that of the center of mass of the two particles; we find

$$\mathbf{v}_1 = \mathbf{v}_g - \mathbf{v}m_2/(m_1 + m_2), \tag{4.103}$$

$$\mathbf{v}_2 = \mathbf{v}_g + \mathbf{v}m_1/(m_1 + m_2). \tag{4.104}$$

The reaction rate then becomes

$$R = \frac{(4\pi)^2 n_1 n_2}{(2\pi kT)^3} (m_1 m_2)^{3/2} \int_{v_g=0}^{\infty} dv_g\, v_g{}^2 \times \int_{v=0}^{\infty} dv\, v^3 \sigma(v) \exp\left(-\frac{m_s v_g{}^2}{2kT} - \frac{m_r v^2}{2kT}\right), \tag{4.105}$$

where

$$m_r = \frac{m_1 m_2}{m_1 + m_2}, \tag{4.106}$$

$$m_s = m_1 + m_2. \tag{4.107}$$

The integral over $\mathbf{v}_g$ can be performed immediately. Usually the cross section is expressed in terms of the energy variable $U = m_1 v^2/2$, so that Eq. 4.105 becomes

$$\overline{\sigma v} = \frac{4}{(2\pi m_1)^{1/2}} \left(\frac{m_r}{m_1 kT}\right)^{3/2} \int_0^{\infty} dU\, U\sigma(U) \exp\left(-\frac{m_r U}{m_1 kT}\right). \tag{4.108}$$

A special precaution is necessary in case the two types of particles are really the same. In the integration, the thought is to sum over each reaction once and only once. For example, in integrating over $\mathbf{v}$ it might be thought that each collision is counted twice, since both $+\mathbf{v}$ and $-\mathbf{v}$ are considered in the integration of Eq. 4.100, and there are particles at both ends of the vector. If the particles 1 and 2 are really different, then the volume elements $d\mathbf{v}_1$ and $d\mathbf{v}_2$ really contain different particles; hence the collisions will not have been counted twice. If, however, the two types of particles are really the same as in the D-D reaction, then each collision will have been counted twice, so the results in Eqs. 4.98, 4.101, and 4.105 must be divided by 2.

For reactions in which the reactants must penetrate a barrier, which is the case with thermonuclear reactions, the cross section can be expressed in the form

$$\sigma = Kv^{-k} \exp(-L/v), \tag{4.109}$$

where k, K, and L are constants. In particular for the D-D reaction[5]

$$\sigma_{DD} = \frac{288}{u} \exp(-45.8/\sqrt{u}); \tag{4.110}$$

here u is the energy in kev, and σ_{DD} is the cross section in barns. The reaction rate for the D-D process is then approximately given by[5]

$$R_{DD} = \frac{n^2}{2}\, 260 \times 10^{-22} T^{-2/3} \exp(-18.76 T^{-1/3}), \quad \text{if } T < 50 \text{ kev}, \tag{4.111}$$

where T stands for kT in kilovolts. The average rates per pair of reacting particles, $\overline{\sigma v}$, for the D-D and D-T reactions are shown in Fig. 4.10. The one-third power in Eq. 4.111 may appear strange; it arises from the

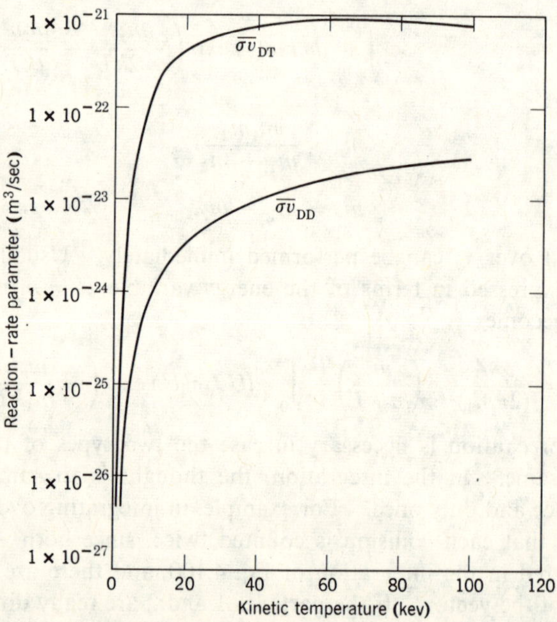

Fig. 4.10. Reaction-rate parameters $\overline{\sigma v}_{DD}$ and $\overline{\sigma v}_{DT}$ for a Maxwellian particle distribution. After R. F. Post, *Revs. Modern Phys.*, **28**, 338–362 (1956).

fact that the most rapidly varying term in the required rate integral is of the form $\exp[-(m_r v^2/2kT) - (B/v)]$; the exponential has its maximum value at $v_{max} = (BkT/m_r)^{1/3}$. Neglecting the slow dependence of the other terms, we conclude that for $v \ll v_{max}$, the cross section σ is very small; and for $v \gg v_{max}$, the Maxwellian distribution is very small.

4.10. Kinetics of the Nuclides and the Power Density

In this section the consumption and formation of various nuclides in the D-D and D-T reactions will be considered, using the averaged coefficients developed in the previous section. From the resulting rate equations, such important parameters as the thermonuclear power density are calculated. Of the reactions listed in Fig. 2.2, only the D-D, D-T, and D-He^3 are important. Assume for simplicity that the branching ratio of the D-D reaction is 0.5. Then with the notation $He^3 \equiv He$, we have at a given temperature

$$\frac{dn_D}{dt} = S_D - \frac{2n_D^2}{2}\overline{\sigma v}_{DD} - n_D n_T \overline{\sigma v}_{DT} - n_D n_{He} \overline{\sigma v}_{DHe}, \tag{4.112}$$

$$\frac{dn_T}{dt} = S_T + n_D^2 \tfrac{1}{4}\overline{\sigma v}_{DD} - n_D n_T \overline{\sigma v}_{DT}, \tag{4.113}$$

$$\frac{dn_{He}}{dt} = S_{He} + n_D^2 \tfrac{1}{4}\overline{\sigma v}_{DD} - n_D n_{He} \overline{\sigma v}_{DHe}, \tag{4.114}$$

where S is the external source, if any, of the constituent denoted by the subscript. The peculiar notation 2/2 in Eq. 4.112 is purposely inserted to remind the reader that the reaction rate for the similar particles is $n_D^2 \overline{\sigma v}/2$ (see Sec. 4.9), but that for each reaction two deuterons are consumed.

Since the equations are somewhat involved to solve, it is better to discuss simplified examples. Consider the case of a steady-state reaction with deuterium source only, under conditions such that the T burns but the He^3 does not. Then

$$n_T = \frac{n_D}{4}\frac{\overline{\sigma v}_{DD}}{\overline{\sigma v}_{DT}}, \tag{4.115}$$

$$S_D = \tfrac{5}{4}(n_D)^2 \overline{\sigma v}_{DD}. \tag{4.116}$$

Because of the high cross section, as may be seen from Fig. 4.10, the amount of T in the mixture will be very small. The power density is readily computed. Each D-D reaction produces on the average

$$U_{DD} = \tfrac{1}{2}(3.27) + \tfrac{1}{2}(4.03) \text{ Mev}$$
$$= 3.65 \text{ Mev} = 5.85 \times 10^{-13} \text{ joule.} \tag{4.117}$$

Each D-T reaction produces

$$U_{DT} = 17.6 \text{ Mev} = 2.82 \times 10^{-12} \text{ joule.} \tag{4.118}$$

The total power density w is given by

$$w = R_{DD}U_{DD} + R_{DT}U_{DT}$$
$$= n_D^2\, \overline{\sigma v}_{DD}\left(\frac{U_{DD}}{2} + \frac{U_{DT}}{4}\right)$$
$$\approx 1.0 \times 10^{-12}\, n_D^2\, \overline{\sigma v}_{DD} \text{ watts/m}^3. \tag{4.119}$$

At a temperature of 60 kev, $w \approx 2 \times 10^{-35}\, n_D^2$. On the assumption that a power density of 10^8 watts/m^3 is reasonable, we find in this example that $n_D = 2.5 \times 10^{21}$/m^3. We can check that most of the power comes from the D-T reaction; however, the D-D cross section is completely controlling, because that is the source of the tritium. We also find that $n_D \gg n_T$ in the plasma.

The mean lifetime of each particle against fusion is the inverse collision frequency $(n\overline{\sigma v})^{-1}$. The mean life of a deuteron in this example is 20 seconds and that of a triton is about 0.5 second. It is important to note from this typical example that the reacting plasma must be confined for a substantial time, in the order of seconds.

As a second example, suppose there is no external source of any nuclides, but that an initial charge of deuterium and the tritium product are burned. Then

$$\frac{dn_D}{dt} = -n_D^2\, \overline{\sigma v}_{DD} - n_D n_T\, \overline{\sigma v}_{DT}, \tag{4.120}$$

$$\frac{dn_T}{dt} = \frac{n_D^2}{4}\, \overline{\sigma v}_{DD} - n_D n_T\, \overline{\sigma v}_{DT}. \tag{4.121}$$

These equations are nonlinear. However, the amount of tritium soon comes to equilibrium, so that after a short initial transient Eq. 4.115 holds, and

$$\frac{dn_D}{dt} = -\tfrac{5}{4} n_D^2\, \overline{\sigma v}_{DD}, \tag{4.122}$$

the solution of which is

$$n_D = n_{D0}(1 + \tfrac{5}{4}\overline{\sigma v}_{DD}\, n_{D0} t)^{-1}, \tag{4.123}$$

where n_{D0} is the initial density of deuterium.

The power density resulting from thermonuclear reactions, as a function of deuteron density, temperature, and conditions, is shown in Fig. 4.11. Because of the radiation losses of the electrons to be discussed later, the D-D reaction is not interesting below 30 kev and the D-T reaction is uninteresting below 5 kev. Operating densities are then restricted to the range of a few times 10^{20} to 10^{22} deuterons/m^3. Although deuterium at 0° C would have a pressure of only 0.1 mm Hg, at operating temperatures the pressure would be 20 to 200 atmospheres.

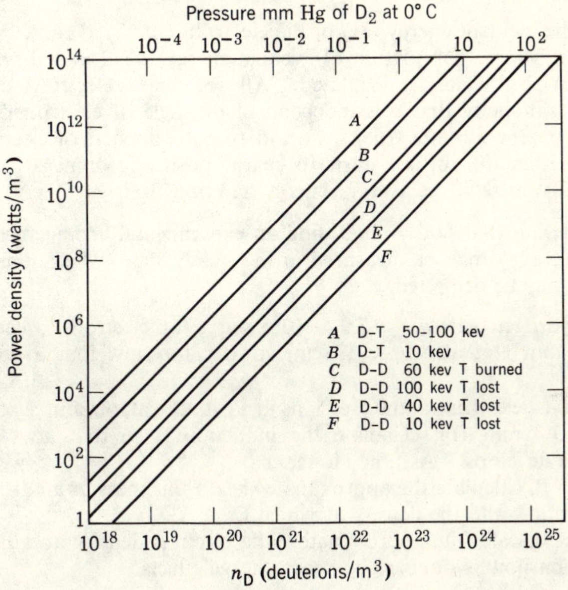

Fig. 4.11. Reaction power density as functions of deuteron particle density; $n_D/n_T = 1$ in D-T mixtures.

The reaction mean free paths are very long, namely, 10^7 to 10^8 m. Impurities even in small amounts can therefore have very serious effects.

For a D-D reaction alone, 66% of the energy goes into charged particles. For the D-T reaction alone, 20% of the energy goes into charged particles. If some practical scheme for direct conversion to electricity were discovered, this charged-particle energy would be the maximum recoverable. The remainder of the energy, 34% for the D-D reaction and 80% for the D-T reaction, must be recovered from the neutrons in a blanket of some sort outside the plasma.

PROBLEMS

1. Suppose that f is expanded in velocity space in spherical harmonics; that is,

$$f = \sum F_{m}^{n}(v, \mathbf{r}, t)\, Y_{m}^{n}(\theta, \varphi),$$

where θ and φ are the usual angles in a spherical system. Which harmonics enter into the calculation of number density, average energy, and drift velocity? Explain why. Write appropriate formulas for the quantities.

2. Assume a secondary-emission coefficient of the form

$$\delta = 8 \times 10^{-3} u \exp(-u/500),$$

where u is the incident energy (ev) of the electrons.

Consider a Maxwellian plasma of average energy $3kT/2 = 800$ ev in contact with the surface, which is insulated. All secondary electrons can escape. Assume for simplicity that δ is independent of angle of electron impact (this assumption is generally not true). Considering the impacts of electrons alone, calculate whether the surface starts to charge positively or negatively. What would the answer be if the average energy had been 10 ev?

3. Considering that $\mathsf{D}/\mu = 2u/3$, find an experimental arrangement whereby the average energy may be measured in the steady state, where ionization and attachment may be neglected.

4. For hydrogen gas, $\nu_m \approx 5.9 \times 10^9 p$ sec^{-1} for electrons, where p is the pressure in mm Hg; also $\mu_i \approx 13$ cm^2/volt-sec for slow ions at atmospheric pressure.

(*a*) For 10-ev electrons and 0.5-ev ions in a cylindrical ion source at a pressure $p = 10^{-2}$ mm Hg, calculate the induction **B** so that no radial space charge develops. Assume H_2^+ ions.

(*b*) If **B** = 0, calculate the approximate sheath thickness at a particle density of 10^{16}/m^3, and the density at which $\mathsf{D}_s \approx 2\mathsf{D}_a$.

(*c*) If **B** = 0, calculate approximately the electron density at which the ion diffusion flow is doubled by space-charge effects.

5. Make reasonable additional assumptions as required, and discuss whether the charges in the following media will move by free or ambipolar diffusion (neglect recombination, attachment, and so on):

(*a*) An ionospheric layer for which $T \approx 500°$ K, $n_e \approx 10^{12}$/m^3, thickness $\approx$ 20 km.

(*b*) A Bunsen burner flame, in which the fractional ionization of the gas is $1/10^8$.

(*c*) A thermonuclear plasma, 1-m diameter.

(*d*) The interior of a concert hall, with charge density 10^6/m^3 generated by natural causes.

6. Suppose that an electron is acted upon by a viscous force, Cv, and an alternating electric field $E_0 e^{i\omega t}$. Find the velocity v of the electron, the current density j, and the resulting conductivity defined as j/E. Show how the viscosity may be related to the collision probability ν_m and how the mobility may be derived in this way.

7. Suppose collisions of electrons of mass m and density n with gas molecules are the only processes taking place. Starting from Newton's second law and characterizing the frequency of collisions by ν_m, write the momentum conservation equation. Noting that the diffusion velocity acquired by a particle has a random direction, so that $\overline{v_z^2} = \frac{1}{3}\overline{v^2}$ independent of position, show that

$$D = \overline{v^2}/3\nu_m.$$

8. Write conservation equations for the electron density n and for the atom density N in a field-free region when both attachment and diffusion are significant. Solve for N in terms of n. Find an equation for n alone.

9. For the time-dependent case, find the solution for the electron density in a nonionizing, nonattaching gas between two infinite parallel plates when diffusion controls the current. Compare the decay rates of the second and third modes with the first.

10. Do Problem 9 with the gas in a cubical cavity.

11. Do Problem 9 with the gas in a right cylinder of radius r and height h.

12. Write the first Townsend coefficient α/p in terms of integrals over the distribution function $f(v)$ of the electrons, the gas density, and the appropriate cross sections.

13. Discuss under what conditions the recombination of electrons with positive ions may be described by an equation $dn_e/dt = -\alpha_r n_e^2$. How is α_r derived from the appropriate two-body cross section?

14. Calculate the collision rate $\bar{\nu}$ of a gas molecule in a Maxwellian distribution, the molecule being a hard sphere of cross section σ.

15. Find the energy of the ions giving the maximum contribution to the D-D reaction rate in terms of the temperature of the Maxwell-Boltzmann distribution according to which the ions are assumed to be distributed.

16. In an idealized thermonuclear plasma consisting of deuterons and electrons only, the density of deuterons is $2 \times 10^{21}/\text{m}^3$ and a deuteron makes a fusion collision at a frequency $\nu_f = 10^{-2}\text{sec}^{-1}$. Calculate the energy generated per m^3-sec by the D-D reaction.

17. A thermonuclear reactor is based on the D-D and D-T cycles, and charged particles are confined. Assume that production of tritium and He^3 are equally likely, and neglect any effects due to He^3. The only source of new fuel is D. By what factor can the total power output be increased by converting the neutrons in a blanket $Li^6(n, \alpha)T$ reaction into T with efficiency $\mathfrak{h}$, and recycling the T? The $Li^6(n, \alpha)T$ reaction yields 4.8 Mev.

Now assume T = 20 kev, $\overline{\sigma v}_{DD} = 3.5 \times 10^{-24}$ m^3/sec, $\overline{\sigma v}_{DT} = 6 \times 10^{-22}$ m^3/sec, $\mathfrak{h} = 0.7$, $n_D = 5 \times 10^{20}/\text{m}^3$.

Calculate n_T and the total power.

18. A mixture of D and T is comprised initially of 10^{21} atoms/m^3 of each material. The mixture is suddenly brought up to a temperature of 10 kev (T = 10 kev) and starts to react. Neglect the infrequent D-D reactions. All charged particles are confined, and any neutrons escape freely. All positive charged particles thermalize in the plasma rapidly compared with the thermonuclear reaction rate per particle. The electrons are very light and do not interchange much energy with the heavy nuclei, hence they remain Maxwellian at 10 kev. The positive ions are also Maxwellian but not necessarily at 10 kev. Neglect all radiation losses.

(*a*) What is the temperature (T in kev) when 1% of the deuterium is burned?
(*b*) What is the final energy of the entire charged debris after all is burned and after all electrons and ions have come to thermal equilibrium?

REFERENCES

1. E. H. Kennard, *Kinetic Theory of Gases*, McGraw-Hill Book Co., New York (1938).
2. W. P. Allis and D. J. Rose, *Phys. Rev.*, **93**, 84 (1954).
3. B. Lehnert, *Proc. 2nd U.N. Conf. on Peaceful Uses of Atomic Energy*, Vol. 32, United Nations, Geneva (1958), p. 349.
4. A. Simon, *An Introduction to Thermonuclear Research*, Pergamon Press, New York (1959), pp. 150 ff.
5. R. F. Post, *Revs. Modern Phys.*, **28**, 338–362 (1956).

GENERAL REFERENCES

W. P. Allis, "Motions of Ions and Electrons," *Handbuch der Physik*, Vol. 21, Springer, Berlin (1956), pp. 383–444.

S. C. Brown, *Basic Data of Plasma Physics*, The Technology Press and John Wiley & Sons, New York (1959).

5

Maxwell's equations and electromagnetic energy

In Chap. 1 we pointed out that a plasma hot enough to experience thermonuclear reactions is so hot that no normal material can contain it. It must be confined by a magnetic field or a combination of such a field and an electric field. The forces that act on a charged particle arise principally from the electromagnetic field at the charge. This field in turn arises from other charges. It behooves us to examine the way in which these fields can be created. The classical electromagnetic theory will suffice for our purposes. We present only a very brief résumé of the more common parts of the theory and a more extensive development of the special features needed in the study of hot plasmas. From the concept of electromagnetic forces we are led very naturally to the concept of electromagnetic stress and pressure. For the reader not skilled in the subject, we present here the basic rudiments of Maxwell's equations and related potentials, which underly the hydromagnetic equations of the next chapter, and which indeed are essential to the development of all our further theory and applications.

In setting forth the concepts, we propose that the forces on a charged particle arise principally from the electric field and the magnetic induction at the charge. The space around a test charge may somewhat arbitrarily be divided into four regions, as shown in Fig. 5.1. These regions may be called (1) the quantum-mechanical region nearest the charge, (2) the Debye region next, (3) the fluid region, and (4) the external field region farthest away.

Region 1 is that in which very close collisions between particles occur. For such close encounters quantum mechanics must be used, and the particles are deflected by relatively large angles. In region 2, classical theory may be used. The field at the test particle fluctuates because of the discrete nature of the sources within both regions 1 and 2. The Debye length, as discussed in Sec. 4.8, defines the periphery of region 2, because the thickness of the sheath shielding the test particle from other parts of the plasma is comparable with this distance. In Sec. 4.2, the fields were somewhat arbitrarily divided into rapidly varying and slowly varying parts. The rapidly varying part arises from interactions within regions 1

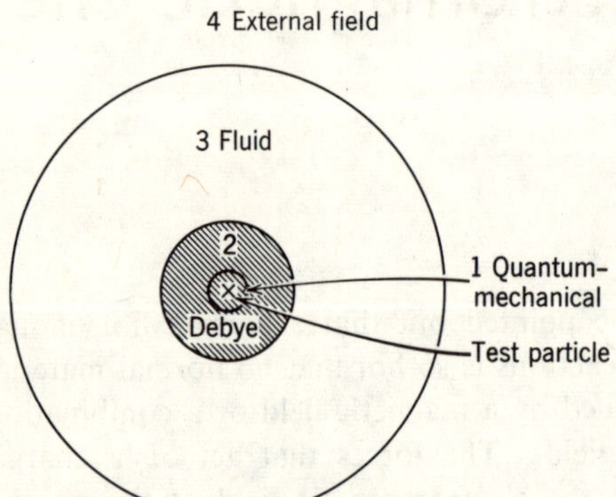

Fig. 5.1. Subdivision of space about a test particle in a plasma. The shaded regions are those within which the discrete nature of the particles is important. The plasma occupies regions 1, 2, and 3.

and 2, and is represented by collision cross sections. These cross sections were discussed in Chaps. 2 and 3, and the case of Coulomb collisions will be discussed in Sec. 8.2. Therefore, the rapidly varying part will not be discussed further here.

In region 3, the sources of the fields are so remote from the test charge that their discrete nature is not distinguishable at the test charge. Therefore, the charges in the fluid region may be considered to form a continuum. In region 4, the charges and currents may *a fortiori* be so considered, and they are prescribed in addition. Any variation of these fields at the test particle does not arise from the discrete character of the sources but rather from macroscopic variations in the fluid or the external region. These fields **E** and **B** are those that appear explicitly in the Boltzmann equation (Eqs. 4.17 and 4.19) of Sec. 4.2. We discuss these fields in the present chapter and approximate the actual distribution of discrete charges by a continuum.

Further, in this chapter discussion is limited to the basic foundations and concepts. The very numerous and elaborate deductions worked out elsewhere will be avoided.[1–3] It is assumed that the reader is familiar with the integral form of the laws of Gauss, Ampère, and Faraday and the solenoidal character of the magnetic induction. The divergence theorem and Stokes's theorem will be used to deduce the differential form of these laws.[4]

5.1. Formulation of Maxwell's Equations

We shall be making intensive use of Maxwell's equations in all of our later work, both in the integral form, with which we assume the reader is familiar, and in the differential form, which in many cases is easier to apply. In this section, we shall deduce the differential form of Maxwell's equations from the integral form. It is assumed that the integral form is given by experiment.

We start with the law of Gauss, which states that the normal component of the electric displacement† **D** integrated over any closed surface s is equal to the total charge q:

$$\int d\mathbf{s}\cdot\mathbf{D} = q. \tag{5.1}$$

The surface is shown in Fig. 5.2. By the divergence theorem the surface integral in Eq. 5.1 may be reduced:[4]

$$\int d\mathbf{r}\nabla\cdot\mathbf{D} = \int d\mathbf{r}\ \rho, \tag{5.2}$$

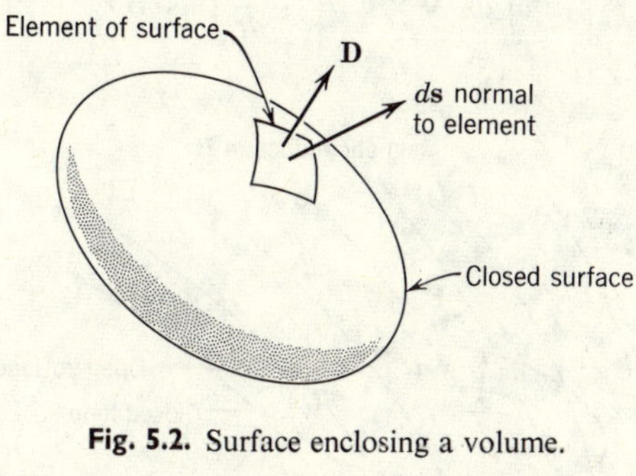

Fig. 5.2. Surface enclosing a volume.

† The displacement vector **D** or its component D_α in the x-direction is readily distinguished from the diffusion coefficient D or its tensor component $\mathrm{D}_{\alpha\beta}$ (see Sec. 5.2).

where ρ is the charge density. Since the element of surface and the element of volume enclosed are perfectly arbitrary, both elements may be taken to be so small that the integrands are substantially constant; therefore,

$$\nabla \cdot \mathbf{D} = \rho. \tag{5.3}$$

The solenoidal law of the magnetic induction states that the integral of the normal component of the magnetic induction over a closed surface is zero. Figure 5.2 applies to the present case by replacing **D** with **B**:

$$\int d\mathbf{s} \cdot \mathbf{B} = 0. \tag{5.4}$$

Again, by the divergence theorem the surface integral can be reduced to a volume integral:

$$\int dr\, \nabla \cdot \mathbf{B} = 0. \tag{5.5}$$

The surface element and the enclosed volume element are perfectly arbitrary; therefore, let the elements be chosen infinitesimally small. It follows that

$$\nabla \cdot \mathbf{B} = 0. \tag{5.6}$$

Faraday's law states that the electromotive force around a closed loop is equal to the negative of the total rate of change with time of the flux Φ of induction linking that loop:

$$\oint d\ell \cdot \mathbf{E} = -\frac{d}{dt}\Phi. \tag{5.7}$$

The closed loop is shown in Fig. 5.3. The line integral may be transformed into a surface integral by means of Stokes's theorem:[4]

$$\int d\mathbf{s} \cdot \nabla \times \mathbf{E} = -\frac{d}{dt}\int d\mathbf{s} \cdot \mathbf{B}. \tag{5.8}$$

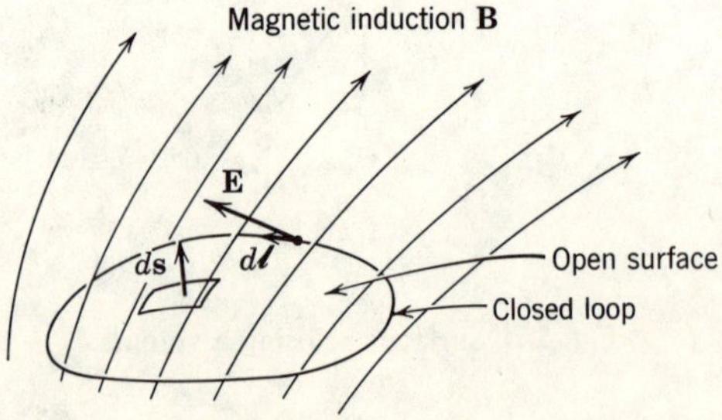

Fig. 5.3. Distribution of magnetic induction in space, and a loop marked in the space.

For the moment the surface will be taken to be fixed with respect to the observer. In that case,

$$\int d\mathbf{s}\cdot\nabla\times\mathbf{E} = -\int d\mathbf{s}\cdot\frac{\partial\mathbf{B}}{\partial t}. \tag{5.9}$$

Since the loop and surface bounded thereby are completely arbitrary, both may be taken very small so that the integrand changes by a negligible amount over the surface of integration. We then find

$$\nabla\times\mathbf{E} = -\frac{\partial\mathbf{B}}{\partial t}. \tag{5.10}$$

Ampère's law states that the magnetomotive force integrated around a closed loop is equal to the enclosed current I plus the rate of change with respect to time of the electric displacement $\mathbf{D}$ linking the loop:

$$\oint d\boldsymbol{\ell}\cdot\mathbf{H} = I + \frac{d}{dt}\int d\mathbf{s}\cdot\mathbf{D}. \tag{5.11}$$

The integral around the closed loop may be transformed into an integral over the bounded surface by means of Stokes's theorem. Figure 5.3 applies to the present case by replacing $\mathbf{E}$ with $\mathbf{H}$, and $\mathbf{B}$ with $\mathbf{j} + \partial\mathbf{D}/\partial t$:

$$\int d\mathbf{s}\cdot\nabla\times\mathbf{H} = \int d\mathbf{s}\cdot\left(\mathbf{j} + \frac{\partial\mathbf{D}}{\partial t}\right), \tag{5.12}$$

where $\mathbf{j}$ is the current crossing a unit area. Since the loop and surface bounded are completely arbitrary, they are taken so small that the integrands change negligibly over the surface of integration. Therefore,

$$\nabla\times\mathbf{H} = \mathbf{j} + \frac{\partial\mathbf{D}}{\partial t}. \tag{5.13}$$

Equations 5.1, 5.4, 5.7, and 5.11 are Maxwell's equations in integral form, and Eqs. 5.3, 5.6, 5.10, and 5.13 are Maxwell's equations in differential form. The electric displacement $\mathbf{D}$ and the magnetic intensity $\mathbf{H}$ may be regarded as the fields created by charges and currents, respectively, and the electric intensity $\mathbf{E}$ and the magnetic induction $\mathbf{B}$ may be regarded as the fields giving the forces acting on a charge.

Not all of the Maxwell equations are independent of each other. The divergence of Eq. 5.10 is zero, so that

$$\frac{\partial}{\partial t}(\nabla\cdot\mathbf{B}) = 0. \tag{5.14}$$

Consequently, if at any instant of time whatsoever $\nabla\cdot\mathbf{B}$ was zero, it must be zero forever. Again, the divergence of Eq. 5.13 is zero. Then, by Eq. 5.3, we have

$$\nabla\cdot\mathbf{j} + \frac{\partial\rho}{\partial t} = 0. \tag{5.15}$$

This equation merely states that charge is neither created nor destroyed. The point is made clearer by integrating Eq. 5.15 over a volume and applying the divergence theorem:

$$\int d\mathbf{s}\cdot\mathbf{j} = -\frac{d}{dt}\int d\tau\ \rho. \tag{5.16}$$

The current of charge leaving the volume is equal to the rate at which the charge within the volume decreases with time. There is nothing unique about the particular sequence of deductions used here; many other permutations exist.

5.2. The Constitutive Relations

As mentioned in the previous section, **E** and **B** measure the force on a charge, and **D** and **H** are the fields created by other charges. We must relate the fields created to the forces exerted on a particle if we wish to determine its motion. The resulting relations, which are called constitutive relations, provide the required connections.

Most media are isotropic, homogeneous, and linear. For such media

$$\mathbf{D} = \epsilon\mathbf{E}, \tag{5.17}$$

where ϵ, the permittivity, is independent of **E**, position, time, and direction. If the medium is a vacuum, the permittivity is denoted by ϵ_0, and

$$\epsilon_0 = 10^7/4\pi c^2 \text{ coulomb}^2\text{-sec}^2/\text{kg-m}^3, \tag{5.18}$$

or

$$\epsilon_0 = 8.854 \times 10^{-12} \text{ farad/m}. \tag{5.19}$$

Further, for isotropic, homogeneous, and linear media,

$$\mathbf{B} = \mu\mathbf{H}, \tag{5.20}$$

where μ, the permeability, is independent of **H**, position, time, and direction. If the medium is a vacuum, the permeability is denoted by μ_0, and

$$\mu_0 = 4\pi \times 10^{-7} \text{ kg-m/coulomb}^2, \tag{5.21}$$

or

$$\mu_0 = 1.257 \times 10^{-6} \text{ henry/m}. \tag{5.22}$$

Finally, the electric field **E** is related to the current density **j** by the resistivity η of the medium:

$$\mathbf{E} = \eta\mathbf{j}. \tag{5.23}$$

Equation 5.23 is a statement of Ohm's law. It is often more convenient to speak of the conductivity. We avoid the use of the normal symbol σ for this quantity to escape from a conflict with the symbol for a cross section, and write the conductivity as s. Thus,

$$\mathbf{j} = s\mathbf{E}. \tag{5.24}$$

Note that the conductivity s is the reciprocal of the resistivity η in an isotropic medium, because η is a simple scalar number.

Plasmas to which a magnetic induction is applied are not isotropic. For these and all other anisotropic media, **E** is usually not collinear with **D**, and similarly for **B** and **H**, and **E** and **j**. The quantities in each of these three pairs are in general related in a more complicated way. We choose Ohm's law (Eq. 5.23) as an example. Suppose that the magnitudes of **E** and **j** are proportional, which, by the way, is not always the case. Then let the medium have different properties in different directions. Most generally, we may write

$$\left.\begin{aligned} E_x &= \eta_{xx}j_x + \eta_{xy}j_y + \eta_{xz}j_z, \\ E_y &= \eta_{yx}j_x + \eta_{yy}j_y + \eta_{yz}j_z, \\ E_z &= \eta_{zx}j_x + \eta_{zy}j_y + \eta_{zz}j_z, \end{aligned}\right\} \tag{5.25}$$

where η_{xx}, η_{xy}, and so on are the appropriate coefficients.

It is convenient at this time to introduce a shorthand notation for Eq. 5.25:

$$E_\alpha = \sum_{\beta=1}^{3} \eta_{\alpha\beta} j_\beta. \tag{5.26}$$

Here, α stands for x, y, or z, and $\beta = 1, 2$, or 3 stands for x, y, and z, respectively. The quantity E_α is any one of the components of **E**. If we define the usual three unit vectors ($\mathbf{i}_1 = \mathbf{i}_x$, $\mathbf{i}_2 = \mathbf{i}_y$, and $\mathbf{i}_3 = \mathbf{i}_z$) in the x, y, and z directions, respectively, then **E** itself is given by

$$\mathbf{E} = \sum_{\alpha=1}^{3}\sum_{\beta=1}^{3} \mathbf{i}_\alpha \eta_{\alpha\beta} j_\beta. \tag{5.27}$$

In this manner, Eq. 5.26 defines the relation between the vectors **E** and **j**. They may not be collinear.† If the medium is in fact isotropic,

† It can be shown[1, 5] that $\eta_{\alpha\beta}$ is a tensor of rank 2. We use the word tensor only to classify and label certain quantities. We do not need to know the precise mathematical definition of a tensor nor any of its algebra or calculus. Accordingly, we shall not go into these subjects in this volume.

$\eta_{xx} = \eta_{yy} = \eta_{zz} = \eta$, and all off-diagonal terms $\eta_{\alpha\beta}$ $(\alpha \neq \beta)$ are zero. This statement may be conveniently summarized by the notation

$$\eta_{\alpha\beta} = \eta\delta_{\alpha\beta} \tag{5.28}$$

for an isotropic medium, where

$$\delta_{\alpha\beta} = \begin{cases} 1, & \text{if } \alpha = \beta, \\ 0, & \text{if } \alpha \neq \beta. \end{cases} \tag{5.29}$$

The indices α and β themselves are dummy indices and may be changed at will, provided only that the proper summation over repeated indices is preserved, as in Eq. 5.26.

The conductivity ς may be related to the mobility of the charges. Thus, if the medium is isotropic, the current $\mathbf{j}_i$ of positive ions is

$$\mathbf{j}_i = q_i\mathbf{\Gamma}_i = q_i\mu_i n_i \mathbf{E} \tag{5.30}$$

by definition; μ is the mobility and is not to be confused with the permeability, Eq. 5.20. If the medium is anisotropic, we have in general

$$j_{\alpha i} = \sum_{\beta=1}^{3} q_i n_i \mu_{\alpha\beta i} E_\beta. \tag{5.31}$$

If both electrons and ions move,

$$j_\alpha = \sum_{\beta=1}^{3} (q_i n_i \mu_{\alpha\beta i} + q_e n_e \mu_{\alpha\beta e}) E_\beta. \tag{5.32}$$

Here, to preserve the symmetry, the mobility is taken with positive or negative sign for positive or negative charges. As a further generalization, for N types of charge,

$$j_\alpha = \sum_{\beta=1}^{3} \left(\sum_{\zeta=1}^{N} q_\zeta n_\zeta \mu_{\alpha\beta\zeta} \right) E_\beta; \tag{5.33}$$

the terms in parentheses in Eq. 5.32 or 5.33 are the components of the conductivity ς. Note that η and ς are not reciprocals of each other if the medium is anisotropic. Both of them are matrices, and their reciprocals are found by the usual method.[6]

Exactly the same formalism may be applied to Eqs. 5.17 and 5.20. Thus,

$$D_\alpha = \sum_{\beta=1}^{3} \epsilon_{\alpha\beta} E_\beta, \tag{5.34}$$

$$B_\alpha = \sum_{\beta=1}^{3} \mu_{\alpha\beta} H_\beta, \tag{5.35}$$

where μ is here the permeability. The same symbol has been used above to designate the mobility. Convention dictates this ambiguity. The context determines the meaning. We shall find little application here for Eq. 5.35.

The notation developed in this section is particularly useful and will be used from time to time as necessity requires.†

5.3. Maxwell's Equations for Uniformly Moving Media

It is a central result of the theory of relativity that the Maxwell equations (Eqs. 5.3, 5.6, 5.10, and 5.13) have exactly the same form for an observer in a uniformly moving system as they do in a system fixed with respect to an observer. The discussion is facilitated by considering Fig. 5.4. An

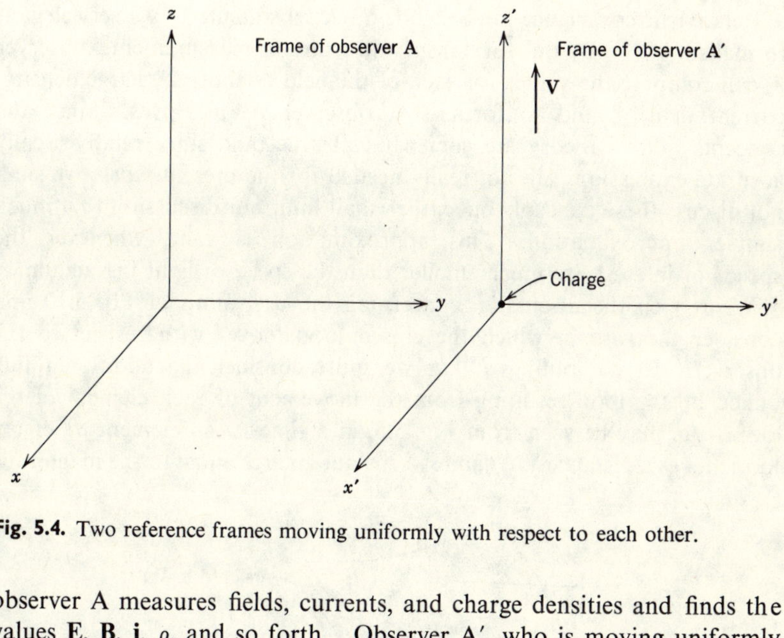

Fig. 5.4. Two reference frames moving uniformly with respect to each other.

observer A measures fields, currents, and charge densities and finds the values **E**, **B**, **j**, ρ, and so forth. Observer A′, who is moving uniformly with respect to observer A and who is stationary with respect to the

† In fact, we have in an earlier context come across a case which was so expressible. In Sec. 4.7, the diffusion coefficient D of charges was stated to be anisotropic in the presence of a magnetic induction. We could have written in that case

$$\Gamma_\alpha = \sum_{\beta=1}^{3} D_{\alpha\beta} \frac{\partial n}{\partial x_\beta}.$$

uniformly moving medium, measures fields, currents, and charge densities and finds the values $\mathbf{E}'$, $\mathbf{B}'$, $\mathbf{j}'$, ρ', and so forth. The values measured by observer A′ are not the same as those measured by observer A. A simple example will support this point. Suppose that a charge moves uniformly with respect to observer A and suppose that observer A′ is at rest with respect to the charge. Then observer A sees an electric field, a current, and a magnetic field. Observer A′, on the other hand, sees only an electric field: to him there is no magnetic field. Each observer now develops a theory relating the various field quantities, the charges, and the currents. Each finds that the Maxwell equations provide the exact relations for the values of the quantities that he measures, in spite of the fact that the other observer measures different values. The Maxwell equations are said to be relativistically covariant because of this property.

Relativistic covariance can be made obviously manifest by a very elegant formalism.[1, 3, 5] By this formalism and from his measurements, observer A can compute the values for each of the field quantities, charge density, current density, and so forth, that observer A′ measures. Since the concepts of this theory are somewhat abstract and since relativistically accurate expressions are not really needed for the present work, we shall not discuss these concepts but rather shall limit our discussion to a much simpler approximation. This approximation is valid whenever the speeds of interest are much smaller than the speed of light in vacuum.

We reopen the discussion concerning the derivation of Eq. 5.10 and consider the case in which the closed loop moves with respect to the observer. In computing $d\Phi/dt$, we must consider the additional flux linked by the loop resulting from the movement of each element of the loop. As may be seen from Fig. 5.5, in a time Δt an element $d\ell$ of the loop moves a distance $\mathbf{V}\,\Delta t$ and sweeps out an area equal to the magnitude

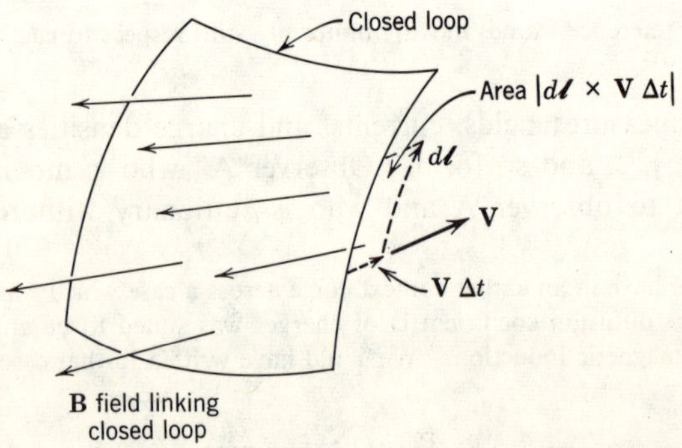

Fig. 5.5. Lines of induction linking a uniformly moving closed loop.

of $d\ell \times \mathbf{V}\,\Delta t$. The additional flux linking the loop is then

$$-\mathbf{B}\cdot d\ell \times \mathbf{V}\,\Delta t. \tag{5.36}$$

The total additional flux is given by the integral around the whole loop of Expression 5.36. Per unit time the added flux is

$$-\int d\ell\cdot\mathbf{V}\times\mathbf{B}. \tag{5.37}$$

We now compute the total electromotive force around the loop. Since the loop is moving, it is the electromotive force in the moving frame that will be obtained. The electric field in the moving frame is denoted by $\mathbf{E}'$. The Faraday law then becomes

$$\int d\ell\cdot(\mathbf{E}' - \mathbf{V}\times\mathbf{B}) = -\int ds\cdot\frac{\partial\mathbf{B}}{\partial t}. \tag{5.38}$$

By Stokes's theorem and the complete arbitrariness in the choice of the closed loop, it follows that

$$\nabla\times(\mathbf{E}' - \mathbf{V}\times\mathbf{B}) = -\frac{\partial\mathbf{B}}{\partial t}. \tag{5.39}$$

The form of Maxwell's equations must be independent of any uniform motion, as is required by the theory of relativity. Therefore,

$$\mathbf{E} = \mathbf{E}' - \mathbf{V}\times\mathbf{B}. \tag{5.40}$$

In other words, the electric field $\mathbf{E}'$ in the moving frame is given by

$$\mathbf{E}' = \mathbf{E} + \mathbf{V}\times\mathbf{B}, \tag{5.41}$$

where $\mathbf{E}$ and $\mathbf{B}$ are measured in the stationary frame.

This result is very plausible. An observer holding onto a charge q at rest with respect to the sources of a magnetic induction will notice a force $q\mathbf{E}$. If he moves and carries the charge with him, he will state that the force is $q\mathbf{E}'$, although his colleague at rest will claim that the force is $q\mathbf{E} + \mathbf{V}\times\mathbf{B}$. Equation 5.41 is a statement of the equality of these two results.

The approximate transformation law for the current density is easily found. Observer A states that the value $\mathbf{j}$ he measures for the current density is related to that $\mathbf{j}'$ measured by observer A′ by

$$\mathbf{j} = \mathbf{j}' + \rho\mathbf{V}. \tag{5.42}$$

Observer A sees an added velocity $\mathbf{V}$ of the charge that observer A′ does not.

The relations in Eqs. 5.12 and 5.42 can be used to show that

$$\mathbf{H}' = \mathbf{H} - \mathbf{V}\times\mathbf{D}. \tag{5.43}$$

5.4. Vector and Scalar Potentials

The use of scalar and vector potentials greatly simplifies the calculation of the fields. Let us begin with the vector potential. In view of the solenoidal character of the magnetic induction, as expressed by Eq. 5.6, the magnetic induction can be deduced from a vector potential[1,4] **A**. This potential is defined, apart from any additive gradients, by

$$\mathbf{B} = \nabla \times \mathbf{A}. \tag{5.44}$$

Clearly, Eq. 5.44 is compatible with Eq. 5.6. If this relation is substituted into Eq. 5.10, there follows that

$$\nabla \times \left(\mathbf{E} + \frac{\partial \mathbf{A}}{\partial t}\right) = 0. \tag{5.45}$$

In view of the irrotational character of $\mathbf{E} + \partial \mathbf{A}/\partial t$, this quantity can be deduced from a scalar potential ϕ according to[1,4]

$$\mathbf{E} = -\nabla\phi - \frac{\partial \mathbf{A}}{\partial t}. \tag{5.46}$$

Should the vector potential be independent of the time, then by Eq. 5.3

$$\nabla^2\phi = -\rho/\epsilon \tag{5.47}$$

if ϵ is a scalar independent of position.

The solution of this last equation is[4]

$$\phi(\mathbf{r}) = \frac{1}{4\pi\epsilon}\int dr' \, \frac{\rho(\mathbf{r}')}{|\mathbf{r} - \mathbf{r}'|}. \tag{5.48}$$

That Eq. 5.48 is the solution of Eq. 5.47 is easily proved as follows. Let

$$\mathbf{r}'' = \mathbf{r} - \mathbf{r}'. \tag{5.49}$$

Then if $\mathbf{r}'' \neq 0$,

$$\begin{aligned} \nabla^2 \frac{1}{|\mathbf{r} - \mathbf{r}'|} &= \frac{1}{r''^2}\frac{d}{dr''}\left[r''^2 \frac{d}{dr''}\left(\frac{1}{r''}\right)\right] \\ &= 0. \end{aligned} \tag{5.50}$$

Accordingly,

$$\begin{aligned} \nabla^2\phi(\mathbf{r}) &= \frac{1}{4\pi\epsilon}\int dr' \, \rho(\mathbf{r}')\nabla^2 \frac{1}{|\mathbf{r} - \mathbf{r}'|} \\ &= \frac{\rho(\mathbf{r})}{4\pi\epsilon}\int dr' \, \nabla^2 \frac{1}{|\mathbf{r} - \mathbf{r}'|}, \end{aligned} \tag{5.51}$$

since the integrand in this equation is zero everywhere, except possibly at the point $\mathbf{r}' = \mathbf{r}$. To evaluate this integral, put a small sphere of radius $\mathfrak{d}$ about the point $\mathbf{r} = \mathbf{r}'$ and apply the divergence theorem. Thus,

$$\nabla^2\phi(\mathbf{r}) = \frac{\rho(\mathbf{r})}{4\pi\epsilon} \int\limits_{\substack{\text{Sphere of} \\ \text{radius } \mathfrak{d}}} d\mathbf{s}\cdot\nabla \frac{1}{|\mathbf{r} - \mathbf{r}'|}$$

$$= -\frac{\rho(\mathbf{r})}{4\pi\epsilon} \int\limits_{\substack{\text{Sphere of} \\ \text{radius } \mathfrak{d}}} d\mathbf{s}\cdot\frac{\mathbf{r}''}{\mathbf{r}''^3}$$

$$= -\frac{\rho(\mathbf{r})}{\epsilon}, \tag{5.52}$$

as was to be proved. The convenience of the scalar potential is to be noted: an algebraic summation is all that is needed, instead of the vector summation that would have been necessary had we calculated the electric intensity itself.

Now let us find the so-called gauge condition for the two potentials. To keep the work simple, we assume that ϵ and μ are independent of position and time. Let Eq. 5.46 be substituted into Eq. 5.3:

$$\nabla^2\phi + \nabla\cdot\frac{\partial\mathbf{A}}{\partial t} = -\rho/\epsilon. \tag{5.53}$$

Next let Eq. 5.44 be substituted into Eq. 5.13:

$$\nabla\times(\nabla\times\mathbf{A}) = \nabla(\nabla\cdot\mathbf{A}) - \nabla^2\mathbf{A} = \mu\mathbf{j} - \mu\epsilon\,\nabla\frac{\partial\phi}{\partial t} - \mu\epsilon\frac{\partial^2\mathbf{A}}{\partial t^2}, \tag{5.54}$$

the intermediate relation resulting from a frequently used vector identity. These equations can be simplified. It is noted that the vector potential is not uniquely determined as yet because the gradient of any scalar function χ can be added to $\mathbf{A}$ without changing the induction $\mathbf{B}'$. By Eq. 5.46, $-\partial\chi/\partial t$ must be added to ϕ in order that the electric intensity $\mathbf{E}$ be unaltered. We can so choose χ that

$$\nabla\cdot\mathbf{A} + \epsilon\mu\frac{\partial\phi}{\partial t} = 0. \tag{5.55}$$

Let χ be so chosen. Then Eq. 5.53 becomes simply

$$\nabla^2\phi - \epsilon\mu\frac{\partial^2\phi}{\partial t^2} = -\rho/\epsilon, \tag{5.56}$$

and Eq. 5.54 becomes

$$\nabla^2\mathbf{A} - \epsilon\mu\frac{\partial^2\mathbf{A}}{\partial t^2} = -\mu\mathbf{j}. \tag{5.57}$$

Equation 5.57 reduces to

$$\nabla^2 \mathbf{A} = -\mu \mathbf{j} \tag{5.58}$$

in case **A** is time independent. The solution of this equation follows from the solution Eq. 5.48 of Eq. 5.47 by duality:[1,3]

$$\mathbf{A}(\mathbf{r}) = \frac{\mu}{4\pi} \int dr' \, \frac{\mathbf{j}(\mathbf{r}')}{|\mathbf{r} - \mathbf{r}'|}. \tag{5.59}$$

This result is very useful for determining the magnetic field from current distributions.

The wave properties of the solutions of the inhomogeneous wave equations (Eqs. 5.56 and 5.57) will be discussed in Chap. 9.

5.5. Electromagnetic Energy Density and Poynting's Vector

We are now prepared to discuss the concept of energy density in an electromagnetic field and the concept of the energy flow across surfaces.[1,3] To this end, let the following scalar products be formed from the Maxwell equations, Eqs. 5.10 and 5.13:

$$\mathbf{H} \cdot \nabla \times \mathbf{E} = -\mathbf{H} \cdot \frac{\partial \mathbf{B}}{\partial t}, \tag{5.60}$$

$$\mathbf{E} \cdot \nabla \times \mathbf{H} = \mathbf{E} \cdot \mathbf{j} + \mathbf{E} \cdot \frac{\partial \mathbf{D}}{\partial t}. \tag{5.61}$$

Note that

$$\mathbf{H} \cdot \nabla \times \mathbf{E} - \mathbf{E} \cdot \nabla \times \mathbf{H} = \nabla \cdot (\mathbf{E} \times \mathbf{H}) \tag{5.62}$$

is a vector identity. If Eq. 5.61 is subtracted from Eq. 5.60, the result is

$$\nabla \cdot (\mathbf{E} \times \mathbf{H}) + \mathbf{E} \cdot \frac{\partial \mathbf{D}}{\partial t} + \mathbf{H} \cdot \frac{\partial \mathbf{B}}{\partial t} + \mathbf{E} \cdot \mathbf{j} = 0. \tag{5.63}$$

If the permittivity and permeability are independent of time, then

$$\nabla \cdot (\mathbf{E} \times \mathbf{H}) + \frac{\partial}{\partial t} \left(\frac{\mathbf{E} \cdot \mathbf{D}}{2} + \frac{\mathbf{H} \cdot \mathbf{B}}{2} \right) + \mathbf{E} \cdot \mathbf{j} = 0. \tag{5.64}$$

For isotropic media

$$\nabla \cdot (\mathbf{E} \times \mathbf{H}) + \frac{\partial}{\partial t} \left(\frac{\epsilon E^2}{2} + \frac{B^2}{2\mu} \right) + \mathbf{E} \cdot \mathbf{j} = 0. \tag{5.65}$$

The physical meaning of Eq. 5.64 or Eq. 5.65 can be found by integrating either equation over a volume of interest. For Eq. 5.64 we find that, upon applying the divergence theorem,

$$\int d\mathbf{s} \cdot \mathbf{E} \times \mathbf{H} + \frac{\partial}{\partial t} \int dr \left(\frac{\mathbf{E} \cdot \mathbf{D}}{2} + \frac{\mathbf{H} \cdot \mathbf{B}}{2} \right) + \int dr \, \mathbf{E} \cdot \mathbf{j} = 0. \tag{5.66}$$

The interpretation is now clear if this equation is merely stated in words. In the last term the factor $\mathbf{E}\cdot\mathbf{j}$ appears. This term represents the resistive work done per unit of time by the field $\mathbf{E}$ on the charges in moving them about within the volume element of interest. The first term then represents the rate at which energy escapes over the surface of the volume of interest; the second term represents the rate of change with time of the electric and magnetic energy within the volume of interest. The equation then states that the rate at which the energy of the electric and magnetic fields increases within the volume element equals the rate at which such energy leaks in over the surface bounding the volume element minus the rate at which the electric field does work on charges within the volume element. The magnetic induction does no work on charges, of course, because it always acts perpendicular to their motion.

The statements made in the previous paragraph can be supported by showing that the work done in so arranging charges to create a static electric field $\mathbf{E}$ is $\int d\mathbf{r}\, \mathbf{E}\cdot\mathbf{D}/2$; the work done by establishing currents such that the magnetic induction $\mathbf{B}$ results is $\int d\mathbf{r}\, \mathbf{H}\cdot\mathbf{B}/2$, where the integrals are taken over the entire extent of the fields. Then by analogy, we associate $\mathbf{E}\cdot\mathbf{D}/2$ and $\mathbf{H}\cdot\mathbf{B}/2$ with the electrostatic and magnetic energies per unit volume, respectively.

The Poynting vector $\mathbf{S}$ is defined to be

$$\mathbf{S} = \mathbf{E} \times \mathbf{H}. \tag{5.67}$$

It may be thought of as representing the rate of flow of energy in the electric and magnetic fields across a closed surface per unit area. We must take care, however, in applying this concept of the Poynting vector. For example, Eq. 5.65 may be looked upon as an energy-conservation equation per unit volume. However, only the integral of $\mathbf{S}$ over a closed surface has been defined, and only the integral over such a surface enters into the energy balance. Incorrect results may be obtained if the Poynting vector is blindly associated with the energy flow per unit area at any particular point. For example, $\mathbf{S}$ may have nonzero values in the region occupied by static electric and magnetic fields; and yet $\nabla\cdot\mathbf{S} = 0$ everywhere. The integral of $\mathbf{S}$ over the surface bounding any such volume is zero. The energy flow would then be zero.

5.6. Electromagnetic Stress

In the next chapter, the hydromagnetic equations relating energy flow and forces in a plasma will be developed. As a prelude, we develop here the concept of electromagnetic stress and a formalism whereby the body forces may be calculated from the electromagnetic field quantities.

To this end, we return again to Maxwell's equations, multiply Eq. 5.10 vectorially by $\epsilon\mathbf{E}$ and Eq. 5.13 vectorially by $\mathbf{B}$. After adding, we find for an isotropic, homogeneous, linear, time-independent medium

$$\epsilon(\nabla \times \mathbf{E}) \times \mathbf{E} + \frac{1}{\mu}(\nabla \times \mathbf{B}) \times \mathbf{B} = \mathbf{j} \times \mathbf{B} + \epsilon\mu\frac{\partial \mathbf{S}}{\partial t}. \tag{5.68}$$

We now rewrite the first term of Eq. 5.68. It can be checked that the x-component of $\epsilon(\nabla \times \mathbf{E}) \times \mathbf{E}$ may be written

$$\epsilon[(\nabla \times \mathbf{E}) \times \mathbf{E}]\cdot\mathbf{i}_x = \epsilon\frac{\partial}{\partial x}\left(E_x^2 - \frac{E^2}{2}\right) + \epsilon\frac{\partial}{\partial y}(E_yE_x) + \epsilon\frac{\partial}{\partial z}(E_zE_x) - \epsilon E_x\nabla\cdot\mathbf{E}. \tag{5.69}$$

The last term of Eq. 5.69 is exactly the x-component of $-\rho\mathbf{E}$. Then considering the y- and z-components of $\epsilon(\nabla \times \mathbf{E}) \times \mathbf{E}$, we may write

$$\epsilon(\nabla \times \mathbf{E}) \times \mathbf{E} = \sum_{\alpha=1}^{3}\sum_{\beta=1}^{3}\mathbf{i}_\alpha\frac{\partial \mathsf{T}^e_{\alpha\beta}}{\partial x_\beta} - \rho\mathbf{E}, \tag{5.70}$$

where the notation of Sec. 5.2 has been used. The components of $\mathsf{T}^e_{\alpha\beta}$ are listed in Table 5.1.

Table 5.1. Components of $\mathsf{T}^e_{\alpha\beta}$ for an isotropic, homogeneous, linear, time-independent medium

	$\beta = 1$	$\beta = 2$	$\beta = 3$
$\alpha = 1$	$E_xD_x - \frac{1}{2}\mathbf{E}\cdot\mathbf{D}$	E_xD_y	E_xD_z
$\alpha = 2$	E_yD_x	$E_yD_y - \frac{1}{2}\mathbf{E}\cdot\mathbf{D}$	E_yD_z
$\alpha = 3$	E_zD_x	E_zD_y	$E_zD_z - \frac{1}{2}\mathbf{E}\cdot\mathbf{D}$

The transformation of $(\nabla \times \mathbf{B}) \times \mathbf{B}$ is similar, and we note that $\mathbf{B}\nabla\cdot\mathbf{B} = 0$. Thus we form

$$\frac{1}{\mu}(\nabla \times \mathbf{B}) \times \mathbf{B} = \sum_{\alpha=1}^{3}\sum_{\beta=1}^{3}\mathbf{i}_\alpha\frac{\partial \mathsf{T}^m_{\alpha\beta}}{\partial x_\beta}, \tag{5.71}$$

where the components of $\mathsf{T}^m_{\alpha\beta}$ are identical with those of Table 5.1, but with $\mathbf{E}$ replaced by $\mathbf{B}$ and $\mathbf{D}$ replaced by $\mathbf{H}$.

Expressions of the form of $\mathsf{T}^e_{\alpha\beta}$ or $\mathsf{T}^m_{\alpha\beta}$ occur very frequently. It is convenient to introduce an abbreviated method of writing a table, such as Table 5.1, of such quantities. The method will be explained by illustration. The abbreviated form for the quantity $\mathsf{T}_{\alpha\beta}$,

$$\mathsf{T}_{\alpha\beta} = \mathsf{T}^e_{\alpha\beta} + \mathsf{T}^m_{\alpha\beta}, \tag{5.72}$$

is obtained by element-by-element addition and displaying the result

$$\begin{bmatrix} E_xD_x + B_xH_x - \tfrac{1}{2}(\mathbf{E}\cdot\mathbf{D} + \mathbf{B}\cdot\mathbf{H}) & E_xD_y + B_xH_y & E_xD_z + B_xH_z \\ E_yD_x + B_yH_x & E_yD_y + B_yH_y - \tfrac{1}{2}(\mathbf{E}\cdot\mathbf{D} + \mathbf{B}\cdot\mathbf{H}) & E_yD_z + B_yH_z \\ E_zD_x + B_zH_x & E_zD_y + B_zH_y & E_zD_z + B_zH_z - \tfrac{1}{2}(\mathbf{E}\cdot\mathbf{D} + \mathbf{B}\cdot\mathbf{H}) \end{bmatrix}. \tag{5.73}$$

In other words, the labeling of α and β is omitted from the table, leaving an ordered array. The elements in such an ordered array are always entered in the same way. Therefore, there is no need to label the entries made in the rows and columns. Such an ordered array is called a matrix.

If the results of Eqs. 5.70 and 5.71 are inserted into Eq. 5.68, we obtain

$$\rho\mathbf{E} + \mathbf{j}\times\mathbf{B} = \sum_{\alpha=1}^{3}\sum_{\beta=1}^{3} \mathbf{i}_\alpha \frac{\partial \mathsf{T}_{\alpha\beta}}{\partial x_\beta} - \epsilon\mu\,\frac{\partial \mathbf{S}}{\partial t}. \tag{5.74}$$

This equation expresses the total forces exerted on the charges and currents in terms of field quantities only. In other words, the field quantities give a complete description; the concepts of charges and currents do not really need to be used.

To emphasize this last point, we state that the whole matter could have been put the opposite way: we might not have considered fields at all. All that really matters are the forces that charges and currents exert on each other. These forces could be described in terms of the positions of the charges and currents only, as a function of time. Because of the finite speed with which the interactions are propagated, the retardation of the interactions would have to be considered in the mathematical formalism. The concept of fields becomes useful at this point as a way of conveniently arranging and subdividing the problems to be considered. We thus introduce the concepts that fields are created by charges and currents, that the energy in the fields takes some time to propagate, and that the fields exert forces in turn upon the charges and currents. These are the ideas underlying an expression such as the left-hand side of Eq. 5.74. But we might go further; we might state that insofar as the mathematics is concerned, the charges and currents can be completely replaced by the fields that they create. Indeed, this point of view is taken in writing the right-hand side of Eq. 5.74. The fields may be interpreted as exerting forces on themselves, much as the charges and currents might be regarded as exerting forces on themselves in the all-particle picture.

To understand Eq. 5.74 better, let it be integrated over some volume of interest. With the application of the divergence theorem we learn that

$$\int d\tau\,(\rho\mathbf{E} + \mathbf{j}\times\mathbf{B}) = \sum_{\alpha=1}^{3}\sum_{\beta=1}^{3} \int ds_\beta\,\mathbf{i}_\alpha \mathsf{T}_{\alpha\beta} - \int d\tau\,\epsilon\mu\,\frac{\partial \mathbf{S}}{\partial t}. \tag{5.75}$$

Consider the static case first. The left-hand side of Eq. 5.74 is the force exerted on the charges and currents by the fields. Equation 5.75 then states that in the static case the forces on the charges within the volume element are equal to the integral of the stresses over the surface bounding the element. From the field concept we should expect such a result. The forces exerted on some body must be calculable in terms of the stress transferred across a surface bounding the volume, as indeed turns out to be the case. In the field concept, forces are transmitted from one point to the next. Action at a distance is excluded from thought. From the right-hand side of Eq. 5.75, we see that the first index of $\mathsf{T}_{\alpha\beta}$ specifies the component of the force under consideration; the second index of $\mathsf{T}_{\alpha\beta}$ specifies the direction of the normal to the surface on which the force is exerted. For example, T_{xy} is the stress along the x-direction exerted on a surface whose normal is along the y-direction. The quantity $\mathsf{T}_{\alpha\beta}$ is called the electromagnetic stress tensor, in view of the interpretation. It, of course, has the dimensions of a stress. The concept of a tensor has many profound implications;[1, 5] we shall not discuss them.

Next consider the time-dependent case. In the interest of simplicity, consider the case in which the charge density and the current density are both zero. The first term of the right-hand size of Eq. 5.74 is in general not zero. Something must be subtracted from it to make the left-hand side zero, namely, the remaining term of the right-hand side. Let us inquire into the nature of this second term. We have noted that **S** may be thought of as the power flux through an area, providing suitable caution is exercised. Since $(\epsilon\mu)^{-1/2}$ is a speed, it is satisfactory to interpret $\epsilon\mu\mathbf{S}$ as momentum density. Since the charges and currents are in no way involved in the expression, the momentum must be that of the fields themselves. Thus, Eq. 5.75 states that the force transmitted across the closed surface in part arises from the electromagnetic field itself. Furthermore, this part of the total force is equal to the rate of change with time of the electromagnetic momentum within the volume element bounded by the closed surface. In other words, in order that Newton's second law be valid, the momentum density of the electromagnetic field must be considered together with that of the particles.

A simple application will illustrate the use of the stress tensor. Suppose that $\mathbf{E} = 0$, that $\mathbf{B}$ points along the z-axis at the point of interest, and that the fields are static. Then $\mathsf{T}_{\alpha\beta}$ becomes

$$\mathsf{T}_{\alpha\beta} = \frac{1}{2\mu}\begin{bmatrix} -B^2 & 0 & 0 \\ 0 & -B^2 & 0 \\ 0 & 0 & +B^2 \end{bmatrix}. \tag{5.76}$$

According to Eq. 5.76, the stress exerted by the induction on a surface whose normal lies along the z-axis is $+B^2/2\mu$. We may say there is a tension in the lines of induction of this magnitude. Again, Eq. 5.76 implies that the stress exerted by the induction on a surface whose normal lies along the x-axis is $-B^2/2\mu$; that is, the induction exerts a pressure on the surface. Likewise, the induction exerts a pressure of magnitude $-B^2/2\mu$ on a surface whose normal lies along the y-axis. For these reasons, we speak of a tension along the lines of induction and a pressure between the lines. The lines of induction exert a sidewise pressure on each other. This pressure may be thought of as preventing the tension along the lines from collapsing all of them into nothing. The circular lines of induction around a current flowing through a long wire are supported by the pressure between them. Otherwise, the tension would cause them to collapse around the current.

If the normal to the surface is at 45° to the x- and z-axes, then there is a pure shear along the surface, as shown in Fig. 5.6.

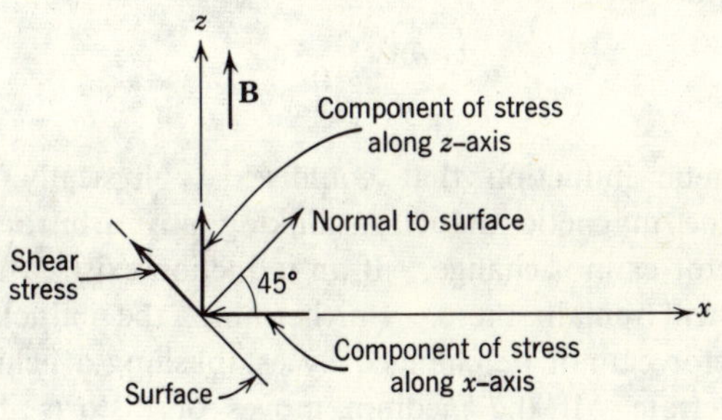

Fig. 5.6. Stress on a surface at 45° to a magnetic induction.

5.7. The Wave Equation and Magnetic Diffusion

In this section, we develop the wave equations that determine the electric and magnetic fields so that we can discuss the trapping of a magnetic induction within a conductor. This subject will be of interest in the next chapter. A detailed discussion of electromagnetic waves will be deferred until Chap. 9, where a number of different types of waves will be discussed together.

It is assumed that the medium is isotropic, homogeneous, linear, and time independent, so that the permittivity, permeability, and resistivity are constants. The wave equation for the induction **B** is obtained by computing the curl of Eq. 5.13:

$$\nabla \times \nabla \times \mathbf{B} = \frac{\mu}{\eta} \nabla \times \mathbf{E} + \epsilon\mu \frac{\partial}{\partial t} (\nabla \times \mathbf{E}). \tag{5.77}$$

By means of the identity Eq. 5.54 and by Eqs. 5.6 and 5.10, the present result may be simplified to

$$\nabla^2 \mathbf{B} - \frac{\mu}{\eta}\frac{\partial \mathbf{B}}{\partial t} - \epsilon\mu \frac{\partial^2 \mathbf{B}}{\partial t^2} = 0. \tag{5.78}$$

Identical equations for **E**, **A**, and ϕ can be derived if the charge density is zero.

If the resistivity is infinite, as in a perfect insulator or vacuum, then Eq. 5.78 reduces to a wave equation. Any function of the form $\mathbf{B}[(\boldsymbol{\kappa}/\kappa)\cdot\mathbf{r} - ct]$ is a solution, where $\boldsymbol{\kappa}/\kappa$ is a unit vector and

$$c = 1/\sqrt{\epsilon\mu}. \tag{5.79}$$

These solutions will be more fully discussed in Chap. 9.

The case of very small resistivity is of interest here. The second term of Eq. 5.78 then becomes all-important. If the resistivity is zero, as in a perfect conductor, then

$$\frac{\partial \mathbf{B}}{\partial t} = 0 \tag{5.80}$$

for any magnetic induction that could exist physically. This result implies that the magnetic induction linking any arbitrary loop of a perfect conductor cannot change. If an induction exists within a perfect conductor, it will remain there. Furthermore, the induction within a perfect conductor cannot be changed by establishing a field outside, for none will penetrate. If the medium moves or distorts its shape, the induction will move correspondingly. The induction will be reduced, for example, if the conductor expands.

The physical reason for the freezing of the lines of induction to a perfect conductor is that any motion of the conductor with respect to the lines is completely countered by an equal and opposite induction created by the currents that are induced by such a motion. For example, the flux linking a loop of infinite conductivity can never be changed by an external induction, because such an induction will induce current to flow in the conductor which would cancel any change of the induction linking the conductor.

If the resistivity is low but not zero, then Eq. 5.78 is a diffusion equation at times t greater than the relaxation time τ of the charge,[1] where

$$\tau = \epsilon\eta. \tag{5.81}$$

This time is very short for good conductors, that is, in the order of 10^{-17} sec. For any possible times of interest, the magnetic induction **B** changes by the diffusion process in good conductors. The diffusion coefficient for the motion is

$$\mathsf{D} = \eta/\mu. \tag{5.82}$$

For good metallic conductors, this diffusion coefficient is in the order of 10^{-4} m^2/sec. The induction can change substantially in a metal plate 10^{-2} m thick in about 1 sec.

We shall show in Chap. 8 that the resistivity of a thermonuclear plasma is much lower than that of metals at room temperature. The diffusion of the magnetic induction will be very slow, especially for large distances.

The diffusion of a magnetic induction within a conducting medium will heat it. High concentrations of the induction will gradually become weaker because of diffusion, thereby decreasing the total magnetic energy stored within the medium. Very little of this energy of the magnetic induction will have escaped over the surface of the medium, and, according to Eq. 5.66, it must therefore have gone into heating the medium.

PROBLEMS

1. Show that the work done in charging a condenser of capacity C to a voltage V is $CV^2/2$. Then for a parallel-plate condenser, show that this energy is equal to $\int d\mathbf{r}\, \epsilon E^2/2$. Neglect edge effects.

2. The plates of a parallel-plate condenser are charged to a potential V. Compute the force acting on the plates from the electrostatic energy density.

3. A material body may be characterized as a medium having a permittivity, permeability, and conductivity; the values of these quantities will differ from those of a vacuum. Alternatively, the body may be considered to be a collection of charges moving about in a vacuum. The body may then be represented by a polarization **P** and a magnetization **M** defined by

$$\mathbf{D} = \epsilon_0 \mathbf{E} + \mathbf{P},$$

$$\mathbf{H} = \frac{\mathbf{B}}{\mu_0} - \mathbf{M}.$$

By substituting these relations into Eqs. 5.3 and 5.13, show that insofar as forces on a test charge are concerned the material body acts like a vacuum having a charge density ρ_b,

$$\rho_b = -\nabla \cdot \mathbf{P},$$

and a current density $\mathbf{j}_b$,

$$\mathbf{j}_b = \nabla \times \mathbf{M} + \frac{\partial \mathbf{P}}{\partial t}.$$

The usual charge density ρ_b and current density $\mathbf{j}_b$ are those arising from charges and currents not considered to form a part of the body. Derive relations between the permittivity and the polarization and between the permeability and the magnetization. These relations connect the two points of view.

4. Suppose that a perfect conductor has excluded all fields within it. Show that if **n** is a unit vector normal to the surface, $\mathbf{n} \times \mathbf{E} = 0$ and $\mathbf{n} \cdot \mathbf{B} = 0$ at the surface. Then derive the surface charge density and surface current for arbitrary fields satisfying the boundary conditions.

5. A rudimentary electric generator consists of a wire of length l, laid out in the x-direction. It moves at speed V in the y-direction across a constant induction B_z. Prove that the voltage across the ends is $B_z l V$. If a current I is drawn from it by sliding contacts, show that the electrical output is just balanced by the rate of work done in moving the wire.

6. A loop of radius r carries a current I. Calculate (*a*) the vector potential **A** at points along the axis and (*b*) the induction **B**.

7. Suppose that the magnetic induction points along the z-axis at the point of interest. Compute the force on a surface whose direction cosines are α, β, and γ.

8. A long cylinder of plasma 0.1-m diameter carries a current of 10,000 amp concentrated on a shell at its surface. The plasma exerts an outward gas pressure of 10^5 newtons/m^2 ($\approx$ 1 atmosphere). Describe the force on the plasma from the point of view of (*a*) the $\mathbf{j} \times \mathbf{B}$ force and (*b*) the magnetic stress tensor. Will the plasma compress or expand? Find the current necessary so that the radial forces are in balance. This geometrical arrangement is that of a simple pinch discharge.

9. A plasma discharge may be approximated by a cylinder. A uniform magnetic induction exists within the discharge. The discharge current I is sufficiently strong to create a magnetic field about itself. Compute an expression for the value of the current I at which the azimuthal field at the surface of the cylinder creates a pressure equal to that created by the axial stabilizing field and the kinetic pressure. (An axial field is often used to stabilize such a discharge.)

10. A liquid-metal coolant moves in a magnetic induction. Compute the ohmic drag on the liquid-metal coolant as a function of its resistivity, speed, and the magnetic induction. Compute the magnetic induction caused by the currents that flow in the metal. Assume that the coolant is an infinite plane and that the magnetic induction is originally uniform everywhere.

11. A long straight wire carries an alternating current $I = I_0 \sin \omega t$. Derive a formula for rate of energy radiation per unit length from the wire. In what direction is the radiation propagated? Hint: Consider the Poynting vector.

REFERENCES

1. J. A. Stratton, *Electromagnetic Theory*, McGraw-Hill Book Co., New York (1941).
2. W. R. Smythe, *Static and Dynamic Electricity*, McGraw-Hill Book Co., New York (1950).
3. W. Panofsky and M. Phillips, *Classical Electricity and Magnetism*, Addison-Wesley, Reading (1955).
4. H. B. Phillips, *Vector Analysis*, John Wiley & Sons, New York (1933).

5. P. G. Bergmann, *Introduction to the Theory of Relativity*, Prentice-Hall, New York (1942).
6. For example, see F. B. Hildebrand, *Methods of Applied Mathematics*, Prentice-Hall, New York (1952), or any equivalent treatise.

GENERAL REFERENCES

N. H. Frank, *Introduction to Electricity and Optics*, McGraw-Hill Book Co., New York (1950).

H. H. Skilling, *Fundamentals of Electric Waves*, John Wiley & Sons, New York (1948).

J. C. Slater and N. H. Frank, *Electromagnetism*, McGraw-Hill Book Co., New York (1947).

6

The hydromagnetic equations

In Chapter 4, averages over velocity distributions were discussed, and the Boltzmann equation was formulated. In this chapter, equations are developed for macroscopic quantities by taking appropriate averages of the Boltzmann equation. These equations are called the hydromagnetic equations and involve the macroscopic currents, fields, charge densities, mass densities, and so forth. Such quantities are more directly related to the experimental observables than are the distributions from which they arose. In this manner, we proceed toward the development of tractable, macroscopic plasma theory. This presentation is short and somewhat mathematical. We shall discuss some of its applications in Chap. 7, which is in fact a direct continuation of our present work. The hydromagnetic equations are necessary in investigating the confinement of the plasma and its stability against small perturbations.

The Boltzmann equation describes the changes of a distribution of particles as a function of the seven variables $(\mathbf{r}, \mathbf{v}, t)$. The macroscopic equations describing the conservation of particles, the conservation of momentum, and the conservation of energy must be derivable from it. The hydromagnetic equations are specific statements of these conservation laws. Here we shall derive these conservation laws by taking appropriate averages or moments of the Boltzmann equation. Further equations, which do not express the conservation of some familiar quantity, could be developed by the method to be discussed; but we do not find them useful and will not derive them.

In the course of the developments just mentioned, certain useful subsidiary concepts appear, such as the heat flux tensor or the pressure exerted by a magnetic field. The concept of magnetic pressure is very useful in discussing confinement, and will be treated in Chap. 7.

Finally, the Boltzmann factor for a particle in a potential well will be derived as an illustration of the methods.

6.1. The Method of Obtaining the Macroscopic Equations

Macroscopic equations may be derived very easily from the Boltzmann equation by multiplying it by $v_\alpha \cdots v_\beta$ and by averaging the result over all possible velocities. Various averages were discussed in Chap. 4; the work there will now be extended. We refer to Eq. 4.5. By integrating by parts, there follows that

$$\int dv\, g(\mathbf{r}, \mathbf{v}, t)\, \frac{\partial f(\mathbf{r}, \mathbf{v}, t)}{\partial t} = \frac{\partial}{\partial t}\, n(\mathbf{r}, t)\, \overline{g(\mathbf{r}, \mathbf{v}, t)} - n(\mathbf{r}, t)\, \overline{\frac{\partial g(\mathbf{r}, \mathbf{v}, t)}{\partial t}}, \tag{6.1}$$

$$\int dv\, g(\mathbf{r}, \mathbf{v}, t) v_\alpha \frac{\partial f(\mathbf{r}, \mathbf{v}, t)}{\partial x_\alpha} = \frac{\partial [n(\mathbf{r}, t)\overline{v_\alpha g(\mathbf{r}, \mathbf{v}, t)}]}{\partial x_\alpha} - n(\mathbf{r}, t)\, \frac{\partial [\overline{v_\alpha g(\mathbf{r}, \mathbf{v}, t)}]}{\partial x_\alpha}, \tag{6.2}$$

and

$$\int dv\, g(\mathbf{r}, \mathbf{v}, t)\, \frac{F_\alpha}{m} \frac{\partial f}{\partial v_\alpha} = \int dv_\beta \int dv_\gamma \left\{ \left[\frac{g(\mathbf{r}, \mathbf{v}, t) F_\alpha f(\mathbf{r}, \mathbf{v}, t)}{m} \right]_{v_\alpha = -\infty}^{v_\alpha = +\infty} - \int_{-\infty}^{\infty} dv_\alpha f(\mathbf{r}, \mathbf{v}, t)\, \frac{\partial}{\partial v_\alpha} \left[\frac{g(\mathbf{r}, \mathbf{v}, t) F_\alpha}{m} \right] \right\}. \tag{6.3}$$

For all quantities g of interest, the integrand of the first term on the right side of Eq. 6.3 vanishes at both limits. In other words, gF_α tends to infinity more slowly than the velocity distribution tends to zero for infinite speeds.

The force laws of present interest are given by Eq. 4.17. For these forces

$$\frac{\partial F_\alpha(\mathbf{r}, \mathbf{v}, t)}{\partial v_\alpha} = 0; \tag{6.4}$$

consequently, Eq. 6.3 becomes

$$\int dv\, g(\mathbf{r}, \mathbf{v}, t)\, \frac{F_\alpha(\mathbf{r}, \mathbf{v}, t)}{m} \frac{\partial f(\mathbf{r}, \mathbf{v}, t)}{\partial v_\alpha} = -\frac{n(\mathbf{r}, t)}{m}\, \overline{F_\alpha(\mathbf{r}, \mathbf{v}, t)\, \frac{\partial g(\mathbf{r}, \mathbf{v}, t)}{\partial v_\alpha}}. \tag{6.5}$$

If now we multiply the Boltzmann equation (Eq. 4.15 or Eq. 4.19) by $g(\mathbf{r}, \mathbf{v}, t)$ and integrate over all velocities, we find

$$\frac{\partial}{\partial t}[n(\mathbf{r}, t)\overline{g(\mathbf{r}, \mathbf{v}, t)}] - n(\mathbf{r}, t)\frac{\overline{\partial g(\mathbf{r}, \mathbf{v}, t)}}{\partial t} + \nabla_r \cdot n(r, t)\overline{\mathbf{v}g(\mathbf{r}, \mathbf{v}, t)}$$
$$- n(\mathbf{r}, t)\overline{\nabla_r \cdot \mathbf{v}g(\mathbf{r}, \mathbf{v}, t)} - \frac{n(\mathbf{r}, t)}{m}\overline{\mathbf{F}(\mathbf{r}, \mathbf{v}, t) \cdot \nabla_v g(\mathbf{r}, \mathbf{v}, t)}$$
$$= \int dv\, g(\mathbf{r}, \mathbf{v}, t)\left[\frac{\partial f(\mathbf{r}, \mathbf{v}, t)}{\partial t}\right]_{\text{coll}}. \tag{6.6}$$

By suitable choice of the function g, various macroscopic equations may be obtained. For example, the particle conservation equation analogous to Eq. 4.72 corresponds to letting $g = 1$ in Eq. 6.6:

$$\frac{\partial n(\mathbf{r}, t)}{\partial t} + \nabla_r \cdot [n(\mathbf{r}, t)\overline{\mathbf{v}(\mathbf{r}, t)}] = \int dv \left[\frac{\partial f(\mathbf{r}, \mathbf{v}, t)}{\partial t}\right]_{\text{coll}}. \tag{6.7}$$

The integral in the right-hand member is zero if there are no sources or sinks of particles, because that term is the net gain in the number of particles in a collision. For example, if new particles are created by ionization, then

$$\int dv \left[\frac{\partial f(\mathbf{r}, \mathbf{v}, t)}{\partial t}\right]_{\text{coll}} = n(\mathbf{r}, t)\bar{\nu}_i. \tag{6.8}$$

That Eq. 6.7 corresponds to particle conservation is easily seen by integrating it over a volume of space and by applying the divergence theorem. See, for example, the derivation of Eq. 5.66 and the subsequent interpretation. The first term of the left-hand side represents the rate of change of the particle density in an element of volume; the second term represents the net rate at which particles within this element flow out of it.

6.2. The Momentum Transport Equation

A macroscopic equation for the conservation of momentum, that is, Newton's second law, results from letting g be the vector quantity

$$\mathbf{g} = m\mathbf{v}. \tag{6.9}$$

The rate of change of momentum density is given by Eq. 6.6, the α-component of which becomes

$$\frac{\partial}{\partial t}[n(\mathbf{r}, t)m\overline{v_\alpha}] + \sum_{\beta=1}^{3}\frac{\partial}{\partial x_\beta}[n(\mathbf{r}, t)m\overline{v_\beta v_\alpha}] - n(\mathbf{r}, t)\overline{F_\alpha(\mathbf{r}, t)}$$
$$= \int dv\, mv_\alpha\left[\frac{\partial f(\mathbf{r}, \mathbf{v}, t)}{\partial t}\right]_{\text{coll}}, \tag{6.10}$$

if the scalar products in Eq. 6.6 are written explicitly. The first term of the left-hand side represents the rate of change of momentum density with time at a fixed point of space; the second term represents the net rate at which momentum is transported across the surfaces bounding an element of volume at the point of interest; and the third term is the rate at which momentum density is created by external forces $\overline{F_\alpha}$. The integral in the right-hand member of Eq. 6.10 expresses the total momentum gained per unit time as a result of collisions by the particles. If all particles are identical, there is only one momentum transport equation for the system. Since the total momentum is conserved at each collision, the integral must be zero in such a case. If there are several different kinds of particles in the system, then there is a momentum transport equation for each species. The integrals in each equation contain contributions arising from collisions of unlike particles, and these contributions are in general not zero.

The second term in the left-hand side of Eq. 6.10 is particularly interesting. It may be written out completely as

$$\sum_{\beta=1}^{3} \frac{\partial}{\partial x_\beta}[n(\mathbf{r}, t)m\overline{v_\beta \mathbf{v}}] = m\left\{\left[\frac{\partial}{\partial x}(n\,\overline{v_x v_x}) + \frac{\partial}{\partial y}(n\,\overline{v_x v_y}) + \frac{\partial}{\partial z}(n\,\overline{v_x v_z})\right]\mathbf{i}_x\right.$$
$$+ \left[\frac{\partial}{\partial x}(n\,\overline{v_y v_x}) + \frac{\partial}{\partial y}(n\,\overline{v_y v_y}) + \frac{\partial}{\partial z}(n\,\overline{v_y v_z})\right]\mathbf{i}_y$$
$$\left. + \left[\frac{\partial}{\partial x}(n\,\overline{v_z v_x}) + \frac{\partial}{\partial y}(n\,\overline{v_z v_y}) + \frac{\partial}{\partial z}(n\,\overline{v_z v_z})\right]\mathbf{i}_z\right\}. \tag{6.11}$$

The physical significance of the quantity $mn\,\overline{v_\alpha v_\beta}$ will be explained in the next section.

Equation 6.10 may be put in a more useful form. Let $\mathbf{v}_r$ be defined by

$$\mathbf{v} = \overline{\mathbf{v}(\mathbf{r}, t)} + \mathbf{v}_r, \tag{6.12}$$

so that the velocity $\mathbf{v}_r$ is the velocity of random motion as seen if we follow the average drift. From the definition of Eq. 6.12 for $\mathbf{v}_r$,

$$\overline{\mathbf{v}_r} = 0. \tag{6.13}$$

From this result, we find in turn that since

$$\overline{v_\alpha v_\beta} = \overline{v_\alpha}\,\overline{v_\beta} + \overline{v_{r\alpha} v_{r\beta}}, \tag{6.14}$$

$$\sum_{\beta=1}^{3} \frac{\partial}{\partial x_\beta}[n(\mathbf{r}, t)m\overline{v_\alpha v_\beta}] = \sum_{\beta=1}^{3} \frac{\partial[n(\mathbf{r}, t)m\overline{v_\alpha}\,\overline{v_\beta}]}{\partial x_\beta} + \sum_{\beta=1}^{3} \frac{\partial}{\partial x_\beta}[n(\mathbf{r}, t)m\overline{v_{r\alpha} v_{r\beta}}]\cdot \tag{6.15}$$

From the usual rule of differentiating a product, we find

$$\sum_{\beta=1}^{3} \frac{\partial}{\partial x_\beta} [n(\mathbf{r}, t) m \overline{v_\alpha} \overline{v_\beta}] = n(\mathbf{r}, t) m \sum_{\beta=1}^{3} \overline{v_\beta} \frac{\partial \overline{v_\alpha}}{\partial x_\beta} + \overline{v_\alpha} \sum_{\beta=1}^{3} \frac{\partial}{\partial x_\beta} [n(\mathbf{r}, t) m \overline{v_\beta}] \quad (6.16)$$

and

$$\frac{\partial}{\partial t} [n(\mathbf{r}, t) m \overline{v_\alpha}] = n(\mathbf{r}, t) m \frac{\partial \overline{v_\alpha}}{\partial t} + m \overline{v_\alpha} \frac{\partial n(\mathbf{r}, t)}{\partial t}. \quad (6.17)$$

In addition, we can multiply Eq. 6.7 by $m\overline{v_\alpha}$ and discover that

$$\frac{\partial}{\partial t} [n(\mathbf{r}, t) m \overline{v_\alpha}] = n(\mathbf{r}, t) m \frac{\partial \overline{v_\alpha}}{\partial t} - \overline{v_\alpha} \sum_{\beta=1}^{3} \frac{\partial}{\partial x_\beta} [n(\mathbf{r}, t) m \overline{v_\beta}] + m \overline{v_\alpha} \int d\mathbf{v} \left[\frac{\partial f(\mathbf{r}, \mathbf{v}, t)}{\partial t} \right]_{\text{coll}}. \quad (6.18)$$

It is useful now to define the quantity $\mathsf{P}_{\alpha\beta}$ by

$$\mathsf{P}_{\alpha\beta}(\mathbf{r}, t) = m \int d\mathbf{v}\, v_{r\alpha} v_{r\beta} f(\mathbf{r}, \mathbf{v}, t) = mn\, \overline{v_{r\alpha} v_{r\beta}}. \quad (6.19)$$

This quantity is called the kinetic stress tensor. It will be more fully treated in the next section. We obtain the final result by inserting Eqs. 6.15, 6.16, 6.18, and 6.19 into Eq. 6.10. The α-component of the vector equation becomes

$$mn(\mathbf{r}, t) \left[\frac{\partial}{\partial t} + \sum_{\beta=1}^{3} \overline{v_\beta(\mathbf{r}, t)} \frac{\partial}{\partial x_\beta} \right] \overline{v_\alpha(\mathbf{r}, t)} = n(\mathbf{r}, t) \overline{F_\alpha(\mathbf{r})} - \sum_{\beta=1}^{3} \frac{\partial}{\partial x_\beta} \mathsf{P}_{\alpha\beta}(\mathbf{r}, t) + \int d\mathbf{v}\, m v_\alpha \left[\frac{\partial f(\mathbf{r}, \mathbf{v}, t)}{\partial t} \right]_{\text{coll}} - m \overline{v_\alpha(\mathbf{r}, t)} \int d\mathbf{v} \left[\frac{\partial f(\mathbf{r}, \mathbf{v}, t)}{\partial t} \right]_{\text{coll}}. \quad (6.20)$$

The left-hand member of Eq. 6.20 is easily interpreted. Imagine a cloud of particles traveling with an average velocity $\overline{\mathbf{v}(\mathbf{r}, t)}$. The changes of a variable h associated with the cloud may be specified in two ways. In the Lagrangian method, the observer rides along with the cloud at its average velocity $\overline{\mathbf{v}(\mathbf{r}, t)}$. In the Eulerian method, which is the point of view taken here, the observer, who is stationary in the laboratory system, looks at the moving cloud. A quantity h describing a property of the cloud will in general be a function $h(\mathbf{r}, t)$ of the space coordinates $\mathbf{r}$ and of the time t. The equations of motion apply to the particles and not to the arbitrarily situated observer. The rate at which the property h is changing with respect to a point moving along with the fluid will be the total derivative dh/dt. But to a fixed Eulerian observer, the total derivative becomes

$$\frac{dh(\mathbf{r}, t)}{dt} = \frac{\partial h(\mathbf{r}, t)}{\partial t} + \sum_{\alpha=1}^{3} \frac{dx_\alpha}{dt} \frac{\partial h(\mathbf{r}, t)}{\partial x_\alpha} = \frac{\partial h}{\partial t} + (\bar{\mathbf{v}} \cdot \nabla)h. \tag{6.21}$$

The second term of the right-hand side of Eq. 6.21 appears only if the quantity h depends explicitly on the coordinates $\mathbf{r}$ of position. The left-hand side of Eq. 6.20 is then merely the total mass of all particles in a unit volume times their acceleration as seen by a stationary observer in the laboratory system.

The last term in Eq. 6.20 represents the rate at which momentum is created by the generation of new particles as a result of collisions. This term will be zero if no net number of particles is created in a collision. All the remaining terms on the right-hand side of Eq. 6.20 must then be the total forces acting on all particles within a unit volume. The first term of the right-hand side of this equation is the rate of change of momentum of the particles in a unit volume caused by the long-range, slowly varying forces discussed in the introduction of Chap. 5 and in Sec. 4.2. The third term on the right-hand side represents the rate of gain of momentum resulting from the rapidly changing, short-range forces (that is, collisions) mentioned in the introduction of Chap. 5 and in Sec. 4.2. In each collision, the total momentum is conserved. Therefore, if the particles are identical, this term is zero. The only term remaining unexplained is the second one on the right-hand side of Eq. 6.20. It will be discussed in the next section.

★ 6.3. The Kinetic Stress Tensor

We now discuss the physical interpretation of the kinetic stress tensor defined by Eq. 6.19 of the previous section. It is first observed that this entity relates to two directions simultaneously, as described by its two indices. The kinetic stress tensor has nine components, of which only six are independent, in view of the fact that

$$\mathsf{P}_{\alpha\beta} = \mathsf{P}_{\beta\alpha}. \tag{6.22}$$

Next, we notice that only the components of the velocity relative to the average drift velocity appear in this tensor. Since $f\,d\mathbf{v}$ is the total number of particles in the velocity volume $d\mathbf{v}$ located at $\mathbf{v}$, $fmv_{rx}\,d\mathbf{v}$ is the x-component of momentum of all particles (relative to the average motion) in the velocity volume $d\mathbf{v}$ and located in a unit volume of space at $\mathbf{r}$. The product $dv\,(fmv_{rx})v_{ry}$ states the rate at which the x-component of momentum is transported in the y-direction for all particles in dv located at $(\mathbf{r}, \mathbf{v})$. A rate of transfer of momentum is by definition force per unit area, that is, a stress. The stress under discussion is a shear directed along the x-axis exerted by the particles on a surface perpendicular to the y-axis. The integration merely computes the sum of the contribution from particles of all velocities. For this reason, the quantity P is properly called a kinetic stress tensor.

The special case where the distribution function f is isotropic in the system drifting with $\bar{\mathbf{v}}$ is very important. Let us drift along with the system and observe it. By definition,

$$\bar{\mathbf{v}}_r = 0, \tag{6.23}$$

and

$$\mathsf{P}_{\alpha\beta} = \begin{cases} nm\overline{v_{r\alpha}^2} = nm\overline{v_{r\beta}^2}, & \alpha = \beta, \\ 0, & \alpha \neq \beta. \end{cases} \tag{6.24}$$

Furthermore, the isotropy requires

$$\overline{v_{r\alpha}^2} = \overline{v_r^2}/3, \tag{6.25}$$

and we associate P—now a simple scalar number—with the rate of transport of momentum in any direction, that is, the pressure p measured by the drifting observer. In this case, we can write finally

$$\mathsf{P} \rightarrow p = nm\overline{v_r^2}/3 = 2n\bar{U}_r/3, \tag{6.26}$$

where $\bar{U}_r$ is the mean random energy. Equation 6.26 is a generalization of Eq. 4.24. The distribution f may or may not be Maxwellian in the drifting frame; but if the distribution is Maxwellian, we see at once that the pressure and temperature are related by

$$\mathsf{P} \rightarrow p = nm\overline{v_r^2}/3 \rightarrow nkT. \tag{6.27}$$

An illustration of these foregoing thoughts can be taken from everyday experience. If we neglect gravitational effects, snowflakes in a blizzard move with the drift speed of the wind and experience only the static pressure. That is, the stress tensor is isotropic. However, as the wind blows around a corner, the stress tensor becomes anisotropic, and the snowflakes experience a net force; so does the corner, and so do we who stand observing it.

In plasmas having a high temperature, the stress tensor may very well not have the scalar form described by Eq. 6.24, because the small collision rate is insufficient to keep the velocity distribution isotropic.

Another quantity, called the heat flux tensor, is sometimes used. It is defined by

$$Q_{\alpha\beta\gamma}(\mathbf{r}, t) = n(\mathbf{r}, t)m \int d\mathbf{v}\; v_{r\alpha}v_{r\beta}v_{r\gamma}f(\mathbf{r}, \mathbf{v}, t). \tag{6.28}$$

It measures the rate at which energy is transported across a surface. For isotropic distributions, $\bar{\mathbf{v}}_r = 0$, and all odd moments of the distribution function f are zero. The heat flux tensor is then zero, as should be the case if $Q_{\alpha\beta\gamma}(\mathbf{r}, t)$ is to be interpreted as a rate of transport of heat energy, that is, random energy. Again, the product

$$mv_{r\alpha}v_{r\beta}f(\mathbf{r}, \mathbf{v}, t)\, d\mathbf{v} \tag{6.29}$$

is clearly proportional to the random energy of the particles in $d\mathbf{v}$ when viewed from a frame moving with the average drift velocity of the particles. The product of Expression 6.29 with $v_{r\gamma}$ is proportional to the rate at which energy is transported across a surface normal to the γ-axis.

Note that Q is related to the rate of transfer of energy through the gas distribution and not necessarily through space. If the distribution function describes a beam, then $f = 0$ except for $\mathbf{v} = \bar{\mathbf{v}}$; and $Q = 0$ in spite of the fact that the beam is transporting kinetic energy. If the particles in the beam have no spread in velocity, then there can be no interchange of energy between them. In other words, there is no transfer of heat among the particles.

6.4. The Energy Transport Equation

We derive a macroscopic equation for the conservation of energy by letting

$$g = \frac{m}{2} v^2 \tag{6.30}$$

in Eq. 6.6. Since

$$\mathbf{v} \times \mathbf{B} \cdot \nabla_v v^2 = 0, \tag{6.31}$$

the result is

$$\frac{\partial}{\partial t}\left(\frac{nm}{2}\overline{v^2}\right) + \nabla_r \cdot \left(\frac{n}{2} m \overline{\mathbf{v}v^2}\right) = \mathbf{j} \cdot \mathbf{E} + \int d\mathbf{v}\, \frac{mv^2}{2}\left(\frac{\partial f}{\partial t}\right)_{\text{coll}}. \tag{6.32}$$

This equation expresses the conservation of energy. The first term of the left-hand side is the rate of change of energy density with time; the second term is the leakage of energy out from a closed surface bounding the element of interest. The sum of these two terms equals the work done on the particles by the electric field per unit time plus any energy that might be created per unit time as a result of collisions. If all particles are identical and if there is no absorption or creation of energy, then this term is zero. If there are several different kinds of particles in the system, then an energy transport equation exists for each species, and the individual collision integrals differ from zero. Their sum is zero, however, unless the internal energy of the particles is altered. The magnetic induction does no work on charged particles since it always acts perpendicular to the direction of motion of the particles. This fact is expressed by Eq. 6.31.

An isotropic distribution in a drifting coordinate system is of particular interest. In that case, we reduce the terms on the left side of Eq. 6.32 by replacing $\mathbf{v}$ everywhere by $\bar{\mathbf{v}} + \mathbf{v}_r$ from Eq. 6.12 and by expanding the products. The first term in Eq. 6.32 involves the term

$$\frac{1}{2}nm\overline{v^2} = \frac{1}{2}nm\bar{\mathbf{v}}^2 + \frac{1}{2}nm\overline{v_r{}^2} = \frac{1}{2}nm\bar{\mathbf{v}}^2 + \frac{3}{2}p, \tag{6.34}$$

which follows immediately from Eqs. 6.14, 6.23, and 6.26.† Equation 6.34 states that the total energy is the sum of the drift energy plus the random energy.

The quantity $\overline{v^2\mathbf{v}}$ is similarly expanded:

$$\begin{aligned}\overline{v^2\mathbf{v}} &= \overline{(\bar{\mathbf{v}} + \mathbf{v}_r)\cdot(\bar{\mathbf{v}} + \mathbf{v}_r)(\bar{\mathbf{v}} + \mathbf{v}_r)} \\ &= \bar{\mathbf{v}}^2\bar{\mathbf{v}} + 2\overline{\bar{\mathbf{v}}\cdot\mathbf{v}_r\bar{\mathbf{v}}} + \overline{v_r{}^2}\bar{\mathbf{v}} + \overline{\bar{\mathbf{v}}^2\mathbf{v}_r} + 2\overline{\bar{\mathbf{v}}\cdot\mathbf{v}_r\mathbf{v}_r} + \overline{v_r{}^2\mathbf{v}_r}.\end{aligned} \tag{6.35}$$

The quantity $\bar{\mathbf{v}}$ is already an average; the second, fourth, and sixth terms on the right side of Eq. 6.35 are odd functions of $\mathbf{v}_r$, and their average vanishes by Eq. 6.23. The fifth term is re-expressed as $2\sum_\alpha \bar{\mathbf{v}}_\alpha\overline{v_{r\alpha}\mathbf{v}_r}$, whence by isotropy (Eqs. 6.24 and 6.25) we can write

$$\begin{aligned}\overline{v^2\mathbf{v}} &= \bar{\mathbf{v}}^2\bar{\mathbf{v}} + \overline{v_r{}^2}\bar{\mathbf{v}} + 2\bar{\mathbf{v}}\overline{v_{r\alpha}{}^2}, \\ &= \bar{\mathbf{v}}^2\bar{\mathbf{v}} + \frac{5}{3}\overline{v_r{}^2}\bar{\mathbf{v}}, \\ &= \bar{\mathbf{v}}^2\bar{\mathbf{v}} + \frac{5p\bar{\mathbf{v}}}{nm}.\end{aligned} \tag{6.36}$$

† The quantity $\bar{\mathbf{v}}$ appearing in Eq. 6.34 is the average of the velocity vector $\mathbf{v}$ $(= \bar{\mathbf{v}} + \mathbf{v}_r)$, that is, the drift velocity $\mathbf{v}$ itself. The quantity $\bar{\mathbf{v}}^2 = \bar{\mathbf{v}}\cdot\bar{\mathbf{v}}$ is to be distinguished from the quantity $\bar{v}^2 = \overline{(\mathbf{v}_r + \bar{\mathbf{v}})}^2$, that is, the square of the average magnitude of the velocity. This latter quantity would be other than zero for an isotropic distribution, but the drift velocity would be zero for such a case.

Now substitute Eqs. 6.34 and 6.36 into Eq. 6.32, differentiate three of the products appearing, and find

$$nm\bar{\mathbf{v}}\cdot\frac{\partial\bar{\mathbf{v}}}{\partial t} + \frac{1}{2}m\bar{\mathbf{v}}^2\frac{\partial n}{\partial t} + \frac{3}{2}\frac{\partial p}{\partial t} + \frac{1}{2}m\bar{\mathbf{v}}^2\nabla\cdot(n\bar{\mathbf{v}}) + \frac{1}{2}nm\bar{\mathbf{v}}\cdot\nabla\bar{\mathbf{v}}^2 + \frac{5}{2}\bar{\mathbf{v}}\cdot\nabla p + \frac{5}{2}p\nabla\cdot\bar{\mathbf{v}} = \mathbf{j}\cdot\mathbf{E} + \int d\mathbf{v}\frac{1}{2}mv^2\left(\frac{\partial f}{\partial t}\right)_{\text{coll}} \cdot \tag{6.37}$$

The second and fourth terms on the left-hand side are now combined, using the conservation equation (Eq. 6.7). After differentiating $\nabla\bar{\mathbf{v}}^2$, we re-express the result in component form as

$$nm\sum_{\alpha=1}^{3}\bar{v}_\alpha\frac{\partial\bar{v}_\alpha}{\partial t} + \frac{3}{2}\frac{\partial p}{\partial t} + \sum_{\alpha\,\beta,=1}^{3}nm\bar{v}_\beta\bar{v}_\alpha\frac{\partial\bar{v}_\alpha}{\partial x_\beta} + \frac{5}{2}\sum_{\alpha=1}^{3}\bar{v}_\alpha\frac{\partial p}{\partial x_\alpha} + \frac{5}{2}p\sum_{\alpha=1}^{3}\frac{\partial\bar{v}_\alpha}{\partial x_\alpha} = \sum_{\alpha=1}^{3}j_\alpha E_\alpha + \int d\mathbf{v}\,\frac{mv^2}{2}\left(\frac{\partial f}{\partial t}\right)_{\text{coll}} - \frac{1}{2}m\sum_{\alpha=1}^{3}\bar{v}_\alpha\bar{v}_\alpha\int d\mathbf{v}\left(\frac{\partial f}{\partial t}\right)_{\text{coll}} \tag{6.38}$$

in order to cancel more of the terms using Eq. 6.20 for momentum conservation. To this end, multiply Eq. 6.20 by $\bar{v}_\alpha$ and sum over α from 1 to 3. Subtract the result from Eq. 6.38 and find

$$\frac{3}{2}\left(\frac{\partial}{\partial t} + \sum_{\alpha=1}^{3}\bar{v}_\alpha\frac{\partial}{\partial x_\alpha}\right)p + \frac{5}{2}p\sum_{\alpha=1}^{3}\frac{\partial\bar{v}_\alpha}{\partial x_\alpha} = \int d\mathbf{v}\,\frac{mv^2}{2}\left(\frac{\partial f}{\partial t}\right)_{\text{coll}} + \frac{1}{2}m\sum_{\alpha=1}^{3}\bar{v}_\alpha\bar{v}_\alpha\int d\mathbf{v}\left(\frac{\partial f}{\partial t}\right)_{\text{coll}} - \sum_{\alpha=1}^{3}\bar{v}_\alpha\int d\mathbf{v}\,mv_\alpha\left(\frac{\partial f}{\partial t}\right)_{\text{coll}} \cdot \tag{6.39}$$

From Eq. 6.21, we recognize the term in parentheses as the total derivative d/dt. Similarly, the conservation equation (Eq. 6.7) can be written as

$$\frac{dn}{dt} + n\nabla\cdot\bar{\mathbf{v}} = \int d\mathbf{v}\left(\frac{\partial f}{\partial t}\right)_{\text{coll}}, \tag{6.40}$$

and the term $\nabla\cdot\bar{\mathbf{v}}$ in Eq. 6.39 can be eliminated. The result is

$$3\frac{dp}{dt} - \frac{5p}{n}\frac{dn}{dt} = m\int d\mathbf{v}\left[v_r^2 - \frac{5}{3}\overline{v_r^2}\right]\left(\frac{\partial f}{\partial t}\right)_{\text{coll}} \tag{6.41}$$

after collecting the collision integrals, expanding the velocity integrands, and canceling terms.

Provided only that the collisions do not change the random energy or create new particles, the collision integral is zero. This provision is trivially satisfied if there is only one kind of particle. In that case, Eq. 6.41 becomes

$$\frac{dp}{dn} = \frac{5p}{3n}; \tag{6.42}$$

the equation can be integrated to give

$$p = p_0 n^{5/3}. \tag{6.43}$$

This law is the familiar one of adiabatic compression. Note that the distribution need not be Maxwellian for Eq. 6.43 to apply.

More elaborate conservation relations for the kinetic stress tensor $\mathsf{P}_{\alpha\beta}$ can be derived in a manner rather similar to that used for energy conservation.[1] The quantity $g = m\mathbf{v}_r\,\mathbf{v}_r$ is inserted in Eq. 6.6; the result involves the heat flow tensor, Eq. 6.28. The derivation and result are complicated and not particularly useful for our purposes. For this reason, we forgo the derivation.

An infinite sequence of higher moments of the Boltzmann equation may be computed by appropriate choices of g in Eq. 6.6. The distribution function itself may be computed in principle if all of its moments are known. The difficulty is that each equation of the sequence is linked to both of its neighbors, so there is no way of solving for any one moment exactly. However, if in the equation containing the highest moment this moment is neglected, then the sequence is terminated and each moment can in principle be found.[2]

6.5. The Boltzmann Factor for Particles in a Potential Well

As a simple and instructive application of Eq. 6.20, consider the case of a number of charges trapped in a potential well. The distribution under equilibrium conditions is to be calculated. The well is to be so deep that negligibly few particles escape over its top.

After equilibrium is established,

$$\bar{\mathbf{v}} = 0. \tag{6.44}$$

It is assumed that there are no sources or sinks. Equation 6.20 becomes

$$nq\mathbf{E} = \boldsymbol{\nabla}(nkT), \tag{6.45}$$

by Eq. 6.27, $\mathbf{E}$ being the electric intensity in the well. Our problem is a time-independent one. By Eqs. 5.46 and 6.45, it follows that

$$n = n_0 \exp(-q\phi/kT), \tag{6.46}$$

since in equilibrium the temperature is a constant. Equation 6.46 will be a good approximation, providing the current of particles over the top of the well is small compared with the random current in the well. This condition will be satisfied, providing

$$q\phi \gg kT, \tag{6.47}$$

for then few particles will be found near the top of the well where the approximations fail. It is interesting to note that a Maxwell-Boltzmann distribution is obtained in potential energy as well as in kinetic energy. Further, Eq. 6.46 is quite accurate, providing $q\phi$ is only two or three times kT.

6.6. Hydromagnetic Equations for a Plasma

The preceding theory involves particles of only one type. The theory may be easily extended to include a number of particles of different types. We illustrate the extension by treating a mixture of electrons and ions of masses m_e and m_i and of charges q_e and q_i, respectively, in combined electric, magnetic, and gravitational fields. The total mass density is denoted by ρ_m, defined by

$$\rho_m = n_e m_e + n_i m_i. \tag{6.48}$$

The mass current $\mathbf{J}$ is defined by

$$\mathbf{J} = n_e m_e \overline{\mathbf{v}}_e + n_i m_i \overline{\mathbf{v}}_i, \tag{6.49}$$

and the average velocity $\mathbf{V}$ by

$$\mathbf{V} = \mathbf{J}/\rho_m. \tag{6.50}$$

As applied to the present considerations, we define the charge density

$$\rho = n_e q_e + n_i q_i, \tag{6.51}$$

and the current density

$$\mathbf{j} = n_e q_e \overline{\mathbf{v}}_e + n_i q_i \overline{\mathbf{v}}_i. \tag{6.52}$$

These definitions are consistent with the previous customs: see Eqs. 5.30 and 4.84.

If no particles are created in an element of volume, then a particle conservation equation like Eq. 6.7 with the right-hand side zero holds for each constituent:

$$\frac{\partial}{\partial t}(n_e m_e) + \boldsymbol{\nabla}_r \cdot (n_e m_e \overline{\mathbf{v}}_e) = 0, \tag{6.53}$$

$$\frac{\partial}{\partial t}(n_i m_i) + \boldsymbol{\nabla}_r \cdot (n_i m_i \overline{\mathbf{v}}_i) = 0. \tag{6.54}$$

If these two equations are added, then

$$\frac{\partial \rho_m}{\partial t} + \nabla_r \cdot (\rho_m \mathbf{V}) = 0. \tag{6.55}$$

This equation expresses the conservation of mass of the system.

A conservation equation for the charge can be found by multiplying Eq. 6.7 written for electrons by the charge q_e and by multiplying Eq. 6.7 for ions by the charge q_i. If the results are added,

$$\frac{\partial \rho}{\partial t} + \nabla_r \cdot \mathbf{j} = 0, \tag{6.56}$$

analogous to the mass conservation equation. We find a momentum conservation equation for electrons results by applying Eqs. 6.14 and 6.19 to Eq. 6.10:

$$\frac{\partial}{\partial t}(n_e m_e \bar{v}_\alpha) + \sum_{\beta=1}^{3} \frac{\partial}{\partial x_\beta}(n_e m_e \bar{v}_\beta \bar{v}_\alpha)$$

$$= n_e q_e [E_\alpha + (\bar{\mathbf{v}} \times \mathbf{B})_\alpha] - n_e m_e \frac{\partial \phi}{\partial x_\alpha} - \sum_{\beta=1}^{3} \frac{\partial \mathsf{P}_{\beta\alpha}}{\partial x_\beta} + \int dv\, m_e v_\alpha \left[\frac{\partial f_e(\mathbf{r}, \mathbf{v}, t)}{\partial t}\right]_{\text{coll } ei}, \tag{6.57}$$

where ϕ is here the gravitational potential and the subscript ei denotes the momentum transferred to the electrons from the ions. We use the symbol f_e to denote the electron distribution function. There is no term arising from electron-electron collisions because there can be no net momentum transfer among them. A similar equation holds for the ions. It is noted by Newton's third law that

$$\int dv\, m_i \mathbf{v} \left(\frac{\partial f_i}{\partial t}\right)_{\text{coll } ie} = -\int dv\, m_e\, \mathbf{v} \left(\frac{\partial f_e}{\partial t}\right)_{\text{coll } ei}, \tag{6.58}$$

where f_i denotes the ion distribution function. In other words, the collisions of ions and electrons within a system cannot change the total momentum of the system. Equation 6.57 is added to a similar equation for ions to yield the equation for total momentum conservation:

$$\rho_m \frac{\partial V_\alpha}{\partial t} + V_\alpha \frac{\partial \rho_m}{\partial t} + \sum_{\beta=1}^{3} \frac{\partial}{\partial x_\beta}(n_e m_e \bar{v}_{\beta e} \bar{v}_{\alpha e} + n_i m_i \bar{v}_{\beta i} \bar{v}_{\alpha i})$$

$$= \rho E_\alpha + (\mathbf{j} \times \mathbf{B})_\alpha - \sum_{\beta=1}^{3} \frac{\partial}{\partial x_\beta} \mathsf{P}_{\beta\alpha} - \rho_m \frac{\partial \phi}{\partial x_\alpha}, \tag{6.59}$$

where

$$\mathsf{P}_{\beta\alpha} = \mathsf{P}_{\beta\alpha e} + \mathsf{P}_{\beta\alpha i} \tag{6.60}$$

is the total kinetic stress tensor.

The result just derived is rather complicated, so we frequently make the approximation that the plasma is nearly isotropic. The drift velocities, electric currents, and gradients are all first-order perturbations, and we neglect second-order perturbations. The assumption of approximate isotropy implies that the kinetic stress tensor is the pressure, that is, that Eq. 6.23 holds. In this case, Eq. 6.59 simplifies to

$$\rho_m \frac{\partial \mathbf{V}}{\partial t} = \rho \mathbf{E} + \mathbf{j} \times \mathbf{B} - \nabla_r p - \rho_m \nabla_r \phi. \tag{6.61}$$

This equation merely states Newton's second law, as indeed it must do. In this result, we note that the left-hand side is itself a first-order quantity by Eq. 6.55.

Charge neutrality is also often assumed. However, the neglect of the first term of Eq. 6.61 must be treated with caution. A very small fractional charge separation gives rise to very large electric fields and very large forces that, in turn, usually tend to restore the neutrality. Since Eq. 6.61 is a force equation, the validity of setting $\rho = 0$ depends on the magnitude of $\rho\mathbf{E}$ compared with the other terms in the equation. The matter can be settled by recalling the discussion of Sec. 4.8. A plasma shields itself from an electric field by means of a sheath in which the charge density $\rho \neq 0$. As it were, the field created by the net charge density within the sheath just cancels that applied from the outside. The thickness of the sheath is about equal to the Debye shielding length, so that the term $\rho\mathbf{E}$ must be retained if Eq. 6.61 is to be applied to a sheath. On the other hand, the $\mathbf{j} \times \mathbf{B}$ forces are likely to be found at greater depths within the plasma. Then in the interior of the plasma $\mathbf{E}$ is small, and we find justification for neglecting $\rho\mathbf{E}$ compared with the other forces.

An equation for the rate of change of the current density $\mathbf{j}$ can be derived by multiplying Eq. 6.57 by q_e/m_e, and the analogous equation for ions by q_i/m_i and adding the two results. We find for the α-component

$$\begin{aligned}
&\frac{\partial j_\alpha}{\partial t} + \sum_{\beta=1}^{3} \frac{\partial}{\partial x_\beta} (n_e q_e \bar{v}_{\beta e} \bar{v}_{\alpha e} + n_i q_i \bar{v}_{\beta i} \bar{v}_{\alpha i}) \\
&\quad = \frac{n_e q_e^2}{m_e} \left(1 + \frac{m_e n_i q_i^2}{m_i n_e q_e^2}\right) E_\alpha \\
&\quad + \frac{n_e q_e^2}{m_e} \left\{\left[\mathbf{V} + \frac{\mathbf{j}}{n_e q_e} + \frac{n_e m_e}{\rho_m} (\bar{\mathbf{v}}_i - \bar{\mathbf{v}}_e) + \left(\frac{m_e n_i q_i^2}{m_i q_e} - \rho\right) \frac{\bar{\mathbf{v}}_i}{n_e q_e}\right] \times \mathbf{B}\right\}_\alpha \\
&\quad - \sum_{\beta=1}^{3} \frac{\partial}{\partial x_\beta} \mathsf{R}_{\beta\alpha} - \rho \frac{\partial \phi}{\partial x_\alpha} + \left(\frac{q_e}{m_e} - \frac{q_i}{m_i}\right) \int dv\, m_e v_\alpha \left(\frac{\partial f_e}{\partial t}\right)_{\text{coll } ei},
\end{aligned} \tag{6.62}$$

where

$$\mathsf{R}_{\beta\alpha e} = q_e \int dv\, v_{r\beta} v_{r\alpha} f = \frac{q_e}{m_e} \mathsf{P}_{\beta\alpha e} \tag{6.63}$$

and

$$\mathsf{R}_{\beta\alpha} = \mathsf{R}_{\beta\alpha e} + \mathsf{R}_{\beta\alpha i}. \tag{6.64}$$

The quantity $\mathsf{R}_{\beta\alpha}$ is called the electrokinetic stress tensor.

The present result is very complicated, and we customarily make the following approximations in the interest of simplicity:

1. The plasma is nearly isotropic, so that the drift velocities, electric currents, and gradients all represent small first-order perturbations. The higher-order perturbations are neglected. Then also

$$\mathsf{R}_{\alpha\beta e} \simeq q_e p_e / m_e. \tag{6.65}$$

2. The plasma is neutral.

3. The ratios m_e/m_i and $Zm_e p_i/m_i p_e$ are very small and may be neglected compared to 1. As a result, we find also that

$$\mathsf{R}_{\alpha\beta i} \ll \mathsf{R}_{\alpha\beta e}. \tag{6.66}$$

4. The interchange of momentum between electrons and ions is proportional to the current. That is, we assume that

$$\int dv m_e \mathbf{v} \left(\frac{\partial f_e}{\partial t}\right)_{\text{coll } ei} = -n_e q_e \eta \mathbf{j}. \tag{6.67}$$

The proportionality factor η is a resistivity, as we shall see. With these approximations, Eq. 6.62 becomes

$$\frac{m_e}{n_e q_e^2} \frac{\partial \mathbf{j}}{\partial t} = \mathbf{E} + \left(\mathbf{V} + \frac{\mathbf{j}}{n_e q_e}\right) \times \mathbf{B} - \frac{1}{n_e q_e} \nabla p_e - \eta \mathbf{j}. \tag{6.68}$$

The resistivity will be discussed in more detail in Chap. 7. However, we note here that if $\partial \mathbf{j}/\partial t$, $\mathbf{B}$, and ∇p are all zero,

$$\mathbf{E} = \eta \mathbf{j}, \tag{6.69}$$

and identification of η with a scalar resistivity follows immediately. On the other hand, η is not necessarily the resistivity of the plasma, principally because **B** is not zero. The plasma resistivity relates **E** to **j**, whereas the phenomenological constant relates **j** to momentum transfer. The plasma conductivity (and resistivity) will be discussed in Secs. 7.6 and 7.7.

PROBLEMS

1. For any distribution f which is isotropic about its drift velocity $\bar{\mathbf{v}}$, prove that

$$\overline{v_{r\alpha} v_{r\beta}} = v_r^2 \delta_{\alpha\beta}/3.$$

2. An electron gun emits a cloud of electrons. There are enough positive ions in the vicinity to neutralize their space charge, and the electrons collide among themselves to produce an approximately Maxwellian distribution at temperature T in front of the cathode. These electrons drift slowly to an accelerating gap and then are accelerated into a field-free drift region. Describe the appearance of the kinetic stress tensor in the drift region, and show that it will not be isotropic. If the electrons thermalize among themselves far down the drift region, will the final temperature be greater or less than T?

3. A velocity distribution of particles has a three-dimensional elliptical shape in velocity space. That is, $\mathsf{P}_{xx} = C_1\mathsf{P}_{yy} = C_2\mathsf{P}_{zz}$, where C_1 and C_2 are constants different from 1; $\mathsf{P}_{xy} = \mathsf{P}_{xz} = \mathsf{P}_{zy} = 0$. Find the temperature of the distribution after the particles thermalize. Discuss qualitatively, in terms of the density, particle mass, and so forth, the rate of thermalization in the case of (*a*) hard-sphere collisions, and (*b*) collisions governed by a potential ϕ_{12} between any two particles 1 and 2 varying as $(r_{12})^{-1}$.

4. For a one-dimensional electron gas, derive the Boltzmann distribution in a potential well (Eq. 6.46) directly from the Boltzmann equation (Eq. 4.19).

5. For an isothermal atmosphere above the earth, derive the density as a function of height, starting with derivatives of pressure or density. Interpret the result in terms of a gravitational potential well.

6. An annular disk of plasma has a potential ϕ applied between its inner and outer radii r_1 and r_2. A uniform induction B exists in the direction of the disk axis. Calculate the angular velocity as a function of radius in the steady state. Neglect viscosity and the effect of electrostatic sheaths at r_1 and r_2.

7. A current I of 1 amp flows along a copper wire whose cross section is 1 mm on a side. An induction B exists normal to the current flow. There is one conduction electron per atom in copper. Calculate the voltage across the two faces of the wire normal to B and I. This phenomenon is known as the Hall effect, and the voltage is called the Hall voltage.

8. Calculate reasonable values of $\mathbf{B}$ and $\mathbf{j}$ so that a liquid Na-K alloy ($\rho_m = 0.8$ g/cm^3, $\eta = 4 \times 10^{-7}$ ohm-m) can be pumped 10 m high. Calculate the electrical loss per unit volume.

9. Show that for a simple plasma, where $\partial\mathbf{j}/\partial t$, $\mathbf{B}$, and ∇p are all zero, Eq. 6.67 can be proved. Hint: Let $f = f_0 + f_1$, where f_0 is isotropic, and f_1 is a small perturbation. Then refer to the resistivity as defined in Eq. 5.33 for an isotropic medium.

REFERENCES

1. J. L. Delcroix, *Introduction to the Theory of Ionized Gases* (translated by M. Clark, Jr., D. J. BenDaniel, and J. M. BenDaniel), Interscience Publishers, Inc., New York (1960).
2. W. P. Allis, "Motions of Ions and Electrons," *Handbuch der Physik*, Vol. 21, Springer, Berlin (1956), pp. 383–444.

7

Macroscopic motions of a plasma

In the previous chapter, a plasma was considered from a macroscopic point of view as consisting of a continuum to which the ideas of hydrodynamics could be applied. Each volume element contains many particles. We are not interested in the motion of each individual particle but only in macroscopic groupings of particles. The motion of each individual particle will be discussed in Chap. 10.

The macroscopic concepts led to the hydromagnetic equations of motion. These equations are very useful in the quantitative treatment of a plasma and have been applied extensively in computing motion of the plasma. In this chapter, we wish to interpret the hydromagnetic equations from a physical point of view. This work will enable us to derive several constraints on the motion of the plasma. To this end, we shall have to use Maxwell's equations to compute the electromagnetic fields arising from the macroscopic motion. Maxwell's equations and the hydromagnetic equations form a complete set, in the sense that from the former the fields can be computed from the plasma fluid motion, and from the latter equations the fluid motion can be computed from the fields. The system is deterministic. This work will lead to the concept of magnetic pressure and give a preliminary basis for the confinement of a plasma by magnetic fields.

We shall derive in Sec. 7.4 the conservation equation for the kinetic energy of the particles and for the electromagnetic fields. The forces exerted by the fields on the charged particles and the creation of the fields by the charged particles do not appear explicitly in this conservation relation. Indeed, it will be seen that only the momentum of *both* the particles and the field is conserved.

Frequently a plasma is compressed by a magnetic induction, not only to increase its density but also to heat it. For this reason, the subject of one-, two-, and three-dimensional adiabatic heating is discussed in the third section.

Since a plasma is confined and may be heated by a magnetic induction, it behooves us to examine the transport of the plasma constituents. Electric fields and pressure gradients lead to particle currents of both electrons and ions. Some of these are in the same direction, so that there is a flow of material but no current. Some of the flows are antiparallel, whence electric currents arise. Momentum lost by electrons colliding with ions is exactly equal to the momentum thereby gained by the ions. This truism and the absence of any third species in a fully ionized plasma profoundly affects the plasma behavior. Such matters are discussed in Sec. 7.6. If fixed scattering centers—neutral particles, for example—are present, the momentum transfer processes are modified; we discuss this case in Sec. 7.7.

Finally, the subject of shock heating is briefly considered in the last sections. To this end, the conservation laws of mass, momentum, and energy are applied to the Hugoniot relations. It will be seen that shock heating can be a very effective method, especially since the ions are preferentially heated.

7.1. Discussion of the Simplified Hydromagnetic Equations

The hydromagnetic equations (Eqs. 6.61 and 6.68) were derived in the previous chapter. We shall start to interpret these equations in this section. The first equation essentially states that the mass in an element of volume times its acceleration equals the total of all forces acting. The forces in this equation are, respectively, those due to the electric field, the magnetic induction, the pressure gradient, and the gradient of the gravitational potential. Equation 6.68 also relates forces to mass and acceleration, but is expressed in terms of the electric current created by the particles, rather than in terms of the particles themselves, as in Eq. 6.61. The first term of the right-hand side is the force caused by the electric field, the fourth term is the force due to the pressure gradient, and the last term is the damping caused by the resistivity. It represents the effect of the scattering of particles as they plow through the plasma. The second and third terms are considered together; the force on the moving particles is $n_e q_e(\mathbf{E} + \mathbf{V} \times \mathbf{B})$ in the light of Eq. 5.41, since

$\mathbf{E} + \mathbf{V} \times \mathbf{B}$ is just the electric field seen by the particle in the moving frame. It is interesting to note in passing that the equation

$$m_e \frac{\partial \mathbf{j}}{\partial t} = q_e(n_e q_e \mathbf{E}' + \mathbf{j} \times \mathbf{B} - \nabla p_e), \tag{7.1}$$

which is deduced by neglecting the resistivity term in Eq. 6.68, is the dual of the equation

$$m\dot{\mathbf{v}} = q(\mathbf{E} + \mathbf{v} \times \mathbf{B}) \tag{7.2}$$

for a charged particle moving in an electric field and a magnetic induction. Thus, the solutions of Eq. 7.1 will be of the same form as those for a particle.

For practical use of the hydromagnetic equations, a number of further approximations are frequently made. The gravitational terms may always be neglected, except in astronomical and some ionospheric applications. Within a plasma, the net charge density is zero. In the steady state, Eq. 6.61 becomes

$$\nabla p = \mathbf{j} \times \mathbf{B}. \tag{7.3}$$

Also in the steady state, the displacement current is zero, so that Ampère's law,

$$\nabla \times \mathbf{B} = \mu_0 \mathbf{j}, \tag{7.4}$$

applies. The third applicable relation is

$$\nabla \cdot \mathbf{B} = 0. \tag{7.5}$$

Equations 7.3 to 7.5 are used together to compute p, $\mathbf{j}$, and $\mathbf{B}$. There are a number of important consequences that will be discussed in the next two sections.

7.2. Magnetic and Isobaric Surfaces and Confinement

From Eq. 7.4, we find that

$$\nabla \cdot \mathbf{j} = 0. \tag{7.6}$$

There are no sources or sinks of current in the present extreme approximation. Furthermore, by Eq. 7.3,

$$\mathbf{B} \cdot \nabla p = 0, \tag{7.7}$$

$$\mathbf{j} \cdot \nabla p = 0. \tag{7.8}$$

Expressed in words, Eqs. 7.7 and 7.8 state that the magnetic induction and the current density are everywhere normal to the pressure gradient. The most general solution of these equations (Eqs. 7.3 *et seq.*) can be visualized as follows. Since the pressure p is a scalar quantity that varies from place to place, a curve $p = C =$ const must define a simple closed surface, characterized by the constant value C of the pressure over it. The gradient of the pressure is everywhere normal to the surface $p = C$, and by Eq. 7.7 no line of induction $\mathbf{B}$ crosses the surface. We can therefore equivalently consider the surface to be made up of the lines of induction. Similarly, by Eq. 7.8 the same surface is described by the lines of current density $\mathbf{j}$. Figure 7.1 illustrates such a surface. Because any closed surface $p = C$ divides all space into two regions, inside and outside, we see that all such closed surfaces must topologically be similar to tori.

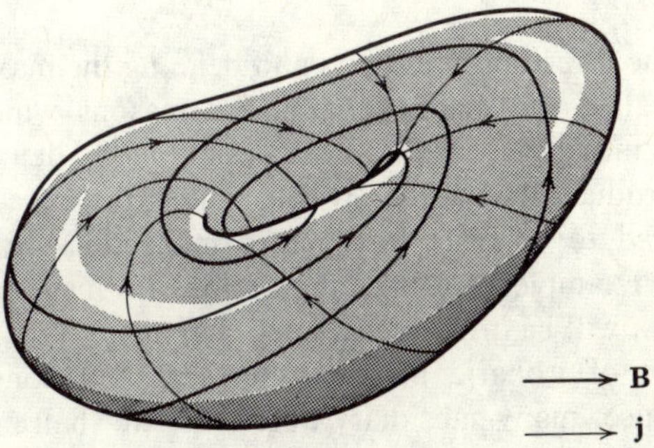

Fig. 7.1. Closed surface $p =$ const showing current density $\mathbf{j}$ and induction $\mathbf{B}$ lying in the surface. Pressure gradient ∇p is normal to the surface.

There are few restrictions on $\mathbf{B}$ and $\mathbf{j}$ other than the ones we have described. Usually, neither $\mathbf{B}\cdot\mathbf{j}$ nor $\mathbf{B}\times\mathbf{j}$ is zero, so the lines of $\mathbf{B}$ and $\mathbf{j}$ intersect at various angles. While the pressure p is constant over any of these surfaces, ∇p is usually not constant and varies from place to place.

Beside forming the basis for interesting topological studies, these considerations are essential to the understanding of a number of plasma-confinement schemes. By suitable disposition of currents either in external conductors or in the plasma, the induction $\mathbf{B}$ forms a set of nested surfaces similar to that shown in Fig. 7.1. With the plasma currents on the surface of Fig. 7.1 arranged so that $\mathbf{j}\times\mathbf{B}$ points inward, the pressure increases as we penetrate the torus from the outside. Then inside the torus of Fig. 7.1, there lie further nested tori; the pressure p increases on these successive nested magnetic surfaces. If $p = 0$ on the outermost one, we can in principle confine a plasma free from any material walls. Figure 7.2 shows a cross section of such a torus with isobaric lines drawn on the cross section. Because each torus $p = C$ is closed and because each contour $p = C$ is closed on the cut, there must

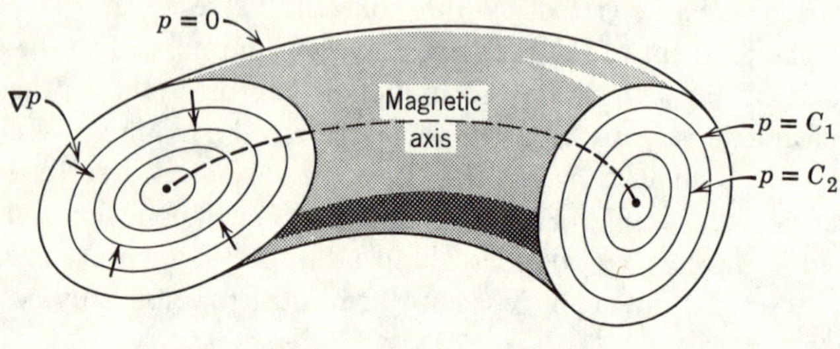

Fig. 7.2. Series of nested tori p = const cut open to show contours of constant pressure.

be a unique line threading the forms and defining the maximum pressure. This line is called the magnetic axis, for reasons which will become apparent in a moment. If the magnetic surface is defined, the current required to produce it is given by Eq. 7.4. Specific configurations of current required to produce satisfactory magnetic surfaces tend to be complicated. Examples of such configurations, for instance, the stellarator and the Astron, will be discussed in later chapters.

The concept of isobaric magnetic surfaces and consequent plasma confinement raises many subsidiary questions; we shall mention a few of them. In the first place, a line of induction **B** or current **j** does not necessarily appear to close on itself on the torus of Fig. 7.1, in spite of the fact that $\nabla \cdot \mathbf{B}$ and $\nabla \cdot \mathbf{j}$ are both zero. There is no inconsistency here, for the closing of each line is a sufficient but not necessary condition for zero divergence. It is necessary only that as many flux lines leave any volume as enter it, and this requirement is evidently satisfied. We find most generally that in Fig. 7.1 the direction of **B** followed indefinitely will generate the entire isobaric surface, in the sense that it will pass arbitrarily near any point on it. To be sure, situations exist wherein lines of force close exactly on themselves after finite extension; but these cases are singular. Figure 7.3 illustrates the point. A circular torus has a perfect solenoidal winding, so that each **B** line is circular and closes on itself. But if, in addition, any current flows along the axial wire shown, then an additional induction is created: the **B** lines in the torus become helices, and a **B** line followed around the torus will usually not close on itself.

In spite of the fact that most lines of induction on a torus generate surfaces and close on themselves in only an ergodic sense, there is one line in Fig. 7.2 which must close on itself. The surfaces whose cross

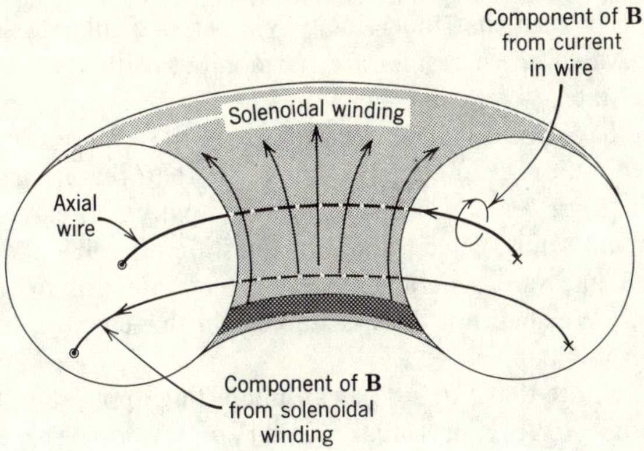

Fig. 7.3. Cut circular torus, showing a solenoidal current and an axial wire carrying current.

sections are shown all contain the magnetic axis in their interior. This line therefore defines the position of the limiting magnetic surface whose cross section approaches zero. The line of induction describing it must close on itself and define a pressure extremum; it is therefore appropriately called the magnetic axis.

The second specific subsidiary question concerns the confinement of particles. These foregoing derivations have all been carried out with the thought that the magnetic forces might confine particles. But permanent confinement of an isolated static system is impossible, for a magnetic induction cannot do work on the particles and hence create an energetic potential well in which particles may be trapped. Only a quasi-equilibrium among the pressure, current, and induction can be established which lasts for a finite time. Alternately, the plasma might be replenished in some way. Indeed, in thermodynamic equilibrium, since the current density is zero, there can be no magnetic forces leading to confinement. The time that an unreplenished confined plasma might last cannot be determined from Eqs. 7.3 through 7.5, because the scattering processes that lead to thermal equilibrium and the disappearance of confinement have not appeared here. The scattering processes are, of course, synonymous with diffusion and resistivity. A connection between all these quantities is deferred until Secs. 7.6 and 7.7. Many discussions of this point may be found in the literature.[1,2]

Although collisions are necessary to establish equilibrium, they are not necessary to maintain the equilibrium state once established. Equations 7.3 through 7.5 must therefore be consistent with the equilibrium condition, and they are. Equation 7.3 implies that $\nabla p = 0$ in equilibrium, since $\mathbf{j} = 0$. These results and Eq. 6.61 imply that $\mathbf{E} = 0$, from which we conclude in turn that the velocity $\mathbf{V}_\perp$ perpendicular to $\mathbf{B}$ is zero by Eq. 6.68. All of these results are harmonious with the concept of an equilibrium state.

Third, we have not discussed the stability of the temporary confinement as described. For example, it is possible that the nested structures described in Figs. 7.1 and 7.2 will deform rapidly and carry the plasma to the vacuum walls. Such stability questions involve the manner in which the $\mathbf{B}$ lines are generated by (and therefore tied to) the external conductors. We shall not discuss stability in this chapter, but return to the question in Chap. 12.

Finally, we note that Eq. 7.8 has an interesting application to diffusion of the plasma. We expect that, since ∇p is normal to the isobaric surfaces, the particle current will be given by

$$\mathbf{\Gamma} = -\mathrm{D}\,\nabla p, \tag{7.9}$$

appropriately calculated for each type of particle. However, Eq. 7.8 states that the current parallel to the gradient is zero. We conclude, therefore, that positive and negative charges must move together: the current is zero, but the mass transport is finite. It can be calculated that the mass flux is in agreement with our notions of diffusion of a plasma across a magnetic induction: if the resistivity $\eta \neq 0$, the diffusion rate (Sec. 5.7) $\neq 0$. But if $\eta \neq 0$, then an electric field $\mathbf{E}$ must exist in the plasma. We may then calculate that the $\mathbf{E}$ and $\mathbf{B}$ forces propel the plasma in the direction of lower pressure, and obtain an alternate but equivalent view of the diffusion process.[2] Diffusion in a multispecies plasma is further considered in Secs. 7.6 and 7.7.

A most unexpected result arises if we consider a plasma made up of only one charged species. Space-charge problems may be avoided if the plasma is sufficiently thin, whereupon Eq. 7.8 should still apply. However, in this case, the particle current $\mathbf{\Gamma}$ and electric current $\mathbf{j}$ are uniquely related by

$$\mathbf{j} = q\mathbf{\Gamma}. \tag{7.10}$$

We therefore come to the surprising conclusion from Eq. 7.8 that $\mathbf{\Gamma}\cdot\nabla p = 0$, and that the diffusion equation (Eq. 7.9) cannot be true in this application. Our conclusion is in fact correct. If the anisotropic kinetic stress tensor had been retained, we should have discovered that a diffusion current of such a plasma across the induction arises only from higher derivatives of density. Diffusion arising from unlike-particle collisions is further discussed in Sec. 7.6. The difficulty with like-particle collisions is most readily stated in terms of individual particle orbits, and we reserve further discussion of this point until Chap. 10.

7.3. Magnetic Pressure

As mentioned in the previous section, the stress exerted on a collection of charges by an induction can under certain circumstances be thought of as a pressure. In this section, we quantify the concept.

If the lines of magnetic induction are straight and parallel, the relation

$$(\mathbf{B}\cdot\nabla)\mathbf{B} = 0 \tag{7.11}$$

is true. Then by inserting Eq. 7.4 into Eq. 7.3, we find

$$\nabla[p + (B^2/2\mu_0)] = 0, \tag{7.12}$$

from which it follows that

$$p + (B^2/2\mu_0) = \text{const.} \tag{7.13}$$

In view of this result, the magnetic energy density $B^2/2\mu_0$ is often called the magnetic pressure and is used as a basis for discussing the magnetic confinement of plasmas. If we immerse a plasma of pressure p in a vacuum and if the induction equals $\sqrt{2\mu p}$, then the magnetic and kinetic pressures are in quasi-equilibrium until the assembly mixes by diffusion. Whether the equilibrium defined by Eq. 7.13 is stable or unstable is another matter.

The concept of $B^2/2\mu_0$ as a magnetic pressure is of value chiefly in a qualitative way. For quantitative deductions, the more exact hydromagnetic equations must usually be used. The energy density or pressure associated with a moderate magnetic induction is high. For example, an induction $B = 1$ weber/m^2 gives an energy density $B^2/2\mu_0 = 4 \times 10^5$ joules/m^3, or a pressure of 4×10^5 newtons/m^2; this pressure is about 58 lb/in.^{2} A pressure of 1 atm requires an induction of about 0.5 weber/m^2. We note particularly that the magnetic pressure increases as the square of the induction. The calculations of Sec. 4.10 show that a pressure in the order of 100 atm might be required in a practical thermonuclear gas. That pressure can be created by the large, but not unreasonable, induction of 5 webers/m^2.

We remark in passing that the pressure exerted by reasonable electric fields is very small compared with that exerted by a magnetic induction. For example, an electric field of 3×10^8 volts/m and an induction of 1 weber/m^2 have the same pressure. Even if such an electric field could be created, its presence near the highly conducting plasma would give rise to many complications. If a steady electric field is externally applied, the plasma moves; if an rf field is applied, the skin losses are immense. For these reasons among others, thermonuclear plasma confinement by electric fields alone is not of interest to us. However, from time to time, we shall have to consider the effect of electric fields produced incidentally by plasma motions and configurations.

In using this qualitative concept of magnetic pressure, it is useful to define the ratio of the kinetic pressure to the magnetic pressure by

$$\beta = \frac{nkT}{(B^2/2\mu_0)}. \tag{7.14}$$

The quantity serves as a measure of the efficiency with which the induction is used. Very often the kinetic and magnetic pressures are not evaluated at the same point. For example, from Eq. 7.13, we can imagine a configuration where a plasma at pressure p excludes the induction B which confines it. Then inside the plasma, $\beta = \infty$, and outside, $\beta = 0$; but the ratio of nkT just inside the boundary to $B^2/2\mu_0$ just outside is unity.

We discover from Eq. 7.13 that the plasma is apparently diamagnetic: where p is high, B is low. The explanation can be found from Eq. 7.3. If $\nabla p \neq 0$, a current flows; the charges rotate in the magnetic induction and create an induction of their own to cancel part of that applied.

Finally, we remark that neutral particles may be swept along by the charges, as a result of collisions. The effect manifests itself by the scattering of charged particles and appears in Eq. 6.68 as the resistivity term. Thus in a moving, weakly ionized plasma, for instance, nearly all the gas may be channeled by a magnetic induction, because the charged particles sweep the neutrals along. This fact is of great importance in studies of hydromagnetic energy conversion from seeded flames, but is not expected to be important in thermonuclear research.

★ 7.4. The Total Stress Tensor

We are now in a position to derive a conservation law for the total momentum in a system. The forces exerted on charged particles are related to the electromagnetic fields by Eq. 5.74 and to the motion of the particles by Eq. 6.59. Equation 6.59 can be written as

$$\frac{\partial}{\partial t}(\rho_m \mathbf{V}) + \sum_{\alpha,\beta=1}^{3} \mathbf{i}_\alpha \frac{\partial}{\partial x_\beta}(n_e m_e \overline{v_{\alpha e} v_{\beta e}} + n_i m_i \overline{v_{\alpha i} v_{\beta i}}) = \rho \mathbf{E} + \mathbf{j} \times \mathbf{B} \tag{7.15}$$

by means of Eqs. 6.14 and 6.19. This result may be added to Eq. 5.74 to yield

$$\frac{\partial}{\partial t}(\rho_m \mathbf{V} + \epsilon\mu \mathbf{S}) + \sum_{\alpha,\beta=1}^{3} \mathbf{i}_\alpha \frac{\partial}{\partial x_\beta}(\rho_m \overline{V_\alpha V_\beta} - \mathsf{T}_{\alpha\beta}) = 0, \tag{7.16}$$

where

$$\rho_m \overline{V_\alpha V_\beta} = n_e m_e \overline{v_{\alpha e} v_{\beta e}} + n_i m_i \overline{v_{\alpha i} v_{\beta i}}. \tag{7.17}$$

The present result states that the rate of change of momentum density equals the rate of inflow of momentum over the surfaces of the bounding volume. This fact can be easily seen by integrating Eq. 7.17 over an element of volume and by use of the divergence theorem, akin to the method used in studying Eq. 5.64. The momentum is of two types: that due to the kinetic energy of the particles and that due to the electromagnetic field. The momentum of the particles alone is not conserved, but changes because of the forces exerted by the electromagnetic field. The quantity conserved is the total momentum of the field plus the particles. The Poynting vector **S** divided by the square of the speed of light may be interpreted as a momentum density in view of the form of Eq. 7.16. This interpretation is in keeping with that of Sec. 5.5. Again, from the form of Eq. 7.16 the tensor $\mathsf{T}_{\alpha\beta}$ is to be interpreted as the momentum flow rate. In static problems, the law of momentum conservation becomes

$$\sum_{\beta=1}^{3} \frac{\partial}{\partial x_\beta} (\rho_m \overline{V_\alpha V_\beta} - \mathsf{T}_{\alpha\beta}) = 0 \tag{7.18}$$

for each coordinate direction α. In words, the momentum transferred from the particles equals that transferred to the field, and conversely.

7.5. Confinement and the Adiabatic Compression of a Plasma

The developments of Secs. 7.2 and 7.3 show that the currents flowing in a magnetic induction give rise to a force on the particles that may be interpreted as a magnetic pressure. At an idealized sharp boundary between a plasma and the induction, the surface currents create a force that balances the pressure and establishes equilibrium. The magnetic induction may be regarded as a piston. However, it is a soft one because the plasma can change its shape, penetrate the induction, and diffuse through it. By increasing the surrounding induction, a plasma may be compressed. The compression may occur rapidly compared with the time required for the particles to come to thermal equilibrium, in which each particle travels as fast in one direction on the average as in any other direction. In other words, if the compression is fast enough, equipartition of energy will lag behind. A change in one velocity component of the particles will not affect the other two components. By suitably controlling the geometry of the magnetic induction as it increases, a compression may be made one-, two-, or three-dimensional.

The changes in the thermodynamic state of the particles when subjected to one-, two-, or three-dimensional adiabatic compressions will now be considered. The adiabatic relation between pressure and volume follows immediately from Eq. 6.42 for three-dimensional compression. In the case of one- or two-dimensional compressions, the averages expressed by Eq. 6.25 must be re-examined. These averages are deduced from the definition of $\mathsf{P}_{\alpha\beta}$ in Eq. 6.24. Equations 6.34, 6.35, 6.37, 6.38, 6.39, 6.41, and 6.42 would be changed. The result is that in the more general case the adiabatic law of compression is given by

$$p\mathscr{V}^{\gamma} = \text{const}, \tag{7.19}$$

where $\mathscr{V}$ is the volume. The quantity γ, which incidentally is the ratio of specific heats, is given by

$$\gamma = (2 + \delta)/\delta, \tag{7.20}$$

and δ is the number of degrees of freedom. This relation is valid in the limit where the compressions are so slow that the reactions are reversible. A gas molecule must hit the bounding walls many times during the time of the compression. The compression must not be so slow, however, that the particles have time to scatter in angle. Further, if the speed of the moving wall were comparable with the speed of the particles, then shock waves would develop. This matter will be treated in Sec. 7.7. From the perfect-gas law (Eq. 6.27) we deduce further that

$$T_2/T_1 = (\mathscr{V}_1/\mathscr{V}_2)^{\gamma-1}. \tag{7.21}$$

The heating expected by the adiabatic compression of a gas depends upon the geometry of the compression. We now develop the general formula for the final total energy of a gas which is compressed adiabatically in δ-directions simultaneously, but in which the degrees of freedom do not couple; that is, the particles do not scatter into or out of the uncompressed directions. In order to avoid use of the word temperature in this nonequilibrium situation, we replace the symbol T in Eq. 7.21 by the energy U appropriate to the directions involved. Then we find that the energy $U_{\parallel}$ in directions parallel to the compression is

$$U_{\parallel 2}/U_{\parallel 1} = (\mathscr{V}_1/\mathscr{V}_2)^{\gamma-1}. \tag{7.22}$$

We now assume for simplicity that the initial distribution was isotropic. Then for δ-directions of compression,

$$U_{\parallel 1} = (\delta/3)U_1, \tag{7.23}$$

where U_1 is the initial total energy. The energy

$$U_{\perp 1} = (3 - \delta)U_1/3 \tag{7.24}$$

in the other directions perpendicular to the compression remains unchanged. We then find that the total energy U_2 after compression is given by

$$U_2/U_1 = [3 - \delta + \delta(\mathscr{V}_1/\mathscr{V}_2)^{\gamma-1}]/3. \tag{7.25}$$

For a one-dimensional compression, $\delta = 1$, $\gamma = 3$, and

$$U_2/U_1 = [2 + (\mathscr{V}_1/\mathscr{V}_2)^2]/3. \tag{7.26}$$

If the compression is two-dimensional, for example, radial compression of a cylinder, $\delta = 2$, $\gamma = 2$, and

$$U_2/U_1 = [1 + 2(\mathscr{V}_1/\mathscr{V}_2)]/3. \tag{7.27}$$

In a three-dimensional compression or in any compression where the particles maintain energy equipartition through collisions, we find

$$U_2/U_1 \equiv T_2/T_1 = (\mathscr{V}_1/\mathscr{V}_2)^{2/3}. \tag{7.28}$$

We note that the energy increase in a one-dimensional compression is greater than that in a two-dimensional compression, and that the energy increase in a two-dimensional compression is greater than that in a three-dimensional compression.

The second law of thermodynamics tells us that the energy increase in an irreversible compression must always be greater than that in a reversible compression. However, irreversible processes are very difficult to treat generally; we limit ourselves here to a few remarks concerning the rate of compression normally found in devices. In many fast plasma-compression experiments, the initial plasma radius is about 0.1 m. A pulsed magnetic induction compresses this plasma to a much smaller radius in a time of about 1 μsec. The inward speed is then about 10^5 m/sec, which is to be compared with the random speed of the plasma particles. For electrons, 10^5 m/sec corresponds to an average energy $\bar{u} \approx 0.04$ ev. The electron energy will be several electron volts at least; we conclude that the electrons will be compressed adiabatically. Therefore, in a radial compression, Eq. 7.27 applies if interparticle collisions are neglected. For a 10-to-1 volume reduction, the total electron energy is increased by a factor of 7.

For ions, the result may be quite different. The speed 10^5 m/sec corresponds to 150-ev deuterons. The deuteron energy at the start of the compression may be much less than 150 ev. Then deuterons will gain *twice* the wall speed on each bounce off the oncoming wall. If all of them were struck only once by the wall, they would acquire an energy of about 600 ev, even if they had negligible energy at the start. Numerous compression experiments are reported[3] in which the final electron energy was about 20 ev, the ion energy was about 400 ev, and the original plasma had an average energy of 1 or 2 ev.

We remark also that many slow compression experiments have been carried out, in which the compression speed was 10^4 m/sec or less. In that case, light ions are compressed adiabatically, once their energy rises above a few electron volts.

7.6. Particle Currents and Conductivity of a Fully Ionized Magnetized Gas

In the previous sections, we have applied the hydromagnetic equations developed in Chap. 6 to the discovery of configurations and principles of confinement and compression. Our development to this point has been specifically formulated to avoid questions of conductivity, diffusion, and the like. Even if the configuration is stable, these processes will lead to a gradual dissolution of the confined plasma. Here, we return to these questions and re-examine the hydromagnetic equation for each species from this point of view.

To this end, consider Eq. 6.57 for electrons in the case of

1. No time variation.
2. No net charge density.
3. An isotropic stress tensor.
4. Negligible drift energy compared with the kinetic energy.
5. No gravitational effects.
6. Setting the momentum transfer rate proportional to the current, as in Eq. 6.67.

Equation 6.57 is a force equation from which we may deduce the close approximation of a plasma to neutrality. Indeed, in all equations but the force equation the net charge density may be zero; in the force equation the term ρE may, of course, be quite large in spite of the small magnitude of ρ. For the examination of plasma sheaths and the propagation of some plasma waves, this term is essential. Its neglect in the present section corresponds to a neglect of the very high frequency oscillations superimposed upon the more gross motions discussed here.

With the convenient substitutions

$$\mathbf{\Gamma}_e = n\bar{\mathbf{v}}_e \quad \text{and} \quad q_e = -e,$$

we find

$$en\eta\mathbf{\Gamma}_i - en\eta\mathbf{\Gamma}_e - \mathbf{\Gamma}_e \times \mathbf{B} = n\mathbf{E} + \nabla p_e/e. \tag{7.29}$$

Similarly for the ions,

$$en\eta\mathbf{\Gamma}_i - en\eta\mathbf{\Gamma}_e - \mathbf{\Gamma}_i \times \mathbf{B} = n\mathbf{E} - \nabla p_i/e. \tag{7.30}$$

These equations can readily be solved for the individual particle flows; however, there is an apparent inconsistency with the components in the direction of **B**, say the z-components. We shall consider this subset first. All the z-components appear in the two component equations

$$en\eta\Gamma_{zi} - en\eta\Gamma_{ze} = nE_z + \frac{\partial(p_e/e)}{\partial z}, \tag{7.31}$$

$$en\eta\Gamma_{zi} - en\eta\Gamma_{ze} = nE_z - \frac{\partial(p_i/e)}{\partial z}. \tag{7.32}$$

which seem inconsistent unless the gradients are zero.

In fact, Eqs. 7.31 and 7.32 are not inconsistent and have real physical significance. Note first that only differences in the axial particle currents are defined. The reason is simple. An axial translation of the whole plasma along the **B** lines at any uniform velocity cannot affect the behavior of the plasma or our formulation of its behavior. If we are to write specific equations for each axial particle flow, we must evidently have further information about the axial mass flow. This information must be obtained from additional inspection of the particular system to be analyzed. If there are no pressure gradients in the direction of the induction, the equations state that

$$e(\Gamma_{zi} - \Gamma_{ze}) \equiv j_z = E_z/\eta; \tag{7.33}$$

Ohm's law is obeyed in the direction of the induction, with a conductivity $1/\eta$.

Consider now the axial pressure gradients. Equations 7.31 and 7.32 state only that

$$\frac{\partial(p_e + p_i)}{\partial z} = \frac{\partial p}{\partial z} = 0. \tag{7.34}$$

This result is consistent with Eq. 7.3, which states that the **B** lines define isobaric surfaces, whence the gradient along them is zero. This result can lead to interesting observations in addition to those described in earlier sections. Because neutrality is preserved, the two particle density gradients must be identical. Suppose then that $\partial n_e/\partial z \equiv \partial n_i/\partial z \neq 0$ and that initially the temperatures are everywhere uniform (a violation of Eq. 7.34). Currents of ions and electrons flow, and in general they will not be the same because of the difference in mass. As a result of the net current, there is a resistive heating of the medium leading to an increase in temperature in the region of lower density. This heating tends to restore the pressure equilibrium. Thus Eq. 7.34 is again satisfied as equilibrium is approached.

It is our intent in this section to express the particle flows as matrix summations over the fields and gradients, in the style adopted for Table 5.1 of Sec. 5.6. What shall we insert for the zz-element of the matrix associated with the pressure gradients, in view of this discussion? We resolve the problem simply by saying that the term is undefined, because no such pressure gradients exist in equilibrium in a closed system: electrons and ions flow along a magnetic surface like that in Fig. 7.1, and the pressure of each will be uniform. For convenience, we adopt a reference frame in which $\Gamma_{zi} = 0$. This is not an approximation, because we could add any arbitrary axial velocity to the system and to our results (for example, the velocity of the center of mass).

The x- and y-components of Eqs. 7.29 and 7.30 present no difficulty; the α-component of the electron particle current, $\Gamma_{\alpha e}$, is found to be

$$\Gamma_{\alpha e} = \sum_{\beta=1}^{3} \mu_{\alpha\beta e} n E_{\beta} - \sum_{\beta=1}^{3} \mathsf{D}'_{\alpha\beta ep} \frac{\partial p}{\partial x_{\beta}} - \sum_{\beta=1}^{3} \mathsf{D}'_{\alpha\beta ee} \frac{\partial p_e}{\partial x_{\beta}}. \tag{7.35}$$

The mobility μ_e is

$$[\mu_{\alpha\beta e}] = \begin{bmatrix} 0 & \dfrac{1}{B} & 0 \\ -\dfrac{1}{B} & 0 & 0 \\ 0 & 0 & -\dfrac{1^*}{ne\eta} \end{bmatrix}; \tag{7.36}$$

the asterisk reminds us that the term reflects our assumption that the ions are axially stationary. The coefficient D'_{ep} is an effective diffusion coefficient, relating the electron particle flow to the total pressure gradient (rather than density), and is

$$[\mathsf{D}'_{\alpha\beta ep}] = \begin{bmatrix} \dfrac{n\eta}{B^2} & 0 & 0 \\ 0 & \dfrac{n\eta}{B^2} & 0 \\ 0 & 0 & \text{—} \end{bmatrix}. \tag{7.37}$$

The coefficient D'_{ee} is a similar coefficient relating to the electron pressure gradient alone. It is

$$[\mathsf{D}'_{\alpha\beta ee}] = \begin{bmatrix} 0 & -\dfrac{1}{eB} & 0 \\ \dfrac{1}{eB} & 0 & 0 \\ 0 & 0 & \text{—} \end{bmatrix}. \tag{7.38}$$

In exactly the same way, we write for the ions

$$[\mu_{\alpha\beta i}] = \begin{bmatrix} 0 & \dfrac{1}{B} & 0 \\ -\dfrac{1}{B} & 0 & 0 \\ 0 & 0 & 0^* \end{bmatrix}. \tag{7.39}$$

Again, the asterisk represents the assumption $\Gamma_{zi} = 0$. Also

$$[\mathrm{D}'_{\alpha\beta ip}] = \begin{bmatrix} \frac{n\eta}{B^2} & 0 & 0 \\ 0 & \frac{n\eta}{B^2} & 0 \\ 0 & 0 & - \end{bmatrix} \tag{7.40}$$

and

$$[\mathrm{D}'_{\alpha\beta ii}] = \begin{bmatrix} 0 & \frac{1}{eB} & 0 \\ -\frac{1}{eB} & 0 & 0 \\ 0 & 0 & - \end{bmatrix}. \tag{7.41}$$

Several results appear at once from these equations. The current

$$j_\alpha = e(\Gamma_{\alpha i} - \Gamma_{\alpha e}) = \sum_{\beta=1}^{3} s_{\alpha\beta} E_\beta \tag{7.42}$$

defines the conductivity s of the plasma. We find at once that

$$s_{\alpha\beta} = \begin{bmatrix} 0 & 0 & 0 \\ 0 & 0 & 0 \\ 0 & 0 & \frac{1}{\eta} \end{bmatrix}. \tag{7.43}$$

An electric field drives a current only parallel to **B** in a fully ionized plasma, in the approximations to which we have solved the equations. The nondiagonal terms in μ_e and μ_i exactly cancel in the conductivity. No current flows, but the entire fluid travels in a direction perpendicular both to **E** and **B**. The velocity is

$$\mathbf{V}_E = \mathbf{E} \times \mathbf{B}/B^2 \tag{7.44}$$

independent of mass of the particles, or sign or magnitude of their charge. Equation 7.44 can be expressed also as

$$\mathbf{E}_\mathrm{T} + \mathbf{V}_E \times \mathbf{B} = 0; \tag{7.45}$$

in other words, the plasma moves as a whole so that there is no electric field in the moving system.

The addition of the flows characterized by Eqs. 7.38 and 7.41 together with Eq. 7.34 yields $\boldsymbol{\nabla} p = \mathbf{j} \times \mathbf{B}$, which is Eq. 7.3. Only a pressure gradient drives a current across **B** lines.

The flows described by Eqs. 7.37 and 7.40 represent the diffusion across the magnetic induction in the direction of decreasing pressure. The net current in this direction is zero from this effect, as stated also in Eq. 7.8. Note that this transverse diffusion of both species varies inversely as B^2 and is proportional to η. The coefficients of D'_{ip} or D'_{ep} are $nkT\eta/B^2kT$.

In most practical plasma systems, we expect that $\beta = nkT/(B^2/2\mu_0)$ will have some more or less fixed value determined by efficiency considerations, independent of the size or temperature. The transverse diffusion coefficient will then vary as η/T. We shall calculate η in Chap. 8 and show that it is proportional to $T^{-3/2}$. Diffusion is therefore drastically reduced at high temperature.

These particle flows and currents have been derived by Allis and Buchsbaum[4] in the case where neutral particles are also present. Their results, equivalent to our Eqs. 7.35 and following, are considerably more complicated because of the presence of the neutrals. Their results in the limit of vanishing neutral density are identical to ours, except that for the *zz*-components of the pressure matrices they propose a density gradient with uniform temperature.

7.7. Particle Currents and Conductivity of a Magnetized Lorentzian Gas

The results derived in the previous section are applicable to a fully ionized plasma in a magnetic induction. That case is the one of primary interest to us. In order to complete the discussion on diffusion and mobility, we consider also a gas in which the charged particles scatter from fixed centers. Such a system is called a Lorentzian gas; it was used as a convenient model for introducing the concept of diffusion and mobility in Sec. 4.4. We enlarge upon that discussion.

The only difference between the developments of this section and those of the previous one is in the handling of the momentum-transfer term. There, momentum of the combined electron-ion system was conserved; here it is not, because both electrons and ions transfer momentum to the fixed scattering centers. The particle current $\mathbf{\Gamma}$ of each species tends to be dissipated at the rate $\nu_m\mathbf{\Gamma}$, because ordered momentum is destroyed at the momentum collision frequency ν_m by definition. The equation analogous to Eq. 6.57 for either particle (or Eqs. 7.29 and 7.30) for the case of a near-isotropic distribution is

$$m\dot{\mathbf{\Gamma}} + m\nu_m\mathbf{\Gamma} = qn\mathbf{E} - \nabla p + q(\mathbf{\Gamma} \times \mathbf{B}). \tag{7.46}$$

An equation of this form exists for both $\mathbf{\Gamma}_e$ and $\mathbf{\Gamma}_i$. The two equations are not coupled because of our assumption about the momentum-transfer process. It is useful now to introduce the cyclotron frequency

$$\boldsymbol{\omega}_b = -q\mathbf{B}/m. \tag{7.47}$$

As before, the equation is solved in the steady state by expressing it in component form, or by the simple expedient of multiplying it from the left by $(\nu_m + \boldsymbol{\omega}_b \times)$, which stratagem extracts the cross-product terms containing $\boldsymbol{\Gamma}$. The solution can be written in component form as

$$\Gamma_\alpha = -\sum_{\beta=1}^{3} \frac{\partial}{\partial x_\beta} (\mathsf{D}_{\alpha\beta} n) + \sum_{\beta=1}^{3} \mu_{\alpha\beta} n E_\beta. \tag{7.48}$$

Using Eq. 6.27, we can express the coefficients D and μ as

$$[\mathsf{D}_{\alpha\beta}] = \frac{\overline{v^2}}{3} \begin{bmatrix} \dfrac{\nu_m}{\nu_m^2 + \omega_b^2} & \dfrac{-\omega_b}{\nu_m^2 + \omega_b^2} & 0 \\ \dfrac{+\omega_b}{\nu_m^2 + \omega_b^2} & \dfrac{\nu_m}{\nu_m^2 + \omega_b^2} & 0 \\ 0 & 0 & \dfrac{1}{\nu_m} \end{bmatrix}$$

$$= \begin{bmatrix} \mathsf{D}_\mathsf{T} & -\mathsf{D}_\mathsf{H} & 0 \\ \mathsf{D}_\mathsf{H} & \mathsf{D}_\mathsf{T} & 0 \\ 0 & 0 & \mathsf{D}_\parallel \end{bmatrix}, \tag{7.49}$$

and

$$[\mu_{\alpha\beta}] = \frac{q}{m} \begin{bmatrix} \dfrac{\nu_m}{\nu_m^2 + \omega_b^2} & \dfrac{-\omega_b}{\nu_m^2 + \omega_b^2} & 0 \\ \dfrac{\omega_b}{\nu_m^2 + \omega_b^2} & \dfrac{\nu_m}{\nu_m^2 + \omega_b^2} & 0 \\ 0 & 0 & \dfrac{1}{\nu_m} \end{bmatrix}$$

$$= \begin{bmatrix} \mu_\mathsf{T} & -\mu_\mathsf{H} & 0 \\ \mu_\mathsf{H} & \mu_\mathsf{T} & 0 \\ 0 & 0 & \mu_\parallel \end{bmatrix}. \tag{7.50}$$

Before discussing the nature of these results, we wish to state them more exactly. In writing Eq. 7.46, we have assumed that ν_m was a constant independent of particle speed. If that is not so, the equivalent of Eq. 7.46 would have to be written as integrals over the velocity distribution function. Analysis[4, 5] shows that the average appearing in Eq. 7.48 for D should be the average of v^2 times the matrix. The mobility terms are in fact given for an isotropic distribution by

$$n\mu_\mathsf{T} = -\frac{4\pi q}{3m} \int dv\, v^3 \frac{\nu_m}{\nu_m^2 + \omega_b^2} \frac{\partial f}{\partial v}, \tag{7.51}$$

$$n\mu_\mathsf{H} = -\frac{4\pi q}{3m} \int dv\, v^3 \frac{\omega_b}{\nu_m^2 + \omega_b^2} \frac{\partial f}{\partial v}, \tag{7.52}$$

and

$$n\mu_{\|} = -\frac{4\pi q}{3m}\int dv\,\frac{v^3}{\nu_m}\frac{\partial f}{\partial v}. \tag{7.53}$$

In case ν_m = const, Eqs. 7.51 to 7.53 reduce to Eq. 7.50 by integration by parts. The elements $n\mathsf{D}_{\|}$ and $n\mu_{\|}$ are exactly those given by Eqs. 4.49 and 4.53 (or 4.54). Therefore, the motion along **B** lines is governed by the induction-free coefficients, and the induction has no effect. Because the induction can have no effect on motion parallel to it, the result is to be expected. If the magnetic induction approaches zero, the diffusion coefficient

$$\mathsf{D}_{\alpha\beta} \to \mathsf{D}_0 = \overline{v^2/3\nu_m}\,\delta_{\alpha\beta}. \tag{7.54}$$

The coefficient is then a scalar number in agreement with Eq. 4.50. For this isotropic case, all the diagonal elements of the diffusion matrix are equal, and all the nondiagonal elements are zero. Identical remarks apply to the mobility.

In discussing the nature of the terms in case $\mathbf{B} \neq 0$, we shall use the approximate form, Eqs. 7.49 and 7.50, for convenience. Consider first the diffusion.

The transverse *xx*- and *yy*-components of the diffusion coefficient are equal to each other and are denoted by the symbol D_T. The coefficient D_T is smaller than $\mathsf{D}_{\|}$ by the factor $1 + (\omega_b/\nu_m)^2$. If the induction is large, that is, if the cyclotron frequency is high compared with the collision frequency, then D_T is reduced by the factor $(\nu_m/\omega_b)^2$. The transverse diffusion of ions is much larger than that of electrons having the same mean free path and temperature, because of the much greater mass of the ions. On the other hand, and for the same reason, the diffusion of electrons along the induction is much higher than that of ions of the same temperature. In the absence of collisions and in a uniform induction, the particles would rotate in circles in the magnetic induction. At each Coulomb collision the center of these circles is displaced. The displacement is in the order of the radius of the cyclotron orbit. Indeed, the transverse diffusion may be computed from considerations such as these.

As a result of the collisions and rotation in a magnetic induction, there is a diffusion of the particles in a direction perpendicular to both the gradient of the pressure and the magnetic induction. These effects are represented by the nondiagonal elements of the diffusion matrix. For example, suppose that the gradient of the pressure is along the *x*-axis only. With the magnetic induction along only the *z*-axis, there is a component of diffusion along the *y*-axis given by the factor $-\omega_b/(\nu_m^2 + \omega_b^2)$ in Eq. 7.49. This flow is conventionally called the Hall diffusion current, and the coefficient is written as D_H. At low fields the

Hall diffusion is proportional to the induction and inversely proportional to the mass, for the same mean free path. At inductions so high that $\omega_b \gg \nu_m$, D_H is proportional to the particle energy and inversely proportional to the induction. In this limit, D_H is independent of both the collision frequency and mass, for particles of the same energy. Note that the Hall diffusion causes electrons and ions to move in opposite directions. Therefore, an electric current is associated with Hall diffusion.

The remarks just made regarding diffusion apply equally well to mobility. Indeed, we see directly from Eq. 7.46 that the coefficients are similar: the body forces $nq\mathbf{E}$ and $-\nabla p$ appear in the same way, and their effects must be identical. From Eqs. 7.49 to 7.53, it is evident that if ν_m is constant

$$\frac{\mathsf{D}_{\alpha\beta}}{\mu_{\alpha\beta}} = \frac{2\bar{U}}{3q}. \tag{7.55}$$

In addition, if the distribution is Maxwellian,

$$\frac{\mathsf{D}_{\alpha\beta}}{\mu_{\alpha\beta}} = \frac{kT}{q}, \tag{7.56}$$

independent of the form of ν_m. Equations 7.55 and 7.56 are generalizations of Eqs. 4.56 and 4.57.

The conductivity s of the plasma is

$$[s_{\alpha\beta}] = \sum_{\zeta} q_\zeta n_\zeta \mu_{\alpha\beta\zeta}, \tag{7.57}$$

as shown in Eq. 5.33 and the discussion following it. Thus for either component

$$[s_{\alpha\beta}] = \frac{nq^2}{m}\begin{bmatrix} \dfrac{\nu_m}{\nu_m{}^2 + \omega_b{}^2} & \dfrac{-\omega_b}{\nu_m{}^2 + \omega_b{}^2} & 0 \\ \dfrac{\omega_b}{\nu_m{}^2 + \omega_b{}^2} & \dfrac{\nu_m}{\nu_m{}^2 + \omega_b{}^2} & 0 \\ 0 & 0 & \dfrac{1}{\nu_m} \end{bmatrix}$$

$$= \begin{bmatrix} s_\mathsf{T} & -s_\mathsf{H} & 0 \\ s_\mathsf{H} & s_\mathsf{T} & 0 \\ 0 & 0 & s_\parallel \end{bmatrix}, \tag{7.58}$$

which appears quite different from the analogous equation (Eq. 7.43) for a fully ionized plasma.

Equations 7.50 for the mobility can be made similar to Eq. 7.36 or 7.39

in the limit of no collisions. The transverse components approach zero as ν_m/B^2; the Hall components approach $1/B$. Here again, no current flows across lines of induction, but the plasma is transported at the velocity $\mathbf{E} \times \mathbf{B}/B^2$.

In order to illustrate the flow in case the collision frequency is not zero, assume for simplicity that the current arises principally from one charged species (for example, the light electrons). Here, an electric field E_T transverse to $\mathbf{B}$ produces a current j_T in the direction of $\mathbf{E}$, plus a Hall current j_H perpendicular both to $\mathbf{B}$ and $\mathbf{E}$. Figure 7.4 shows the currents

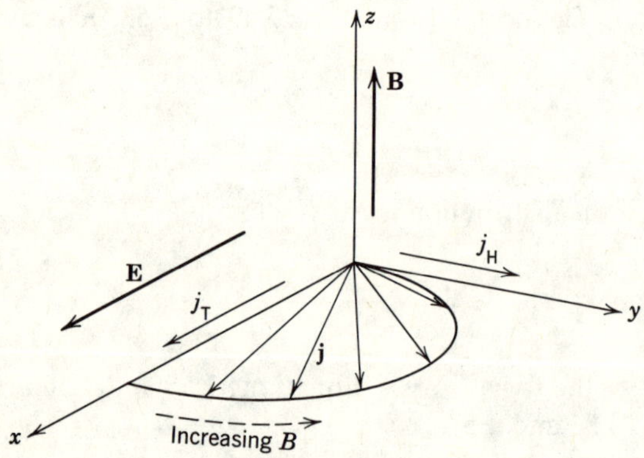

Fig. 7.4. Directions of transverse and Hall currents arising from an induction B_z and electric field E_x.

j_T and j_H flowing, for the case of an induction B_z and a transverse electric field E_x. The directions of j_T and j_H are shown by the detached arrows. As B_z is increased with other parameters kept constant, the locus of $\mathbf{j}$ moves around a semicircle in the x-y plane. Note that a negative sign of one component of the Hall conductivity does not indicate a negative conductivity in the usual sense of a scalar. A negative nondiagonal component of conductivity merely relates two perpendicular vectors; $\mathbf{j}_\mathrm{H}\cdot\mathbf{E} = 0$, and no work is involved.

In discussing the resistivity, we must invert the conductivity matrix. Again, if one charged species contributes principally to the current, Eq. 7.58 can be formally inverted to give

$$\eta = \mathfrak{s}^{-1} = \frac{m}{nq^2}\begin{bmatrix} \nu_m & \omega_b & 0 \\ -\omega_b & \nu_m & 0 \\ 0 & 0 & \nu_m \end{bmatrix}$$

$$= \begin{bmatrix} \eta_T & \eta_H & 0 \\ -\eta_H & \eta_T & 0 \\ 0 & 0 & \eta_\parallel \end{bmatrix}. \tag{7.59}$$

The diagonal components in Eq. 7.59 appear independent of the induction, but this conclusion is actually valid only if ν_m is independent of particle speed. If ν_m is a function of speed, we recall that the terms in Eq. 7.58 should be replaced by the integrals of Eqs. 7.51 to 7.53. The quantities are no longer so simple, and the induction appears in the transverse resistivity. It can be shown that for any speed dependence of ν_m the transverse resistivity $\eta_T \geq \eta_\parallel$. The ratio $(\eta_T - \eta_\parallel)/\eta_\parallel$ is called the magnetoresistance, and for Lorentzian systems in which $\nu_m = \text{const}$, the magnetoresistance is zero.

A difference between the fully ionized plasma and the Lorentzian gas is apparent if we consider the resistance derived from the conductivity of the fully ionized plasma, Eq. 7.43: the matrix cannot even be inverted. Physically, no transverse current j_T can be generated in the direction of a transverse field E_T. Thus we could have assigned an infinite transverse resistivity to a fully ionized plasma. Because of difficulties such as these, the concept of conductivity is a much more useful one than the concept of resistivity. It is much more realistic to apply an electric field (or a pressure gradient) and ask what currents flow than to presuppose a current and ask what mechanism produced it.

The dynamic conductivity for a Lorentzian gas can be derived by a simple extension of the results just obtained. The subject is of importance in the heating of a plasma by a rapidly oscillating electric field. Let the electric intensity depend on time as

$$\mathbf{E} = \mathbf{E}_0 \exp i\omega t. \tag{7.60}$$

The equation of motion for a plasma component is then given by inserting Eq. 7.60 into Eq. 7.46, with $\nabla p = 0$ for convenience. A solution of the inhomogeneous equation is

$$\mathbf{\Gamma} = \mathbf{\Gamma}_0 \exp i\omega t. \tag{7.61}$$

If this trial solution is substituted into Eq. 7.46, we find

$$m(\nu_m + i\omega - \omega_b \times)\mathbf{\Gamma} = nq\mathbf{E}. \tag{7.62}$$

From this relation, it is evident that the dynamic equation will be identical to the static equation, except that everywhere that ν_m appears it is replaced by $\nu_m + i\omega$. We utilize this correspondence to write a conductivity tensor for dynamic conductivity by inspection of the static conductivity (Eq. 7.58). The result is

$$[s_{\alpha\beta}] = \frac{nq^2}{m}\begin{bmatrix} \dfrac{\nu_m + i\omega}{(\nu_m + i\omega)^2 + \omega_b{}^2} & \dfrac{-\omega_b}{(\nu_m + i\omega)^2 + \omega_b{}^2} & 0 \\ \dfrac{\omega_b}{(\nu_m + i\omega)^2 + \omega_b{}^2} & \dfrac{\nu_m + i\omega}{(\nu_m + i\omega)^2 + \omega_b{}^2} & 0 \\ 0 & 0 & \dfrac{1}{\nu_m + i\omega} \end{bmatrix}. \tag{7.63}$$

The conductivity tensor is complex. The real part leads to dissipation; the imaginary part of the conductivity can be thought of as a real permittivity, modifying that of free space. Thus the propagational speed of electromagnetic waves will be affected by a plasma. We shall discuss such propagation in Chap. 9.

★ 7.8. Simple Shocks: the Hugoniot Relations

The simplest properties of shock waves[6] will be investigated in this section, since some of the thermonuclear devices proposed have been dependent on shocks for heating. Shocks may be used to achieve high temperatures and densities simultaneously. We shall discuss only one-dimensional shocks. The phenomena are somewhat different in two dimensions.

In a shock, the motions are so rapid that very little momentum or energy is transported across the flow lines. For this reason, the viscosity and heat conduction may be neglected if the structure of the shock itself is not of interest. A shock is rather thin—in the order of a few mean free paths. For example, at the front of a bullet, the thickness may be 10^{-4} cm. For a violent shock in air, the thickness is even less.

The development of a shock can be understood physically quite easily. The velocity of sound is given by

$$c = \sqrt{\gamma p/\rho_m} \tag{7.64}$$

and is related to the mean speed of the particles. Since in an adiabatic compression

$$p \propto \rho_m{}^{\gamma}, \tag{7.65}$$

we find that

$$c \propto (\gamma \rho_m{}^{\gamma-1})^{1/2}. \tag{7.66}$$

Because $\gamma > 1$, the speed of sound increases with density under adiabatic conditions; that is, the temperature and particle speed increase. Let there be a perturbation of the density, as shown in Fig. 7.5, corresponding

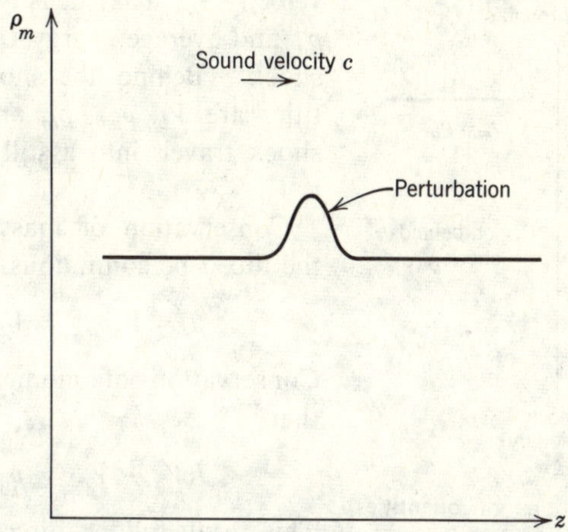

Fig. 7.5. Perturbation in the density of a gas leading to a shock.

to a pressure increase. The speed of sound is greater at the maximum of the density than at the plateaus on either side. Consequently, the wave front gets steeper and steeper as the maximum catches up with its own base.

The steepness of the shock front grows until the adiabatic conditions assumed above in the expression for the speed of sound fail. Eventually a significant fraction of the heat is conducted away, and the density changes become so rapid that viscous forces produce significant damping. These latter two processes fix the shape of the shock wave. It is clear that the entropy will increase through the shock because of viscosity and heat conduction, and the increase depends on the strength of the shock.

In a thermonuclear plasma, the magnetic induction usually present causes the particles to rotate. The finite size of these orbits limits the thinness of the shock front and will determine the structure. We shall not go into the complicated problem of determining the structure of a shock. Much useful information can be derived from the conservation laws; these will enable us to deduce connecting relations across the shock for the more important physical quantities.

In our approximation, there is a sharp discontinuity in various macroscopic parameters across a shock. Material flows across this discontinuity. Conservation equations facilitate the determination of the values of various quantities on each side of the shock. A shock wave is schematically represented in Fig. 7.6. We travel with the shock (the Lagrangian system) and observe a mass velocity V_a, mass density ρ_{ma}, pressure p_a, and average energy U_a ahead of the shock. Behind the shock, the quantities are V_b, ρ_{mb}, p_b, and U_b. If the shock travels into a still gas, its speed is V_a.

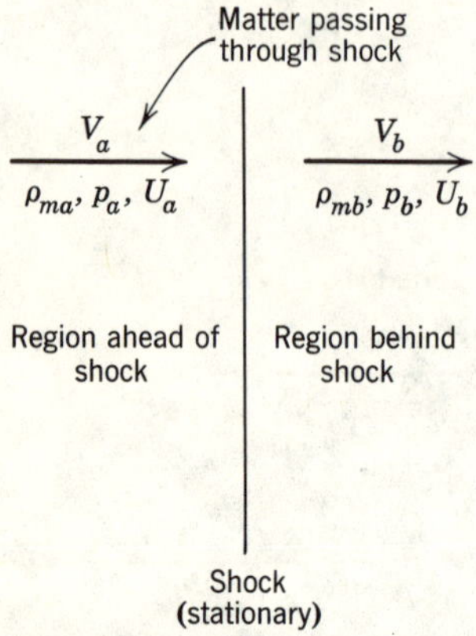

Fig. 7.6. Definitions of quantities associated with a shock wave. The shock is stationary in this moving reference frame.

Conservation of mass requires that the flux J be continuous, so that

$$J = V_a \rho_{ma} = V_b \rho_{mb}. \tag{7.67}$$

Conservation of momentum requires that

$$J(V_b - V_a) = p_a - p_b. \tag{7.68}$$

This result follows directly from Newton's second law, for the left-hand side of the equation is the change in momentum flow rate per unit surface area of the shock as a mass of gas passes through it; the right-hand side is the net force per unit area of the shock acting on this mass of gas.

Conservation of energy states that

$$J\left(\frac{V_b^2}{2} + U_b\right) - J\left(\frac{V_a^2}{2} + U_a\right) = V_a p_a - V_b p_b. \tag{7.69}$$

In words, the rate of change of energy equals the work done on the fluid per unit time per unit area. The kinetic energy is $JV^2/2$, and the internal energy is JU.

Equations 7.67 through 7.69 are four equations relating nine quantities. We may eliminate J, V_a, and V_b and find

$$\frac{p_b + p_a}{2} = \frac{U_b - U_a}{\rho_a^{-1} - \rho_b^{-1}}. \tag{7.70}$$

These equations are known as the Hugoniot relations.

★ 7.9. Application of the Hugoniot Relations

The equations developed in the previous section have a number of important applications. We shall illustrate the use of these relations with several applications to an ideal gas. For such a gas,

$$\mathscr{C}_p - \mathscr{C}_{\mathscr{V}} = Ak, \tag{7.71}$$

where $\mathscr{C}_p$ and $\mathscr{C}_{\mathscr{V}}$ are the specific heats of the gas at constant pressure and volume, respectively, and A is Avogadro's number. Because

$$U = \mathscr{C}_{\mathscr{V}} T, \tag{7.72}$$

we find that

$$U = \frac{p}{\rho_m(\gamma - 1)}. \tag{7.73}$$

We now define the ratio of mass densities behind and ahead of the shock,

$$\Theta = \rho_{mb}/\rho_{ma}, \tag{7.74}$$

and the pressure ratio,

$$\xi = p_b/p_a. \tag{7.75}$$

In terms of these quantities, Eq. 7.70 may be re-expressed as

$$\Theta = \frac{(\gamma - 1) + (\gamma + 1)\xi}{(\gamma + 1) + (\gamma - 1)\xi}. \tag{7.76}$$

This expression for Θ agrees through terms to order $(\xi - 1)^2$ with the expression for Θ in the absence of a shock and any change in entropy. For moderate shocks, the change of entropy obviously cannot be great.

For intense shocks, ξ is large, and the density ratio is

$$\Theta \approx \frac{\gamma + 1}{\gamma - 1}. \tag{7.77}$$

A graph of Θ as a function of γ is shown in Fig. 7.7 for strong shocks. There is a very definite limit of the density ratio, regardless of the pressure ratio, as the intensity of the shock increases. As the pressure ratio increases, so do the entropy and the associated heating. The contrast with reversible compressions is to be noted, since in such compressions the density ratio may be anything.

The entropy change can be readily computed from the expression for the entropy of an ideal gas. This expression is[7]

$$\Delta\mathscr{S} = \mathscr{C}_{\mathscr{V}} \ln (p\rho_m^{-\gamma}) + \mathscr{S}_0. \tag{7.78}$$

The change of entropy with the passing of a shock is

$$\begin{aligned} \Delta\mathscr{S} &= \mathscr{C}_{\mathscr{V}} \ln (p_b\rho_{ma}^{\gamma}/p_a\rho_{mb}^{\gamma}) \\ &= \gamma\mathscr{C}_{\mathscr{V}} \ln (\xi^{1/\gamma}/\Theta). \end{aligned} \tag{7.79}$$

Entropy increases if the fluid flows from a region of low density to a region of high density. The fluid flows from a region of low pressure to a region of high pressure.

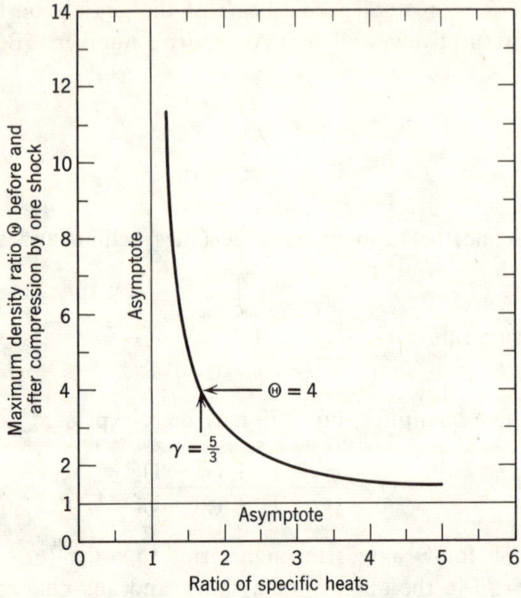

Fig. 7.7. Density ratio as a function of the specific heat ratio for strong shocks.

The entropy must always increase or remain constant. Therefore, if matter flows from a region of low density to a region of higher density, so that Θ exceeds 1, a shock is possible. On the other hand, if matter flows from a region of high density to a region of lower density, no shock is possible; otherwise the entropy would decrease. A fluid passing through a shock does so irreversibly and becomes hotter than it would be in the absence of any shock.

PROBLEMS

1. Prove that no magnetic surface as defined in Sec. 7.2 can be simply connected, that is, a topological egg. Hint: Consider Ampère's law.

2. In Figs. 7.1 or 7.2, prove that both the total magnetic flux Φ and the total current I through the following surfaces are constants:

(*a*) Any surface defined by a closed curve on $p = C$ which encircles the magnetic axis, that is, the cut shown in Fig. 7.2.

(*b*) Any surface defined by a closed curve on $p = C$ which encircles the hole in the torus and the magnetic axis.

3. It has been proposed that "force-free" coils be built, in which the conductors are so arranged that the magnetic force on them is zero. From the equations of Sec. 7.1, find a simple proof that the coils cannot be force-free everywhere, and that a force-free plasma confined by its own currents cannot be created. A more elegant proof, based on the electromagnetic stress tensor, is given by E. N. Parker, *Phys. Rev.*, **109**, 1440 (1958).

4. Calculate the θ-current in amp/m in a solenoid required to produce a magnetic pressure of 100 atm inside.

5. A plasma is supposed to be confined in the region of the magnetic axis of a simple torus. Show that the plasma is not in equilibrium there and experiences a net force toward the periphery.

6. For the configuration of Problem 6, Chap. 6, suppose that the plasma consists of protons and electrons. Calculate the stored energy per meter of axial length in such a plasma. From this stored energy, calculate the effective dielectric constant. The device is sometimes called a hydromagnetic capacitor.

7. We develop simple scaling laws for a very long solenoid that confines a thermonuclear plasma. All radial dimensions r will be scaled by the same factor. The current density j in the coils limits the design, and the magnetic pressure limits the particle density.

(*a*) Show that the induction scales as rj.
(*b*) Show that the total magnetic energy U stored per unit length scales as r^4j^2.
(*c*) Find the scaling law for the magnet power W_M in terms of the resistivity η of the coils, j and r.
(*d*) For a fixed operating temperature, show that the thermonuclear power W_T/unit length scales as r^6j^4.
(*e*) Assume that the capital investment C is proportional to the volume of the magnet. Find the scaling law for W_T/C.
(*f*) If, as is probably not the case, the confinement time τ is limited by diffusion of the plasma to the walls, show that $\tau \propto r^2$.

8. Consider a plasma contained by a magnetic induction of 2.0 webers/m^2.

(*a*) Find the pressure created if the field lines are straight.
(*b*) Find the corresponding density of particles if the temperature is 10 kev, and $\beta = 1$.
(*c*) Find the power density in a D-T reactor.

9. The energy balance in a thermonuclear D-T plasma is to be considered. The plasma is confined in a very long solenoid. The copper coils producing the induction have inner radius 2.0 m and outer radius 3.0 m. Plasma temperature $T = 30$ kev, and $n_D = 10^{20}/\text{m}^3$. The resistivity η of copper at its operating temperature of 300° C is 1.7×10^{-8} ohm-m. Insulation and cooling channels occupy 25% of the coil volume. The quantity $\beta = 0.1$ throughout the plasma, which fills all but a negligible annulus inside the coils, and nkT is independent of position.

Calculate:

(*a*) The axial induction which confines the plasma.
(*b*) The induction just inside the coils.
(*c*) The power dissipated in the coils per meter of length with constant current density.
(*d*) The thermonuclear power produced per meter of length.

10. In a long solenoid, coils lie between an inner radius r_1 and an outer radius r_2. Find the radial dependence of the current density $j_\theta(r)$ if the total resistive loss is to be a minimum for fixed B_z inside, and fixed r_1 and r_2.

11. Prove that under appropriate simplifying assumptions, Eq. 7.18 reduces to the pressure balance equation (Eq. 7.13).

12. Assume that the degrees of freedom of a gas do not couple. The gas is initially at a temperature T_1 and is to be compressed as follows: first, an axial compression from volume $\mathscr{V}_1$ to volume $\mathscr{V}_2$; second, a radial compression from $\mathscr{V}_2$ to $\mathscr{V}_3$, where $(\mathscr{V}_2/\mathscr{V}_3) = (\mathscr{V}_1/\mathscr{V}_2)^2$. In this way, the compression in each direction is the same.

(*a*) Calculate the final temperature T_3, and show that it is greater than $(\mathscr{V}_1/\mathscr{V}_3)^{2/3}$.

(*b*) Explain, using the kinetic stress tensor, why the temperature is high, and that the work done in compressing the gas equals the increase in energy of the gas.

13. A plasma is stably confined in a long solenoid by an axial field B_z. The gas is to be heated by magnetic pumping, using collisional effects. To this end, the induction B_z is pulsed as shown. Time t_1 is very short compared with the collision time but is long enough so that the plasma response is adiabatic. Time t_2 is very long compared with the collision time. For a given average energy U_0 at time t_0, calculate the energy 1 cycle later, in terms of U_0 and the parameters defined in the figure.

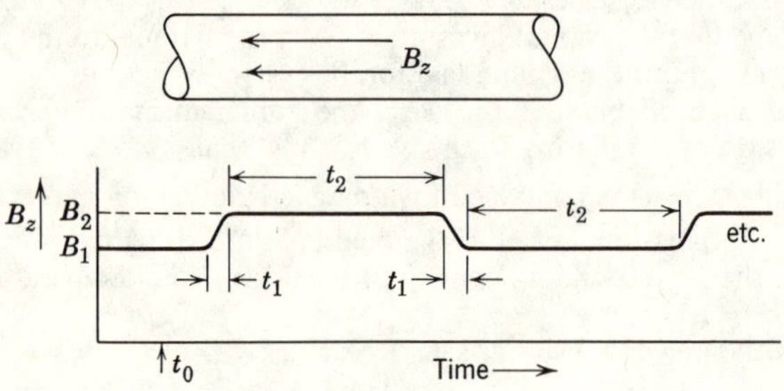

14. A deuterium plasma at a temperature of 10 ev lies inside a solenoid with diameter 0.1 m. The solenoid is a heavy, one-turn coil and is to be pulsed from a condenser bank charged to 50 kv. Energy stored in the bank is 10^6 joules, and the inductance of the system is 10^{-8} henry. Estimate whether the electrons or ions are heated adiabatically.

15. Prove Eq. 7.56 for a Maxwellian gas.

16. Derive the resistivity tensor, Eq. 7.59, for a Lorentzian gas from the conductivity tensor.

17. In the absence of collisions and a magnetic induction, derive the ac resistivity of a Lorentzian plasma directly from the equation of motion of the charges.

18. Using the results of Sec. 7.7 for diffusion and mobility, write expressions for the currents for the case of ambipolar diffusion of a cylindrical plasma in an axial induction B_z. Show that a current j_θ flows, which cancels part of the applied induction. See B. Lehnert, *Proc. 2nd U.N. Conf. on Peaceful Uses of Atomic Energy*, Vol. 32, United Nations, Geneva (1958), p. 349.

19. Why may Eq. 7.69 be considered in the moving frame? Is there any difference in this equation if considered in a stationary frame?

20. Prove from Eq. 7.79 and the Hugoniot relations that the entropy change $\Delta\mathscr{S} \geqslant 0$ in a shock.

21. A shock wave passes through a quiescent helium gas ($\gamma = \frac{5}{3}$), whose pressure and temperature ahead of the shock are 10 mm Hg and 500° C. The temperature behind the shock is 3000° C. Calculate its speed and the pressure behind it.

REFERENCES

1. E. S. Weibel, *Phys. Fluids*, **2**, 52 (1959).
2. A. Simon, *An Introduction to Thermonuclear Research*, Pergamon Press, New York (1959), pp. 173 ff.
3. E. P. Butt, R. Carruthers, J. T. D. Mitchell, R. S. Pease, P. C. Thonemann, M. A. Bird, J. Blears, and E. R. Hartell, *Proc. 2nd U.N. Conf. on Peaceful Uses of Atomic Energy*, Vol. 32, United Nations, Geneva (1958), pp. 42–64.
4. W. P. Allis and S. J. Buchsbaum, "The Conductivity of an Ionized Gas in a Magnetic Field," in A. B. Cambel and J. B. Fenn (Eds.), *Dynamics of Conducting Gases*, Northwestern University Press, Evanston, Ill. (1960), pp. 3–14.
5. W. P. Allis, "Motions of Ions and Electrons," *Handbuch der Physik*, Vol. 21, Springer, Berlin (1956), pp. 383–444.
6. H. A. Bethe, K. Fuchs, J. von Neumann, R. Peierls, and W. G. Penney, *Shock Hydrodynamics and Blast Waves*, AECD-2860, U.S. Atomic Energy Commission, Technical Information Service, Oak Ridge (1944).
7. P. S. Epstein, *Thermodynamics*, John Wiley & Sons, New York (1937), p. 63.

GENERAL REFERENCE

L. Spitzer, Jr., *Physics of Fully Ionized Gases*, Interscience Publishers, Inc., New York (1956).

8

Coulomb interactions and their consequences

Until now we have mentioned Coulomb scattering without either calculating it or deducing many of its consequences. We have mentioned the conductivity of a plasma but have discussed nothing of its actual behavior or magnitude. These points will be discussed in this chapter, together with their implications.

In the calculation of Coulomb scattering we encounter serious difficulty, in that the Coulomb cross section for an isolated charge by another charge is infinite.† In practice, a charged scattering center is always surrounded by charges of opposite sign that shield the scattered particle from the scatterer at sufficiently great distances. As a result, the actual cross section is finite, as it must be. It behooves us, however, to investigate the distance in which an individual charge may be considered to be shielded by other charges and to determine the approximate potential in such a shielded charge distribution. This matter is taken up first in this chapter and leads to the Debye length. The Coulomb cross section can then be approximately computed.

With the Coulomb cross section in hand we may compute the time required to scatter particles through substantial angles. This time is a measure of the containment time in a certain class of thermonuclear machines. Unless it is sufficiently long, the plasma will be lost before a

† We neglect radiation effects, which would make the cross section finite without the screening to be discussed here.

substantial fraction has reacted. We shall conclude from such calculations that the plasma temperature must be very high.

From the Coulomb cross sections we can also compute the rate of transfer of energy between ions and electrons. This matter is of importance in determining the radiation losses from a thermonuclear machine, as we shall see in Chap. 11. The electron temperature is determined in part by the condition that the rate of transfer of energy from the ions to the electrons equal the rate of energy loss by radiation of the electrons. The consequences of this energy balance will be examined in Chap. 11, and further in Chap. 13.

Finally we shall compute the resistivity and conductivity, and show that the resistivity is very low at the operating temperatures required. As a result, the confining magnetic induction will require a relatively long time to diffuse into the plasma. The high conductivity also prevents electric fields from penetrating the plasma and greatly aggravates the heating problem. For the most part, the specific methods of heating will be discussed in connection with particular devices in Chaps. 14 to 16.

Next we shall develop the concept of diffusion and attraction in velocity space and use it to show that a constant electric field applied to a plasma will cause some electrons to be heated indefinitely. The energy absorbed by these electrons prevents the main body of electrons from being heated. This method of heating is therefore restricted to relatively low temperatures. The preferential heating of the electrons has other very serious effects, which will be discussed in Chap. 11.

8.1. Screening of Electrostatic Fields

We have seen that a plasma may exclude electrostatic fields, as well as magnetic fields. The electrostatic shielding led us to discover the Debye length of a plasma; we wish to enlarge upon the concept here.

We now compute the small variations of electron density in an infinite plasma, which will, of course, be neutral to a very high order of approximation. In the interest of simplicity, the following assumptions are made: (1) the electrons are in a Maxwell-Boltzmann distribution at a temperature T_e; (2) $|q_e\phi| \ll kT_e$; (3) positive ions are uniformly distributed throughout the plasma with the average density $|n_{e0}q_e/q_i|$, where n_{e0} is the average density of the electrons; and (4) the microscopic variations of potential arising from the discrete nature of electrons surrounding the ions may be neglected.

The electron density varies from one point to another according to the Maxwell-Boltzmann distribution:

$$n_e = n_{e0} \exp(-q_e\phi/kT_e), \tag{8.1}$$

or

$$n_e \approx n_{e0}[1 - (q_e\phi/kT_e)]. \tag{8.2}$$

The mean potential is taken as zero. The electrons tend to congregate in regions of high potential.

Poisson's equation must also be satisfied, so that

$$-\epsilon_0\nabla^2\phi = q_e(n_e - n_{e0}), \tag{8.3}$$

since $n_e - n_{e0}$ is the difference in density that leads to a net charge. If Eq. 8.2 is substituted into Eq. 8.3, we find that

$$\nabla^2\phi - \phi/\lambda_e^2 = 0, \tag{8.4}$$

where

$$\lambda_e = (\epsilon_0 kT_e/n_{e0}q_e^2)^{1/2}, \tag{8.5}$$

or

$$\lambda_e = 7.45 \times 10^3(kT_e/en_{e0})^{1/2} \text{ m}. \tag{8.6}$$

The quantity λ_e is called the Debye length of electrons. This quantity was derived approximately in Sec. 4.8.

If a disturbance is localized to one region of a plasma, the potential ϕ must approach zero at infinity. The spherically symmetric solution of Eq. 8.4 is

$$\phi = (C/r)\exp(-r/\lambda_e). \tag{8.7}$$

Thus, electrostatic disturbances of the plasma are shielded by electrons in a distance of the order of λ_e. We note that λ_e is a function only of the parameters of electrons, and so we suspect that Eq. 8.5 may have a greater range of validity than the derivation given above.

The constant C in Eq. 8.7 may be evaluated from the observation that there is an excess of electrons in the vicinity of an ion in a plasma. The total extra charge of electrons in the vicinity must be $-q_i$ in order to neutralize the charge on the ion. The solution of Eq. 8.4 that goes as $q_i/4\pi\epsilon_0 r$ at small r and that approaches zero as r approaches infinity is required, so that

$$\phi = (q_i/4\pi\epsilon_0 r)\exp(-r/\lambda_e). \tag{8.8}$$

The total charge within an infinite sphere not including the origin itself is just $-q_i$. Equation 8.8 represents a screened Coulomb potential that extends to a distance of approximately $r = \lambda_e$. It is this screening that keeps the Coulomb scattering cross section from being infinite.

The approximate derivation of the Debye length given in Sec. 4.8 may now be recast. Poisson's equation gives a potential

$$\phi = (nq/2\epsilon_0)L^2 \tag{8.9}$$

for a slab of thickness $2L$, particle density n, and charge q per particle. The origin of coordinates is at the middle of the slab. If the potential energy of a charge in the electric field is about equal to the average particle energy attributable to motion perpendicular to the slab, then the electrical forces will cause an appreciable alteration of the energy distribution. The Debye length is the half-thickness for which this condition is obtained, or

$$\lambda^2 nq^2/2\epsilon_0 = q\phi = kT/2. \tag{8.10}$$

An assembly of electrons exhibits little self-interaction if its linear dimensions are much smaller than the Debye length, and behaves like a true plasma if the linear dimensions are large compared with the Debye length. Beams of lateral dimensions greater than the Debye length develop sheaths.

Figure 8.1 displays contours of equal Debye length for various densities

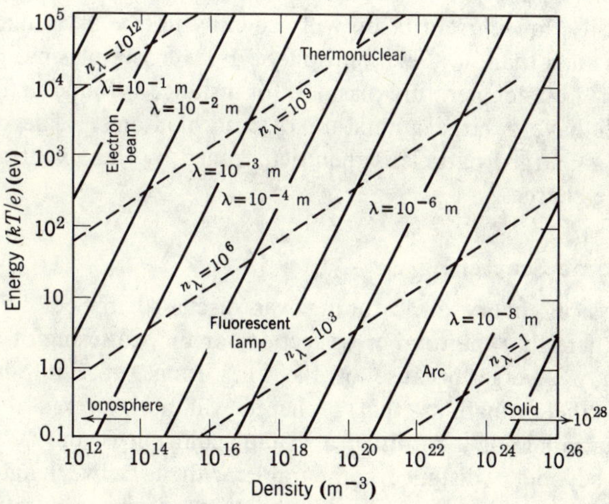

Fig. 8.1. Debye length λ (meters) and number per Debye sphere for singly charged particles. Several typical regions are marked.

n and temperatures T for singly ionized particles. If the particles are ionized Z times, then λ must be divided by Z. The number n_λ of charges contained within a sphere the radius of which equals the Debye length is shown as dotted lines, and is

$$n_\lambda = \frac{4\pi}{3}\frac{(\epsilon_0 kT/q^2)^{3/2}}{n^{1/2}}. \tag{8.11}$$

The present derivation has been given for electrons, and we see from the form of the Debye length λ_e that it depends only upon their density and temperature. The properties of the ions do not enter. Thus we may also define in the same way a Debye length λ_i for the ions. In an isothermal neutral plasma, $\lambda_e = \lambda_i$. Any given charge interacts simultaneously with approximately its n_λ nearest neighbors. In such Coulomb collisions, a charge moves in the fluctuating potential of its many neighbors.

For plasmas of interest to us, $n_\lambda \gg 1$; that is, the Debye length is much greater than the interparticle distance. Furthermore, the Debye length is much less than the minimum dimension of the ionized medium. If this condition were not satisfied, the electrostatic interaction of the charges would be negligible. The medium should then be viewed as a collection of free charges, rather than as a plasma.

Several characteristic charge collections have been marked in Fig. 8.1, according to their electron density and energy. At high density, we note that in solids $n_\lambda < 1$, and quantum statistics is necessary as well. A low-density, low-current beam will generally not be a plasma, because its radius is less than λ. The ionospheric F-layer is a plasma, because it is so large. Low-temperature plasmas, for instance as found in fluorescent lamps, will have a large admixture of neutral atoms. The collisional history of a charged particle is then determined by these neutrals, rather than by the charges.

8.2. Coulomb Scattering

The classical theory of scattering was discussed in Sec. 2.5. It was indicated there that the total cross section for the scattering of a charged particle may diverge because of the long range of Coulomb forces. Physically, the probability that a charge will be scattered through an infinitesimal angle as a result of a distant collision is unity. However, collisions beyond a distance comparable with a Debye length will be shielded from the scattering center, so that an upper cutoff distance exists. Therefore, the total Coulomb cross section in practice is finite.

An exact treatment of the shielded Coulomb scattering does not exist, essentially because the interaction of many bodies is involved. We shall use very approximate derivations of the scattering in this section and obtain answers that differ by a factor close to 1 from more accurate ones. The more accurate though still approximate answers will be quoted.

For a two-body Coulomb collision, the scattering angle χ may be obtained from Eq. 2.33. The result can be expressed as

$$\cot(\chi/2) = 4\pi\epsilon_0 m_r b v_r^2/q_1 q_2 \tag{8.12}$$

in the notation of Sec. 2.5. From Eq. 2.35, which defines $\sigma(\chi)$, we find that

$$\sigma(\chi) = \frac{(q_1 q_2)^2}{[8\pi\epsilon_0 m_r v_r^2 \sin^2(\chi/2)]^2}. \tag{8.13}$$

Equation 8.13 is the Rutherford scattering relation.

The total cross section for momentum transfer produced by scattering is defined by Eq. 2.37. In our present case, we find

$$\sigma_m = \frac{1}{4\pi}\int_{\chi=\delta}^{\pi} d\left(\sin\frac{\chi}{2}\right)\left(\frac{q_1 q_2}{\epsilon_0 m_r}\right)^2 \frac{1}{v_r^4 \sin(\chi/2)}. \tag{8.14}$$

The difficulty is now apparent if we let $\delta \to 0$, for the integral diverges logarithmically.

Our resolution of this difficulty will be direct but naive: we shall limit the scattering to particles within a Debye length of the test charge and hence find a nonzero lower limit for δ. We shall discuss the validity of the approximation after writing the answer. To this end, we set the impact parameter $b = \lambda_e$ in Eq. 8.12 and thus define the minimum scattering angle $\chi_{\min}$. Since $\chi_{\min}$ is very small, we find very closely

$$\delta = (\sin\chi/2)_{\min} = \frac{q_1 q_2}{4\pi\epsilon_0 m_r v_r^2}\left(\frac{q_e^2 n_e}{\epsilon_0 k T_e}\right)^{1/2}. \tag{8.15}$$

The Coulomb scattering cross section is then

$$\sigma_m = \frac{(q_1 q_2)^2}{4\pi\epsilon_0^2 (m_r v_r^2)^2} \ln\left[\frac{4\pi\epsilon_0 m_r v_r^2}{q_1 q_2}\left(\frac{\epsilon_0 k T_e}{n_e q_e^2}\right)^{1/2}\right]. \tag{8.16}$$

We find that the quantity appearing in square brackets in Eq. 8.16 is greater than 10^6 in all our thermonuclear applications. Approximations for its magnitude therefore will not affect σ_m appreciably. The factor $m_r v_r^2$ in the logarithm is replaced by $3kT$, from Eq. 4.32, and all temperatures in the argument of the logarithm are set equal to T_e. In addition, we set $q_1 = q_2 = e$, and write finally

$$\sigma_m = \frac{(q_1 q_2)^2 \ln\Lambda}{4\pi\epsilon_0^2 (m_r v_r^2)^2}, \tag{8.17}$$

where

$$\Lambda = \frac{12\pi(\epsilon_0 k T/e^2)^{3/2}}{n_e^{1/2}}. \tag{8.18}$$

We note that the quantity Λ is just nine times the number n_λ of electrons in a Debye sphere, as given by Eq. 8.11. Thus, from Fig. 8.1, we find further evidence that Λ is a large quantity. In our illustrations, we shall use $\ln\Lambda \approx 20$.

Having derived Eq. 8.17, we wish to discuss briefly the physical nature of the scattering process. The final approximations made in expressing the quantity Λ are trivial, and we shall not dwell on them.

In view of Debye shielding, our entire derivation of Eq. 8.17 suffers from an initial flaw. We should have calculated the scattering in a screened Coulomb potential, rather than in an unscreened one with a sharp cutoff. The equations would have become very complicated, but the logarithmic divergence would not have appeared. That calculation has been performed,[1] and the result obtained therefrom differs negligibly from Eq. 8.17, where we cut off the scattering sharply at a Debye distance. We might from this correspondence gain some reassurance about the validity of the result, but the real difficulty is even more basic. It stems from our insistence upon defining a collision cross section at all. The necessity of our doing so arose in Sec. 4.2, where the Boltzmann equation was first derived: an effective two-body collision term is required, and we have manufactured one. The inadequacy of the Boltzmann formulation in this respect was first discussed in the paragraph following Eq. 4.19. In actual fact, isolated Coulomb collisions in a plasma cannot physically be distinguished. With the exception of rare very close encounters, a charge moves in the fluctuating potential of its many neighbors. The physically meaningful quantity is the average scattering of a charge; a cross section is a derived concept and should properly be used only when two-body interactions are involved. Analyses much more sophisticated than any we shall discuss[2] take this point of view and deal with the plasma as a very generalized ensemble. The physical quantities that appear naturally in such analyses are correlations in space between the charges, and the rates of change of momentum and energy. The space correlation leads naturally to phenomena that we may interpret as Debye shielding. The momentum rate of change $m\dot{v}$ may be interpreted *a posteriori* in terms of a cross section defined as $(m\dot{v})/(mv)(nv)$, but the interpretation is purely *ad hoc*.

In spite of these considerations, our approximate derivation yields answers close to those obtained by more sophisticated means. This fact is not overly surprising. In principle, we should have calculated the effect of the motion of the test charge on the motions of the other charges. However, this effect is negligible. This fact may be seen by noting that the deflection of the test charge at nearly all points in a Debye sphere is minute and by observing that the behavior of the test charge is representative of that of the other charges in a Debye sphere. Only in very close collisions with other electrons would an electron be substantially deflected and the mutual effects of each charge on the other have to be considered. Such collisions are extremely rare and can be neglected, as has been done.

We may get a feeling for the magnitudes of these minute scattering angles by computing the angle of scattering of a particle that has an impact parameter of a Debye length. If the plasma temperature is 10 kev and if the density of electrons is $10^{20}/\text{m}^3$, then the angle through which a singly charged ion of average energy is scattered is only 7.4×10^{-8} degree. Thus, even at 1 Debye length away the potential of the scattering ion is very small. For this reason, the fluid model is a good approximation at distances larger than a Debye length.

8.3. Relaxation Times

The characteristic time for some property of a charge distribution to change substantially, called the relaxation time, is of importance in several connections, such as the computation of the loss rate from a device.

As a result of Coulomb collisions, a particle may be deflected from its original trajectory. We have seen from Eq. 8.13 and the previous discussion that the probability of a large-angle collision is very low. Because of the very slow manner in which the Coulomb interaction decreases with distance, the small-angle collisions are very numerous. The probability of a sequence of small-angle collisions that will result in a substantial deflection is higher than the probability of a deflection through the same angle caused by one large-angle collision. We are, therefore, encouraged to examine the effect of a large number of small-angle collisions.

In a small-angle collision, a particle will lose very little energy. Therefore, as in Fig. 8.2, a sequence of small-angle collisions may be pictured as a sequence of trajectories randomly directed and lying approximately on a sphere in velocity space. As a result of the collisions through small angles, there is a kind of diffusion of the particles in velocity space. In all diffusion processes, the mean-square displacement is proportional to the elapsed time. The displacement is here an angular one, and we let $1/\tau_\theta(v)$ be the proportionality factor. Thus

$$\overline{\theta^2} = t/\tau_\theta(v). \tag{8.19}$$

The relaxation time, as τ_θ is called, must be in the order of the Coulomb collision time. Consequently, to lowest order of approximation,

$$\tau_\theta(v) = (n\sigma_m v_r)^{-1} = \frac{4\pi\epsilon_0{}^2 m_r{}^2 v_r{}^3}{(q_1 q_2)^2 n \ln \Lambda}. \tag{8.20}$$

Here, n is the density of scattering centers. If the test charge is being

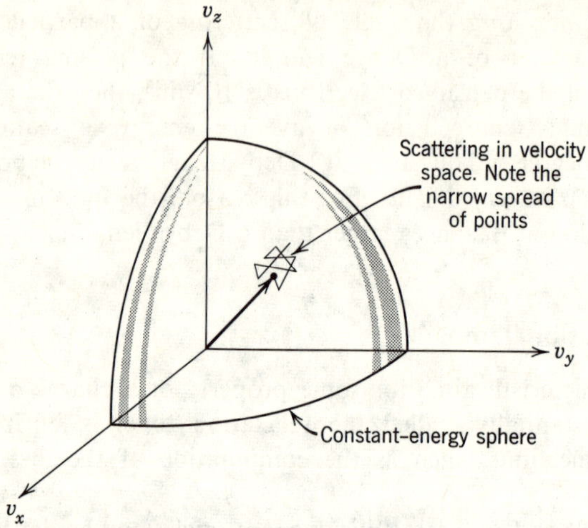

Fig. 8.2. Scattering in velocity space.

scattered by more than one species, the appropriate frequencies τ^{-1} must be summed.

The expression must be averaged over the possible values of speed. We shall not be overly concerned about exact averaging, because the derivation is already approximate. But we shall compute the average in two cases, in order to make comparisons. In the first case, we consider the time $\tau_{\theta ee}$ or $\tau_{\theta ii}$ for scattering of an electron or ion by its own species. In this case, $m_r = m/2$, and v_r^3 must be computed for all possible pairs. The development of Sec. 4.9 applies here. We may formally replace σ by v_r^2 in the notation of that section, change variables to $U = mv_r^2/2$, and apply Eq. 4.108. Because all particles are alike, we divide the integral by 2, for the reasons stated in the sentences following Eq. 4.108. The answer is

$$v_r^3 = \frac{16}{\sqrt{\pi}}\left(\frac{kT}{m}\right)^{3/2}, \tag{8.21}$$

and from Eq. 8.20

$$\left.\begin{matrix}\tau_{\theta ee}\\ \\ \tau_{\theta ii}\end{matrix}\right\} = \frac{16\sqrt{\pi}\epsilon_0^2 m^{1/2}(kT)^{3/2}}{q^4 n \ln \Lambda}. \tag{8.22}$$

Subscripts for electrons or ions must be inserted on the right side of Eq. 8.22. A more accurate result given by Spitzer[3] is

$$\left.\begin{matrix}\tau_{\theta ee}\\ \\ \tau_{\theta ii}\end{matrix}\right\} = \frac{25.8\sqrt{\pi}\epsilon_0^2 m^{1/2}(kT)^{3/2}}{q^4 n \ln \Lambda}, \tag{8.23}$$

or 1.61 times our approximate time. Note that $\tau_{\theta ii}/\tau_{\theta ee} = (m_i/m_e)^{1/2}$ for singly charged particles at the same temperature. This ratio is merely the inverse ratio of speeds: the faster particles thermalize more quickly. Electrons will relax in angle about 61 times as fast as deuterons at the same energy. For deuterons, Eq. 8.23 reduces to

$$\tau_\theta = 6.4 \times 10^{17} T^{3/2}/n_{\mathrm{D}} \ln \Lambda, \tag{8.24}$$

where $T \equiv kT/e$ is measured in kev, and n_{D} in m^{-3}.

It will appear that a certain class of confinement devices (mirrors) can contain particles only if the particles move at a sufficiently large angle with respect to the axis of the magnetic induction. The time τ_θ is a measure of the time that particles can be contained in such a device, for in that time the particles will have been so deflected that $\overline{\theta^2} \approx 1$. At low energies, the Coulomb cross section is large, the time τ_θ is short, and the charges will be quickly lost. For example, if the charge density is $10^{20}/\mathrm{m}^3$ and if $T = 10$ ev, $\tau_\theta \approx 0.3$ μsec. On the other hand, if $T = 300$ kev, then $\tau_\theta \approx 1.7$ sec for the same density. This time is very reasonable. In these illustrations, we have taken $\ln \Lambda = 20$ for simplicity.

The second case of much interest is the angular scattering time $\tau_{\theta ei}$ for scattering of electrons by the ions. In this case, $m_r \approx m_e$; in addition, $v_r \approx v_e$ unless the ion energy greatly exceeds the electron energy; with these approximations, Eq. 8.20 gives

$$\tau_{\theta ei} = \frac{32\sqrt{2\pi}\epsilon_0{}^2 m_e{}^{1/2}(kT_e)^{3/2}}{(Ze^2)^2 n_i \ln \Lambda}, \tag{8.25}$$

where Ze is the ionic charge. This time is $2\sqrt{2}$ times as long as $\tau_{\theta ee}$, as Eq. 8.22 for electron-electron relaxation shows. We therefore conclude that the electrons relax in angle almost as fast by colliding with singly charged ions as they do by colliding among themselves. If $Z \geqslant 2$, $\tau_{\theta ei} < \tau_{\theta ee}$.

On the other hand, the time $\tau_{\theta ie}$ for ions to be scattered by electrons is very long; we could check that $\tau_{\theta ie} \approx (m_i/m_e)\tau_{\theta ei}$. Thus by all angular scattering processes, the electrons relax in angle much faster than the ions.

In those devices in which particles are lost if they move at too small an angle with respect to the magnetic axis, the electrons tend to pull the ions out, and the ions tend to restrain the electrons from leaving, by virtue of the space-charge electric field that is built up. This field causes rather complicated drifts.

Having disposed of the angular scattering, we turn to the problem of relaxation in energy, which we may characterize by a time τ_u. We can obtain at once a fairly good estimate of the times $\tau_{uee}(v)$ or $\tau_{uii}(v)$: for like-particle scattering, the time must be comparable with the corresponding angular relaxation time τ_θ. The time τ_{uee} or τ_{uii} for an entire Maxwellian distribution has no meaning. But such times are a rough measure of the time in which a non-Maxwellian distribution approaches a Maxwellian one.

If the electrons and ions are at different temperatures, they will interchange energy; we now address ourselves to the calculation of the rate of transfer of energy between electrons and ions. The approximate method used here is similar to that used in Sec. 8.2 on Coulomb scattering. For the purposes of calculation, we assume that the electrons are more energetic than the ions. After our derivation, we shall display an equation that applies even if this restriction is removed. The electrons in passing an ion give it a number of random impulses. On the average, the change $\delta\mathfrak{p}$ of momentum $\mathfrak{p}$ in any particular direction will be zero if the velocity distribution of the electrons is isotropic, as is assumed. Since the collisions may be assumed independent, isotropy of the velocity distribution also implies that $\delta\mathfrak{p}_\alpha\delta\mathfrak{p}_\beta$ is zero if $\alpha \neq \beta$. The average $\overline{(\delta\mathfrak{p}_\alpha)^2}$ will not, however, be zero, since for each collision $(\delta\mathfrak{p}_\alpha)^2$ is positive. For this reason, the effect of the collisions may be characterized by this quantity. It will now be estimated.

In any one collision of a particle of charge q_e and speed v with one of charge q_i,

$$\begin{aligned}\delta\mathfrak{p} &\approx (b/v)q_eE,\\ &\approx q_eq_i/4\pi\epsilon_0 bv,\end{aligned} \tag{8.26}$$

where b is the impact parameter.

We must now compute the number of particles with impact parameter between b and $b + db$. To this end, we derive the number of electrons that enter a sphere of radius b centered on the ion per unit time. If the electrons of speed v are isotropically distributed, then $n_ev/4$ per unit time cross a unit surface of the sphere headed inward. The total number entering per unit time is then

$$(n_ev/4)4\pi b^2 = \pi b^2 n_e v. \tag{8.27}$$

Thus, the number of electrons entering a spherical shell of thickness db per unit time is

$$2\pi b n_e v\, db. \tag{8.28}$$

The change of $(\delta\mathfrak{p})^2$ in time δt is given by

$$\sum (\delta\mathfrak{p})^2 = (n_ev)(2\pi b\, db)(q_eq_i/4\pi\epsilon_0 bv)^2\, \delta t \tag{8.29}$$

summed over all particles between b and $b + db$ of speed v.

This expression must be integrated over all possible impact parameters to find the total deflection of the particle. As a result of the many collisions, a particle will be headed in a different direction from that in which it was originally going. Further, it will have a different momentum. From an over-all viewpoint, we define $\Delta(\mathfrak{p}^2)$ as the change of the square of the momentum as a result of these many collisions:

$$\Delta(\mathfrak{p}^2) = \int_{b_{\min}}^{b_{\max}} \sum (\delta\mathfrak{p})^2. \tag{8.30}$$

The same argument is used here to establish the limits of integration as was used in Sec. 8.2. Accordingly, for an ensemble of particles

$$\Delta(\mathfrak{p})^2 = \frac{n_e q_e^2 q_i^2 \ln \Lambda}{8\pi\epsilon_0^2}\, \delta t(\overline{1/v}), \tag{8.31}$$

where Λ was defined by Eq. 8.18. The average

$$\overline{v^{-1}} = (2m/\pi kT)^{1/2} \tag{8.32}$$

in a Maxwell-Boltzmann distribution. As a result of the many collisions, the over-all change ΔU in the kinetic energy of the ions is

$$\Delta U = \Delta(\mathfrak{p}^2)/2m_i. \tag{8.33}$$

By Eqs. 8.31, 8.32, and 8.33, we find that the rate of gain of energy by all ions n_i per unit volume is

$$n_i \frac{dU_i}{dt} = \frac{q_e^2 q_i^2 n_e n_i m_e \ln \Lambda}{8\pi\epsilon_0^2 (2\pi m_e k T_e)^{1/2} m_i}. \tag{8.34}$$

A factor $q_e^2 q_i^2/16\pi^2\epsilon_0^2$ in Eq. 8.34 represents the effect of Coulomb interactions.

Equation 8.34 is the approximate result that was to be derived. A much more exact solution has been derived by Chandrasekhar[4] and discussed at length by Spitzer.[3] The more exact relation is a function of an exponential integral, and we shall not display it. Instead, we give a good approximation to their results, then discuss the connection with Eq. 8.34. The good approximation is

$$n_i \frac{dU_i}{dt} = \frac{q_e^2 q_i^2 n_e n_i m_e \ln \Lambda\, [1 - (2U_i/3kT_e)]}{2\pi\epsilon_0^2 (2\pi m_e k T_e)^{1/2} m_i [1 + (4/3\sqrt{\pi})(m_e U_i/m_i k T_e)^{3/2}]}. \tag{8.35}$$

In making the connection, recall first that Eq. 8.34 was derived under the assumption that $kT_e \gg U_i$. With that assumption, both terms in square brackets in Eq. 8.35 approach unity, and the correct answer is four times larger than Eq. 8.34. The factor 4 arises principally from our cavalier treatment of the problem, and we shall not discuss it further. Now let us imagine that the ratio $3kT_e/2U_i$ approaches 1 or even becomes less than 1, but that v_i is still much less than v_e. In that case, the term in square brackets in the denominator of Eq. 8.35 is still close to 1. The additional term in the numerator expresses the reduced rate of energy transfer as equilibrium is approached. If $U_i = 3kT_e/2$, the ion on the average will not change its energy by colliding with the electrons. If $U_i > 3kT_e/2$, the ion on the average will lose energy. Note that if $m_i kT_e/m_e \gg U_i \gg 3kT_e/2$, then $dU_i/dt \propto T_e^{-3/2}$, indicative of a v_e^{-4} Coulomb cross section and a collision frequency proportional to v_e. The factor in square brackets in the numerator of Eq. 8.35 is accurate at energies close to equilibrium and for extreme ratios of the energies. Being phenomenological, it is inaccurate at intermediate energy ratios.

Finally, if the electron energy is so low that the ions are actually much faster than the electrons, we should have set the relative speed equal to the ion speed, instead of the electron speed. In that case, $U_i/kT_e \gg 1$; the second term in square brackets in the denominator is a measure of $(v_i/v_e)^3$ and is much greater than 1. From Eq. 8.35, an ion loses energy at a rate proportional to $U_i^{-3/2}$, independent of the electron energy. This dependence reflects the product of the v_i^{-4} Coulomb cross section with the collision frequency proportional to v_i.

Equation 8.35 is correct both for $v_e \gg v_i$ and for $v_e \ll v_i$. In the intermediate range $v_e \approx v_i$, Eq. 8.35 is a good approximation ($\pm 10\%$ at worst) to the more exact exponential integral dependence derived by Chandrasekhar.[4] This circumstance is fortuitous: it allows us to express the rate of ion-electron energy transfer in a form convenient for practical calculations. Equation 8.35 is a very important result, and we shall return to it many times.

From Eq. 8.35, we note that the addition of heavy impurities would increase the pressure and disproportionately raise the rate of energy transfer between ions and electrons.

If the electron temperature were held constant, as by heating, then the ratio of the total rate of transfer of energy from ions to electrons with impurities present to the rate with no impurities would be given by

$$\sum_\beta (Z_\beta n_{i\beta}) \sum_\alpha Z_\alpha^2 \left(\frac{n_{i\alpha}/m_{i\alpha}}{Z_1^3 n_1^2/m_1}\right), \tag{8.36}$$

where the subscript 1 denotes the original constituent. A numerical example is illuminating. Suppose that enough fully ionized oxygen impurity is added to deuterium gas to make the number of impurity atoms 3% of the number of deuterium atoms. The power transferred is increased by the factor 1.54, although the electron density increased by only the factor 1.24. If the electrons and ions were at the same temperature, which they will not be, then the pressure would increase by the factor

$$\sum_\alpha \frac{(Z_\alpha + 1)n_{i\alpha}}{(Z_1 + 1)n_1}. \tag{8.37}$$

In the example, the pressure increases by a factor of 1.135.

In cases of interest where the ion energy U_i is greater than the electron energy U_e, Eq. 8.35 can be conveniently reduced. In terms of temperature,

$$n_i \frac{dT_i}{dt} = -\frac{n_i T_i}{\tau_{uie}}, \tag{8.38}$$

where

$$\frac{1}{\tau_{uie}} = \frac{1.0 \times 10^{-19} Z_i^2 n_e \, [1-(T_e/T_i)] \ln \Lambda}{A_i T_e^{3/2}[1 + \sqrt{6/\pi}(m_e T_i/m_i T_e)^{3/2}]}. \tag{8.39}$$

Here, A_i is the mass number of the ion, temperatures are measured in kev, and τ_{uie} is the energy time constant relating to loss of ion energy because of the presence of colder electrons.

In steady-state applications, where the ions are kept at constant temperature by heating, and the electrons lose energy either by radiation or escaping, Eq. 8.38 may be applied to calculate the rate of transfer of energy to the electrons. Applications are discussed in Chaps. 11 and 13.

In time-varying applications, Eq. 8.38 can be recast into other useful forms to fit the circumstances. We discuss two simple examples, both pertaining to the usual case where $\bar{v}_e \gg \bar{v}_i$ (that is, the square bracket in the denominator of Eq. 8.39 is equal to 1). If the electron temperature is constant (determined, for example, by radiation) and if the ions are cooled from a higher temperature by the electrons, Eq. 8.38 can be integrated by inspection. The answer is

$$T_i - T_e = (T_i - T_e)_0 \exp(-t/\tau_1), \tag{8.40}$$

where

$$(1/\tau_1) = \frac{1.0 \times 10^{-19} Z_i^2 n_e \ln \Lambda}{A_i T_e^{3/2}}. \tag{8.41}$$

As an example of Eq. 8.40, if deuterons of density $10^{20}/m^3$ lose energy to 10 ev electrons at the same density, the time constant for the deuterons losing 63% of their excess energy would be 10 μsec. Under these circumstances, the electrons would certainly be heated, thus invalidating our assumption. If the electrons were at 40 kev, the time constant would be about 2.6 sec.

If the total energy of both electrons and ions remains constant, for example, the electrons do not radiate, the problem may also be solved. We consider singly charged ions for convenience. In that case,

$$\frac{dT_i}{dt} = -\frac{dT_e}{dt}. \tag{8.42}$$

It is most convenient to recast Eq. 8.38 in the form

$$\frac{dT_i}{dt} = -\frac{C}{T_e^{3/2}}(T_i - T_e). \tag{8.43}$$

We solve these equations explicitly. From Eq. 8.42 we learn that

$$(T_i + T_e)/2 = T_f, \tag{8.44}$$

where T_f is the equilibrium temperature approached asymptotically. This result may be used to eliminate either T_e or T_i in Eq. 8.43. The variables can be separated and the result integrated by standard procedures. We find that

$$Ct = T_f^{3/2}\left[\frac{1}{2}\ln\left(\frac{1 + \sqrt{T/T_f}}{1 - \sqrt{T/T_f}}\right) - \left(\frac{T}{T_f}\right)^{1/2} - \frac{1}{3}\left(\frac{T}{T_f}\right)^{3/2}\right]_{T=T_{e0}}^{T=T_e}, \tag{8.45}$$

where the initial temperature of the electrons is denoted by T_{e0}. The temperature of the ions may then be found from Eq. 8.44.

8.4. Resistivity of a Fully Ionized Plasma

From the foregoing discussions, we are now in a position to determine the disordering effects of collisions. We may thus determine the resistivity of a fully ionized plasma or the phenomenological constant relating the change of momentum induced by Coulomb scattering. The relation of resistivity to this disordering is manifest by our noting that, were there no scattering and no magnetic induction, a current could be made to flow across a vacuum with no electric field whatever.

We consider the current and electric field parallel to the induction; the conductivity in this direction is just $1/\eta$, and the resistivity is η, as shown in Eq. 7.43. Equally well, we could consider an induction-free plasma, in which the resistivity is the scalar η. We state now that the quantity η is $m\nu_m/nq^2$. To see this point, we note that each momentum transfer collision produces a large deflection. If the collisions are small-angle ones, a large number of them are summed until a large deflection is produced. This thought was applied in our derivation of τ_θ for a fully ionized plasma in Sec. 8.3. Under these circumstances, the momentum gained by an electron in a collision on the average is

$$m_e(\overline{\mathbf{v}}_i - \overline{\mathbf{v}}_e). \tag{8.46}$$

The number of such collisions per unit volume is $\nu_m n_e$. The current density is

$$n_e q_e \overline{\mathbf{v}}_e + n_i q_i \overline{\mathbf{v}}_i \approx ne(\overline{\mathbf{v}}_i - \overline{\mathbf{v}}_e). \tag{8.47}$$

From Eqs. 8.46 and 8.47, the resistivity parallel to **B** follows:

$$\eta_{\parallel} = m_e \nu_m / n_e e^2. \tag{8.48}$$

This result agrees with that deduced from the mobility, Eq. 7.59.

In order to obtain a value for the resistivity, we must substitute an appropriate momentum transfer frequency ν_m in Eq. 8.48. We assume that currents parallel to **B** arise principally from motion of the mobile electrons. They are scattered by Coulomb collisions with the slow ions, for which the time constant $\tau_{\theta ei}$ was derived in Eq. 8.25. If we now assume that $\nu_m = (\tau_{\theta ei})^{-1}$ and that $n_e = Zn_i$, the resistivity is

$$\eta_{\|} = \sqrt{\frac{m_e}{2\pi}} \frac{Ze^2 \ln \Lambda}{32\epsilon_0{}^2(kT)^{3/2}}. \tag{8.49}$$

Note that for a fully ionized plasma the resistivity η is independent of charge density. An increase of the electrons density is exactly counteracted by a proportional increase in the density of ions that scatter the electrons. In practical units, Eq. 8.49 reduces to

$$\eta_{\|} = 1.92 \times 10^{-9} Z \ln \Lambda / T_e{}^{3/2} \quad \text{ohm-m}, \tag{8.50}$$

where T_e is expressed in kev. A more accurate derivation has been carried out by Spitzer,[3] who finds

$$\eta_{\|} = 1.65 \times 10^{-9} Z \ln \Lambda / T_e{}^{3/2} \quad \text{ohm-m}, \tag{8.51}$$

or 0.86 of our approximate value. A numerical example is illuminating: if $T_e = 1$ kev, the resistivity given by Eq. 8.51 is a little greater than that of copper at room temperature. At $T_e = 50$ kev, however, the resistivity is a factor of 350 lower.

8.5. Dynamical Friction

The form of the Coulomb scattering cross section permits the development of an elegant physical duality to potential theory in ordinary space. Consider the interaction of a single test charge of velocity **v** with those in the plasma having velocities in the element $d\mathbf{v}$ about the zero of velocity space. The interaction is shown in Fig. 8.3. According to the development of Secs. 8.2 and 8.3, we could describe the incremental frequency $d\nu_m$ for momentum transfer interaction of the charge with this collection as

$$d\nu_m = v\sigma_m(v) f(0)\, d\mathbf{v}, \tag{8.52}$$

where $f(0)$ is the distribution function at $v = 0$. The relative momentum of the test charge is $m_r \mathbf{v}_r$ and is altered by the

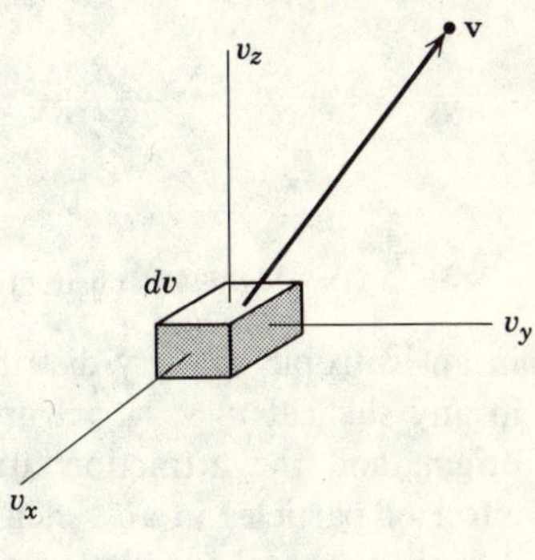

Fig. 8.3. Interaction of a charge at velocity **v** with those in the volume element $d\mathbf{v}$ at the origin.

collisions at the rate $m_r \mathbf{v}_r \, dv_m$ in the center of mass system. By symmetry, the change of momentum on the average is in the direction $-\mathbf{v}_r$. Thus, by combining Eqs. 8.17 and 8.52, the incremental acceleration is

$$\Delta\left[\frac{d}{dt}(m\mathbf{v})\right] = -\frac{1}{4\pi}\left(\frac{q_1 q_2}{\epsilon_0 m_r}\right)^2 \ln \Lambda \frac{m_r \mathbf{v}_r}{v_r^3} f(0)\, d\mathbf{v}. \tag{8.53}$$

The duality to the problem of attraction of a particle in a $1/r$ potential in ordinary space is immediate, for the force acting on a particle in such a potential is proportional to $\mathbf{r}/r^3$. As it were, the particle is attracted to the origin of the velocity space as a result of the forces exerted by all other particles that are situated there. The force is inversely proportional to the square of the relative speed. This force is called the dynamical friction, because it tends to reduce all particles to one common velocity. Since only the relative speed is pertinent, the choice of an attractive center at the origin of Fig. 8.3 is purely one of convenience.

Potential theory developed for an inverse-square-law force can now be applied to advantage. Figure 8.4 illustrates a single particle of velocity $\mathbf{v}_1$

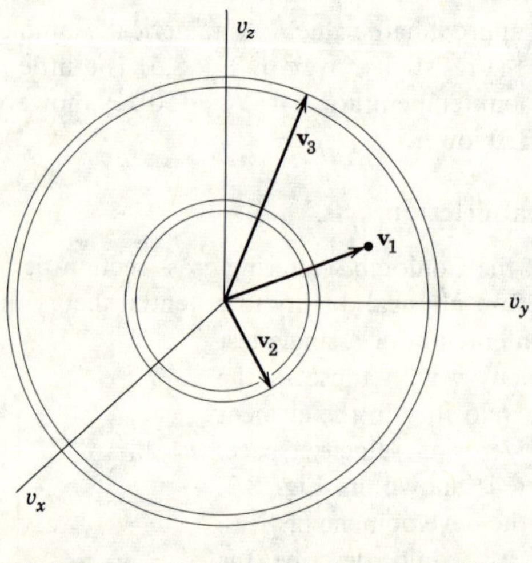

Fig. 8.4. Construction to illustrate dynamical friction.

in an isotropic velocity distribution. By Gauss's theorem, the particles in any shell at $v_2 < v_1$ act as though they were all concentrated at the origin, and the attraction arises solely from these particles. The net effect of particles in any shell at $v_3 > v_1$ is zero.

Since the total energy and momentum of the system are conserved, the dynamical friction, which urges particles toward the origin $\mathbf{v} = 0$, must be counterbalanced on the average by some outward flow in velocity space. The outward flow would have appeared in Eq. 8.53 if we had preserved the transverse increments of velocity imparted at each collision. The effect of these increments leads to diffusion in velocity space, similar to that shown in Fig. 8.2. Because the particles occupy a region in velocity space about the origin, their diffusion in this space urges them outward, on the average. In an equilibrium distribution, the combination of inward dynamical friction and outward diffusion leads to zero net flow in velocity space, and a Maxwellian distribution results.[5]

Finally, note that the v_r^{-3} dependence in Eq. 8.53 reflects the same dependence as the time constant derived from Eq. 8.35, in the limit where the scattering centers are stationary.

8.6. Runaway Electrons

The concepts of dynamical friction and potentials in velocity space can be usefully applied to show that relatively fast electrons in a fully ionized plasma will acquire an increasingly larger velocity in an electric field.[6,7] Figure 8.5 illustrates the case of a charge far outside the main part of

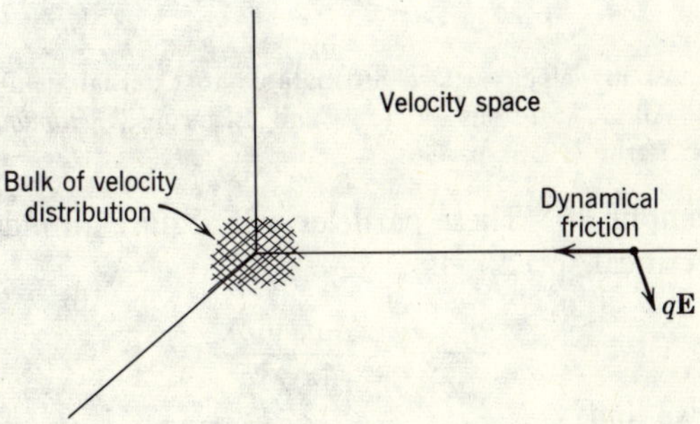

Fig. 8.5. Fast charge acted upon by dynamical friction from a velocity distribution and an electric field.

the distribution, which may be anisotropic. To a good approximation, the total dynamical friction force on the test particle acts directly toward the origin and arises from all the particles of the distribution, which are concentrated there. Diffusion in velocity is negligible for the test charge, because it has very few neighbors in velocity space. If the particle is an electron, its interaction with the electron cloud is much greater than that with the ion cloud. Then in Eq. 8.53, $m_r = m_e/2$, and the total force acting on this fast electron is

$$\mathbf{F} = -e\mathbf{E} - \frac{e^4 n_e \mathbf{v} \ln \Lambda}{2\pi \epsilon_0^2 m_e v^3}. \tag{8.54}$$

Equation 8.54 is the dual of the sum of a uniform field plus the field of a point charge. The directions of flow in velocity space for such a system are shown in Fig. 8.6. Particles for which v lies within the contour passing through the point v_c tend to stay within the main body of the distribution; those lying outside the contour gain speed indefinitely so long as the

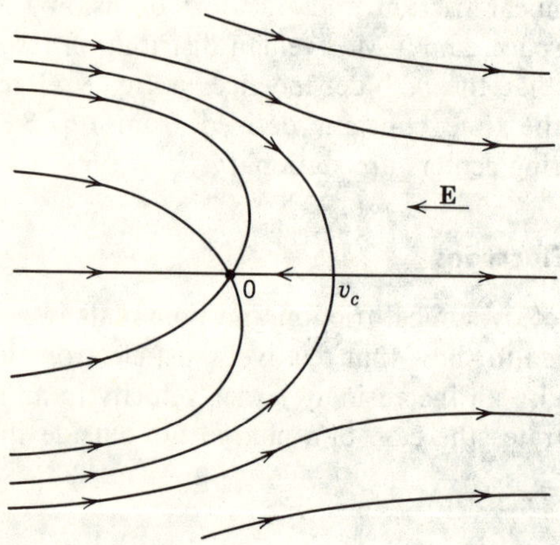

Fig. 8.6. Flow lines in velocity space attributable to external field and dynamical friction. W. P. Allis, "Motions of Ions and Electrons," *Handbuch der Physik*, Vol. 21, Springer, Berlin (1956), p. 436.

electric field is applied. These particles will acquire an indefinitely great energy. The critical speed v_c is

$$\frac{m_e v_c^2}{2e} = \frac{e^2 n_e \ln \Lambda}{4\pi\epsilon_0^2 E}. \tag{8.55}$$

If $\ln \Lambda = 20$, we find

$$\frac{m v_c^2}{2e} = 5.2 \times 10^{-16} n_e/E \tag{8.56}$$

in mks units. A very small electric field will cause a significant fraction of the electrons to be accelerated freely. For example, in a plasma for which $n_e = 10^{20}/\text{m}^3$ and in which $E = 10$ volts/m, electrons with an energy $u_c > 5$ kev will run away. The gradual heating of the remainder in the body of the distribution by ohmic losses brings more and more of them up to the speed v_c. In addition, the body of the distribution, which attracts the electrons of less than runaway speed, is eroded by the escape of those that are brought up to the critical speed. Finally, if the field could be applied for a long enough time, all the electrons would run away in energy.

The runaway process will be modified if a magnetic induction **B** is present. However, if **E** and **B** are collinear, the runaway condition is not altered.

As a result of these various considerations, plasma heating by means of static electric fields does not for several reasons appear feasible at electron temperatures exceeding 100 ev:

1. Conservation of linear momentum implies that the electrons will acquire most of the energy. It is desired to heat the ions, not the electrons, since it is the ions that undergo fusion. The electrons will dissipate most of their energy at high temperatures in radiation, as will be discussed in Chap. 11. Further, the transfer of energy from the electrons to the ions is most inefficient because of the huge disparity in masses.

2. The fast electrons in the distribution run away, carrying most of the energy from the heating device with them. The faster these electrons go, the less likely they are to give any of their energy at all to the rather cold ions.

3. As the plasma heats up, its conductivity becomes so great that a direct electric field is completely canceled at the surface. Few of the charged particles see any of the heating field.

The unfortunate results just enumerated have all been observed in attempts to heat a plasma by unidirectional electric field pulses.

The entire phenomenon of runaway electrons arises because of the long-range character of the Coulomb scattering. It may be readily checked that for short-range forces the electrons will not run away.

PROBLEMS

1. A deuterium plasma at kinetic temperature $T_D = 20$ kev is confined in a magnetic mirror. Assume that the deuterons are lost from the system in a time equal to three angular relaxation times. Calculate the approximate fractional burnup, that is, the fraction of deuterons that react before the gas escapes.

2. Do Problem 1 for a deuterium temperature $T_D = 100$ kev.

3. Do Problem 1 for a 50% D–50% T plasma at 20 kev and 100 kev, neglecting the difference in ionic masses.

4. A simple approach to the problem of Coulomb scattering is to be considered. A particle of mass m, speed v, and charge Z_1e collides with an impact parameter b with a nucleus of charge Z_2e. The time of interaction will be of the order of $2b/v$.

(*a*) Show that the angular deflection $\Delta\theta$ is approximately

$$\Delta\theta = \frac{Z_1Z_2e^2}{2\pi bmv^2\epsilon_0}.$$

(*b*) Let the density of nuclei be n. Show that

$$\overline{\theta^2} = \int_{b_{\min}}^{b_{\max}} db\, 2\pi bnl\left(\frac{Z_1Z_2e^2}{2\pi bmv^2\epsilon_0}\right)^2,$$

where l is the mean free path.

(*c*) Justify the selection of an upper limit of

$$b_{\max} = \sqrt{\epsilon_0 kT/ne^2},$$

and a lower limit of

$$b_{\min} = \frac{e^2}{2\pi mv^2\epsilon_0},$$

or

$$b_{\min} = h/2\pi mv.$$

(*d*) Indicate why below an energy of 200 kev the second lower limit should be taken.

(*e*) Show that an effective cross section for deflection through 90° is given by

$$\frac{1}{2\pi}\left(\frac{Z_1Z_2e^2}{mv^2\epsilon_0}\right)^2 \ln\left(\frac{b_{\max}}{b_{\min}}\right).$$

(*f*) Compute the ratio of the number of Coulomb scatterings through 90° each by multiple scatterings, to the number of thermonuclear reactions for the D-D reaction.

5. Compare the angular relaxation time $\tau_{\theta DD}$ of deuterons with temperature $T_D = 20$ kev among themselves with the angular relaxation time $\tau_{\theta\alpha D}$ of α-particles with these same deuterons, if $T_\alpha = 2.4$ Mev. Justify the use of the temperature T_α in the relaxation rate of the α-particles for an approximate derivation.

6. A plasma of ions at initial temperature T_{i0} and electrons at initial temperature T_{e0} thermalize. The electrons radiate at a rate $CT_e^{1/2}$ per unit volume. Calculate the rate of thermalization, analogous to Eq. 8.45.

7. We may regard the total current in a plasma as the sum of the displacement current and the current arising from the motion of the charges comprising the medium. We can characterize the medium by a permittivity $\epsilon_{\alpha\beta}$ in place of the conductivity plus the vacuum permittivity. Show that

$$\epsilon_{\alpha\beta} = \frac{s_{\alpha\beta}}{i\omega} + \epsilon_0\delta_{\alpha\beta},$$

where ω is the angular frequency of the applied field.

8. Calculate the resistivity of a fully ionized proton plasma at $T_e = 5$ ev. Compare this result with the resistivity of a plasma in H_2 gas, in which 0.1% of the gas is ionized, and the molecular density is $10^{22}/m^3$. Assume that ν_m = const for H_2, from Fig. 3.1.

9. A deuterium plasma at temperature $T_D = 1$ kev has density $n_D = 10^{19}/m^3$. Calculate the frequency ω that equals ν_m. Calculate also the real and imaginary parts of the conductivity $s_\parallel$ at a frequency of 10^{10}cps.

10. Do Problem 9 for a fully ionized deuterium plasma at $T_D = 5$ ev, $n_D = 10^{19}/\text{m}^3$.

11. Calculate the current density and the power absorption per unit volume if an electric field $E = 5$ volts/m is maintained in a deuterium plasma at $T_e = 60$ kev. If the deuteron density is $10^{20}/\text{m}^3$, approximately what fraction of the electrons is above the runaway threshold?

12. In considering the average energy of α-particles in a thermonuclear D-T reactor, we adopt a simplified model. The α-particles are each created with an energy $U_{\alpha 0} = 3.5$ Mev. They cool by dynamical friction to the ion temperature, which we set equal to zero. When the α-particles reach zero energy, they are magically removed by a scattering process. Show that the distribution function $f(v)$ in velocity space of the α-particles is constant out to $U_{\alpha 0}$ and that $\overline{U}_\alpha = 0.6U_{\alpha 0}$. Compute $\overline{U}_\alpha$ on the supposition that the α-particles are scattered out when their energy is reduced to $U_{\alpha 0}/2$.

REFERENCES

1. E. Everhart, *Phys. Rev.*, **99**, 1287 (1955).
2. C. M. Tchen, *Phys. Rev.*, **114**, 394 (1959).
3. L. Spitzer, Jr., *Physics of Fully Ionized Gases*, Interscience Publishers, Inc., New York (1956), p. 78.
4. S. Chandrasekhar, *Principles of Stellar Dynamics*, University of Chicago Press, Chicago (1942), p. 89.
5. W. P. Allis, "Motions of Ions and Electrons," *Handbuch der Physik*, Vol. 21, Springer, Berlin (1956), pp. 432–439.
6. E. R. Harrison, *Phil. Mag.*, **3**, 1318 (1958).
7. H. Dreicer, *Phys. Rev.*, **115**, 238 (1959).

GENERAL REFERENCES

W. P. Allis, "Motions of Ions and Electrons," *Handbuch der Physik*, Vol. 21, Springer, Berlin (1956), pp. 383–444.

L. Spitzer, Jr., *Physics of Fully Ionized Gases*, Interscience Publishers, Inc., New York (1956).

9

Plasma waves of small amplitude

The theory of the propagation of waves of small amplitude in a plasma, which is the subject of this chapter, is based on the electromagnetic theory presented in Chap. 5. These waves are important for heating plasmas, for stability studies, and for the measurement of the properties of plasmas.

A combination of electric and/or magnetic fields may be superimposed in the plasma. The attenuation and speed of propagation of several of the various types of waves and the rotation of the plane of polarization of the waves, if any, are among the topics of interest.

There are many types of waves; they may be classified and derived from the conservation law of total momentum, Eq. 7.16. We shall not engage in this complicated activity, but rather shall restrict ourselves to the consideration of three of the more important limiting types: electromagnetic, hydromagnetic, and electroacoustic. The electromagnetic waves have high frequency and travel at very high speed. The simplest of them are transverse, in that the electric and magnetic fields are perpendicular to the direction of propagation. The phase speed is changed by the electrons of the plasma. Propagation cannot take place above a certain density of electrons unless a magnetic field is present. Hydromagnetic or Alfvén waves are the low-frequency limiting form of electromagnetic waves traveling along lines of induction in a plasma. The plasma and lines of induction are strongly coupled and move together. Electroacoustic waves are very much like sound waves, in that their speed of propagation is temperature dependent and that they are longitudinal. That is, the electric field is parallel to the direction of propagation. We shall derive the mode in which the electrons oscillate and the heavy ions are stationary.

Only waves of small amplitudes are treated, in order that the nonlinear terms that arise may be neglected compared with linear ones. The solutions may therefore be superimposed. The results to follow can be derived in a number of different ways. Sometimes we utilize the conductivity tensor, and sometimes we use a single-particle calculation. The approaches are entirely equivalent, and the choice in any particular case depends upon convenience.

9.1. Electromagnetic Waves in a Charge-Free Region

The simplest example of propagation is that of a wave in a charge-free vacuum. A wave equation (Eq. 5.78) for the magnetic induction was developed in Sec. 5.7. An identical equation can be developed for the electric intensity by computing the curl of Eq. 5.10 and by using Eqs. 5.13 and 5.20 to eliminate **B**. By using vector identities and Eqs. 5.3, 5.17, and 5.23, we find

$$\nabla^2\mathbf{E} - \mu\epsilon\frac{\partial^2\mathbf{E}}{\partial t^2} - \frac{\mu}{\eta}\frac{\partial\mathbf{E}}{\partial t} = 0 \tag{9.1}$$

in any charge-free, time-independent, linear, isotropic medium.

Let $\boldsymbol{\kappa}$ be a vector in the z-direction. We seek solutions of the wave equations (Eqs. 5.78 and 9.1) of the form $\mathbf{E}(\boldsymbol{\kappa}\cdot\mathbf{r}, t)$ and $\mathbf{H}(\boldsymbol{\kappa}\cdot\mathbf{r}, t)$. Since the z-direction is arbitrary, the choice is not restrictive. We observe that

$$\nabla = \boldsymbol{\kappa}\frac{\partial}{\partial(\boldsymbol{\kappa}\cdot\mathbf{r})}. \tag{9.2}$$

This equation and the scalar product of Eq. 5.10 with $\boldsymbol{\kappa}$ imply that

$$\boldsymbol{\kappa}\cdot\frac{\partial\mathbf{H}}{\partial t} = 0. \tag{9.3}$$

From $\nabla\cdot\mathbf{B} = 0$, it follows that

$$\boldsymbol{\kappa}\cdot\frac{\partial\mathbf{H}}{\partial(\boldsymbol{\kappa}\cdot\mathbf{r})} = 0. \tag{9.4}$$

By adding the results (Eqs. 9.3 and 9.4), we find

$$\boldsymbol{\kappa}\cdot d\mathbf{H} = 0; \tag{9.5}$$

in other words, either the component of **H** parallel to $\boldsymbol{\kappa}$ is a constant or is zero. For fields of the form sought, the magnetic intensity must be perpendicular to the direction of propagation, that is, perpendicular to $\boldsymbol{\kappa}$.

From Eqs. 5.13 and 9.3, it follows that

$$\frac{dE_\kappa}{dt} + \frac{1}{\epsilon\eta} E_\kappa = 0 \tag{9.6}$$

for a charge-free space where E_κ is the component of $\mathbf{E}$ along $\boldsymbol{\kappa}$. This result implies that

$$E_\kappa = E_{0\kappa} \exp(-t/\epsilon\eta). \tag{9.7}$$

If the resistivity is other than infinite, the electric intensity along $\boldsymbol{\kappa}$ decays away. For copper at room temperature the electric intensity decays in about 10^{-19} sec. If the resistivity is finite, then for fields dependent on position and time the electric intensity must be perpendicular to the direction of propagation.

A solution of Eq. 9.1 is

$$\mathbf{E} = \mathbf{E}_{T1} \exp[i(\boldsymbol{\kappa}\cdot\mathbf{r} + \omega t)] + \mathbf{E}_{T2} \exp[i(-\boldsymbol{\kappa}\cdot\mathbf{r} + \omega t)], \tag{9.8}$$

where $\mathbf{E}_{T1}$ and $\mathbf{E}_{T2}$ are two coupled constant vectors perpendicular to $\boldsymbol{\kappa}$. Likewise, a solution of Eq. 5.78 is

$$\mathbf{H} = \mathbf{H}_{T1} \exp[i(\boldsymbol{\kappa}\cdot\mathbf{r} + \omega t)] + \mathbf{H}_{T2} \exp[i(-\boldsymbol{\kappa}\cdot\mathbf{r} + \omega t)], \tag{9.9}$$

where $\mathbf{H}_{T1}$ and $\mathbf{H}_{T2}$ are two complex constant vectors perpendicular to $\boldsymbol{\kappa}$. Since all waves in this section are transverse, we shall suppress the subscript T.

We can find the magnitude of the propagation vector $\boldsymbol{\kappa}$ by the method of separation of variables or by direct substitution of Eqs. 9.8 and 9.1 into the differential equations. The result is

$$\kappa^2 = \epsilon\mu\left(\omega^2 - \frac{i\omega}{\eta\epsilon}\right). \tag{9.10}$$

If the resistivity η of the medium is infinite, the wave propagates without attenuation; the speed of propagation is

$$c = (\epsilon\mu)^{-1/2}. \tag{9.11}$$

For a vacuum, $c = 2.998 \times 10^8$ m/sec is the velocity of light and is independent of frequency.

The imaginary term in Eq. 9.10 leads to an attenuation of the wave. Its amplitude is reduced to 37% of its initial value in a distance d, known as the skin depth. If $\eta\epsilon\omega \ll 1$, which is the case for good conductors at radio frequencies,

$$d = (2\eta/\mu\omega)^{1/2}. \tag{9.12}$$

If the resistivity depends on frequency, care must be used in substituting into Eqs. 9.10 and 9.12. In particular, if the resistivity is complex, the term $i\omega/\eta\epsilon$ in Eq. 9.10 contains both real and imaginary terms. The resistivity of a plasma to alternating fields does depend on frequency, as shown in Sec. 7.7. This case is discussed in the next section.

Relations between the constant vectors $\mathbf{E}_1$ and $\mathbf{H}_1$, and $\mathbf{E}_2$ and $\mathbf{H}_2$ may be found by substituting Eqs. 9.8 and 9.9 into Eq. 5.10:

$$i(\boldsymbol{\kappa} \times \mathbf{E}_1 + \omega\mu\mathbf{H}_1) \exp[i(\boldsymbol{\kappa}\cdot\mathbf{r} + \omega t)] - i(\boldsymbol{\kappa} \times \mathbf{E}_2 - \omega\mu\mathbf{H}_2) \exp[i(-\boldsymbol{\kappa}\cdot\mathbf{r} + \omega t)] = 0. \tag{9.13}$$

Since the expression must hold at all times and positions, the coefficient of each exponential must vanish. Consequently,

$$\mathbf{H}_1 = -\frac{1}{\mu\omega}\boldsymbol{\kappa} \times \mathbf{E}_1, \tag{9.14}$$

$$\mathbf{H}_2 = \frac{1}{\mu\omega}\boldsymbol{\kappa} \times \mathbf{E}_2. \tag{9.15}$$

Further, from Eq. 9.13,

$$\mathbf{E}\cdot\mathbf{H} = 0. \tag{9.16}$$

Figure 9.1 pictures the waves. The electric intensity, the magnetic intensity, and the direction of propagation are all mutually perpendicular.

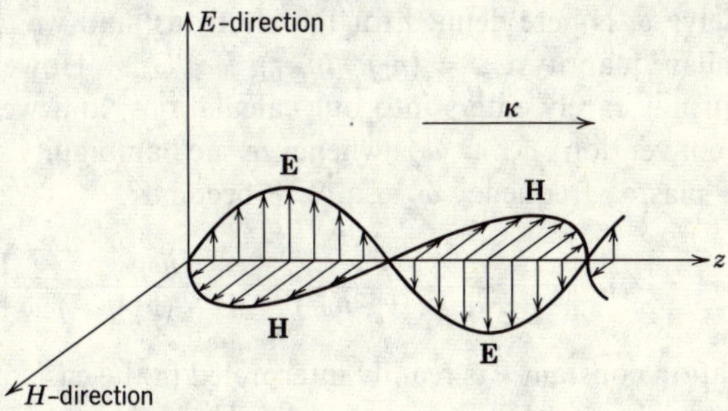

Fig. 9.1. The vectors **E**, **H**, and **κ** for a plane electromagnetic wave in a uniform, isotropic, lossless medium.

9.2. Electromagnetic Waves in a Plasma and Plasma Resonance

We have seen that a wave of any frequency may propagate in a medium of infinite resistivity. We now examine the propagation in a vacuum having charges; we shall find that only waves of sufficiently high frequency can propagate in the plasma. The reason is physically obvious. If the frequency is too low, the charges will have time to rearrange themselves to screen the interior of the plasma from the oscillating field.

Consider propagation of a plane wave in a uniform plasma in which no induction **B** exists independent of time. The nature of the propagation is determined from Eq. 9.10, where we must insert the conductivity for an alternating field. The heavy ions are almost stationary, and they are the scattering centers for the electrons. Therefore, in this derivation we may use the Lorentzian gas model of Sec. 7.7, and write

$$\frac{1}{\eta} = \frac{n_e e^2}{m(\nu_m + i\omega)}, \tag{9.17}$$

from Eq. 7.63. The propagation vector becomes

$$\kappa^2 = \epsilon_0\mu_0\left[\omega^2 - \frac{n_e e^2\omega^2}{\epsilon_0 m_e(\nu_m^2 + \omega^2)} - \frac{i n_e e^2 \nu_m \omega}{\epsilon_0 m_e(\nu_m^2 + \omega^2)}\right]. \tag{9.18}$$

An equation for the propagation constant is called a dispersion relation. It is very convenient at this time to define a quantity ω_p, called the plasma frequency, where

$$\omega_p^2 = n_e e^2/m_e\epsilon_0. \tag{9.19}$$

For the density expressed in mks units, Eq. 9.19 becomes

$$\omega_p = 56.4 n_e^{1/2}\ \text{rad/sec}. \tag{9.20}$$

Most precisely, ω_p is here defined for the electrons, and we should call it ω_{pe}. A similar quantity $\omega_{pi} = (n_i q_i^2/m_i\epsilon_0)^{1/2} \ll \omega_{pe}$. However, the ion plasma frequency rarely enters into our calculations, and we shall adopt the usual convention $\omega_{pe} \equiv \omega_p$ whenever no ambiguity results. In terms of the plasma frequency ω_p, Eq. 9.18 becomes

$$\kappa^2 = \epsilon_0\mu_0\left[\omega^2 - \frac{\omega_p^2}{1 + (\nu_m^2/\omega^2)} - \frac{i\omega_p^2}{1 + (\nu_m^2/\omega^2)}\frac{\nu_m}{\omega}\right]. \tag{9.21}$$

The propagation constant κ is readily interpreted in the case of a collisionless plasma or in any case where $\omega \gg \nu_m$. The propagation constant is pure real for $\omega_p < \omega$ and is pure imaginary if $\omega_p > \omega$. That is, propagation of an electromagnetic wave is possible provided the density n_e is low enough so that ω_p (Eq. 9.20) does not exceed the applied frequency. The phase velocity of the wave

$$c_p = \omega/\kappa \tag{9.22}$$

exceeds the velocity of light in vacuum and approaches infinity as $\omega_p \to \omega$. On the other hand, the group velocity of the wave $c_g = \partial\omega/\partial\kappa$

is the speed of transmission of a signal and is always less than the speed of light in vacuum. In fact, as $\omega_p \to \omega$, the group velocity $c_g \to 0$. The plasma is dispersive; that is, the speed of propagation depends upon the frequency.

At frequencies for which $\omega < \omega_p$, only attenuated waves exist. A wave incident upon a plasma whose thickness $L \gg \kappa^{-1}$ is completely reflected by a collisionless plasma. Equations 9.14 and 9.15 show that **E** and **H** are 90° out of phase. Since the Poynting's vector averaged over a cycle is zero, no net energy flows. If $\omega \ll \omega_p$, then $\kappa L = 1$ when

$$L = 5.32 \times 10^6/n_e^{1/2}\ \text{m}. \tag{9.23}$$

For $n_e = 10^{21}/m^3$, $L = 0.017$ mm. In general, the wave damps out in a distance

$$d' = c/(\omega_p^2 - \omega^2)^{1/2} \tag{9.24}$$

to 37% of its initial value.

There is a simple relation between the plasma frequency ω_p and the Debye length λ. From Eqs. 8.5 and 9.19, we find

$$\lambda\omega_p = (kT/m)^{1/2} \approx \bar{v} \tag{9.25}$$

if the appropriate quantities are calculated for either electrons or ions. The interpretation of Eq. 9.25 is that the charges can move a Debye distance in a time about equal to $1/\omega_p$ and can therefore shield the plasma from any lower frequency.

If the collision frequency ν_m is not zero, two effects occur, as may be seen from Eq. 9.21. First, the cutoff frequency beyond which no propagation occurs is modified, but this effect is generally of little importance. Second, there is attenuation, indicated by the imaginary part of κ^2. The term $n_e e^2 \nu_m/m_e(\nu_m^2 + \omega^2)$ appearing there and in Eq. 9.18 makes up the real part of the conductivity. The case usually of interest is the one where real propagation is possible and the attenuation per wavelength is small. In that case, $\nu_m/\omega \ll 1$, and the attenuation distance d of the wave is

$$d = 2c\omega(\omega^2 - \omega_p^2)^{1/2}/\nu_m\omega_p^2. \tag{9.26}$$

We may readily check that $\nu_m \approx (\tau_{\theta ei})^{-1}$ from Eq. 8.25 is indeed much smaller than ω for any fully ionized plasma of thermonuclear interest. Then if $\omega > \omega_p$, that is, if the wave propagates at all, it propagates with virtually no attenuation.

The attenuation and frequency shift of electromagnetic structures have been used with great advantage in plasma diagnostics. Small plasmas may be inserted into microwave cavities and waveguides. Their presence is made known by the shift in resonant frequency of the system and by the increased loss. From the frequency shift, the electron density can be calculated; from the loss, the collision frequency is calculated.[1–6] In addition, an electromagnetic wave propagated through a plasma has different phase shift from one propagated through a vacuum. The phase shift is a measure of the density, and the attenuation is a measure of the collision frequency.

In our present approach, we have chosen to treat the conductivity explicitly. We could equally well have characterized the medium by a dielectric constant $\epsilon \neq \epsilon_0$. Thus for a collisionless plasma, from Eq. 9.21,

$$\epsilon/\epsilon_0 = \kappa/\kappa_0 = [1 - (\omega_p{}^2/\omega^2)]^{1/2}. \tag{9.27}$$

Here, the dielectric constant ϵ/ϵ_0 is less than 1, whereas for most un-ionized materials the dielectric constant is greater than 1. The dielectric constant is zero at $\omega = \omega_p$. If collisions exist, the dielectric constant ϵ is complex.

9.3. Electromagnetic Waves in a Magnetized Plasma

The general case of the propagation of electromagnetic waves in a plasma immersed in a magnetic induction is rather complicated and will not be treated here. However, two limiting cases are sufficiently simple and important to merit discussion. In the first one, the electric intensity vector is parallel to the magnetic induction, and the propagation direction is perpendicular to the magnetic induction. In the second case, the direction of propagation is taken parallel to the applied magnetic induction.

Figure 9.2 displays the relation of various vectors with respect to each other in the first case. We can readily show from Eqs. 9.14 or 9.15 that the magnetic force of a wave on a charge moving slowly compared with the speed of light is negligible compared with the electric force exerted by the wave. Therefore, we neglect the magnetic induction associated with the traveling wave, as indeed has implicitly been done in the previous section.

If the electric intensity is parallel to the applied constant induction, then the motion of the particle is composed of two independent parts: one along the induction and one perpendicular to it. The parallel motion is the same as that of a particle moving in only an electric field. The perpendicular motion is the same as that of a particle in only a magnetic field. The two components of motion are uncoupled and may be solved separately. To the order of our approximation, in this case the electron motion generated by the wave is independent of the presence of a static induction $\mathbf{B}_0$. Then the propagation and attenuation are exactly as described in Sec. 9.2.

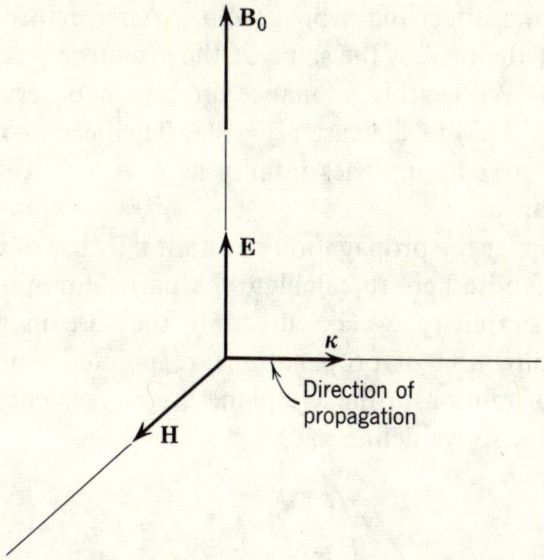

Fig. 9.2. Propagation of an electromagnetic wave in a magnetized plasma. Wave travels across applied induction.

Propagation in this manner across the induction has been used to measure the density and other characteristics of large magnetized plasmas.[7, 8] Care must be taken to ensure that diffraction around the plasma and reflection at the plasma-vacuum interface are properly considered.

If the electric field is not parallel to the constant induction **B**, the charge motion is more complicated. Figure 9.3 shows an electron and

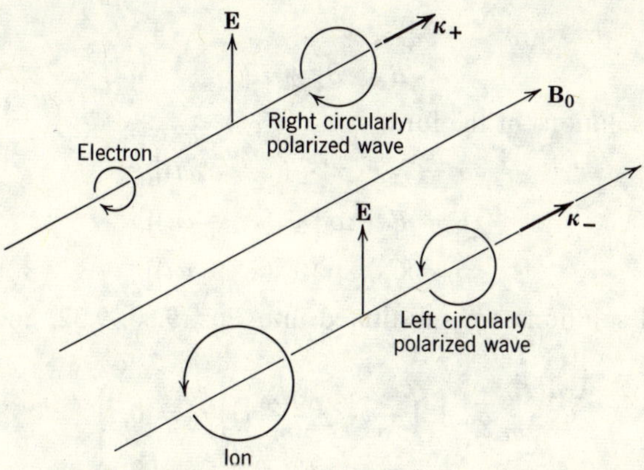

Fig. 9.3. Propagation of an electromagnetic wave in a magnetized plasma. Wave travels along applied induction.

an ion gyrating about lines of induction. The electric field **E** can couple to this motion and affect the propagation. A resonance can exist when the frequency of the wave is the same as the cyclotron frequency of either type of particle. We see this resonance directly by observing the conductivity tensor (Eq. 7.63) in the case $\nu_m = 0$. Then each conductivity term in the x-y submatrix approaches infinity as $\omega \to \omega_b$. Collisions tend to damp the resonance.

We could discover the propagation constant $\boldsymbol{\kappa}$ by use of the conductivity tensor, but we choose here to calculate the particle motions directly. In the interest of simplicity, we discuss only the case in which both the induction and direction of propagation are pointed along the $+z$-axis. The equations of motion in the x-y plane are conveniently expressible in complex notation, so we define

$$r = x + iy, \tag{9.28}$$

$$E = E_x + iE_y. \tag{9.29}$$

With these variables, the equations for the incremental motion are

$$\ddot{r} + \frac{iqB_0}{m}\dot{r} = \frac{q}{m}E, \tag{9.30}$$

$$\ddot{z} = 0. \tag{9.31}$$

The electric and magnetic fields are related by the equations

$$\frac{\partial E}{\partial z} - i\frac{\partial B}{\partial t} = 0, \tag{9.32}$$

$$\frac{\partial B}{\partial z} + i\epsilon_0\mu_0\frac{\partial E}{\partial t} = -inq\mu_0\dot{r}, \tag{9.33}$$

where

$$B = B_x + iB_y. \tag{9.34}$$

Let us try solutions of the form

$$r = r_0 \exp\left[\pm i(\kappa z - \omega t)\right], \tag{9.35}$$

$$E = E_0 \exp\left[\pm i(\kappa z - \omega t)\right], \tag{9.36}$$

$$B = C_0 \exp\left[\pm i(\kappa z - \omega t)\right]. \tag{9.37}$$

These trial solutions are substituted into Eqs. 9.30, 9.32, and 9.33, and we find

$$\left.\begin{aligned} -\frac{q}{m}E_0 + \left(-\omega^2 \pm \frac{qB_0}{m}\omega\right)r_0 &= 0, \\ \kappa E_0 + i\omega C_0 &= 0, \\ \frac{\omega}{c^2}E_0 + i\kappa C_0 + nq\omega\mu_0 r_0 &= 0. \end{aligned}\right\} \tag{9.38}$$

The three algebraic equations (Eqs. 9.38) involve the three quantities E_0, C_0, and r_0. If they are not to be zero, the determinant of the coefficients must vanish. By setting the determinant equal to zero, we find a dispersion relation for the propagation constant κ, which is

$$\kappa_{\pm}{}^2 = \frac{\omega^2}{c^2}\left(1 - \frac{\omega_p{}^2}{\omega^2 \mp \omega\omega_b}\right). \tag{9.39}$$

Equation 9.39 expresses the effect of one component of the plasma on the propagation constant, and would denote by suitable choice of subscripts the effect of electrons or ions. The only term in Eq. 9.39 that depends upon the nature of the particle is the term $\omega_p{}^2/(\omega^2 \mp \omega\omega_b)$. Had we considered electrons and ions together from the beginning, we should have written two force equations in the form of Eq. 9.30. As a consequence, we should find an appropriate resonance term for each species; that is,

$$\kappa_{\pm}{}^2 = \frac{\omega^2}{c^2}\left(1 - \frac{\omega_{pe}^2}{\omega^2 \mp \omega\omega_{be}} - \frac{\omega_{pi}^2}{\omega^2 \mp \omega\omega_{bi}}\right) \tag{9.40}$$

for a plasma composed of electrons and one species of ion. For several ion species, terms appropriate to each would appear.

Equation 9.40 exhibits resonances ($\kappa^2 \to \pm\infty$) and cutoffs ($\kappa^2 \to 0$) depending upon the particular circumstances. In order to illustrate most simply the behavior of the propagation, we shall temporarily neglect the ion motions and return to Eq. 9.39. In other words, we shall consider frequencies high enough so that $\omega \gg \omega_{pi}$ or $|\omega_{bi}|$; because $\omega_{pe} \gg \omega_{pi}$ and $\omega_{be} \gg |\omega_{bi}|$, the phenomena are then determined by the electrons. In that case, the phase speed c_p is given by

$$c_p = \frac{c}{\left(1 - \dfrac{\omega_p{}^2}{\omega^2 \mp \omega\omega_b}\right)^{1/2}}. \tag{9.41}$$

Two values of the propagation constant are permitted, corresponding to two different speeds of propagation in the $+z$-direction. Therefore, two different waves exist, and we must discover their nature. Let us choose the plus sign in Eq. 9.36 and substitute into Eq. 9.29. After equating real and imaginary parts, we find

$$\left.\begin{aligned} E_{x+} &= E_0 \cos(\kappa_+ z - \omega t), \\ E_{y+} &= E_0 \sin(\kappa_+ z - \omega t). \end{aligned}\right\} \tag{9.42}$$

Equations 9.42 represent E_+, a vector whose magnitude is constant, and equal to E_0, but whose direction rotates. That is, the angle θ defining the orientation of E_+ is

$$\theta = \tan^{-1}(E_{y+}/E_{x+}) = \kappa_+ z - \omega t. \tag{9.43}$$

At any fixed position z, θ decreases as t increases; the vector rotates clockwise, looking in the direction of $\boldsymbol{\kappa}$. Such waves are called circularly polarized, and in particular, this one is right circularly polarized. It is shown in the upper part of Fig. 9.3; the **E** vector in the figure rotates in the direction of the circular arrow; its phase angle θ propagates according to the constant κ_+. Sometimes the symbol $\kappa_r(\equiv\kappa_+)$ is used.

The other wave is found by substituting κ_- into Eq. 9.36, taking the corresponding minus sign of the exponent, and solving as before. This wave is left circularly polarized, as shown in the lower part of Fig. 9.3. The symbol κ_l is sometimes applied.

The physical content of our manipulations is now evident. The harmonically oscillating plane wave **E** (and also **B**) has been decomposed into two circularly polarized waves. Figure 9.4 shows the vector

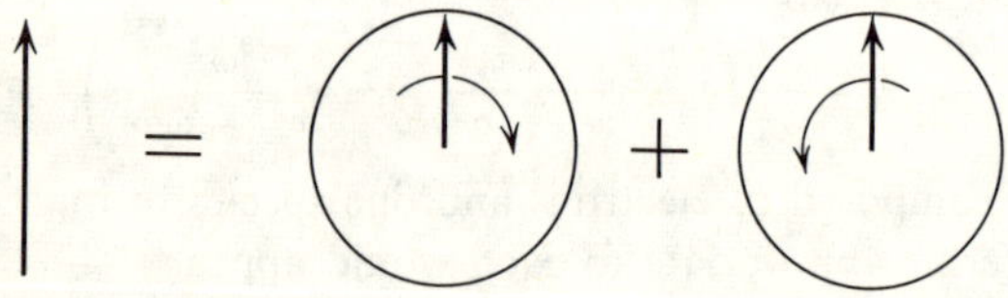

Fig. 9.4. Resolution of a harmonically oscillating plane-polarized vector into two circularly polarized vectors.

decomposition. The actual resolution is performed by the insertion of the $\pm$ sign in Eqs. 9.35 to 9.37. If there were no induction independent of time, this decomposition would tell us nothing new. If a time-independent induction is present, we find that the two waves propagate at different speeds. Physically, the two waves couple differently to the oppositely gyrating electrons and ions.

Note from Fig. 9.3 that if we reverse the induction $\mathbf{B}_0$ or the direction of $\boldsymbol{\kappa}$, the right circularly polarized wave (looking in the direction $\boldsymbol{\kappa}$) would couple to the ions, and the left to the electrons.

If the medium is lossless, as we have assumed, there are two principal differences between right and left circularly polarized waves. First, the right circularly polarized waves will resonate with negative charges. The left circularly polarized wave would resonate with positive ions, had we retained the propagation constant in the form of Eq. 9.40. At resonance, the magnitude of the propagation vector becomes infinite, and the phase speed becomes zero. Second, the angles of the resultant vectors **E** and **B** are rotated as the wave propagates through the plasma. This phenomenon is known as Faraday rotation. Unlike the case of many simple optically active crystals, the Faraday rotation is not unwound if the wave is reflected back through the medium. On the contrary, the plane of polarization is rotated still further. Faraday rotation is also exhibited by ferrites. A quantum-mechanical analogue of our derivation applies to such materials. If the dissipation of the medium differs for the two circularly polarized modes, an incident plane wave becomes elliptically polarized, because one mode is attenuated more than the other.

Note that not all frequencies can be propagated and that the induction B_0 permits waves to propagate at frequencies otherwise forbidden. If we consider the motion of electrons only, we find that propagation is possible under the following circumstances:

$$\text{Left circular polarization:} \quad \begin{cases} \omega_b/\omega > (\omega_p^2/\omega^2) - 1, \\ c_p > c. \end{cases} \tag{9.44}$$

$$\text{Right circular polarization:} \quad \begin{cases} \omega_b/\omega < 1 - (\omega_p^2/\omega^2), \\ c_p > c. \end{cases} \tag{9.45}$$

or

$$\begin{cases} \omega_b/\omega > 1, \\ c_p < c. \end{cases} \tag{9.46}$$

These regions are conveniently shown in Fig. 9.5, in the (ω_b/ω) vs. (ω_p^2/ω^2) space, where the shorthand l and r are used for the waves, and the speed is shown greater or less than c.

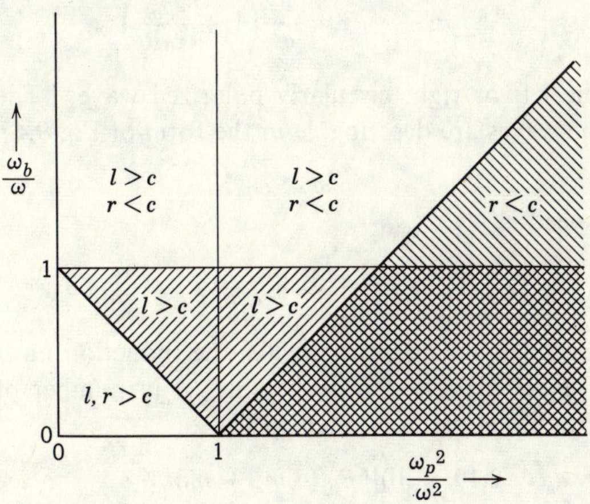

Fig. 9.5. Regions in which left and right circularly polarized waves propagate in the direction of **B**. Phase speed is shown greater than or less than the velocity of light in vacuum.

Propagation at arbitrary angles to the time-independent induction **B** could also be considered. Various hybrid resonances occur in these cases. The task is a complicated one, and we refer the interested reader to the literature.[9–11]

It is evident that propagation of waves in these modes through a plasma will yield information on both the electron and ion densities. Diagnostic techniques can thereby be developed.

In addition, these waves have resonances at $\omega = \omega_{be}$ for the right circularly polarized wave, and at $\omega = \omega_{bi}$ for the left circularly polarized wave. These resonances are evident from Eq. 9.40 by inspection and are those described also in Eq. 7.63. At each resonance, the associated charge gyrates in phase with the electric field and gains energy indefinitely in the absence of collisions. Thus an electric field at the ion cyclotron resonance frequency can be applied to heat the ions in a plasma preferentially. Modifications of this scheme find application to thermonuclear devices, particularly the stellarator, which we shall discuss in Chap. 16.

9.4. Hydromagnetic Waves

Hydromagnetic or Alfvén waves are the waves described in the previous section, in the limit of very low frequency. Here, the ion motions are important, and we evaluate Eq. 9.40 in the limit $\omega \to 0$. From the definition of the ω_p's and ω_b's and the fact that $n_e = n_i$, we find

$$\kappa_{\pm}{}^2 \to \kappa^2 = \frac{\omega^2}{c^2}\left(1 + \frac{\rho_m}{\epsilon_0 B^2}\right) \tag{9.47}$$

for either the left or right circularly polarized wave. The derivation is valid and the waves are describable in the form of Eq. 9.47 if and only if

$$\omega \ll \omega_{bi}. \tag{9.48}$$

The phase speed is

$$c_p = c\left(1 + \frac{\rho_m}{\epsilon_0 B^2}\right)^{-1/2}. \tag{9.49}$$

The phase speed is greater, the greater the induction and the lower the density. If B_0 is expressed in webers/m^2, n in number of particles/m^3, and m in absolute mass units, then

$$c_p = 2.19 \times 10^{16} B_0/(n_e m_e + n_i m_i)^{1/2} \qquad \text{m/sec.} \tag{9.50}$$

If $n_e = n_i = 10^{21}/\text{m}^3$, $m_i = 2$, and $B = 1$ weber/m^2, then $c_p = 4.9 \times 10^5$ m/sec. These waves propagate very slowly at any frequency, provided that the density is high enough so that our derivation is accurate.

The same result could have been derived from the conductivity tensor. If we assume that the wave propagates along the static induction and take the electric intensity parallel to the x-axis, then the conductivity tensor indicates a coacting current that travels along the x-axis. From Eq. 5.10, we see immediately that the oscillating component of the induction must be parallel to the y-axis. The relation of these vectors to each other is shown in Fig. 9.6. The propagation is nondispersive, that is, independent of frequency. The group speed is therefore equal to the phase speed.

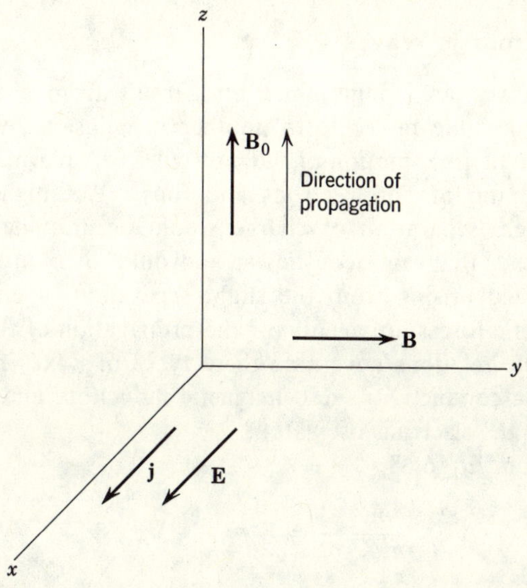

Fig. 9.6. Relations among various vectors used in studying Alfvén waves.

The equations assert that Alfvén waves are controlled by the mass density and inertial effects. The lines of induction and the plasma move together. Indeed, Alfvén waves may be derived for the case of infinite conductivity from the observation that lines of induction are subjected to a tension $B^2/2\mu$ and a sidewise pressure of $B^2/2\mu$. Insofar as the forces on the lines are concerned, we may just as well consider them to be subjected to a uniform hydrostatic pressure $B^2/2\mu$ everywhere and to a tension B^2/μ. Now consider a unit area normal to the lines. The number of lines passing through this area will be B. The tension exerted on any one line is then B/μ. The mass that must be associated with this line is ρ_m/B, since, as the line moves about, this amount of mass moves with the line. The line of induction may now be treated as an ordinary string having a mass ρ_m/B per unit length and subjected to a tension B/μ. The equation of motion for this line or string is

$$\mu \frac{\partial^2 B_y}{\partial t^2} = \frac{B_0^2}{\rho_m} \frac{\partial^2 B_y}{\partial z^2} \tag{9.51}$$

in the absence of any resistivity. The existence of the hydrostatic pressure is irrelevant. The speed $(B_0^2/\mu\rho_m)^{1/2}$ of wave propagation deduced from Eq. 9.51 is approximately that of Eq. 9.49.

Further details concerning this subject may be found in the literature.[12, 13]

9.5. Electroacoustic Waves

Acoustic waves are longitudinal and result from the transfer of momentum from one molecule to another in collisions, which in turn result from the thermal motion of the molecules. A plasma consists of a nearly neutral mixture of electrons and ions. We might expect that some very slight separation of charge would occur under appropriate stimulation and that an acoustic wave would then propagate. The electrostatic field arising from the slight separation of charge and the usual gas kinetic forces are agents for the propagation of the wave. We shall derive the relations governing these types of waves for the simple case of infinite conductivity, zero magnetic induction, and frequency so high that only the electrons oscillate.[12]

We start from Eq. 6.68,

$$\frac{m_e}{n_e q_e^2} \frac{\partial \mathbf{j}}{\partial t} = \mathbf{E} - \frac{1}{n_e q_e} \nabla p_e, \tag{9.52}$$

and the adiabatic gas law (Eq. 7.19), to find the gradient of the electron partial pressure:

$$\nabla p_e = \gamma_e k T_e \nabla n_e. \tag{9.53}$$

Since

$$\nabla n_i = 0, \tag{9.54}$$

then

$$\nabla p_e = \frac{\gamma_e k T_e}{q_e} \nabla \rho. \tag{9.55}$$

Because the density is assumed to be very low, the pressure is highly anisotropic; the compression is then one-dimensional.

Let the divergence of Eq. 9.52 be computed. With Eqs. 5.3 and 5.15 we learn that

$$\frac{\partial^2 \rho}{\partial t^2} = -\omega_p^2 \rho + \frac{3kT_e}{m_e} \nabla^2 \rho. \tag{9.56}$$

Solutions of this equation are of the form $\exp[i(-\kappa z + \omega t)]$. Then we find that

$$\kappa^2 = \frac{m_e(\omega^2 - \omega_p^2)}{3kT_e}; \tag{9.57}$$

the phase speed of propagation c_p is

$$c_p = \frac{\omega}{\kappa} = \left\{\frac{3kT_e}{m_e[1 - (\omega_p^2/\omega^2)]}\right\}^{1/2}; \tag{9.58}$$

the group speed c_g is

$$c_g = \frac{\partial\omega}{\partial\kappa} = \frac{3kT_e}{m_e c_p}. \tag{9.59}$$

Thus the product of the two speeds is related to the average thermal energy of the electrons. The propagation is very dispersive.

The result (Eq. 9.58) is intriguing because the velocity of propagation becomes pure imaginary when $\omega^2 < \omega_p^2$. In other words, there is a cutoff in the propagation frequency of the waves. No waves can be propagated if $\omega^2 < \omega_p^2$, for the electrons will then have time to rearrange themselves to shield the ions completely. To put the matter another way, an electric field arising from space charge cannot exist for longer than ω_p^{-1}. In Sec. 9.2, a similar cutoff was observed for simple electromagnetic waves.

The present subject has been studied extensively.[12] It is found that waves are damped as their wavelength approaches λ, the Debye length.

PROBLEMS

1. Starting from Eq. 5.24 and by means of the equation

$$m_e\ddot{z} = q_e E_0 \exp[-i(\omega t - \kappa z)],$$

derive the propagation constant as given by Eq. 9.10.

2. Prove that in any system for which a dispersion relation $\kappa^2 = (\omega^2 - A)/c^2$ is valid, where A and c are independent of frequency, the product of the phase and group velocities is c^2.

3. Calculate the attenuation of 300 kMc radiation through 1 m of plasma at a density of $10^{20}/\text{m}^3$ and a temperature of 1 kev. How do you reconcile this high transmission with the high dc conductivity of the plasma?

4. A 3-cm electromagnetic wave is incident on sea water having a conductivity of 4.3 mhos/m. Compute the depth to which the wave penetrates.

5. A uniform-density D^+ plasma is confined by an external magnetic induction of 0.5 weber/m², which at time $t = 0$ is completely excluded from the plasma. The radius of curvature of the plasma surface is very large compared with the Larmor radius. It is found that microwave probe signals of free-space wavelength $\lambda \leqslant 3$ mm are transmitted through the plasma, but waves $\lambda > 3$ mm are not. Assuming that the plasma is thermalized, compute the diffusion coefficient D relating to diffusion of the plasma across the magnetic induction. Neglect heating of the plasma by the process.

6. Consider a plasma with zero resistivity having an electron pressure independent of position. A wave propagates parallel to a static, uniform induction B_0 along the z-axis. Show that the speed of propagation is given by

$$c_p{}^2 = \frac{c^2}{1 - (\omega_p/\omega)^2[1 \pm (\omega_{be}/\omega)]^{-1}}.$$

The analysis is easiest in terms of the variables $E_x \pm iE_y$. Neglect ω_{bi} compared with ω^2, note that the current density is parallel to the wave front, and that $\nabla \cdot \mathbf{j} = 0$. Plot the phase speed vs. ω/ω_p for $0 < \omega/\omega_p < 7$ and $\omega_p/\omega_{be} = 0.2$ and 5.0.

7. In principle, we can construct a nonreciprocal microwave circuit (an isolator) by inserting a circular section between two rectangular waveguides, as illustrated. The dimensions of the rectangular guide are such that the

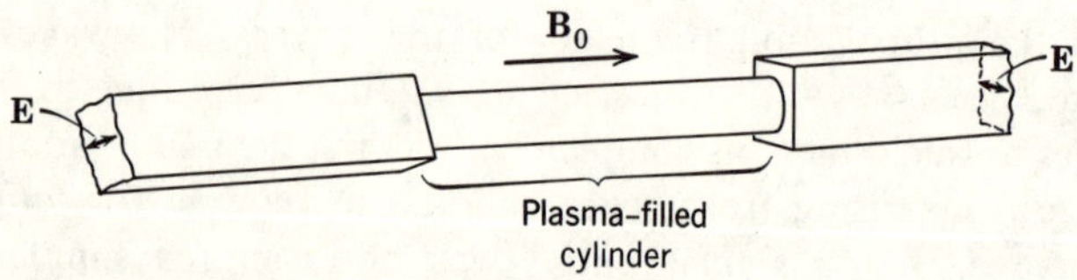

wave cannot propagate with the **E** vector parallel to the wide side. The magnetized plasma rotates the plane of polarization. Neglect matching problems at the cross-section discontinuities. Propose appropriate plasma density, induction $\mathbf{B}_0$, and length of the circular section so that propagation is possible in only one direction at 3000 Mc. Assuming the plasma to be weakly ionized H_2, discuss the losses in the system.

8. A lightning discharge can generate in the earth's ionosphere a local disturbance that travels as an Alfvén wave along lines of the earth's magnetic induction. The disturbance may therefore follow induction lines through the exosphere and return to the earth at the conjugate point. Explain why these signals are detected at some geomagnetic latitudes but not at others. Support your discussion with order-of-magnitude calculations. As a simplified model of the ionosphere and exosphere, use an ion density that is negligible below $h = 150$ km, $10^{12}/\text{m}^3$ up to 200 km, and decreasing higher up as $\exp(-h/500)$. The earth's induction is that of a dipole at the center of the earth; the induction at the earth's surface at the geomagnetic equator is 5×10^{-5} weber/m^2.

9. The Alfvén speed as applied to the preceding problem is independent of frequency. But it is observed, nevertheless, that signals generated by single impulses are received over an extended time. Consider a source of finite size at the surface of the earth, and explain the phenomenon for a monochromatic source.

10. The equations for the motion of hydromagnetic waves and their propagation speed are to be derived for an incompressible fluid of finite conductivity. Assume a uniform, static induction B_0 pointed along the z-axis; neglect gravitational forces and displacement current. Assume that all vectors depend only on z and the time t. Choose all arbitrary constants other than B_0 to be zero. Orient the coordinate system so $j_y = 0$. Show that

$$\frac{\partial^2 B_y}{\partial t^2} = \frac{B_0^2}{\mu_0 \rho_m} \frac{\partial^2 B_y}{\partial z^2} + \frac{\eta}{\mu_0} \frac{\partial^3 B_y}{\partial z^2\, \partial t}.$$

Assuming that the conductivity is infinite and that

$$B_y = C \sin \omega(t - z/c_p),$$

where c_p is a constant, find the current density along the x-axis, the electric intensity along the x-axis, the pressure, and the mass velocity along the y-axis.

REFERENCES

1. S. C. Brown and D. J. Rose, *J. Appl. Phys.*, **23**, 711 (1952).
2. D. J. Rose and S. C. Brown, *J. Appl. Phys.*, **23**, 719 (1952); **23**, 1028 (1952).
3. L. Gould and S. C. Brown, *J. Appl. Phys.*, **24**, 1053 (1953).
4. K.-B. Persson, *Phys. Rev.*, **106**, 191 (1957).
5. S. J. Buchsbaum and S. C. Brown, *Phys. Rev.*, **106**, 196 (1957).
6. S. C. Brown, *Proc. 2nd U.N. Conf. on Peaceful Uses of Atomic Energy*, Vol. 32, United Nations, Geneva (1958), p. 394.
7. R. F. Whitmer, *Phys. Rev.*, **104**, 572 (1956).
8. T. Coor and co-workers, *Phys. Fluids*, **1**, 411 (1958).
9. J. E. Drummond, *Proc. 2nd U.N. Conf. on Peaceful Uses of Atomic Energy*, Vol. 32, United Nations, Geneva (1958), p. 379.
10. R. S. Sagdeyev and V. D. Shafranov, *Proc. 2nd U.N. Conf. on Peaceful Uses of Atomic Energy*, Vol. 31, United Nations, Geneva (1958), p. 118.
11. W. P. Allis, to be published.
12. L. Spitzer, Jr., *Physics of Fully Ionized Gases*, Interscience Publishers, Inc., New York (1956), Chap. 4.
13. H. Alfvén, *Cosmical Electrodynamics*, Clarendon Press, Oxford (1950), Chap. 4.

GENERAL REFERENCES

D. Bohm and E. P. Gross, *Phys. Rev.*, **75**, 1851, 1864 (1949).

J. A. Stratton, *Electromagnetic Theory*, McGraw-Hill Book Co., New York (1941).

10

Motion of individual charges

In Chaps. 4 through 9, we studied a plasma mainly from a continuum point of view. The charges were characterized by a distribution function from which appropriate averages could, in principle, be calculated to find some macroscopic quantity of interest. The detailed motion of each individual particle was of secondary interest. Indeed, we were able to find a number of conservation laws: the details of electromagnetic wave propagation through a plasma, the conductivity of a plasma, and many other physical phenomena of interest. In later chapters, we shall apply this point of view in discussing further the properties of plasmas and the associated thermonuclear experiments.

The continuum approach has disadvantages also, particularly in the analytical treatment of highly anisotropic distributions. The concept of pressure is useless for such distributions; instead we must use the stress tensor with all its complexity. A beam of particles, for instance, is highly anisotropic.

It is sometimes helpful to consider the detailed motions of each individual charge. This approach is the antithesis of that of the continuum. If the motion of each charge were known, then in principle we could average over all the trajectories to find the properties of a continuum. Such a procedure is usually highly impractical. Nevertheless, a number of very general and useful conclusions can be drawn from the individual-particle approach. A number of confinement schemes depend upon simple analyses of single-particle orbits. Some of the properties already found from the fluid model may also be derived from this complementary attack, such as the heating caused by compression with a magnetic field.

In the present chapter we assume that the electromagnetic fields are known from the work of Chap. 5. The motion of a charged particle is then computed by means of Newton's second law and the force law for a charge in an electromagnetic field.

Each approach has its own particular area of application.

10.1. Orbits in Uniform, Static Fields

The simplest possible combination of fields comprises a superposition of uniform, static electric and magnetic fields. This problem is studied first.

A particle of mass m and charge q moves in an orbit determined by Newton's second law:

$$m\ddot{\mathbf{r}} = q(\mathbf{E} + \dot{\mathbf{r}} \times \mathbf{B}), \tag{10.1}$$

where $\mathbf{r}$ is the distance from an arbitrary reference point. A moving charge constitutes a current; this equation is the dual to that for a current (Eq. 7.1). Let $\mathbf{B}$ be directed along the z-axis; let $\mathbf{E}$ lie in the x-z plane, as shown in Fig. 10.1. Equation 10.1 is resolved into two components, one parallel to $\mathbf{B}$ and one transverse to $\mathbf{B}$:

$$m\ddot{\mathbf{r}}_{\parallel} = q\mathbf{E}_{\parallel}, \tag{10.2}$$

$$m\ddot{\mathbf{r}}_{\mathrm{T}} = q(\mathbf{E}_{\mathrm{T}} + \dot{\mathbf{r}} \times \mathbf{B}). \tag{10.3}$$

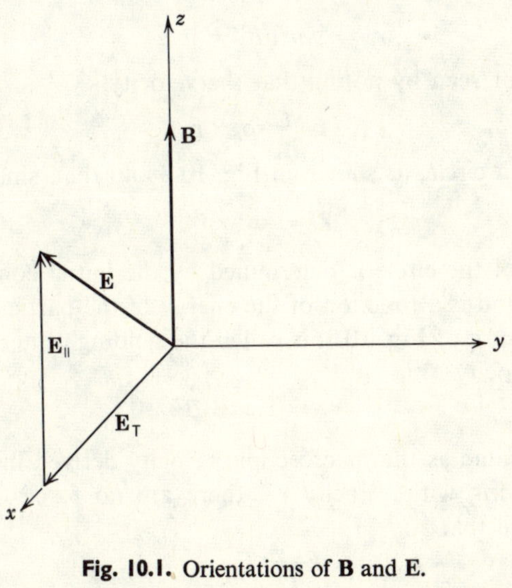

Fig. 10.1. Orientations of **B** and **E**.

The subscripts $\|$ and T denote the components parallel and transverse to **B**, respectively. The magnetic induction is of no consequence insofar as motion along **B** is concerned.

The remaining component of the motion takes place in the x-y plane. Simplifications result if the following transformation is introduced:

$$\mathbf{r}' = \mathbf{r} - [(\mathbf{E} \times \mathbf{B})/B^2]t. \tag{10.4}$$

The velocity $\mathbf{E} \times \mathbf{B}/B^2$ is a constant and is directed along the negative y-axis. Insofar as the transformation is concerned, it is immaterial whether **E** or $\mathbf{E}_\mathrm{T}$ is used, since in either case a cross product is computed with **B**. We make the transformation in order to simplify the problem; in the moving frame the perpendicular component of the electric intensity is zero as a result of the transformation. It yields

$$\dot{\mathbf{r}}' = \dot{\mathbf{r}} - \mathbf{v}_E = \dot{\mathbf{r}} - (\mathbf{E} \times \mathbf{B}/B^2), \tag{10.5}$$

$$\ddot{\mathbf{r}}' = \ddot{\mathbf{r}}. \tag{10.6}$$

In terms of $\mathbf{r}'$, Eq. 10.3 becomes

$$m\ddot{\mathbf{r}}_\mathrm{T}' = q(\dot{\mathbf{r}}_\mathrm{T}' \times \mathbf{B}). \tag{10.7}$$

Let

$$\boldsymbol{\omega}_b = -q\mathbf{B}/m, \tag{10.8}$$

the cyclotron frequency, be substituted into Eq. 10.7; then

$$\ddot{\mathbf{r}}_\mathrm{T}' = \boldsymbol{\omega}_b \times \dot{\mathbf{r}}_\mathrm{T}'. \tag{10.9}$$

The solution of Eq. 10.9 consists of a radius vector, which rotates about the origin with an angular frequency ω_b, of constant length

$$r_b = |mv/qB| = |v/\omega_b|. \tag{10.10}$$

This fact can be seen by noting that the velocity

$$\dot{\mathbf{r}}_\mathrm{T} = \boldsymbol{\omega}_b \times \mathbf{r}_b \tag{10.11}$$

describes such a circle, as shown in Fig. 10.2 and that, since $\dot{\omega}_b = 0$,

$$\ddot{\mathbf{r}}_\mathrm{T} = \boldsymbol{\omega}_b \times \dot{\mathbf{r}}_b. \tag{10.12}$$

The radius r_b of the circle is determined by the initial conditions, which are arbitrary, and is a function of the energy of the particle. The center of motion, point g of Fig. 10.2, is called the guiding center.

The drift velocity

$$\mathbf{v}_E = \mathbf{E} \times \mathbf{B}/B^2 \tag{10.13}$$

is exactly the same as the macroscopic velocity derived in Eq. 7.44. In the frame moving with velocity $\mathbf{v}_E$, there are no forces, because from Eq. 5.41 we find that

$$E' = E_\|. \tag{10.14}$$

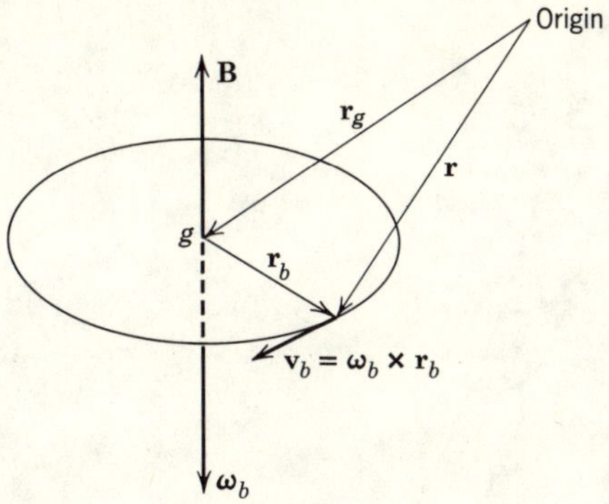

Fig. 10.2. Motion of an ion in a magnetic induction.

If a particle is injected into the crossed-field system with velocity $\mathbf{E} \times \mathbf{B}/B^2$, it travels in a straight line. This fact is utilized in the construction of charged-particle velocity filters. The velocity is independent of the particle's mass and charge, both in sign and magnitude. The associated cancellation of the Hall currents has been discussed at the end of Sec. 7.6.

In summary, the motion of a particle is composed of three components:

1. A circular motion about the guiding center at an angular frequency ω_b, the cyclotron frequency.
2. A uniform drift of the circular orbit and its guiding center in the direction perpendicular to **E** and **B**, the Hall direction, with the velocity $\mathbf{v}_E$.
3. A uniformly accelerated motion along **B**, the acceleration being $qE_{\|}/m$.

A typical trajectory is shown in Fig. 10.3. The particle is initially rotating about the origin when the electric intensity **E** is applied suddenly. The guiding center remains in the y-z plane.

10.2. Inhomogeneous Magnetic Induction

We now examine the motion of a particle moving in a magnetic induction directed everywhere parallel to the z-axis, but having a gradient normal to the z-axis. The radius of curvature of the particle's orbit will be smaller where the induction is large, and conversely. The guiding center moves as a result; charges of opposite sign drift in opposite directions, as shown in Fig. 10.4. If $\mathbf{j} = 0$, a gradient of the induction also implies that the lines curve. A charge is subjected to a centrifugal force as a result of following the curved line of induction. This force causes a cross drift.

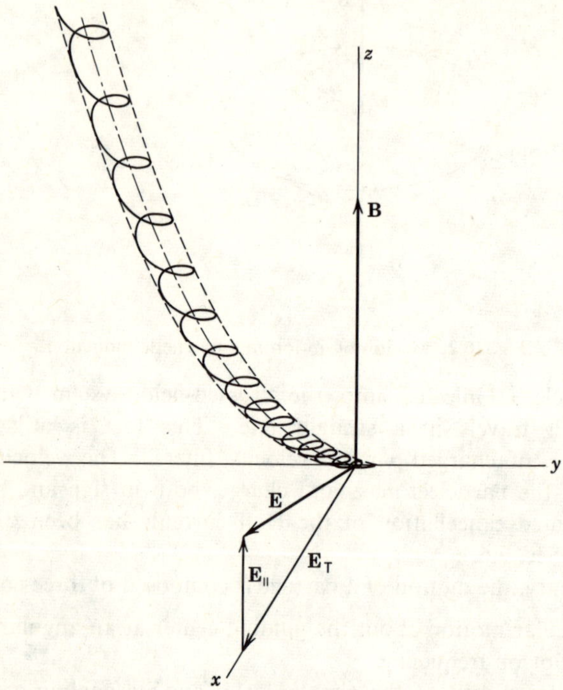

Fig. 10.3. Typical trajectory of an ion in electric and magnetic fields.

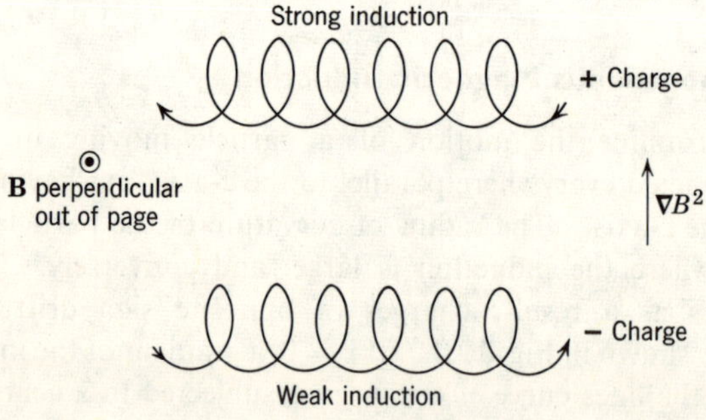

Fig. 10.4. Drift of charged particles in an inhomogeneous magnetic induction.

The rigorous derivation of the drift velocity $\mathbf{v}_g$ of the guiding center is somewhat involved. First we state the result:[1]

$$\mathbf{v}_g = \frac{mv_b^2}{4q}\frac{\mathbf{B} \times \nabla B^2}{B^4} + \frac{mv_{\|}^2}{q}\frac{\mathbf{B} \times (\mathbf{B}\cdot\nabla)\mathbf{B}}{B^4} + \mathbf{v}_{\|}, \tag{10.15}$$

where $\mathbf{v}_{\|}$ is the component of the velocity along the induction $\mathbf{B}$, and $\mathbf{v}_b$ is the component of the velocity of the particle that is perpendicular to the induction $\mathbf{B}$. The velocity $\mathbf{v}_b$ represents the motion of the particle in a circle about the guiding center.

The present resúlt, Eq. 10.15, can be made very plausible. The first term represents the drift speed resulting from the gradient of the induction, and the second term represents the drift speed caused by the force on the particle as it tries to follow a curving line. The position vector $\mathbf{r}$ of a particle and the position vector $\mathbf{r}_g$ of the guiding center are related by

$$\mathbf{r} = \mathbf{r}_g + \mathbf{r}_b \tag{10.16}$$

as in Fig. 10.2, where

$$\mathbf{r}_b = \frac{m}{qB^2}(\mathbf{B} \times \mathbf{v}). \tag{10.17}$$

As the particle drifts transverse to the induction and its gradient, the direction of the induction is hypothesized to be constant. Therefore $\mathbf{B}/B$ will be a constant along the path of the particle. The guiding center $\mathbf{r}_g$ will move at a velocity

$$\dot{\mathbf{r}}_g = \mathbf{v} - \frac{m}{qB^2}\left(\mathbf{B} \times \frac{d\mathbf{v}}{dt}\right) + \frac{m}{qB^3}\frac{dB}{dt}(\mathbf{B} \times \mathbf{v}). \tag{10.18}$$

By Newton's law, Eq. 10.1, applied to the term $d\mathbf{v}/dt$ and expansion of the resulting triple cross product, we find that the guiding center moves at a velocity

$$\dot{\mathbf{r}}_g = \mathbf{v}_{\|} + \frac{m}{qB^3}\frac{dB}{dt}(\mathbf{B} \times \mathbf{v}). \tag{10.19}$$

The time average of the second term over one cycle of the particle's motion is

$$\begin{aligned}\overline{\frac{m}{qB^3}\frac{dB}{dt}(\mathbf{B} \times \mathbf{v})} &= \overline{\frac{m}{qB^3}(\mathbf{v}\cdot\nabla B)(\mathbf{B} \times \mathbf{v})} \\ &= \frac{1}{4}\frac{m}{qB^4}v_b^2\mathbf{B} \times \nabla B^2. \end{aligned}\tag{10.20}$$

The factor $\frac{1}{4}$ appears because the average of one of the two components of v_b^2 over a cycle is $\frac{1}{2}v_b^2$ (∇B lies in one direction), and because $\nabla B = \frac{1}{2}\nabla B^2$. The velocity of the guiding center is then

$$\dot{\mathbf{r}}_g = \mathbf{v}_{\|} + \frac{1}{4}\frac{mv_b^2}{qB^4}\mathbf{B} \times \nabla B^2. \tag{10.21}$$

The second term in the result, Eq. 10.15, can be made plausible rather easily. To this end, the radius of curvature R must be related to the gradient of the induction. Figure 10.5 displays a line of induction together with its osculating circle. From this figure it is obvious that

$$\frac{\Delta B}{B\,\Delta z} = \frac{\theta}{\Delta z} = \frac{1}{R}. \tag{10.22}$$

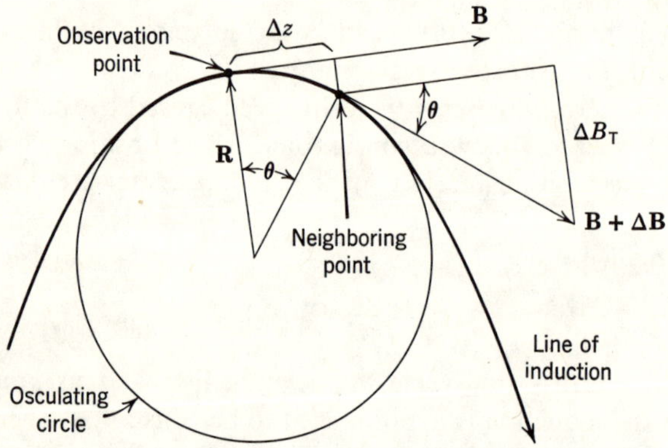

Fig. 10.5. Curving line of induction and its osculating circle. The radius of curvature is **R**.

This relation can be expressed as

$$\frac{R_\alpha}{R^2} = -\frac{\mathbf{B}\cdot\boldsymbol{\nabla}B_\alpha}{B^2}. \tag{10.23}$$

In Fig. 10.5, the coordinate system is chosen so that all components of **R** but one are zero. We measure the radius of curvature of a line from the center of curvature out to the line itself. Incidentally,

$$\frac{R_\alpha}{R^2} = -\frac{\mathbf{B}}{B}\cdot\boldsymbol{\nabla}\left(\frac{B_\alpha}{B}\right), \tag{10.24}$$

as can be seen by writing both Eqs. 10.23 and 10.24 in component form. The last result can also be derived from geometrical considerations, as was Eq. 10.23.

We can now see that the second term of Eq. 10.15 represents the drift due to the centrifugal force exerted on the particle as it attempts to follow a curving line of induction. This force is, of course, given by

$$\mathbf{F} = mv_{\|}^2\,\frac{\mathbf{R}}{R^2}. \tag{10.25}$$

The force causes a drift velocity

$$\frac{mv_{\parallel}^2(\mathbf{R} \times \mathbf{B})}{qR^2B^2}, \tag{10.26}$$

as may be seen from Eq. 10.13. If Eq. 10.23 is substituted into Eq. 10.26, the second term in Eq. 10.15 becomes obvious.

There is a temptation to identify $\mathbf{v}_g$ with a drift current flowing in a direction perpendicular to both $\mathbf{B}$ and ∇B^2. However, we should not yield to such a temptation because the identification of a current with $\mathbf{v}_g$ is often false. In fact, in Sec. 7.2 equilibrium was defined by the condition that the current density be zero, quite independent of the configuration of the static induction.

There is really no contradiction. A current arises from the drift velocities of many particles; the drift velocity is the instantaneous average of the velocities of all particles in a small volume of space, regardless of where their guiding centers may lie. On the other hand, the motion of a guiding center represents the time average of a single particle. Thus, there is no *a priori* reason to expect the two averages to be the same. Although in some cases they are the same, in many such cases they are not.

Tonks has given further insight into this somewhat perplexing matter.[2] Figure 10.6 clarifies the problem. It shows a small volume element with

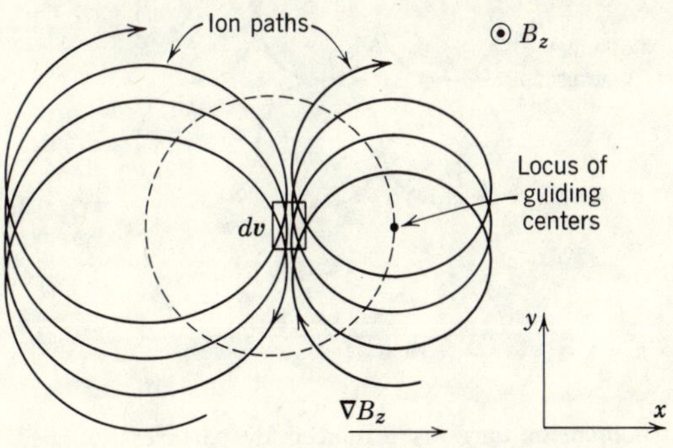

Fig. 10.6. Ion paths, guiding centers, and volume element for inhomogeneous magnetic induction B_z. The locus of guiding centers is approximately circular but not centered on dv.

ions passing through it. Ions that penetrate the element will have their guiding centers disposed in a small region about the locus shown. Tonks then sums over all possible orbits. If the density of guiding centers is uniform, that is, if the density of particles is uniform to a high approximation, it is found that the net particle flux in the volume element is identically zero. Because of the drift velocity of particles as given by Eq. 10.15, the particles in the large orbits go downward through the element of volume more slowly than the particles in the small orbits go upward. However, because of their larger orbits there are more particles traveling downward than traveling upward. The net current is then found to be zero. Further, if there is a gradient of density dn_i/dx in the distribution of Fig. 10.6, Tonks finds that a current j_y exists:

$$j_y = \frac{mv_T^2}{2B_z}\frac{dn_i}{dx}. \tag{10.27}$$

This result is in exact accord with our expectation based on the hydromagnetic model. Equation 7.3 applied to the present problem predicts that

$$\mathbf{j}_y = \nabla p/B_z. \tag{10.28}$$

In view of Eqs. 6.24 and 6.26 for an isotropic distribution, the results agree.

An actual enclosure must have physical walls. Figure 10.7 is used to illustrate two limiting cases. First, if the walls are perfectly reflecting,

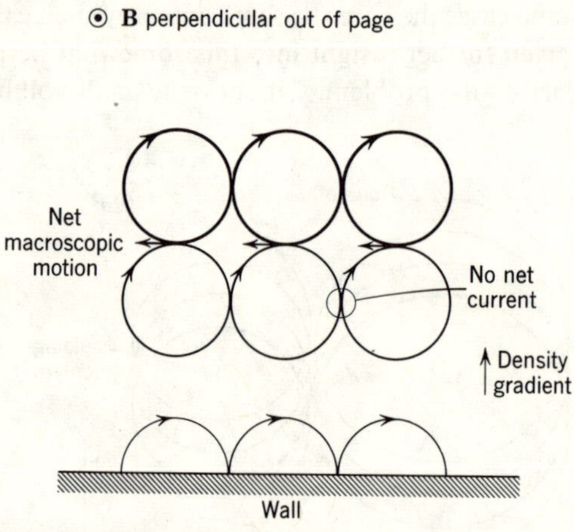

Fig. 10.7. Macroscopic ion currents produced by particles bounded by a wall. The line thickness denotes number of particles in each circle.

then there is no density gradient in equilibrium. Particles hop along the walls and exactly cancel the $\mathbf{B} \times \nabla B$ drift in the body of the plasma. On the other hand, if the walls are perfectly absorbing, a density gradient exists, no particles are reflected from the walls, and the drift velocity of the particles in this case equals the velocity of the guiding centers.

Confinement of a plasma within a torus is impossible in view of the inhomogeneous induction, as follows from the results of this section or from hydromagnetic principles. A cross section of a simple torus is shown in Fig. 10.8. A solenoidal winding (not shown) around the torus generates an induction B_z as illustrated. From Ampère's law, we see that $2\pi B_z r/\mu_0$ must be constant inside the torus, for this quantity is the current that threads the loop. Consequently,

$$B_z = B_a a/r, \tag{10.29}$$

where B_a is the value of the induction at the minimum radius of the plasma. The current destroying the confinement can be seen qualitatively with the approximation that $|\nabla n|$ is constant over the cross section of the plasma. By Eq. 10.27, ions flow upward at $r = b$ and downward at $r = a$, as shown. The current at $r = b$ is larger than that at $r = a$ because B_b is smaller than B_a. The current parallel to the midplane at the top of the discharge cancels that at the bottom. Ions therefore move upward on the average. From Eq. 10.29, which is independent of the sign of the charge, we conclude by the same argument that electrons must on the average move downward. Therefore, we deduce that the charges separate with respect to the midplane.

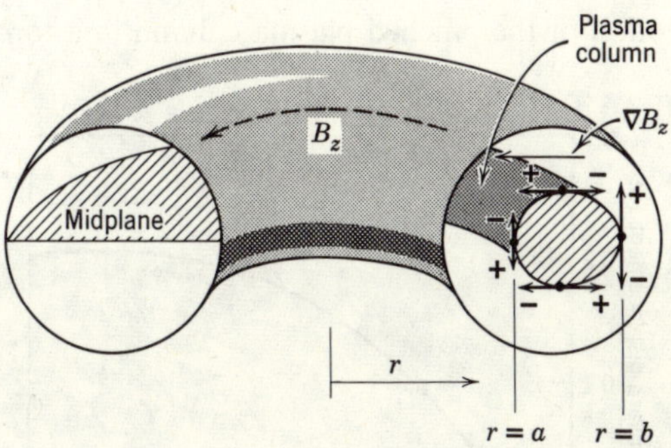

Fig. 10.8. Currents flowing in a plasma confined in a simple torus.

The separation is independent of our assumptions about ∇n. Suppose, for convenience, that the ions are at the temperature T_i. Then from Eq. 10.27 or from Eq. 7.3, we find an ion current traveling up across the midplane,

$$I_i = -kT_i \int_{r=a}^{r=b} dr\,(r/aB_a)(dn_i/dr) \tag{10.30}$$

per unit length around the axis of the torus. This equation may be integrated by parts. Since $n_i = 0$ at $r = a$ and $r = b$,

$$I_i = (n_i k T_i/aB_a)(b - a). \tag{10.31}$$

The electron current is similarly described.

This flow in itself would cause the charges to strike the top and the bottom of the torus were it not for the space-charge electric field that develops across the plasma as a result of the separation. The Debye length in the plasma is small compared with the size of the plasma. Thus the separation establishes a downward field in Fig. 10.8. An **E** x **B** drift results that drives the plasma to the outer periphery of the torus. The same result can be obtained from the concept of magnetic pressure by noting that the magnetic pressure at $r = a$ is higher than at $r = b$. A net outward force is created; the plasma travels across the field to the periphery. In the process, a motional electric field is created, according to Eq. 5.41, equal to **V** x **B**.

In passing it is interesting to note that the driving force is ∇n and not ∇B, as might be expected from Eq. 10.15.

The effect described in this section is observed in practice. Figure 10.9 displays the motion of the pinched plasma column in a toroidal machine

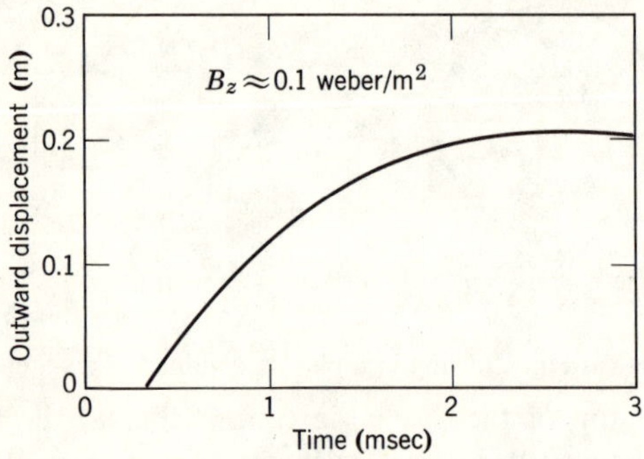

Fig. 10.9. Displacement of the current from the axis threading the toroid as a function of time in Zeta. After E. P. Butt and co-workers, *Proc. 2nd U.N. Conf. on Peaceful Uses of Atomic Energy*, Vol. 32, United Nations, Geneva (1958), p. 42.

called Zeta. The minor radius of the torus, that is, the radius of the pipe of which the torus is made, is 0.5 m. The current j_z is 150,000 amp and is carried by a plasma pinched to a radius of 0.15 or 0.20 m. The solenoidal induction B_z trapped in the plasma is about 0.1 weber/m^2. From the observed drift and magnetic induction, we calculate an electric field perpendicular to the torus plane of only 15 volts/m. The field may be much higher than this value; the outward speed is reduced by at least two effects. First, as the plasma drifts outward, the B_θ induction created by j_z is compressed against the conducting wall, which is aluminum 2.5 cm thick. Second, the magnetic energy trapped inside increases as the plasma moves toward the periphery unless the plasma expands. If it does not expand, the volume of the induction trapped inside and unchanged in magnitude increases.

The stellarator is designed to overcome the difficulty of the toroidal geometry. In the stellarator, a line of magnetic induction twists as it follows around the torus and does not necessarily close on itself, as shown in Fig. 10.10. A magnetic surface is generated as discussed in Sec. 7.2. The plasma tends to follow the lines of induction, and the radial drift in one part of the torus is canceled by a radial drift in the opposite direction in another part of the torus. The twist of the magnetic lines is produced by a set of helical windings, as discussed in Chap. 15.

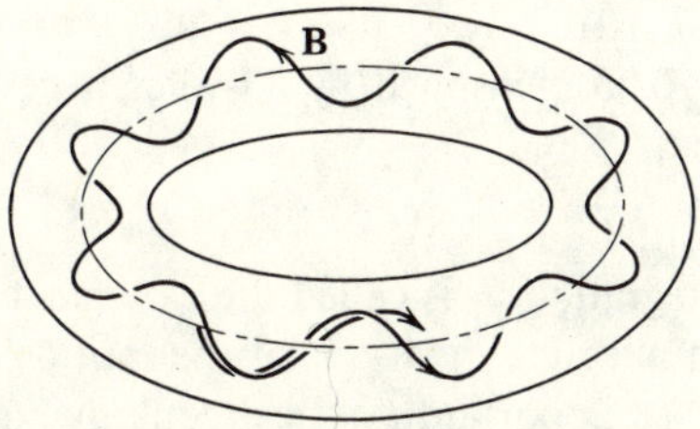

Fig. 10.10. Twisted magnetic induction in the stellarator.

★ 10.3. Invariance of Angular Momentum in a Cylindrically Symmetric Induction

In this and subsequent sections, we generally restrict our remarks to systems with an axis of cylindrical symmetry. Most thermonuclear devices possess such an axis or may be closely approximated by such a system. Some of the results of our derivations apply also to systems of arbitrary shape; we shall point out such applications.

A number of properties of the particle motion which remain invariant or almost invariant will be derived here and in subsequent sections. The notion of invariance will purposely be left vague in this preamble, because we intend to make quite clear in each derivation and associated discussion the property of the motion that is involved and the physical consequences.

We begin by considering a very straightforward case: the conservation of a quantity called the canonical angular momentum. In any cylindrically symmetric system, currents generating the magnetic induction **B** flow in the θ-direction only. Thus, the vector potential has a θ-component only by Eq. 5.59. At this point, we allow arbitrary variations in r, z, or t but note by symmetry or Eq. 5.44 that $B_\theta = 0$. We assume that no scalar electric potential exists; from Eqs. 5.44 and 5.46, the equation of motion of a charged particle is

$$m\ddot{\mathbf{r}} = q[-\dot{\mathbf{A}}_\theta + \dot{\mathbf{r}} \times (\nabla \times \mathbf{A}_\theta)]. \tag{10.32}$$

We are interested in the angular acceleration of the particle; the θ-component of Eq. 10.32 becomes, after some algebraic manipulations,

$$m(r\ddot{\theta} + 2\dot{r}\dot{\theta}) = -\frac{q}{r}\left(r\dot{A}_\theta + \dot{r}A_\theta + r\dot{r}\frac{\partial A_\theta}{\partial r} + r\dot{z}\frac{\partial A_\theta}{\partial z}\right). \tag{10.33}$$

We assert that the quantity in parentheses on the right side of Eq. 10.33 is the total time derivative of the quantity rA_θ. By this statement, we mean that $d(rA_\theta)/dt$ is the rate of change of rA_θ as the particle is followed in its orbit. To see this point, apply Eq. 6.21 to the quantity rA_θ, taking cognizance of the facts that $\partial A_\theta/\partial\theta = 0$ and $\partial r/\partial t \equiv 0$ at a fixed point in space. Equation 10.33 multiplied by r is then a perfect time derivative, and we discover that

$$mr^2\dot{\theta} + qrA_\theta = \mathfrak{p}_\theta = \text{const.} \tag{10.34}$$

The constant of integration $\mathfrak{p}_\theta$ is called the canonical angular momentum, and is evidently an absolute invariant of the motion.

This invariance has important physical consequences, as we shall now see. We can evaluate the quantity $\mathfrak{p}_\theta$, and we consider first the case where the particle does not encircle the axis, as in Fig. 10.11a. The quantity $\dot{\theta}$ is

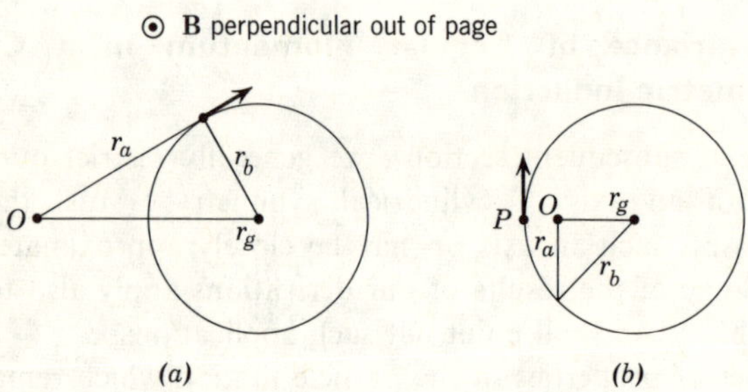

Fig. 10.11. Gyration of a positive ion (a) where it does not encircle an axis O of magnetic symmetry and (b) where it encircles the magnetic axis.

measured with respect to the axis O, and when the particle is at r_a, $\dot{\theta} = 0$. Then most generally

$$\mathfrak{p}_\theta = qr_aA_\theta(r_a). \tag{10.35}$$

The special case of a uniform induction is important; B may vary with time. If so,

$$A_\theta = Br/2, \tag{10.36}$$

whence

$$\mathfrak{p}_\theta = qBr_a{}^2/2 = (q/2\pi)(\pi r_g{}^2 - \pi r_b{}^2)B. \tag{10.37}$$

For a positively charged particle, $\mathfrak{p}_\theta > 0$ if it does not encircle the axis. The manipulations at the right of Eq. 10.37 afford a simple geometrical interpretation: the flux linked by a circle of radius r_a, or the difference of the fluxes in the r_g and r_b circles, is constant.

A similar interpretation is available for particles that encircle the axis, as in Fig. 10.11*b*. For such orbits, $\dot\theta$ is never zero; we can consider the particle at point *P*, for simplicity, note that $\dot\theta = \omega_b r_b/(r_b - r_g)$ there, and write

$$\begin{aligned}\mathfrak{p}_\theta &= -m(r_b - r_g)^2[r_b/(r_b - r_g)](qB/m) + q(r_b - r_g)^2 B/2 \\ &= -qB(r_b{}^2 - r_g{}^2)/2 = -qBr_a{}^2/2. \end{aligned} \tag{10.38}$$

In this case, we define r_a in Fig. 10.11*b* as the half-chord shown. Here $\mathfrak{p}_\theta$ is negative for a positively charged particle. The two classes of orbits have canonical angular momenta of opposite sign. They are topologically distinct and do not mix, regardless of the variations of the axially symmetric induction.

The results displayed in Eq. 10.34 may be applied to show that a particle which encircles the axis of symmetry can be confined forever in a suitably shaped time-independent induction **B**. In this case, the kinetic energy of the particle

$$U_0 = m[\dot r^2 + (r\dot\theta)^2 + \dot z^2]/2 \tag{10.39}$$

is constant. Then

$$\dot r^2 + \dot z^2 = (2U_0/m) - [(\mathfrak{p}_\theta/mr) - (qA_\theta/m)]^2, \tag{10.40}$$

where we have applied Eq. 10.34. The right-hand side of Eq. 10.40 is a known function of *r* and *z*. As the particle moves around in its orbit, the term $[(\mathfrak{p}_\theta/mr) - (qA_\theta/m)]^2$ fluctuates according to the oscillations of $r^2\dot\theta^2$. For encircling particles, $\dot\theta$ is never zero. Indeed, if the induction or vector potential is sufficiently great at the ends of the system, an encircling particle must stop moving at the point for which

$$(2U_0/m) - [(\mathfrak{p}_\theta/mr) - (qA_\theta/m)]^2 \geqslant 0. \tag{10.41}$$

The particle cannot penetrate beyond this point, for to do so would require imaginary velocity components. The particle must be reflected. The reflection of a particle that encircles the axis symmetrically is shown in Fig. 10.12. At the reflection plane, $\dot z = \dot r = 0$. A particle that encircles the axis asymmetrically is reflected earlier, because $\dot r^2 > 0$ in general for this particle; hence, $\dot z = 0$ at a lower value of A_θ, that is, at a lower value of the converging B_z.

In order to confine this class of particles forever in the absence of scattering, we need only construct a magnetic mirror from two current

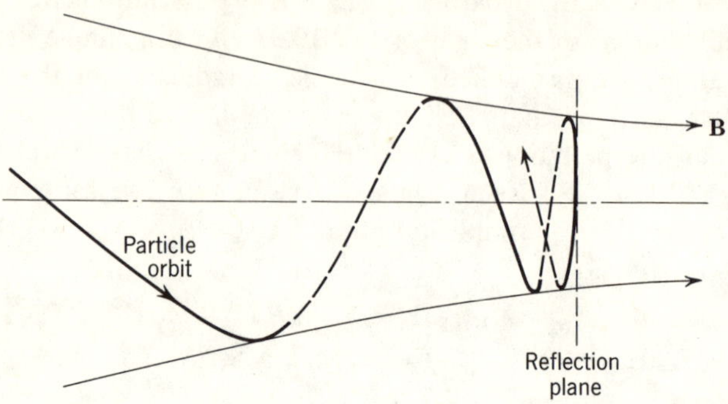

Fig. 10.12. Motion of a particle in a converging magnetic induction.

coils, as shown in Fig. 10.13. A positive particle that encircles the axis is permanently confined if

$$qA_{\theta\,\max} > (2mU_0)^{1/2} + (\mathfrak{p}_\theta/r), \tag{10.42}$$

where $A_{\theta\,\max}$ corresponds to the maximum induction under the mirrors. Encircling particles whose energy and/or canonical angular momentum do not satisfy Eq. 10.42 are not necessarily reflected inside the mirrors and may escape. The confinement condition for these latter particles will be discussed in Sec. 10.6.

The above conclusions are not applicable to a particle that does not encircle the axis, Fig. 10.11*a*. For such particles, $\mathfrak{p}_\theta > 0$, and $r^2\dot{\theta}^2 = [(\mathfrak{p}_\theta/mr) - (qA_\theta/m)]^2$ is zero at some point along the trajectory, regardless of the magnitude of A_θ at other points in the *r-z* plane. While such

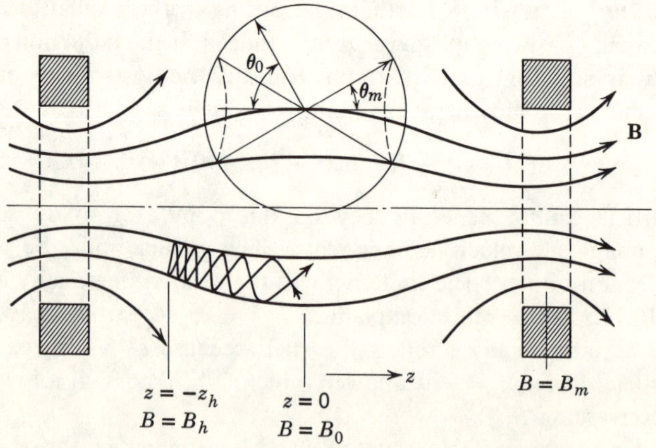

Fig. 10.13. Particle in a magnetic mirror.

particles may not leak from the system, we can no longer prove by the above argument that they will not. However, in the next two sections, we shall show that many of these particles as well as encircling ones not satisfying Condition 10.41 will be confined for many transits. We turn now to this problem.

10.4. Orbits in a Slowly Time-Varying Induction

We consider the case of a magnetic induction that changes slowly with time. Again we restrict ourselves to a cylindrically symmetric system and develop an approximate solution for the motion of the particle. There is no static electric field **E**; if one existed, it could be transformed away by the addition of a drift velocity $(\mathbf{E} \times \mathbf{B})/B^2$ and an acceleration of the guiding center, as discussed in Sec. 10.1. Figure 10.11 shows the guiding center of the particle at $\mathbf{r}_g$ and the radius of gyration $\mathbf{r}_b$. Both distances will change in magnitude with time.

The changing **B** produces an electromotive force that cannot be transformed away. The azimuthal electric field may be found from Eq. 5.7 by integrating around a circle centered at the axis of the cylinder. This component of the electric field is constant over the circle by symmetry. By Eq. 5.3, the radial and axial components of the electric field are zero in the absence of any charge anywhere. It follows that

$$\mathbf{E} = \mathbf{r} \times \dot{\mathbf{B}}/2. \tag{10.43}$$

With the usual sign conventions the equation of motion becomes

$$\ddot{\mathbf{r}} = \boldsymbol{\omega}_b \times \dot{\mathbf{r}} + (\dot{\boldsymbol{\omega}}_b \times \mathbf{r}/2). \tag{10.44}$$

This equation is rather intractable to solve exactly, so we use an approximate method.[1,3] Since only motion in a plane normal to **B** is of interest, we use the complex notation

$$\mathbf{r} = x + iy. \tag{10.45}$$

Equation 10.44 may be written as

$$\ddot{r} = i[\omega_b \dot{r} + (\dot{\omega}_b r/2)]. \tag{10.46}$$

The solution is advisedly assumed to be of the form

$$r = \rho \exp i\psi. \tag{10.47}$$

Let

$$\omega = \dot{\psi}; \tag{10.48}$$

the trial solution, Eq. 10.47 is substituted into Eq. 10.46. It is assumed that both ρ and ψ are real; the real and imaginary parts are separated. The two following relations result:

$$\omega\rho(\omega - \omega_b) = \ddot{\rho}, \tag{10.49}$$

and

$$\frac{d}{dt}\rho^2\left(\omega - \frac{\omega_b}{2}\right) = 0. \tag{10.50}$$

We use a method of successive approximations to compute the solution. Assume that $\ddot{\rho}$ in Eq. 10.49 is small compared with the other terms. Two solutions are found: $\omega = 0$ and $\omega = \omega_b$. In either case, Eq. 10.50 yields

$$\rho^2\omega_b = C, \tag{10.51}$$

where C is a constant. The general solution in the first order of approximation is then a linear superposition of these two solutions:

$$\mathbf{r} = (\mathbf{C}_1/\sqrt{\omega_b}) + (\mathbf{C}_2/\sqrt{\omega_b}) \exp i \int dt\, \omega_b, \tag{10.52}$$

where the coefficients $\mathbf{C}_1$ and $\mathbf{C}_2$ are determined from the initial conditions. The first term of Eq. 10.52 is just the position vector $\mathbf{r}_g$ to the guiding center, and the second term is the position vector $\mathbf{r}_b$ from the guiding center to the particle. The latter rotates in an almost periodic circle as ω_b slowly changes.

10.5. The Adiabatic Invariants

In the approximate solution (Eq. 10.52) of Eq. 10.44, two constants appeared that were determined by the initial conditions. To the degree that the approximation is accurate, these constants do not change during the motion of the particle. We have associated the first term of Eq. 10.52 with the distance to the guiding center, so that

$$\mathbf{r}_g = \mathbf{C}_1/\sqrt{\omega_b}; \tag{10.53}$$

hence,

$$C_1^2 = \omega_b r_g^2. \tag{10.54}$$

Note that this quantity is proportional to the flux linked by a circle of radius r_g. Thus, the guiding center of the particle moves in such a way that to first approximation the flux linked by the circle of radius r_g is constant as the induction changes. Physically, as B is increased, the lines move in toward the center; and the guiding centers of the particles follow along, thus increasing the density of the particles.

We have associated the second term of Eq. 10.52 with $\mathbf{r}_b$. Since

$$\mathbf{r}_b = (\mathbf{C}_2/\sqrt{\omega_b}) \exp i \int dt\, \omega_b, \tag{10.55}$$

$$C_2^2 = \omega_b r_b^2, \tag{10.56}$$

and the flux linking the circle having the guiding center as center also remains constant. This linked flux is proportional to the magnetic moment M of the moving charge, which in turn is given by the current traveling around the r_b circle times the area of the circle. Thus,

$$
\begin{aligned}
M &= (q\omega_b/2\pi)\pi r_b^2 \\
&= q\omega_b r_b^2/2 \\
&= U_\mathrm{T}/B \qquad (10.57)
\end{aligned}
$$

is an approximate constant of the motion. As the magnetic induction is increased, the energy in the rotational motion increases proportionally. The compression by the magnetic induction is the same as a radial compression with real walls. The radial compression heats the plasma; since two degrees of freedom are involved, $\gamma = 2$. This matter is more fully discussed in Sec. 7.5 on the basis of the hydromagnetic model. We find here an equivalent picture.

The invariance derived is not quite exact because our solution of Eq. 10.44 is not exact. The invariants derived here are true constants only in the limit of very slow changes, that is, in the limit $\dot{\omega}_b/\omega_b^2 \ll 1$.

Since the two adiabatic invariants may be expressed as $\pi B r_g^2$ and $\pi B r_b^2$, we see immediately from Eqs. 10.37 or 10.38 that their difference is truly constant and is related to the canonical angular momentum $\mathfrak{p}_\theta$. The approximate invariance of these components of $\mathfrak{p}_\theta$, especially of the magnetic moment M itself, is useful in discussing confinement of particles. The adiabatic invariance of M applies to all particles, whether they encircle the axis or not, and we shall use this invariance in discussing motion in a space-varying induction.

If the rate of change of induction is not slow, we can by analogy with the discussion of Sec. 7.5 discover which way the invariants vary. In a fast compression, the average energy is increased by more than that for a slow change. Since the energy is measured by v_h, we expect that in a fast compression the magnetic moment and the fluxes linking the two types of circles mentioned above will increase, as is indeed the case. In the next order of approximation, we should find that as B changes, the vector $\mathbf{r}_g$ also precesses slowly.[3]

10.6. Spatially Varying Inductions and Magnetic Mirrors

In Sec. 10.3, we showed that a particle that encircles the magnetic axis of Fig. 10.13 could be permanently confined if its energy and canonical angular momentum were suitably chosen. We are interested here in all the charged particles, whether they encircle the axis or not. The adiabatic invariance of the magnetic moment applies to all particles, even if the induction is not cylindrically symmetric. Therefore in this section we base our derivation of confinement upon this adiabatic invariance. A much more sophisticated discussion than ours is given by Northrop and Teller.[4]

Consider the orbit shown in Fig. 10.13. The moving particle sees a time-varying induction; the results of Secs. 10.4 and 10.5 are then applicable. Because the flux linking the circle generated by the rotation of the particle about its guiding center is constant, the particle follows a flux line for gradual changes. From Eqs. 10.56 and 10.57,

$$v_{\mathrm{T}}^2/B = \text{const.} \tag{10.58}$$

Since the energy U is constant,

$$v^2 = v_0^2 = 2U/m = v_{\mathrm{T}}^2 + v_{\|}^2 \tag{10.59}$$

is also constant, and we see that the gyrating particles tend to be reflected from regions of increasing magnetic induction.

Refer again to the magnetic mirror shown in Fig. 10.13. Let $z = 0$ at the midplane, and let a particle of speed v_0 move at an angle θ_0 with respect to the z-axis at the midplane, so that

$$v_{0\|} = v_0 \cos \theta_0, \tag{10.60}$$

$$v_{0\mathrm{T}} = v_0 \sin \theta_0. \tag{10.61}$$

As the particle moves into a region of higher induction, then by Eq. 10.58 and the constancy of v_0

$$v_{\mathrm{T}}/v_0 = \sin \theta = (B/B_0)^{1/2} \sin \theta_0. \tag{10.62}$$

Then we find immediately that

$$v_{\|} = v_0\left(1 - \frac{B}{B_0} \sin^2 \theta_0\right)^{1/2}. \tag{10.63}$$

This result states that the particle cannot penetrate into a region of arbitrarily great induction, unless it is directed exactly along the line of **B**. It will be reflected at a point z_h for which

$$B(z_h) = B_0/\sin^2 \theta_0. \tag{10.64}$$

The confinement condition is independent of mass, speed, and sign of charge of the particle. It is often convenient to speak of the mirror ratio

$$R(z) = B(z)/B_0, \tag{10.65}$$

which measures the increasing induction away from the midplane. Then Eq. 10.64 becomes

$$R_h = \csc^2 \theta_0 \tag{10.66}$$

at the reflection point. It is evident that particles for which θ_0 is too small will not be confined in the mirror of Fig. 10.13. The minimum angle θ_m for which confinement is possible is determined by the mirror ratio $R_m = B_m/B_0$ pertaining to the maximum induction just under the mirrors. Two loss cones in velocity space exist, and particles are immediately lost if

$$\theta_m < \csc^{-1} \sqrt{R_m}; \qquad (10.67)$$

particles whose vector velocities lie in these cones at the midplane are immediately lost.

We showed in Sec. 10.3 that properly chosen particles which encircle the axis will be confined in a mirror forever if collisions are neglected. In this section, we have shown that a much larger class of particles is confined for many transits. This class includes all those encircling particles that satisfy Eq. 10.41, some encircling particles that do not satisfy Eq. 10.41, and in addition all nonencircling particles that are not traveling in the loss cones.

Numerical and analytical calculations have been made to determine the constancy of the magnetic moment. We have discussed in Sec. 10.3 the case of particles that encircle the axis. For them, the magnetic moment M varies slightly about a mean value, but particles are not lost on this account. For particles that do not encircle the axis, as shown in Fig. 10.13, the space outside the mirrors is accessible to them, and we expect on the premise of ergodicity that sooner or later they will escape from the mirror. This matter is not yet finally settled; calculations indicate[5] that the small nonadiabaticity should cause the end point z_h in Fig. 10.13 to vary slightly from transit to transit.

We now consider the effect of this variation. Define now the distance L_m as the axial extent over which the mirror is developed. The distance L_m is shown in Fig. 10.14 for a mirror with long central section of length L. It is tentatively concluded that the relative size of the variation $\Delta z_h/L_m$ arising from the nonadiabaticity is proportional to $(r_b/L_m)^2$. The size

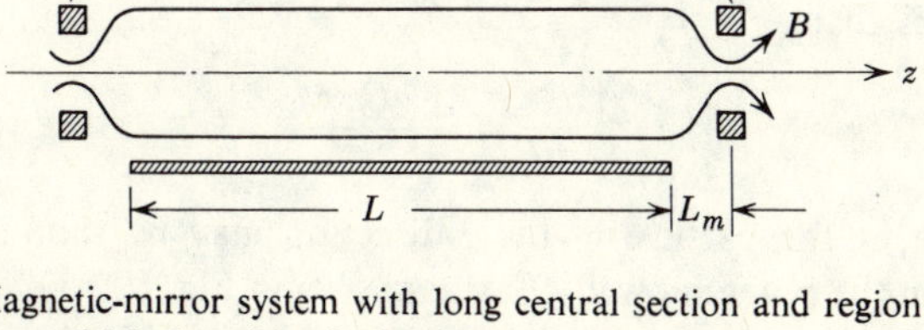

Fig. 10.14. Magnetic-mirror system with long central section and region of length L_m at each end over which the mirror is developed.

of the variation also depends upon the phase of the particle in its gyrating motion. If now we assume that the central section of the mirror is long, as in Fig. 10.14, minute inhomogeneities in the induction or slight scattering will cause the particle to lose memory of its phase between successive transits. As a result, the end point z_h performs a small random walk at each transit; that is, it diffuses slowly in the z-direction. Simple diffusion theory then shows that the particle diffuses a distance L_m through the mirror after approximately N transits, where

$$N = \left(\frac{L_m}{\Delta z_h}\right)^2 = C\left(\frac{L_m}{r_b}\right)^4. \tag{10.68}$$

Calculation shows that the number $C \approx 1$. For plasmas of thermonuclear interest, L_m/r_b is a large number for most of the particles. Then nonadiabatic effects are usually small compared to the effects of Coulomb scattering in the plasma, as discussed in Chap. 8. For very fast ions, the nonadiabatic loss given by Eq. 10.68 could be significant. Should Eq. 10.68 be essentially correct, the effect could be put to advantage in expelling α-particles from a magnetic mirror. We speculate upon this matter in Chap. 13.

The transit time of a particle between two points along the axis of the mirror is easily found. Since

$$v_{\parallel} = \frac{dz}{dt} \tag{10.69}$$

very closely in our approximation,

$$t = \frac{1}{v_0}\int dz\,[1 - R(z)\sin^2\theta_0]^{-1/2}. \tag{10.70}$$

Another invariant called the action integral exists for those types of motion that are separable.[3] A separable motion is one for which the potential energy $Q(x, y, z)$ of a particle can be written as the sum of three separate terms, each dependent on only one spatial variable:

$$Q(x, y, z) = Q_x(x) + Q_y(y) + Q_z(z). \tag{10.71}$$

The equation of motion describing the movement of the particle in the z-direction is then

$$m\frac{d^2z}{dt^2} = -\frac{\partial Q_z(z)}{\partial z}. \tag{10.72}$$

The motion of the particle in the z-direction may be thought of as an oscillation in the potential well $Q(z)$, as shown in Fig. 10.15. The particle may move in the x- and/or y-directions also, but, if it does, motion in

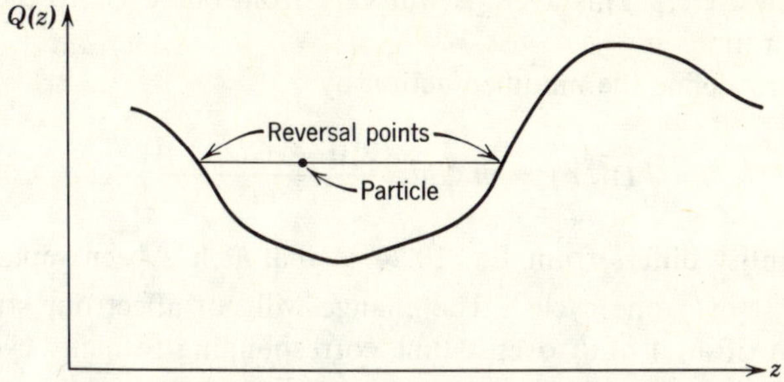

Fig. 10.15. Motion of a particle in a potential well.

these directions is independent of the motion in the z-direction and does not influence the z-component.

The equation of motion (Eq. 10.72) implies that

$$\frac{m}{2}\left(\frac{dz}{dt}\right)^2 + Q(z) = \mathfrak{U} = \text{const}, \tag{10.73}$$

which may be found by multiplying Eq. 10.72 by dz/dt and integrating over time. The quantity $\mathfrak{U}$ is a measure of the constant total energy of the particle arising from its motion in the z-direction.

The action integral for the z-motion is defined as

$$J_z = \oint dz\, mv_z, \tag{10.74}$$

where the integral is to be evaluated over a complete cycle of the motion. This definition is the normal one for the action; but we shall develop a modified one in the following discussion. A particle completes a cycle of motion in a period τ given by

$$\tau = \oint dt. \tag{10.75}$$

The potential energy Q will be determined by one or more parameters. One (let it be called Y) or more of them may vary with time, causing Q to change slowly with time; that is,

$$\frac{d}{dt} \ln Q(z, Y) \ll \frac{1}{\tau}. \tag{10.76}$$

The potential energy $Q(z, Y)$ and the total energy $\mathfrak{U}(Y)$ will then also vary. The total energy will have an average value of

$$\overline{\mathfrak{U}}(Y) = \frac{1}{\tau}\oint dt\left[\frac{m}{2}\left(\frac{dz}{dt}\right)^2 + Q(z, Y)\right] \tag{10.77}$$

over each cycle. This average will vary from one cycle to the next and thus with time.

We now define the modified action by

$$J_z(\bar{u}, Y) = m \oint dz \left\{ \frac{2[\bar{u} - Q(z, Y)]}{m} \right\}^{1/2}; \tag{10.78}$$

this quantity differs from Eq. 10.74 in that u has been replaced by its average $\bar{u}$ over one cycle. The change will not affect our study of the variation (if any) of J_z over a time corresponding to many cycles. Our purpose is now to show that even though $\bar{u}$ and Y both change gradually with time, Eq. 10.78 is independent of time to the degree that Inequality 10.76 is satisfied. The demonstration is made by showing that the total derivative of the modified action

$$\frac{dJ_z}{dt} = \left(\frac{\partial J_z}{\partial \bar{u}}\right)_Y \frac{d\bar{u}}{dt} + \left(\frac{\partial J_z}{\partial Y}\right)_{\bar{u}} \frac{dY}{dt} \tag{10.79}$$

is zero to first order in small quantities. Now from Eqs. 10.78 and 10.73, we learn that

$$\left(\frac{\partial J_z}{\partial \bar{u}}\right)_Y = \oint dz \left\{ \frac{m}{2[\bar{u} - Q(z, Y)]} \right\}^{1/2}; \tag{10.80}$$

but the right-hand side of Eq. 10.80 is the period τ, within the approximation that $\bar{u}$ changes but little over a cycle (and Inequality 10.76 is well satisfied). Next, we observe from Eqs. 10.77 and 10.72 and from Inequality 10.76 that

$$\frac{d\bar{u}}{dt} = \frac{1}{\tau} \oint dt \left(m \frac{dz}{dt} \frac{d^2z}{dt^2} + \frac{\partial Q}{\partial z} \frac{dz}{dt} + \frac{\partial Q}{\partial Y} \frac{dY}{dt} \right) = \frac{1}{\tau} \frac{dY}{dt} \oint dt \frac{\partial Q}{\partial Y}. \tag{10.81}$$

However, by Eq. 10.78,

$$\left(\frac{\partial J_z}{\partial Y}\right)_{\bar{u}} = -\oint dz \frac{\partial Q}{\partial Y} \left\{ \frac{m}{2[\bar{u} - Q(z, Y)]} \right\}^{1/2}, \tag{10.82}$$

which becomes

$$\left(\frac{\partial J_z}{\partial Y}\right)_{\bar{u}} = -\oint dt \frac{\partial Q}{\partial Y} \tag{10.83}$$

by Eq. 10.73. If Eqs. 10.79, 10.80, 10.81 and 10.83 are combined, we find directly that

$$\frac{dJ_z}{dt} = 0. \tag{10.84}$$

The invariance of the action integral may be applied to the heating induced by bringing two mirrors nearer together. Perhaps the simplest case is that in which the magnetic induction increases very rapidly at the mirrors but not so rapidly that the motion of the particle is nonadiabatic (that is, irreversible). Then Eqs. 10.74 and 10.84 yield

$$v_{\|} \propto 1/L, \tag{10.85}$$

where L is the distance between mirrors, as shown in Fig. 10.14. Therefore, we find

$$mv_{\|}^2/2 \propto 1/L^2 \propto n^2, \tag{10.86}$$

where n is the density. If we assume that the compression is one-dimensional and that the transfer of energy to the other two dimensions is negligible, the temperature $T_{\|}$ is given by

$$T_{\|} \propto n^2, \tag{10.87}$$

in agreement with our expectation from Sec. 7.5.

Heating by axial compression has the disadvantage that the velocity of a charged particle becomes more nearly parallel to the magnetic axis as the axial compression proceeds. The particle rides further up on the mirrors. The adiabatic invariance of the action integral may be easily applied to calculate the extent of the effect. For sufficiently great compression, a particle will be lost through the loss cone. Conversely, axial expansion improves the confinement at the expense of reducing the energy of the trapped particles.

10.7. Diffusion of Like Particles across a Magnetic Induction

At the end of Sec. 7.2 and in Secs. 7.6 and 7.7, we noted that like-particle collisions apparently led to no diffusion across a magnetic induction. We return now to this point, and rederive the conclusion from consideration of the individual orbits.

The position of two particles 1 and 2 before a collision are described by the position vectors $\mathbf{r}_1$ and $\mathbf{r}_2$:

$$\mathbf{r}_1 = \mathbf{r}_{g1} + m_1(\mathbf{v}_1 \times \mathbf{B})/q_1 B^2, \tag{10.88}$$

$$\mathbf{r}_2 = \mathbf{r}_{g2} + m_2(\mathbf{v}_2 \times \mathbf{B})/q_2 B^2. \tag{10.89}$$

After the collision the positions are

$$\mathbf{r}_1' = \mathbf{r}_{g1}' + m_1(\mathbf{v}_1' \times \mathbf{B})/q_1 B^2, \tag{10.90}$$

$$\mathbf{r}_2' = \mathbf{r}_{g2}' + m_2(\mathbf{v}_2' \times \mathbf{B})/q_2 B^2. \tag{10.91}$$

At the instant of collision, $\mathbf{r}_1 = \mathbf{r}_2 = \mathbf{r}_1' = \mathbf{r}_2'$. In addition, the total momentum is conserved in the collision, so that

$$m_1\mathbf{v}_1 + m_2\mathbf{v}_2 = m_1\mathbf{v}_1' + m_2\mathbf{v}_2'. \tag{10.92}$$

Then we find that

$$q_1(\mathbf{r}'_{g1} - \mathbf{r}_{g1}) + q_2(\mathbf{r}'_{g2} - \mathbf{r}_{g2}) = 0. \tag{10.93}$$

If the particles are identical, the average position of the guiding centers does not move. Therefore, no net transport of material can take place. If the particles have opposite charge, the guiding centers of the two particles are both displaced by the same amount. In this latter case, a transport of material exists, and the material can diffuse across the induction. We note that the equal displacement of positive and negative charge is in accord with our discussion of Sec. 7.2 and with Eqs. 7.8 and 7.9, where it was shown that no electric current flows in the direction of the gradient.

While no diffusion arises in first order, Simon[6] and Longmire and Rosenbluth[7] have performed a more exact calculation, taking into account higher-order effects.† They find that a particle current exists, arising from higher space derivatives of the density. Longmire and Rosenbluth find the particle current to be

$$\Gamma = \frac{8}{15\pi}\left(\frac{kT}{m}\right)^2 \frac{n}{\tau_\theta \omega_b{}^4} \frac{d}{dx}\left(\frac{1}{n}\frac{d^2 n}{dx^2}\right), \tag{10.94}$$

where τ_θ is the relaxation time given by Eq. 8.22 and x is linear distance. Equation 10.94 is valid if $\omega_b \tau_\theta \gg 1$. If the variation in density is small over a Larmor radius (which is usually the case), this like-particle current is small compared with that arising from unlike-particle collisions.

In view of the fact that momentum is conserved in any collision, whether a magnetic induction is present or not, our orbit demonstration that no particle current arises from like-particle collisions requires some further interpretation. We inquire therefore into the connection, if any, between diffusion and collisions. We must return to our basic concept of diffusion. In Sec. 4.4, we showed that a density gradient and associated particle current required an asymmetry of the velocity distribution. In the simple case of no magnetic induction, we found that each particle has on the average a drift velocity in the direction of decreasing density. Equivalently, the velocity distribution was required to have an anisotropy. Microscopically, this drift arises from the particles of an initially dense blob of gas sorting out in space according to their various velocities; collisions merely impede the process. In a simple gas, the position vectors $\mathbf{r}_1$ and $\mathbf{r}_2$ of two particles before a collision move on the average

† Note that Eq. 7.3, which originally led to no first-order particle current, contains the scalar pressure p. More accurately, the stress tensor P should be used. Simon uses the stress tensor appropriate for the plasma in a magnetic induction. Longmire and Rosenbluth derive their results on the basis of the Fokker-Planck equation.

with the drift motion. If, however, the particles are tied to fixed lines of induction, this average motion of the position vectors $\mathbf{r}_1$ and $\mathbf{r}_2$ between collisions (essential for diffusion in the absence of collisions) vanishes entirely. Since this fact is true for all particles or pairs of particles, whether they are identical or not, we see that the normal diffusion mechanism is entirely absent in the presence of an induction. Any diffusion found must arise from some other mechanism. We have shown that unlike-particle collisions *do* provide one of these other mechanisms; like-particle collisions do not. The same conclusions were stated in Secs. 7.2, 7.6, and 7.7. Diffusion across the induction arises because of collisions and not in spite of them. See Eq. 7.49 and its transverse components.

★ 10.8. Meaning of the Lines of Induction

In this section, we wish to examine the meaning of a line of induction. In previous work, we have assumed that these lines exist and that charged particles will follow the motion of these lines in a perfect conductor. Our purpose is to point out that the concept can sometimes lead to error when applied to imperfect conductors, and to state some of the simpler conditions relating to validity of the concept. Insofar as the Maxwell equations are concerned, the concept of a line of induction is quite extraneous. The Maxwell equations do not require the concept; they do constrain the concept once it is invented, however. This subject has been rather extensively studied.[8]

A line of induction is not an observable in itself. Only certain consequences of the concept can be observed. For example, we may study the manner in which the flux through a closed loop changes with time. If we ascribe a velocity $\mathbf{v}$ to a line of induction, we may say that this velocity preserves the flux if the flux linking a closed loop moving with this velocity is constant. It may be that the velocity $\mathbf{v}$ is a function of position, in which case different parts of the loop move with different velocities.

We can easily show that a velocity is flux-preserving if and only if

$$\nabla \times (\mathbf{E} + \mathbf{v} \times \mathbf{B}) = 0, \qquad (10.95)$$

by computing the total change with time of the flux through a loop. This rate of change is

$$\frac{d\Phi}{dt} = \int d\mathbf{s} \cdot \dot{\mathbf{B}} + \oint d\boldsymbol{\ell} \times \mathbf{v} \cdot \mathbf{B}. \qquad (10.96)$$

The first term represents the change of flux linking the closed loop, caused by the change of the induction. The second term represents the change due to the motion and change of shape of the loop. Application of Eq. 5.9 and Stokes's theorem reveals that

$$\frac{d\Phi}{dt} = -\int d\mathbf{s}\cdot\nabla \times (\mathbf{E} + \mathbf{v} \times \mathbf{B}). \tag{10.97}$$

This expression will vanish for every possible closed loop if and only if

$$\nabla \times (\mathbf{E} + \mathbf{v} \times \mathbf{B}) = 0. \tag{10.98}$$

Many flux-preserving velocities will exist in general, for Eq. 10.98 requires only that

$$\mathbf{E} + \mathbf{v} \times \mathbf{B} = \nabla\phi, \tag{10.99}$$

where ϕ is any scalar function. For a perfect conductor, this equation is trivially true. The left-hand side is the electric field in the moving system. Even in an imperfect conductor, flux preservation exists if this field is derivable from a scalar potential. If part of the electric field arises from a vector potential (see Eq. 5.46), the system must be radiating, and flux of induction is obviously not preserved. From Eq. 10.98, it is easy to see that the drift velocity $\mathbf{v}_E$ of particles in electric and magnetic fields will be flux-preserving if and only if

$$\nabla \times [\mathbf{B}(\mathbf{E}\cdot\mathbf{B})/B^2] = 0. \tag{10.100}$$

If this condition is satisfied, and only if it is, charges may be thought of as moving with the lines of induction. Clearly, whenever the electric intensity is perpendicular to the induction, the condition of Eq. 10.100 is satisfied. It may also be satisfied under other conditions. However, it will not be satisfied, for example, by a plane wave having an electric intensity component along an induction applied externally and traveling across the induction. This case is again one in which the system radiates.

If a moving line is tangent to the induction at all times, the velocity of this line is said to be line-preserving. A flux-preserving velocity must be line-preserving, for otherwise the flux would change through a loop moving with the flux-preserving velocity. This point may be easily seen by considering Fig. 10.16. This figure shows a tiny loop through which there is no flux initially and which moves with a flux-preserving velocity. Line preservation does not imply flux preservation, however.

Lines of induction moving with a velocity $\mathbf{v}$ must then satisfy the conditions (1) that only one line pass through each point of space, (2) that the lines remain tangent to the induction at all times, (3) that the density of lines equal the induction, and (4) that the electromotive force around a closed loop be equal to minus the rate of change of the number of lines

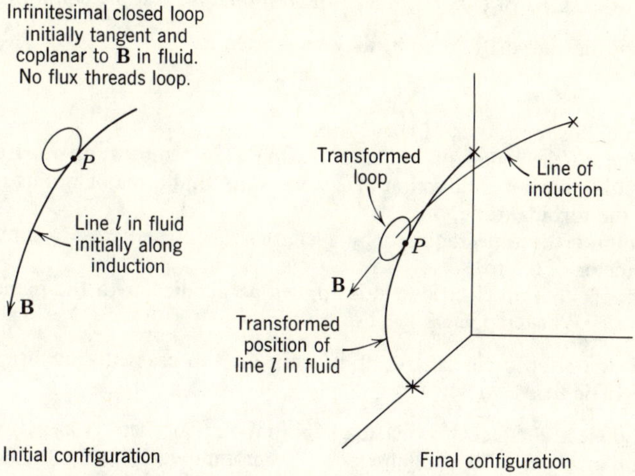

Fig. 10.16. Flux preservation implies line preservation. We show a *reductio ad absurdum.* If a flux-preserving velocity were not line-preserving, then a loop moving with the flux-preserving velocity would exist, having zero initial flux, and that at later time would link flux. To illustrate the three-dimensional aspects, we show a coordinate system in the final configuration.

linking the loop. Conditions 2 and 3 imply that the velocity is flux-preserving. Conversely, if $\mathbf{v}$ is flux-preserving, conditions 1 and 2 are implied by any lines initially tangent to the induction, since flux preservation implies line preservation. Flux preservation implies condition 3, and that the electromotive force induced in a closed loop is

$$\oint d\boldsymbol{\ell}\cdot(\mathbf{E} + \mathbf{v}_l \times \mathbf{B}) = \oint d\boldsymbol{\ell}\cdot(\mathbf{E} + \mathbf{v} \times \mathbf{B}) + \oint d\boldsymbol{\ell} \times (\mathbf{v}_l - \mathbf{v})\cdot\mathbf{B}, \quad (10.101)$$

where $\mathbf{v}_l$ is the velocity of the loop. By Eq. 10.98, the first term vanishes, so the answer obtained agrees with that deduced from condition 4. The velocity $\mathbf{v}_l$ of the loop relative to that of the lines is just $\mathbf{v}_l - \mathbf{v}$. Thus, the flux preservation of the velocity $\mathbf{v}$ implies the consequences of lines of induction moving with velocity $\mathbf{v}$, and conversely.

Numerous other theorems relating to this subject have been proved. We refer the interested reader to the literature.[8]

PROBLEMS

1. Show that the velocity of a charged particle moving in a uniform induction $\mathbf{B}$ satisfies the equation

$$\dot{\mathbf{v}} = (q/m)(\mathbf{v} \times \mathbf{B}),$$

where m is the relativistic mass of the particle.

2. The drift speed of charges in a toroidal induction is to be computed.

(*a*) Show that the drift speed v_g may be written as

$$v_g = 2kT/qBR,$$

where R is the radius of curvature.

(*b*) Compute the value of v_g for a plasma at a temperature of 10 kev, a toroidal magnetic induction of 2 webers/m^2, and a major radius of 1.0 m for the toroidal tube.

(*c*) Compute the time required by a charge to diffuse across a minor tube diameter of 0.5 m.

(*d*) Suppose that an electric field is applied perpendicular to the plane of the torus. What happens?

3. Find the electric current resulting from drift in crossed gravitational and magnetic fields.

4. The dielectric effect of a plasma of initial electron density n_0 between two plane-parallel plates of a condenser is to be computed by orbit theory. The plates are perpendicular to the x-axis; a uniform induction B is parallel to the z-axis. The spacing between the plates is large compared with the Larmor radius.

(*a*) Show that the particles are displaced a distance $mE(x)/qB^2$ along the x-axis as an electric field $E(x)$ is turned on slowly. Hint: Use Eq. 10.13 and equate the work done by the electric field to the energy gain of the particles.

(*b*) By charge conservation, Eq. 5.3, and the result of (*a*), show that the electron density at the positive plate is $n_0\epsilon_0/[\epsilon_0 + (m_e n_0/B^2)]$.

(*c*) By considering the number of electrons that move into the anode and by requiring neutrality, show that the positive charge supplied per unit area from external sources is $[\epsilon_0 + (n_0 m_e/B^2)]E_0$, where E_0 is the field at the anode.

(*d*) Show that the effective capacity of the condenser per unit plate area is

$$\frac{\epsilon_0 + (n_0 m_e/B^2)}{\frac{1}{2}(m_e E_0/q_e B^2) + (q_i m_e/q_e m_i)d},$$

where d is the distance between the plates. Note that the capacitance is very great and dependent upon the field strength.

(*e*) Compute the energy stored in this condenser when the potential ϕ exists across the plates.

(*f*) In what manner is this energy stored?

5. Prove that as an off-axis particle gradually escapes from a magnetic-mirror confinement system because of nonadiabatic effects, its guiding center r_g moves toward the magnetic axis. In terms of the canonical angular momentum, what is the minimum value of r_g?

6. A particle rotates in an initially static induction. Compute its energy change for a linear rise of the induction as a function of time and rate of change of the induction. Suppose that the induction changes by 10^4 webers/m^2-sec. Compute the factor by which the energy of an electron changes.

7. Considering that **B** is a vector, the magnetic moment of a particle

$$\mathbf{M} = -(q^2 r_b{}^2/2m)\mathbf{B}$$

is also a vector. Its interaction with a changing induction is governed by tensor forces. Show that the force on a particle with magnetic moment **M** is $\mathbf{F} = \mathbf{M}\cdot(\nabla\mathbf{B})$ and that the particle is therefore reflected by an increasing induction.

8. Particles are created at the midplane of a symmetric double magnetic-mirror system with an isotropic velocity distribution. Calculate the fraction lost immediately through the mirrors if the mirror ratio is R.

9. From the results of Problem 8:

(*a*) Estimate the time for the loss of an α-particle of 2 Mev; of 0.5 Mev. The deuteron density is $10^{20}/\text{m}^3$, and the energy is $T_D = 5$ kev.

(*b*) If $B = 2$ webers/m^2, and each mirror is 1 m long, estimate whether the α-particles escape by Coulomb scattering or nonadiabatic effects.

10. Suppose that a cylindrically symmetric induction varies parabolically with axial distance, for example, $B_0[1 + (z/a)^2]$. Show that $B_0 z_h{}^4 = \text{const}$ as the induction is made greater, where z_h is the value of z at which a particle is reflected.

11. An electromagnetic wave travels through a static, uniform induction. Show that a unique flux-preserving velocity can be specified to first order if it is proportional to $\exp i(\boldsymbol{\kappa}\cdot\mathbf{r} - \omega t)$, the oscillatory factor in the electromagnetic wave. Assume that the induction of the wave is much smaller than that of the static, uniform induction.

12. Show that if an electromagnetic wave travels at an angle other than 0° to a uniform induction and if the electric intensity is not perpendicular to this induction, the lines of induction and the particles do not move with the same velocity.

13. Show that particles in a stabilized straight pinch type of device do not travel with a flux-preserving velocity. A straight pinch consists of a current so strong that the azimuthal induction created compresses the plasma. A sizable electric field is required to drive the current. An axial induction may be used to stabilize the discharge.

REFERENCES

1. W. P. Allis, "Motions of Ions and Electrons," *Handbuch der Physik*, Vol. 21, Springer, Berlin (1957), pp. 383–444.
2. L. Tonks, *Phys. Rev.*, **97**, 1443 (1955).
3. S. Chandrasekhar, "Adiabatic Invariants in the Motion of Charged Particles," in R. K. M. Landshoff (Ed.), *The Plasma in a Magnetic Field*, Stanford University Press, Stanford (1958), pp. 10 ff.
4. T. G. Northrop and E. Teller, *Phys. Rev.*, **117**, 215 (1960).
5. S. Yoshikawa, S.M. Thesis, Department of Nuclear Engineering, Massachusetts Institute of Technology (1960).
6. A. Simon, *Phys. Rev.*, **100**, 1557 (1955).
7. C. L. Longmire and M. N. Rosenbluth, *Phys. Rev.*, **103**, 507 (1957).
8. W. A. Newcomb, *Ann. Phys.*, **3**, 347 (1958).

11

Radiation losses from a plasma

Up to this point we have been silent about the electromagnetic energy radiated from a plasma. The radiation from a charged particle increases very rapidly with its acceleration. For this reason and because electrons are so much lighter than ions, electrons will radiate copiously if their temperature is in the thermonuclear range. It is desirable in heating a thermonuclear plasma to heat the ions and to keep the electrons relatively cool. There is little else that can be effectively done to prevent the radiation losses. Unfortunately these losses are severe. The optimization of the design parameters, particularly the choice of the temperatures of the constituents and the size of the plasma, is in part determined by these losses.

We shall examine the problem of radiation losses in this chapter. Radiation losses are of two general types: bremsstrahlung and cyclotron radiation. The two are really manifestations of the same phenomenon, namely, radiation caused by the acceleration of the charges. Bremsstrahlung occurs when an electron passes close to a nucleus and is accelerated by the nuclear charge. Quantum mechanics must be used to calculate correctly the bremsstrahlung. Most of the radiant energy has short wavelength, and the average energy of the photons corresponds roughly to the electron energy. Thus in our applications, the bremsstrahlung lies principally in the x-ray region. Cyclotron radiation emerges as charged particles gyrate in the applied magnetic induction. The radiant energy is found principally at the first 10 to 30 harmonics of the cyclotron frequency, depending upon the electron energy. Thus the cyclotron radiation exists largely in the millimeter wavelength region. It may be computed with classical mechanics. The division between the two types of radiation is somewhat arbitrary, because accelerated electrons cause both types of emission. However, the frequency spectra are so disparate that we can with negligible error consider each separately.

Because the electrons are the principal radiators of energy, a kind of competition takes place between the radiative losses in cooling them and the transfer of energy to them by collisions with ions. In this simple view, the temperature of the electrons and hence the radiative losses of the system are determined by the condition that the rate of radiation by the electrons be equal to the rate of gain by collisions with the ions. Fortunately, the transfer of energy from the ions to the electrons is inefficient, because of the great difference in masses. For this reason, the electrons will in general exist at a somewhat lower temperature than the ions. We shall see that the bremsstrahlung of the electrons is very sensitively dependent on the atomic number of the ions. Consequently, impurities have a severe effect in reducing the temperature of the electrons and, in turn, of the ions themselves. This rate of energy transfer between ions and electrons has been discussed in Sec. 8.3.

11.1. Emission, Absorption, and Black-Body Radiation

If a plasma were a black body throughout the spectrum, it would radiate according to the Stefan-Boltzmann law at a rate

$$S_B = aT^4, \tag{11.1}$$

where $a = 5.67 \times 10^{-8}$ watt/m^2-°K^4 is the Stefan-Boltzmann constant. At $T = 3 \times 10^8$ °K, $S_B = 4.5 \times 10^{26}$ watts/m^2 over the exposed surface. Fortunately, this stupendous flux of power is incorrect, since the plasma is not a black body. However, we shall shortly discover that the plasma appears as a black body at low frequency, and we shall from time to time compare the actual radiation $S(\omega)\, d\omega$ in some frequency range with the black-body radiation $S_B(\omega)$ over the same range.

We begin our investigation by summarizing some applicable thermodynamic laws. For a material in which the particles are in kinetic equilibrium, the emissivity and energy absorption are related by

$$\frac{w(\omega, \mathbf{\Omega}, \mathbf{r}, T)}{\alpha(\omega, \mathbf{\Omega}, \mathbf{r}, T)} = S_B(\omega, \mathbf{\Omega}, T). \tag{11.2}$$

Here, $w(\omega, \mathbf{\Omega}, \mathbf{r}, T)$ is the power generated per ünit volume, unit frequency interval at ω, and unit solid angle in a direction $\mathbf{\Omega}$, at a position $\mathbf{r}$ where the temperature is T. The quantity $\alpha(\omega, \mathbf{\Omega}, \mathbf{r}, T)$ is the energy absorption coefficient for a ray of frequency ω propagating in the direction $\mathbf{\Omega}$ at $\mathbf{r}$. The quantity $S_B(\omega, \mathbf{\Omega}, T)$ is the spectral intensity of black-body radiation at temperature T per unit solid angle. Thus if either the emission w or the absorption coefficient α is known, the other can be computed. Equation 11.2 is strictly true only if the temperature is sensibly constant over an absorption distance; we shall ignore any consequent corrections.

In cases of interest to us, the frequency is sufficiently high and the plasma size is sufficiently great that a propagation path may be calculated by ray tracing. For such cases, the radiation from the surface can in principle be obtained from Eq. 11.2 in a straightforward way if either w or α is known. Figure 11.1 shows a plasma bounded by surfaces; we wish to compute the radiation leaving it per unit area at point P. First, consider radiation in the direction $\boldsymbol{\kappa}$ and trace the emergent ray backward through the plasma until it again reaches the surface, at point O. Then choose any point $\mathbf{r}$ along the ray and forward propagation in the direction $d\ell$ toward P. The intensity $S(\omega, \mathbf{\Omega}, \mathbf{r}, T)$ of the radiation along the ray is determined by

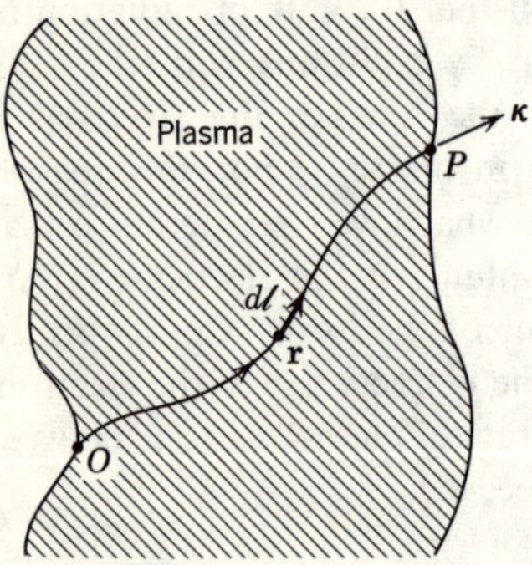

Fig. 11.1. Ray path through a plasma.

$$\frac{dS(\omega, \mathbf{\Omega}, \mathbf{r}, T)}{d\ell} = -\alpha(\omega, \mathbf{\Omega}, \mathbf{r}, T)S + w(\omega, \mathbf{\Omega}, \mathbf{r}, T). \tag{11.3}$$

Equation 11.3 states that the ray is attenuated by the αS term and reinforced by generation along its path. In the interest of brevity, we temporarily omit some of the parenthetical dependences.

From Eq. 11.2, Eq. 11.3 can be written

$$\frac{dS(\omega, \mathbf{\Omega}, \mathbf{r}, T)}{d\ell} = -\alpha(S - S_B). \tag{11.4}$$

For the purpose of illustration, let us assume that the temperature is uniform, so that S_B is independent of position. We also assume that no radiation falls upon the plasma from the outside, so that $S(O) = 0$. (We discuss the effect of reflectors in Sec. 11.6.) The solution of Eq. 11.4 is then

$$S(P, \omega, \boldsymbol{\kappa}) = S_B(\omega, \mathbf{\Omega}) \left[1 - \exp\left(-\int_O^P d\ell\, \alpha\right)\right] \tag{11.5}$$

per unit solid angle in direction $\boldsymbol{\kappa}$ at point P. In order to find the total radiation from point P, we repeat the procedure leading to Eq. 11.5 for all frequencies ω and all directions $\boldsymbol{\kappa}$. Then we erect a hemisphere over the point P and integrate over all frequencies and directions. The total radiation from a unit area of magnitude ds at P is

$$S(P)\,ds = \int_0^\infty d\omega \int_{\text{hemisphere}} d\boldsymbol{\Omega}\, S(P, \omega, \boldsymbol{\kappa})\, d\mathbf{s}\cdot\boldsymbol{\kappa}/|\boldsymbol{\kappa}|. \tag{11.6}$$

The factor $d\mathbf{s}\cdot\boldsymbol{\kappa}/|\boldsymbol{\kappa}|$ accounts for the changing aspect of the area $d\mathbf{s}$ as a function of angle.

Although the difficulties involved in computation of Eqs. 11.5 and 11.6 might be astronomical, we can obtain insight from simple examples. Equation 11.5 states that

$$S \approx S_B, \qquad \text{if} \int d\ell\,\alpha \gg 1; \tag{11.7}$$

that is, the plasma radiates as a black body if the absorption is high. On the other hand,

$$S \approx S_B \int d\ell\,\alpha, \qquad \text{if} \int d\ell\,\alpha \ll 1. \tag{11.8}$$

The radiation is always less than that of a black body, and for low absorption the radiation is small.

For x radiation characteristic of bremsstrahlung, the absorption in a fully stripped plasma is very small, corresponding to a mean free path of cosmical distances. The power density w_x of bremsstrahlung is calculated directly, and we assume that all of it escapes. Since $\int d\ell\,\alpha$ is very small for this radiation, the radiation in any spectral range is much less than that of a black body. On the other hand, the spectral range is huge, so that the power radiated is of much concern.

For cyclotron radiation, we shall find that the quantity $\int d\ell\,\alpha$ is large for low harmonics of the cyclotron frequency and small at high harmonics. Therefore, the developments of this section will be applied in toto.

Further to these considerations, we have already remarked in Chap. 9 that the plasma is virtually opaque to radiation at frequencies ω less than the plasma frequency ω_p. Here, a comparison of the cyclotron and plasma frequencies is useful:

$$\begin{aligned}(\omega_b/\omega_p)^2 &= B^2\epsilon_0/m_e n_e \\ &= (B^2/2\mu_0)(2/n_e m_e c^2) \\ &= 2\overline{v_e{}^2}/3\beta_e c^2,\end{aligned} \tag{11.9}$$

where β_e is the ratio of electron to magnetic pressure. For plasmas of thermonuclear interest, we expect that $\beta_e \approx 0.1$ and $\overline{v^2}/c^2 \approx 0.1$. In such a case, ω_b and ω_p are comparable, and the fundamental of the cyclotron frequency will not appreciably affect the radiation balance. However, the electrons are slightly relativistic, and, in addition, the magnetic induction may not be uniform. In either case, harmonics of the cyclotron frequency are generated; the previous discussion then leads us to expect that the plasma radiates more like a black body through the lower range of these harmonics.

Throughout this chapter, ordinary excitation and recombination radiation are neglected, as befits our study here of a plasma having completely stripped ions.

11.2. Bremsstrahlung

As pointed out in the introduction, bremsstrahlung originates from an accelerated charge. Collisions between electrons produce no radiation in lowest order because the acceleration of one electron is equal and opposite to that of the other. Neither a net charge displacement nor a current is created; no dipole radiation can be produced, and only higher-order effects occur with very low probability. To put the matter another way, the incipient electromagnetic waves of the two electrons destructively interfere. On the other hand, the interaction of an electron with an ion does result in unequal accelerations, and we shall discuss the first-order bremsstrahlung arising on this account.†

Quantum mechanics must be used to compute the bremsstrahlung of electrons. The accurate theory is quite complicated, but for our purposes a simplified version will suffice. According to classical electromagnetic theory, the total power $\dot{U}$ radiated from a charge q moving with a velocity $\mathbf{v}$ and an acceleration $\dot{\mathbf{v}}$ is[1]

$$\frac{dU}{dt} = \frac{q^2}{6\pi\epsilon_0 c^3} \frac{\dot{v}^2 - (\mathbf{v} \times \dot{\mathbf{v}})^2/c^2}{(1 - v^2/c^2)^3}. \tag{11.10}$$

The actual calculation of the radiation is difficult because $\mathbf{v}$ and $\dot{\mathbf{v}}$ are not well known. We therefore assume that an electron of speed v has an impact parameter b with respect to an ion of charge q_i, which is assumed to be infinitely massive. The electron acceleration is

$$\dot{v} \approx \frac{q_e q_i}{4\pi\epsilon_0 m b^2} \tag{11.11}$$

† If the electron temperature is very high (for example > 50 kev), relativistic effects produce unequal accelerations, and electron-electron bremsstrahlung is not entirely negligible. However, since it is small compared with electron-ion bremsstrahlung, we shall neglect it.

and lasts for a time of about b/v. With the assumption that $v^2 \ll c^2$, we neglect the cross product in Eq. 11.10 and write

$$\frac{dU}{dt} = \frac{q_e^2 \dot{v}^2}{6\pi\epsilon_0 c^3}; \tag{11.12}$$

whence, by insertion from Eq. 11.11,

$$\frac{dU}{dt} = \frac{q_i^2 q_e^4}{96\pi^3\epsilon_0^3 c^3 m_e^2 b^4}. \tag{11.13}$$

The total energy-loss rate by the electron colliding with ions at all impact parameters is

$$\frac{dU}{dt} = \frac{q_i^2 q_e^4 (n_i v_e)}{96\pi^3\epsilon_0^3 c^3 m_e^2} \int_{b_{\min}}^{b_{\max}} \frac{db}{b^4}\,(2\pi b)\,\frac{b}{v_e}, \tag{11.14}$$

where n_i is the ion density.

Unlike the case of the Coulomb-scattering cross section, Eq. 8.14, the integral here does not diverge as $b_{\max} \to \infty$. We therefore take $b_{\max} = \infty$ and would find that the somewhat more logical procedure of setting $b_{\max}$ equal to the Debye length introduces a negligible correction. However, the integral will diverge as $b_{\min} \to 0$. The minimum distance cannot be less than the Compton wavelength of the electron, namely,

$$b_{\min} = \frac{h}{2\pi m_e v_e}. \tag{11.15}$$

With this insertion, the power density w_x radiated by all electrons becomes

$$w_x = \frac{q_i^2 q_e^4 n_i n_e}{24\pi\epsilon_0^3 c^3 m_e h} \sqrt{\frac{8kT_e}{\pi m_e}}. \tag{11.16}$$

This expression has been averaged over a Maxwell-Boltzmann distribution by replacing the speed by its average value given by Eq. 4.30. The correct result is larger than Eq. 11.16 by a factor $\sqrt{3/2}$. In terms of T_e expressed in kev, the correct result is

$$w_x = 4.8 \times 10^{-37} Z^2 n_i n_e T_e^{1/2} \text{ watts/m}^3. \tag{11.17}$$

The radiation is proportional to the number of collisions per unit time and hence to the product of the electron and ion densities.

The spectral distribution of the radiation could also be discovered by Fourier analysis of the proper motion approximated by Eq. 11.11. In that case, we should be more careful in treating the acceleration and the time over which it lasts. The acceleration is principally a small transverse one. Fourier analysis shows that the energy radiated per unit frequency interval is approximately constant up to a critical frequency $\omega_{max} \approx v/b$ and decreases precipitously thereafter (see Problem 3). From Eq. 11.15, we discover (within a factor of 2) that the energy spectrum of the photons is sensibly constant up to the initial energy of the electron. This upper limit corresponds to the electron losing all its energy in one photon.

For a Maxwellian electron gas at temperature T_e, the bremsstrahlung spectrum lies in an energy range comparable with T_e. Thus we reassert our previous statement that the bremsstrahlung from a thermonuclear plasma lies principally in the x-ray spectrum and assign the power density symbol w_x to it. Further discussion of these derivations, including proper quantum-mechanical ones, are available in the literature.[1-3]

These x rays escape freely from the plasma, and are strongly absorbed in most solids. Therefore, the bremsstrahlung becomes a source of heat on the first material surface surrounding the plasma.

Impurities greatly increase the radiation losses. If the electron temperature were unchanged by the addition of impurities, Eq. 11.17 would show that the bremsstrahlung power density w_x' with the impurities present is increased over the power density w_x for the original constituent only by the ratio

$$w_x'/w_x = \sum_{\xi} (Z_\xi n_{i\xi}) \sum_{\zeta} (Z_\zeta^2 n_{i\zeta}/Z_1^3 n_{i1}^2). \tag{11.18}$$

Here, $n_e = \Sigma Z_\xi n_{i\xi}$ for charge neutrality, and n_{i1} is the density of the original ions. For example, suppose that a fully stripped oxygen impurity is added to a deuterium plasma so that the density of impurity nuclei is 3% of the deuteron density. The bremsstrahlung increases by a factor of 3.6. For this reason, among others, the charged particles in the plasma must be kept from bombarding the walls and sputtering heavy atoms into the plasma. The problems of sputtering and the necessity for outgassing the vacuum walls were discussed in Sec. 3.10.

In actual fact, the electron temperature will be altered by the presence of impurities. Equations 8.35 and 8.39 display the rate of transfer of energy from ions to electrons, and Relation 8.36 displays the increase in this rate (at constant temperature) as impurities are added. Note that the form of Relation 8.36 is identical to Eq. 11.18, except for the mass factor.

In considering a somewhat more realistic case, suppose that the electron temperature is determined by the balance between bremsstrahlung and energy transfer from the ions. The radiation is given by Eq. 11.17 summed over the ion species. The rate of transfer $3n_ikT_i/2\tau_{uie}$ from ions to electrons is given by Eq. 8.39, and again we must sum over the species. By equating these rates, we find that the electron temperature T_e (kev) is given by

$$T_e^2 = \frac{50(\ln \Lambda) \sum_{\zeta} Z_{i\zeta}^2 n_{i\zeta}(T_{i\zeta} - T_e)/A_\zeta}{\sum_{\xi} Z_{i\xi}^2 n_{i\xi}} \text{ kev}^2. \tag{11.19}$$

The bremsstrahlung power is calculated from Eq. 11.17 summed over the species. Equation 11.19 may be adapted for use in various circumstances. For example, if all the ions are at the same temperature T_i, then

$$\frac{T_e}{T_i} = G\left(\sqrt{1 + \frac{2}{G}} - 1\right), \tag{11.20}$$

where

$$G = \frac{25 \ln \Lambda}{T_i} \frac{\sum_{\zeta} n_{i\zeta} Z_{i\zeta}^2 / A_\zeta}{\sum_{\xi} n_{i\xi} Z_{i\xi}^2}. \tag{11.21}$$

For example, consider a deuterium plasma with ion temperature $T_i =$ 100 kev and for which $\ln \Lambda = 20$. The electron temperature, if determined solely by bremsstrahlung, will be 85 kev. Let us now add 3% of stripped oxygen atoms, as in the previous example. The electron temperature drops slightly to about 74 kev; the bremsstrahlung loss increases by a factor of about 3.4. Thus the cooling of the electrons in this example is small; the bremsstrahlung increases, but so does the rate of energy transfer from the ions. Even without any impurities, the radiated power density w_x is large. If $n_{\text{D}} = n_e = 10^{21}/\text{m}^3$ in this example, $w_x =$ 4.4 Mw/m³.

In a simple thermonuclear plasma, the densities n_e and n_i are approximately equal. Thus both the bremsstrahlung power density w_x and the thermonuclear power density w_{DD} or w_{DT} are proportional to n^2, and their ratio is a function of T. Since the radiated energy can at best be recovered as heat and the plasma must be heated electrically, we require that the thermonuclear power output should exceed the bremsstrahlung loss if net power is to be produced. The loss w_x increases only as $T_e^{1/2}$, but the thermonuclear reaction rate increases very much faster with temperature in the region of a few tens of kilovolts (see Sec. 4.9). Consequently, a minimum temperature exists for both reactions below which operation is not useful. The exact temperature depends strongly upon the fate of the reaction products, which can themselves burn in the D-D reaction and which can increase the bremsstrahlung loss in any case. Very approximately, these temperatures are 35 kev for the D-D reaction and 4 kev for the D-T reaction. Problem 4 illustrates the calculation. In actual fact, this criterion is much too naïve, and we shall discover in due course that other restrictions are more severe for a hydrogenic plasma.

For a nonhydrogenic plasma, the bremsstrahlung loss will be very great. Complete replacement of hydrogen ions by ions of charge number Z at the same temperature and density increases this radiation by a factor of Z^3, as may be seen from Eq. 11.18. Thus a lithium plasma radiates 27 times as much as a hydrogenic plasma if $n_{\mathrm{Li}} = n_{\mathrm{D}}$ at the same temperature. Considering also that the fusion cross sections are much lower for lithium, we see that controlled thermonuclear power from any but hydrogenic plasmas is impossible.

★ 11.3. Cyclotron Radiation from an Isolated Electron

As we mentioned in Sec. 11.1, the cyclotron radiation is partly absorbed in the plasma. The problem is rather more difficult than that of the bremsstrahlung, and our analysis divides into four stages:

1. Calculation of the radiation from a single gyrating electron (Sec. 11.3).
2. The justification for deriving a power density in the plasma as a sum over noncorrelated electrons (Sec. 11.4).
3. Calculation of the harmonic spectrum and energy radiated by the plasma (Sec. 11.5).
4. Effect of reflectors on the energy loss from the plasma (Sec. 11.6).

These problems have been intensively studied in recent years.[4–9]

The total energy radiated by a rotating charged particle is easily calculated from Eq. 11.10. In this case,

$$(v/c)^2 \ll 1, \tag{11.22}$$

and

$$\dot{v} = \dot{v}_\perp. \tag{11.23}$$

The electrons are slightly relativistic. Their angular frequency of rotation ω_0 is determined by their relativistic mass, so that

$$\omega_0 = \frac{q_e B[1 - (v^2/c^2)]^{1/2}}{m} = |\omega_b|\left(1 - \frac{v^2}{c^2}\right)^{1/2}. \tag{11.24}$$

We reserve the symbols m and ω_b for the rest mass and the corresponding cyclotron frequency, and show the relativistic corrections explicitly. The acceleration $\dot{v}_\perp = \omega_0 v_\perp$ substituted into Eq. 11.10 gives

$$\frac{dU}{dt} = \frac{q_e{}^4 v_\perp{}^2 B^2}{6\pi\epsilon_0 m^2 c^3[1 - (v^2/c^2)]}. \tag{11.25}$$

This equation is of limited use to us, because we are vitally interested in the frequency spectrum of the radiation. However, Eq. 11.25 shows that all but a negligible amount of the cyclotron radiation arises from the light electrons. Radiation from the ions is henceforth ignored.

We begin our detailed investigation by calculating the radiation from a charge moving uniformly in a circle with angular frequency ω_0.[10–12] Figure 11.2 shows the configuration. The charge moves in a circle of

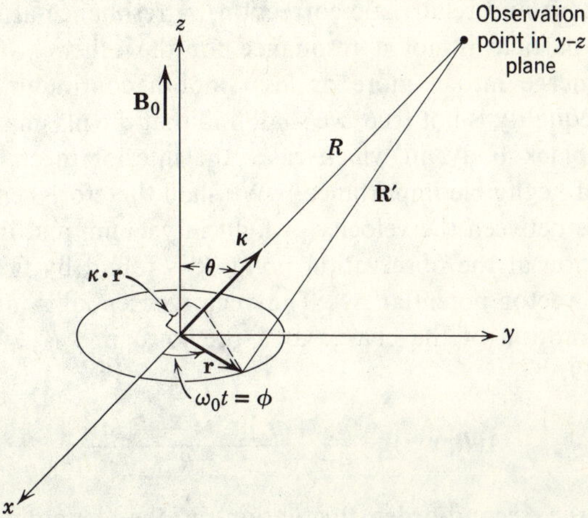

Fig. 11.2. Geometrical constructions for an electron rotating in the x-y plane and radiating in the direction $\boldsymbol{\kappa}$.

radius r in the x-y plane. We are situated at a point at distance R remote from the center of the orbit and define the propagation vector $\boldsymbol{\kappa}$ in our direction. The vector $\boldsymbol{\kappa}$ makes an angle θ with the z-axis, which is in the direction of the steady induction $\mathbf{B}_0$. The instantaneous distance of the charge from the observer is

$$R' = R - (\boldsymbol{\kappa}\cdot\mathbf{r}/|\kappa|). \tag{11.26}$$

Now define the phase angle φ of the particle as

$$\varphi = \omega_0 t; \tag{11.27}$$

then

$$x = r\cos\omega_0 t = r\cos\varphi, \tag{11.28}$$

$$y = r\sin\varphi, \tag{11.29}$$

$$\boldsymbol{\kappa}\cdot\mathbf{r} = \kappa r\sin\theta\sin\varphi, \tag{11.30}$$

and so forth. We shall be concerned with the gth harmonic of the fundamental frequency ω_0, and we note that

$$\kappa = g\omega_0/c = gv/cr. \tag{11.31}$$

Here, c stands for the velocity of propagation in the medium, which for an isolated charge is a vacuum. For a plasma, the velocity of light should be modified. The correction should be small if $g\omega_0 \gg \omega_p$, unless $g\omega_0$ is near some resonant frequency of the plasma. However, because of Doppler shift and relativistic corrections, a resonance at $g\omega_0 \gg \omega_p$ for one plasma particle is not a resonance for the others. Therefore the resonances merge into a more or less smooth continuum. For cases where the inequality is not true, we shall find that the plasma radiates very nearly as a black body, in which cases the interior mechanisms of the plasma are of negligible importance. We shall therefore generally ignore the difference between the velocity of light in vacuum and in the plasma.

The radiation at the observation point R is found by first calculating the retarded vector potential **A**. The vector potential **A** at R at time t arises from motion of the charge at a previous time $t - (R'/c)$. Then from Eq. 5.59,

$$\mathbf{A}(R, \mathbf{r}, t) = \frac{\mu_0}{4\pi} \int \frac{dr\, \mathbf{j}[\mathbf{r}, (t - R'/c)]}{|R'|}. \tag{11.32}$$

We ignore the second-order fluctuation in the denominator and set $R' = R$; but we preserve the fluctuation in **j**, because its effect there appears in first order.

The current density $\mathbf{j}(t)$ is a delta function, running around the orbit circle. Let **n** be a unit vector in the direction of the instantaneous velocity at an arbitrary point on the orbit. Then the current $\mathbf{I}(t)$ is

$$\mathbf{I}(t) = \mathbf{n} q_e \sum_{g=0}^{\infty} \delta(t - g\tau), \tag{11.33}$$

where

$$\tau = 2\pi/\omega_0 = 2\pi r/v \tag{11.34}$$

is the period. Of course, the average current over a cycle is

$$\bar{I}(t) = \frac{q_e}{\tau} \int_0^{\tau} dt \sum_g \delta(t - g\tau) = \frac{q_e}{\tau}. \tag{11.35}$$

We expand the periodic delta function in a cosine series:

$$\sum_{g=0}^{\infty} \delta(t - g\tau) = \frac{a_0}{2} + \sum_{g=1}^{\infty} a_g \cos g\omega_0 t, \tag{11.36}$$

where

$$a_g = \frac{2}{\tau}\int_{-\tau/2}^{\tau/2} dt \sum_{g=0}^{\infty} \delta(t - g\tau) \cos g\omega_0 t = 2/\tau. \tag{11.37}$$

Then we find that

$$\sum_{g=0}^{\infty} \delta(t - g\tau) = \frac{1}{\tau} + \frac{2}{\tau}\sum_{g=1}^{\infty} \cos g\omega_0 t$$

$$= \frac{1}{\tau}\left(1 + 2\mathscr{Re} \sum_{g=1}^{\infty} e^{ig\omega_0 t}\right), \tag{11.38}$$

where $\mathscr{Re}(\ldots)$ indicates that the real part is to be taken. We change now to the more convenient complex notation, and we shall extract the real part at the conclusion of the calculation. The constant term in Eq. 11.38 represents the average current around the orbit ring. This dc contribution will lead to no radiation, and we discard it.

We substitute Eqs. 11.33 and 11.38 into Eq. 11.32 and find for the gth harmonic of the vector potential

$$\mathbf{A}_g = \frac{\mu_0 q_e}{2\pi\tau} \oint d\ell \, \frac{\exp\{ig\omega_0[t - (R/c) + (\boldsymbol{\kappa}\cdot\mathbf{r}/\kappa c)]\}}{R}$$

$$= \frac{\mu_0 q_e \exp(-i\kappa R)}{2\pi\tau R} \oint d\ell \exp[i(g\omega_0 t + \boldsymbol{\kappa}\cdot\mathbf{r})]. \tag{11.39}$$

From Eqs. 11.28, 11.30, and 11.34, the x-component of the vector potential can be written

$$A_{xg} = -\frac{\mu_0 q_e v \exp(-i\kappa R)}{4\pi^2 R} \int_0^{2\pi} d\varphi \sin\varphi \exp\left[ig\left(\varphi + \frac{v}{c}\sin\theta\sin\varphi\right)\right]. \tag{11.40}$$

We evaluate the integral by noting[13] that the Bessel function $J_g(z)$ can be represented as

$$J_g(z) = \frac{i^{-g}}{2\pi}\int_0^{2\pi} d\psi \exp(iz\cos\psi)\exp(ig\psi). \tag{11.41}$$

By differentiating this expression with respect to z, replacing ψ by $\varphi - (\pi/2)$, and by comparing the result with Eq. 11.40, we find the x-component of the vector potential

$$A_{xg} = -(-1)^g \frac{i\mu_0 q_e v \exp(-i\kappa R)}{2\pi R} J_g'\left(\frac{gv}{c}\sin\theta\right), \tag{11.42}$$

where $J'(Y) = dJ(Y)/dY$. The y-component is treated similarly:

$$A_{yg} = \frac{\mu_0 q_e v \exp(-i\kappa R)}{4\pi^2 R}\int_0^{2\pi} d\varphi \cos\varphi \exp\left[ig\left(\varphi - \frac{v}{c}\sin\theta\sin\varphi\right)\right]. \tag{11.43}$$

We evaluate this integral by adding and subtracting $i \sin \varphi$ in the integrand, by applying Demoivre's formula, by using Eq. 11.41 in the same way as in deducing Eq. 11.42, and by using the relation[13]

$$J_g'(z) = (g/z)J_g(z) - J_{g+1}(z). \tag{11.44}$$

We finally obtain

$$A_{yg} = \frac{\mu_0 q_e c \exp(-i\kappa R)}{2\pi R \sin\theta} J_g\left(\frac{gv}{c}\sin\theta\right). \tag{11.45}$$

The component A_{zg} is zero.

The radiation is given by the Poynting vector. The power radiated into an element $d\mathbf{\Omega}$ of solid angle at angle θ and distance R is

$$\dot{U}_g(\theta)\, d\mathbf{\Omega} = |\mathbf{E}_g \times \mathbf{H}_g^*| R^2\, d\mathbf{\Omega}. \tag{11.46}$$

Note that Eq. 11.46 has the dimensions of power and represents the rate of change of electron energy arising from the radiation. The asterisk denotes the complex conjugate. Thus, $\dot{U}_g$ is real. From Eq. 9.14, the fact that $\mathbf{H}\cdot\boldsymbol{\kappa} = 0$ for a plane wave (see Eq. 9.5), $\mathbf{B} = \mu_0\mathbf{H}$, and $\mathbf{B} = \nabla \times \mathbf{A}$, we reduce Eq. 11.46 to

$$\dot{U}_g(\theta) = c|\boldsymbol{\kappa} \times \mathbf{A}_g|^2 R^2/\mu_0. \tag{11.47}$$

Now, from Fig. 11.2 and a familiar vector identity, it follows that

$$|\boldsymbol{\kappa} \times \mathbf{A}_g|^2 = |A_x|^2\kappa^2 + |A_y|^2\kappa^2 \cos^2\theta. \tag{11.48}$$

The vector potential varies sinusoidally with time, and we must average the radiated energy over a cycle. This averaging of Eqs. 11.47 and hence 11.48 introduces a factor of $\frac{1}{2}$. We now combine Eqs. 11.24, 11.42, 11.45, and 11.48 into Eq. 11.47 and find

$$\dot{U}_g(\theta, v, \omega_b) = \frac{g^2 q_e^2 \omega_b^2 [1 - (v^2/c^2)]}{8\pi^2 \epsilon_0 c} \left\{\frac{v^2}{c^2}\left[J_g'\left(\frac{gv}{c}\sin\theta\right)\right]^2 + \cot^2\theta \left[J_g\left(\frac{gv}{c}\sin\theta\right)\right]^2\right\}. \tag{11.49}$$

Equation 11.49 is the result desired. The full dependence $\dot{U}_g(\theta, v, \omega_b)$ is written to show that we have calculated the power radiated in the gth harmonic per unit solid angle at an angle θ relative to the orientation of $\mathbf{B}$ by an electron of speed v in an induction whose magnitude $B = m\omega_b/e$.

Investigation of Eq. 11.49 would show that for $v/c \ll 1$ the intensity of the harmonics declines rapidly as a function of harmonic number. On the other hand, for $v/c \to 1$, the harmonic spectrum is huge. The intensity of the successive harmonics increases until the harmonic number $g \approx [1 - (v^2/c^2)]^{-3/2}$. Thereafter, the harmonic intensities decrease almost exponentially. At energies of principal interest to us, for example, 50 kev, the harmonic intensity decreases with harmonic number g, but at such a rate that about the first 20 harmonics must be considered.

★ 11.4. Cyclotron Radiation by Electrons in a Plasma

It is not *a priori* obvious that we can legitimately compute the generated power density of cyclotron radiation $w_c(\omega, \theta, \mathbf{r})$ by merely summing the contribution of Eq. 11.49 in the range $\omega + d\omega$, $\theta + d\theta$ over all the electrons per unit volume. The various waves may reinforce or cancel. The power density calculated in this straightforward way is an excellent approximation, as will now be shown.

We discuss first the possibility of interference between the many waves and hence of correlations among the particles. The radiation intensity at the observation point could be calculated by summing the contributions to the vector potential $\mathbf{A}$ from all the individual sources. Knowing $\mathbf{A}$, we compute the radiation from $\mathbf{E} \times \mathbf{H}^*$. If interference effects (that is, correlations) exist in the behavior of $\mathbf{E} \times \mathbf{H}^*$, they will also exist in $\mathbf{A} \cdot \mathbf{A}^*$. Thus we choose the simpler task of examining the behavior of AA^*, where A is any one of the components of $\mathbf{A}$.

Consider now N particles in a region of space and the contribution A_ξ to the vector potential at the observation point ascribable to the ξth particle without interference effects. We shall assume that each particle radiates a spectrum consisting of harmonics of its particular angular frequency $\omega_{0\xi}$. Specification of the coefficients $A_{g\xi}$ of the harmonics is not necessary. Indeed, each component of the wave may be gradually attenuated by some process on its way to the observation point. Thus

$$A_\xi = \sum_{g=0}^{\infty} A_{g\xi} \exp(i\gamma_{g\xi}), \tag{11.50}$$

where

$$\gamma_{g\xi} = g\omega_{0\xi}t + \psi_{g\xi} \tag{11.51}$$

at the observation point. The phase angle $\psi_{g\xi}$ is characteristic of each harmonic of each particle. We wish to calculate AA^*, which is

$$AA^* = \left[\sum_{\xi=1}^{N} \sum_{g=1}^{\infty} A_{g\xi} \exp(i\gamma_{g\xi})\right]\left[\sum_{\zeta=1}^{N} \sum_{h=1}^{\infty} A_{h\zeta} \exp(-i\gamma_{h\zeta})\right]. \tag{11.52}$$

The product is real because all imaginary components cancel in pairs.

The product may be separated into terms with $\zeta \neq \xi$ (two different particles) and terms with $\zeta = \xi$ (the same particle). The terms representing one particle are further split into terms with $g \neq h$ (different harmonics) and terms with $g = h$ (the same harmonic). The terms involving the same particle then provide a contribution to AA^* of

$$\sum_{\xi=1}^{N}\left\{\sum_{g=1}^{\infty} A_{g\xi}^2 + \sum_{g\neq h}^{\infty} A_{g\xi}A_{h\xi}\exp\left[i(\gamma_{g\xi} - \gamma_{h\xi})\right]\right\}. \tag{11.53}$$

Each term in this expression representing two different harmonics ($g \neq h$) averages to zero over a cycle, regardless of phase. Our time of observation extends over many cycles, and we therefore find that the harmonics of any one particle do not interact in an observable way.

We turn now to an examination of terms involving different particles ($\zeta \neq \xi$). These terms express the interference between particles. As before, we discover that a term representing two different harmonics averages to zero in about a cycle. We assume that the harmonics are so numbered that the degenerate cases $g\omega_{0\xi} = h\omega_{0\zeta}$ are included with the set $g = h$.

The sole correlation over long times for a term involving different particles may then arise only if it also represents one harmonic ($g = h$), if indeed such correlation arises even then. From Eqs. 11.52 and 11.53, we find

$$AA^* = \sum_{g=1}^{\infty}\left\{\sum_{\xi=1}^{N} A_{g\xi}^2 + \sum_{\xi=1}^{N}\sum_{\zeta\neq\xi}^{N} A_{g\xi}A_{g\zeta}\exp\left[i(\gamma_{g\xi} - \gamma_{g\zeta})\right]\right\}. \tag{11.54}$$

The average value of the second term in Eq. 11.54 is zero if $\gamma_{g\xi}$ and $\gamma_{g\zeta}$ are uncorrelated. Therefore, we are interested in determining if that term yields a significant contribution to AA^* over any appreciable time, for various assumptions about the correlation.

The greatest possible correlation exists if particles denoted by ξ and ζ are so disposed that their radiation amplitudes are the same. If $A_\xi \neq A_\zeta$, less correlation must exist. In the limiting case, we may write

$$AA^* = \sum_{g=1}^{\infty} A_g^2\left\{N + \sum_{\xi=1}^{N}\sum_{\substack{\zeta=1\\ \zeta\neq\xi}}^{N}\exp\left[i(\gamma_{g\xi} - \gamma_{g\zeta})\right]\right\}. \tag{11.55}$$

The double summation in the braces consists of $N(N-1) \approx N^2$ unit vectors in the complex plane, correlated in pairs. If the number N of particles is large, the mean amplitude of this double summation is N. The rate of change of this pair summation is given by the relative rates of change of the γ's. We note that γ depends on both time and position of the particle through the phase $\psi = \mathbf{\kappa}\cdot\mathbf{r}$ + initial phase angle. Thus

$$\frac{d}{dt}\sum_{\xi=1}^{N}\sum_{\substack{\zeta=1\\ \zeta\neq\xi}}^{N} \exp\left[i(\gamma_{g\xi}-\gamma_{g\zeta})\right]$$

$$=\sum_{\xi=1}^{N}\sum_{\substack{\zeta=1\\ \zeta\neq\xi}}^{N}\left[\frac{\partial}{\partial t}(\gamma_{g\xi}-\gamma_{g\zeta})+\boldsymbol{\kappa}_{g\xi}\cdot\mathbf{v}_{\xi}-\boldsymbol{\kappa}_{g\zeta}\cdot\mathbf{v}_{\zeta}\right]\exp\left[i(\gamma_{g\xi}-\gamma_{g\zeta})\right]. \quad (11.56)$$

If the frequencies $\omega_{0\xi}$ and $\omega_{0\zeta}$ are unequal, as they will be in a relativistic velocity distribution, the time derivatives in Eq. 11.56 fluctuate, and the correlation lasts a time in the order of

$$\tau_1 = \{g\omega_b[\sqrt{1-(v_\xi^2/c^2)} - \sqrt{1-(v_\zeta^2/c^2)}]\}^{-1}. \quad (11.57)$$

This time is usually long compared with the time for losing correlation by the first-order difference in velocities. The mean rate of change of the sum ascribable to the $\boldsymbol{\kappa}\cdot\mathbf{v}$ term in Eq. 11.56 is about equal to

$$2\pi g\omega_b(v_\xi - v_\zeta)N/c. \quad (11.58)$$

If the velocities are uncorrelated, the double sum, which is itself of order N, fluctuates at a frequency close to $g\omega_b v/c$. Thus again interference effects average to zero in about a cycle.

We shall have shown then that

$$AA^* = \sum_{g=1}^{\infty}\sum_{\xi=1}^{N} A_{g\xi}^2 \quad (11.59)$$

and that the radiation of each electron is independent of that of any other electron, providing we show that the motions of the electrons are uncorrelated. Two arguments exist. For the first argument, we note that electrons are within each other's Debye sphere in the order of $1/\omega_p$ units of time (see Eqs. 4.94 and 9.19). Since frequencies are also in this order, we see that the particles will be inside each other's sphere for about a cycle. Their motions will therefore be uncorrelated for longer times, so that over several cycles the sums over unlike particles must average to zero.

The second argument is based upon the thought that the kinetic energy of the particles is quite sufficient to randomize them quickly, since this energy is huge compared with that residing in the electromagnetic field. The particles cannot then be constrained by the electromagnetic field and lead to some cooperative motions. We now show that the electromagnetic energy is very small compared with the kinetic energy. The maximum intensity of radiation into a solid angle $d\boldsymbol{\Omega}$ per unit frequency interval is the black-body intensity $S_B(\omega, \boldsymbol{\Omega}, T)\, d\boldsymbol{\Omega}$. At the frequencies of interest for cyclotron radiation, the energy of the photons is much less than the mean energy of the electrons, so that the Rayleigh-Jeans approximation for S_B is valid:

$$S_B(\omega, \mathbf{\Omega}, T)\, d\omega\, d\mathbf{\Omega} = \omega^2 kT_e\, d\omega \frac{d\mathbf{\Omega}}{8\pi^3 c^2}. \tag{11.60}$$

By integrating over a hemisphere, we find that the spectral one-way flux per unit area is

$$S_B(\omega, T)\, d\omega = \omega^2 kT_e \frac{d\omega}{8\pi^2 c^2}. \tag{11.61}$$

This flux in the range ω, $\omega + d\omega$ is related to the energy density $(dp_B/d\omega)\, d\omega$ in the same range by

$$S_B(\omega, T) = \left(\frac{c}{4}\right) \frac{dp_B(\omega, T)}{d\omega}. \tag{11.62}$$

Note that $p(\omega, t)$ has the dimensions of pressure and is in fact related to the radiation pressure existing in the plasma. We are interested in the total radiation density per unit volume (or pressure) over the frequency range below some value ω^*, where ω^* corresponds to the highest cyclotron harmonic of practical interest. Integration of $dp_B(\omega, t)/d\omega$ over the range below ω^* gives

$$p_B(\omega^*, T) = \frac{kT_e \omega^{*3}}{6\pi^2 c^3}, \tag{11.63}$$

and this quantity is to be compared with the kinetic energy density $n_e kT_e$. The ratio is

$$\frac{\text{Radiation energy density}}{\text{Kinetic energy density}} = \frac{4\pi}{3}\left(\frac{d}{\lambda^*}\right)^3 \ll 1; \tag{11.64}$$

here d is the interelectronic distance, and $\lambda^* = c/\omega^*$ is the shortest wavelength of interest. In our applications, $d \leqslant 10^{-6}$ m and $\lambda^* \geqslant 10^{-5}$ m; thus insufficient energy is available to correlate the motions, and we conclude that all fluctuations averaged over several cycles will be close to zero. Therefore, we neglect the summation over unlike particles in Eq. 11.54, and Eq. 11.59 follows.

Having discussed the matter of correlations at some length, we find it instructive to present in addition a simple argument that for a plasma of thermonuclear interest the free-electron model of radiation cannot fail to yield the correct results. We assert that for a thermonuclear plasma the following hierarchy of inequalities exists:

$$\lambda_b < c/\omega_p \ll \lambda' \ll L. \tag{11.65}$$

Here λ_b is a wavelength characteristic of the cyclotron harmonics. Since $g\omega_b > \omega_p$ in general, the inequality on the left results, in view of the argument following Eq. 11.31. The distance L characterizes the size of the plasma. Since ω_{pe} in a thermonuclear plasma corresponds to wavelengths shorter than 10^{-2} m and since $R \approx 1$ m, we can evidently choose a distance λ' so that all the inequalities of Eq. 11.65 are satisfied.

Suppose now that a generated power density w_c of cyclotron radiation is computed, using the free-electron model, and that a corresponding absorption coefficient α is obtained by use of Eq. 11.2 or some other appropriate scheme. At any particular frequency, direction, and position, we shall find that either

$$(1/\alpha) \geqslant \lambda', \tag{11.66}$$

or

$$(1/\alpha) < \lambda'. \tag{11.67}$$

If $(1/\alpha) \geqslant \lambda'$, the absorption of the radiation occurs far from its site of emission. Canceling currents are situated remote from the source, and the electron launches a wave that appears in its vicinity to be undisturbed. Ergo, the free-electron model is appropriate in this circumstance, even though the plasma might appear as a black body at this frequency. On the other hand, if $(1/\alpha) < \lambda'$, we cannot make such categorical statements. But in this case, $(1/\alpha) \ll L$ from Eq. 11.65, and $\alpha L \gg 1$. From Eq. 11.5, the plasma appears at this frequency and direction as a black body. In that case, whatever the assumption about the interior mechanisms of the plasma, the result will not be affected.

Our assumption that the generated power density w_c of cyclotron radiation is obtained by summing over the individual electrons has been justified. In the next section, we calculate this power density and the corresponding attenuation coefficient α for a slightly relativistic distribution.

★ 11.5. Cyclotron-Radiation Power Density and Emission

Although the use of a free-electron model for computing cyclotron radiation in a plasma has been fairly well justified, application of the model in real cases is mathematically difficult. To date no complete calculation has been made. Therefore, we shall point out the physical aspects of such an application and summarize typical calculations that have been made.

It may be checked that the plasma electrons interchange energy at a rate much faster than they lose energy by radiation. Therefore, this energy distribution is approximately Maxwellian. We require the relativistic distribution function, which is most readily derived in momentum space $\mathfrak{p}$, where

$$\mathfrak{p} = \frac{m\mathbf{v}}{[1 - (v^2/c^2)]^{1/2}}. \tag{11.68}$$

The Maxwellian distribution function is[14]

$$f(\mathfrak{p}) = C_{\mathfrak{p}} \exp(-U/kT), \tag{11.69}$$

where

$$U = (\mathfrak{p}^2c^2 + m^2c^4)^{1/2} = mc^2[1 - (v^2/c^2)]^{-1/2} \tag{11.70}$$

is the total energy of the particle. The normalization constant $C_{\mathfrak{p}}$ appropriate to momentum space may be found from the requirement that

$$4\pi \int_0^\infty d\mathfrak{p}\, \mathfrak{p}^2 C_{\mathfrak{p}} \exp\left\{-\frac{mc^2}{kT}[1 + (\mathfrak{p}/mc)^2]^{1/2}\right\} = 1. \tag{11.71}$$

This integral can be evaluated by setting

$$z = \left[1 + \left(\frac{\mathfrak{p}}{mc}\right)^2\right]^{1/2}, \tag{11.72}$$

and by using the Hankel function $H_1^{(1)}(z)$ of the first kind and first order, the derivative relation for these functions, and the definition of the modified Bessel functions $K_2(z)$ of second order.[15] Eventually we find that

$$C_{\mathfrak{p}} = \left[4\pi m^2ckTK_2\left(\frac{mc^2}{kT}\right)\right]^{-1}. \tag{11.73}$$

The function K_2 can be expanded in an asymptotic series. It is easier, however, to make the substitution

$$y^2 = \left[1 + \left(\frac{\mathfrak{p}}{mc}\right)^2\right]^{1/2} - 1 \tag{11.74}$$

in Eq. 11.71, because the resulting integrals can be evaluated directly. In either case,

$$C_{\mathfrak{p}} = \frac{[1 + (15kT/8mc^2) + \cdots] \exp(mc^2/kT)}{(2\pi kT/mc^2)^{3/2}(mc)^3}. \tag{11.75}$$

The predicted mathematical difficulties now begin. We are interested in finding the cyclotron power density $w_c(\omega, \mathbf{\Omega}, \mathbf{r}, T)$ generated per unit volume. Equation 11.49 correctly estimates the power radiated by an electron if it has no velocity $v_\parallel$ along the **B** lines. For electrons in a plasma, $v_\parallel$ is not zero, and two corrections are necessary. First, the proper time of the particle is individual to it and leads to a further relativistic correction in the amplitude of the radiation. For our purposes, this correction is generally small. Second, and more important, the harmonics of each particle will be shifted in frequency by the Doppler effect, except for radiation normal to the **B** lines.

The distribution of electrons in momentum space, Eq. 11.69, can be expressed as a function of velocity. The intensity of the gth harmonic at a frequency ω is obtained from Eq. 11.49 as

$$w_c(\omega, \theta, \mathbf{r}, T, g)\, d\omega = \int_{\omega=\text{const}} dv\, \dot{U}(\theta, g, \omega_b, v_\perp) f(\mathbf{r}, \mathbf{v}, T)\, d\omega, \tag{11.76}$$

where the integral is carried out over a surface of constant frequency.

Neglect of the Doppler shift, corresponding to propagation perpendicular to **B**, simplifies the calculation considerably. All contributions to the gth harmonic lie at or below frequency $g\omega_b$. Inclusion of the Doppler shift at other angles of observation has the effect of smearing the spectral distribution. The approximate shape of the spectral distribution is sketched in Fig. 11.3. These harmonics are also sketched in the references cited.[5, 6, 9]

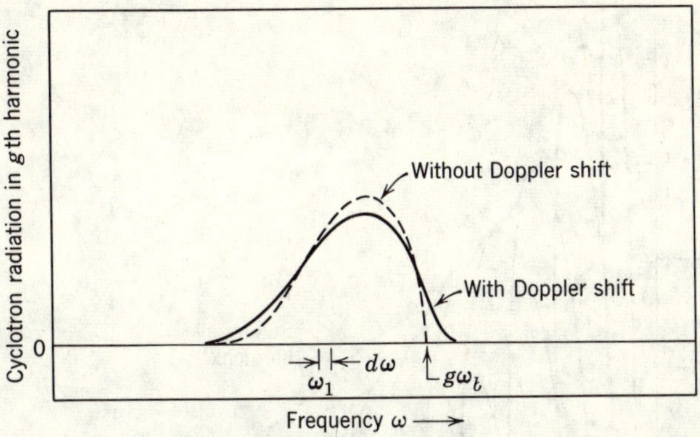

Fig. 11.3. Spectral distribution of the gth harmonic of power density of cyclotron radiation $w_c(\omega, \mathbf{\Omega}, \mathbf{r}, T, g)$ in a plasma.

Detailed calculations made to date refer specifically to propagation perpendicular to **B**, with the assumption of either a two- or three-dimensional relativistic distribution in velocity space. Quantitative results are best expressed in terms of dimensionless ratios. For a plane uniform plasma slab of thickness L, as shown in Fig. 11.4, we can define the absorption exponent αL for a perpendicular ray from Eq. 11.2 as

$$\alpha L = w_c L / S_B. \tag{11.77}$$

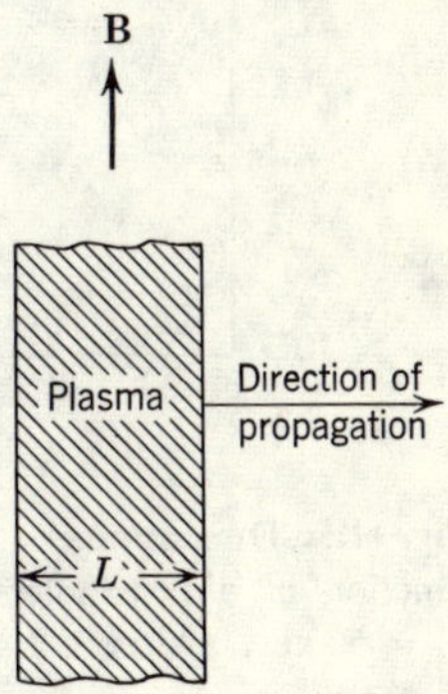

Fig. 11.4. Plane plasma slab of thickness L with propagation normal to the slab.

The black-body radiation intensity S_B is displayed by Eq. 11.61. The exponent αL may be divided by the dimensionless length $\mathscr{L}$, where†

$$\mathscr{L} = \omega_p{}^2 L / \omega_b c; \tag{11.78}$$

† Trubnikov and co-workers and Hirshfield and Baldwin use the symbol Λ for this quantity. We have used Λ elsewhere.

the quantity $\alpha L/\mathscr{L}$ does not contain the length L and is a universal function of ω/ω_b and T_e, for propagation normal to $\mathbf{B}$, as may be seen from a dimensional study of Eqs. 11.76 and 11.77. Figure 11.5 shows Hirshfield and Baldwin's calculation[9] of the quantity $\alpha L/\mathscr{L}$ for the numbered harmonics at an electron temperature $T_e = 50$ kev. In that

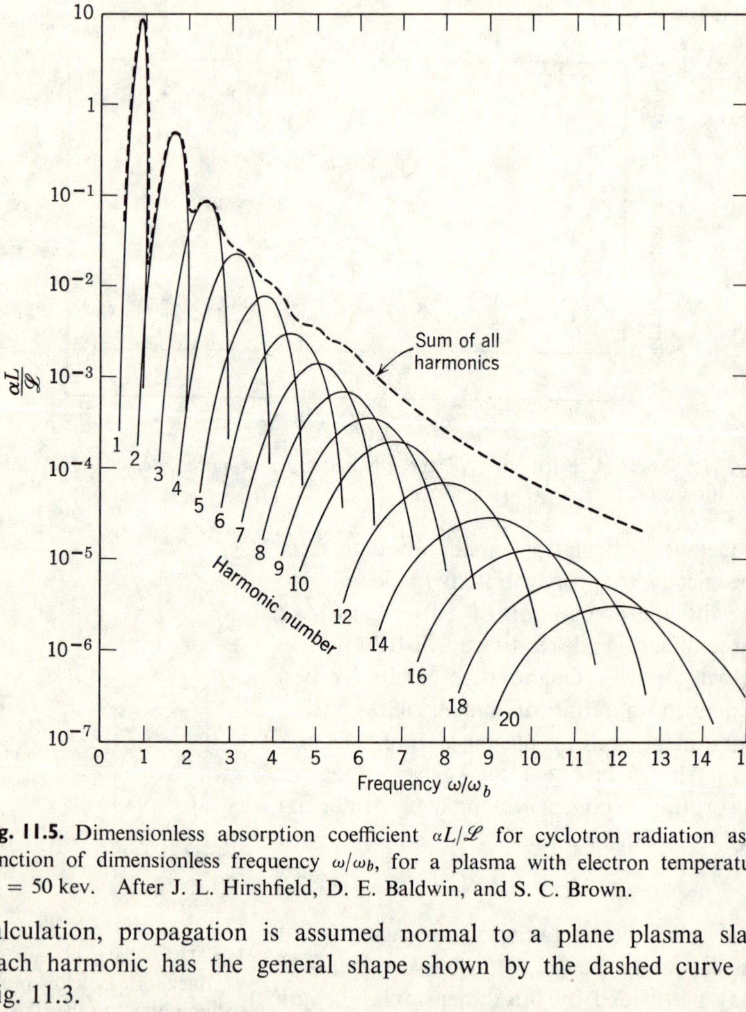

Fig. 11.5. Dimensionless absorption coefficient $\alpha L/\mathscr{L}$ for cyclotron radiation as a function of dimensionless frequency ω/ω_b, for a plasma with electron temperature $T_e = 50$ kev. After J. L. Hirshfield, D. E. Baldwin, and S. C. Brown.

calculation, propagation is assumed normal to a plane plasma slab. Each harmonic has the general shape shown by the dashed curve of Fig. 11.3.

The total radiation at a particular frequency ω is the sum over all harmonics that give any contribution at ω. The sum is shown by the dashed curve of Fig. 11.5. From the summed curve the normal radiation per unit area from the surface of the slab is readily calculated using

Eqs. 11.5 and 11.61. For example, let $L = 1$ m, $B = 5$ webers/m^2, and $n_e = 7.3 \times 10^{20}$/m^3, giving $\mathscr{L} = 9 \times 10^3$. The ordinate $\alpha L/\mathscr{L} = 1.1 \times 10^{-4}$ in Fig. 11.5 then corresponds to $\alpha L = 1$; we find that $\alpha L = 1$ at $\omega/\omega_b \approx 9.5$. The exponent $\alpha L(\omega)$ can be read directly through the frequency range of interest.

Following this procedure, we construct the solid curves of Fig. 11.6 for the radiated spectral intensity $S(\omega, \mathbf{\Omega})$ of cyclotron radiation. The

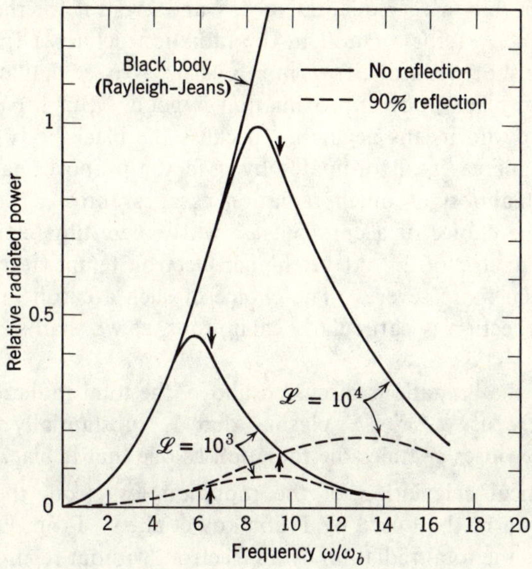

Fig. 11.6. Relative cyclotron radiation loss from a plasma slab at $T_e = 50$ kev for $\mathscr{L} = \omega_p{}^2 L/\omega_b c = 10^3$ and 10^4 with and without 90% reflecting walls. Short arrows indicate the frequency ω^* at which $\alpha L = 1$.

dashed curves of Fig. 11.6 correspond to the case of reflecting walls, which will be discussed in the next section. On the basis of Figs. 11.5 and 11.6, we can now make a number of detailed observations:

1. The total radiated power is large. An absolute scale of ordinates cannot be placed on Fig. 11.6, because ω_b is not specified. For the example quoted above ($B = 5$ webers/m^2, $\mathscr{L} \approx 10^4$), the total radiation is about 5×10^4 watts/m^2 per unit solid angle.

2. For a plasma in which $\mathscr{L} \gg 1$, the vast majority of the radiation is reabsorbed. The large peaks on Fig. 11.5, corresponding to the fundamental and the first few harmonics, are largely constrained by the Rayleigh-Jeans limit. Were this not so, the radiation loss would be of the order of megawatts/m^2.

3. For $\mathscr{L} \gg 1$, the radiation escapes at high harmonics of the cyclotron frequency. The peak in Fig. 11.6 for $\mathscr{L} = 10^4$ appears at $\omega \approx 9\omega_b$; reference to Fig. 11.5 shows that harmonic numbers $g = 11$ to 17 contribute principally at this frequency.

4. The quantity $\alpha L/\mathscr{L}$ decreases rapidly with harmonic number, as might be predicted from the discussion at the end of Sec. 11.3. Thus we might be tempted to approximate Eq. 11.5 by setting $S = S_B$ for $\alpha L > 1$ and setting $S = 0$ for $\alpha L < 1$. We shall succumb to this temptation in only a very qualitative way and avoid it for the most part. The choice $\alpha L = 1$ for a cutoff in the radiation defines a frequency ω^*, shown by the short vertical arrows in Fig. 11.6. We see that at frequencies below ω^* the black-body approximation is good. But for $\omega > \omega^*$, the radiation is by no means negligible, because the black-body function S_B in Eq. 11.5 increases all ordinates by a factor proportional to ω^2. It appears that almost as much radiation escapes above ω^* as below it, and hence the choice of a cutoff at ω^* will underestimate the radiation by almost a factor of 2. At yet higher electron temperature, the effect would be even more severe. The choice of such a cutoff in the case of significant reflection is particularly calamitous, as we shall see in the next section.

5. Within the limitations discussed above, the total radiation increases approximately as $(\omega^*)^3$. A plasma that is substantially black to a frequency $g\omega_b$ loses g^3 times the radiation as one that is black only to ω_b.

The practical calculation of the radiation loss from the plasma is facilitated by a method used by Trubnikov et al.[5,6] From Eq. 11.25, we first compute the total radiation by an electron, without regard to whether it escapes or not. Expressed in terms of the momentum, Eq. 11.25 is

$$\frac{dU}{dt} = \frac{q_e^4 B^2 \mathfrak{p}_\perp^2}{6\pi\epsilon_0 m^4 c^3}, \tag{11.79}$$

which is to be integrated over the relativistic distribution $f(\mathfrak{p})$. Since $f(\mathfrak{p})$ is isotropic and since there are two directions associated with $\mathfrak{p}_\perp$, but only one associated with $\mathfrak{p}_\parallel$, we have

$$\mathfrak{p}_\perp^2 = 2\overline{\mathfrak{p}^2}/3. \tag{11.80}$$

The average of $\mathfrak{p}^2$ is easily computed by the same techniques outlined for $C_\mathfrak{p}$. The exact answer appears in terms of the modified Bessel function $K_3(z)$ of third order. The integral definition of the Hankel functions $H_2^{(1)}(z)$ of first kind and second order and the derivative relations for these functions are used to evaluate the integrals. The total power radiated by all electrons is then

$$w_c = \frac{e^4 B^2}{3\pi\epsilon_0 m^2 c}\left(\frac{n_e k T_e}{mc^2}\right)\frac{K_3(mc^2/kT_e)}{K_2(mc^2/kT_e)}. \tag{11.81}$$

From the asymptotic expansion of the modified Bessel function, $K_3(z)$ of third order, or by the same technique as used previously (see Eq. 11.74), the total power w_c radiated by electrons of all speeds is

$$w_c = \frac{e^2\omega_b^2}{3\pi\epsilon_0 c}\left(\frac{n_e kT_e}{mc^2}\right)\left(1 + \frac{5}{2}\frac{kT_e}{mc^2} + \cdots\right). \tag{11.82}$$

This equation in our units of $T =$ kev, webers/m^2, and number/m^3, becomes

$$w_c = 6.2 \times 10^{-17}\, B^2 n_e T_e[1 + (T_e/204) + \cdots] \quad \text{watts/m}^3. \tag{11.83}$$

More restrictively, if $\beta = 2\mu_0 n_e k(T_e + T_i)/B^2 = 1$, and $T_e = T_i$, the equation becomes

$$w_c = 5.0 \times 10^{-38} n_e^2 T_e^2[1 + (T_e/204) + \cdots] \quad \text{watts/m}^3. \tag{11.84}$$

From these equations, the total generated power density $w_c\mathscr{V}$ over a volume $\mathscr{V}$ can be computed. The actual radiation loss through the plasma surface is given by the integral of S (Eq. 11.6) over the surface. Thus we can define a dimensionless coefficient

$$K_{\mathscr{L}} = \frac{\displaystyle\int_S d\mathbf{s}\cdot\mathbf{S}}{\displaystyle\int_{\mathscr{V}} d\mathbf{r}\, w_c} \tag{11.85}$$

whose value is less than unity and which is a measure of absorption in the plasma. For propagation normal to a plane plasma slab of uniform electron temperature T_e, as in Fig. 11.4, $K_{\mathscr{L}}$ is a function of T_e and the dimensionless plasma size $\mathscr{L}$. Figure 11.7 shows calculated values of the parameter $K_{\mathscr{L}}$.

We present a simple illustration by assuming an electron temperature $T_e = 40$ kev, ion temperature $T_i = 100$ kev, density $n_e = n_i = 10^{20}$/m^3, $\beta = 0.25$, and a slab 1 m thick. For this example, $B = 4.75$ webers/m^2, $\omega_b = 8.35 \times 10^{11}$/sec, and the cyclotron power density w_c from Eq. 11.83 is 6.7×10^6 watts/m^3. The parameter $\mathscr{L} = 6 \times 10^3$; from Fig. 11.7, we find $K_{\mathscr{L}} \approx 0.02$. Then the total radiation actually escaping is about 1.3×10^5 watts/m^2 of the slab (over both surfaces). We may compare this figure with the thermonuclear power density. If the plasma is a D-T mixture, the power density will be about 6×10^6 watts/m^2 of slab. We conclude that for a reasonably large D-T plasma the thermonuclear power density considerably exceeds the radiation loss. However, if the illustrative plasma were pure deuterium, the radiation loss would slightly exceed the thermonuclear power density. Partly because of the low reaction rate and consequent low power compared with bremsstrahlung

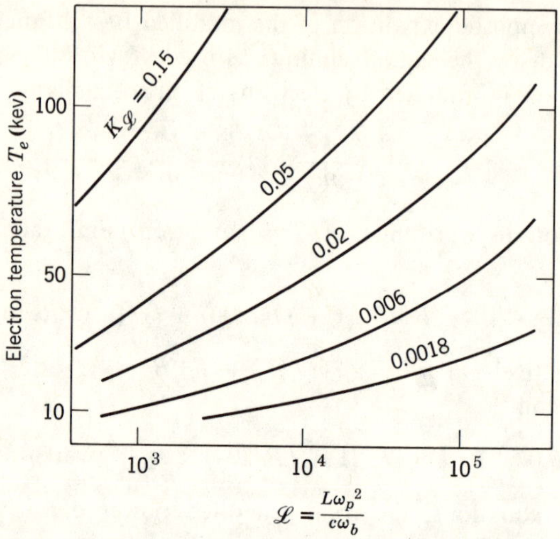

Fig. 11.7. Transparency coefficient $K_{\mathscr{L}}$ for cyclotron radiation in a plasma slab. After B. A. Trubnikov and V. S. Kudryavtsev.[6]

and cyclotron-radiation losses, we provisionally conclude that a controlled D-D thermonuclear reaction appears unprofitable.

As in the case of bremsstrahlung, the electron temperature is determined in part by the cyclotron radiation. In this case, however, the calculation is made difficult by the dependence of the absorption upon frequency (hence electron temperature) and electron density. For example, suppose that the electron energy is maintained in equilibrium through gain from hotter ions and loss to bremsstrahlung plus cyclotron radiation. From Eqs. 8.39, 11.17, 11.83, and 11.85, we find for a plasma of singly charged ions

$$\frac{50(T_i - T_e)\ln\Lambda}{A_i} = T_e^2 + 1.3 \times 10^{20} B^2 T_e^{5/2}\left(1 + \frac{T_e}{204}\right)\frac{K_{\mathscr{L}}}{n_i}. \qquad (11.86)$$

If T_i, n_i, B, and the size are known, the electron temperature may be found from Eq. 11.86.

At the end of Sec. 11.1, we noted that harmonics of the cyclotron frequency are generated by charges gyrating in a nonuniform induction. The magnitude of this effect has not been estimated. Presumably, a proper analysis of the motion described by Eq. 10.15 could yield a set of harmonics whose intensities decrease with harmonic number; then a radiation calculation similar to that described in the previous sections could be performed. It appears likely that such a calculation would be complicated by the fact that the high-energy electrons have the least-circular orbits (see Eq. 10.15). Therefore, both the inhomogeneous induction and the relativistic effects must be considered simultaneously. Finally, we note that if a plasma completely excludes the magnetic induction from its interior, the charges do not gyrate there, and no cyclotron radiation is generated. However, the induction near the plasma surface will be highly nonuniform. Harmonics of the cyclotron frequency will be generated copiously near the surface, and the associated energy will escape with little attenuation. Whether this radiation loss is more severe than that from a plasma in a sensibly uniform induction is not known.

★ 11.6. The Effect of Reflectors

The bremsstrahlung, consisting of x rays in the kilovolt range, is strongly absorbed on any reasonable vacuum wall. Thus nothing of practical value can be done to reduce the bremsstrahlung loss; the radiation becomes a heat source at the vacuum walls surrounding the plasma. Removal of this heat may be difficult, and some very rough estimates of the problem are made in Chap. 13.

The entire cyclotron radiation lies in the frequency range below (let us say) 10^{13} cycles/sec, that is, at wavelengths greater than those corresponding to the far infrared. Radiation at such frequencies is well reflected by metallic surfaces; hence we expect that some reduction of the cyclotron radiation loss can be achieved. Our inability to calculate the cyclotron radiation for other than very simple cases prevents us from estimating properly the effect of such reflectors. For example, diffuse reflection from a surface necessarily produces rays that are not perpendicular to the induction, thus complicating the analysis. Nevertheless, some useful comments can be made.

We consider the case of specular reflection from parallel surfaces lying outside a plane plasma slab, as shown in Fig. 11.8. For simplicity, we assume the plasma to be symmetric about its midplane, so that a ray enters the plasma at O and O' with the same angle. The reflection coefficient at the surface is Γ, which may be a function of the angle of incidence. We further assume that no interference effects take place at the surface, so that geometric optics applies. Equation 11.4 is then applied to the propagation of

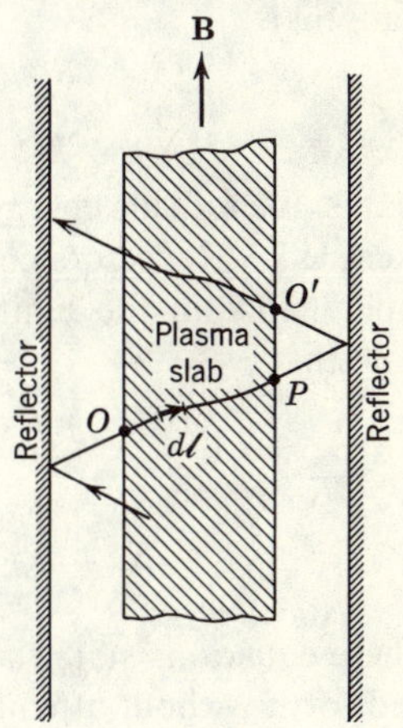

Fig. 11.8. Ray through a plasma reflected by walls with reflection coefficient Γ.

the ray, but with the boundary condition

$$S(O) = \Gamma S(P). \tag{11.87}$$

The solution is

$$S(P, \omega, \boldsymbol{\kappa}) = \frac{S_B(\omega, \boldsymbol{\Omega})\left[1 - \exp\left(-\int_O^P d\ell\, \alpha\right)\right]}{1 - \Gamma \exp\left(-\int_O^P d\ell\, \alpha\right)}. \tag{11.88}$$

This expression is to be compared with Eq. 11.5, for the case of no reflection. We see that the radiant flux has increased because of the reflection. On the other hand, the fraction of the flux absorbed is $1 - \Gamma$. The actual radiant loss S_l is therefore

$$S_l = \frac{S_B\left[1 - \exp\left(-\int_O^P d\ell\, \alpha\right)\right](1 - \Gamma)}{1 - \Gamma \exp\left(-\int_O^P d\ell\, \alpha\right)}, \tag{11.89}$$

which is less than that given by Eq. 11.5.

The loss S_l may be calculated for simple cases. Figure 11.6 shows the relative loss for radiation perpendicular to the slab with 90% reflection. The results were calculated by the method in the paragraphs following Eq. 11.78. We note from the figure that the loss with 90% reflecting walls is much more than 10% of the loss without reflecting walls, and we present the following discussion in support of that fact.

At low frequencies, the absorption is high, and Eqs. 11.88 and 11.89 state that

$$\left.\begin{aligned} S &\approx S_B \\ S_l &\approx S_B(1 - \Gamma) \end{aligned}\right\} \text{ if } \int d\ell\, \alpha \gg 1; \tag{11.90}$$

that is, the radiant energy density is that of a black body, which cannot be exceeded. Reflectors cut down the absorption proportionately. On the other hand, in the limit of plasma transparency, Eqs. 11.88 and 11.89 state that

$$\left.\begin{aligned} S &\approx S_B \int d\ell\, \alpha/(1 - \Gamma) \\ S_l &\approx S_B \int d\ell\, \alpha \end{aligned}\right\} \text{ if } \int d\ell\, \alpha \ll 1. \tag{11.91}$$

These equations state the physical fact that photons merely bounce back and forth without attenuation in the plasma until they are absorbed, and reflectors do not inhibit the energy loss.

We also state an equivalent thermodynamic argument. To the degree that the reflection is perfect, the radiant flux S in the space approaches

S_B at all frequencies and is determined by the temperature T_e of the plasma emitter. Thus, in the limit $\Gamma \to 1$, Eq. 11.88 yields $S = S_B$ at all frequencies. The effect can be seen in Fig. 11.6: the dashed loss curves, multiplied by $1/(1 - \Gamma)$, approximate the black-body limit to higher frequencies than do the solid curves.

Observe that in the case of substantial reflection, the bulk of the radiation appears at frequencies in excess of ω^* at which $\int d\ell\, \alpha = 1$. Thus we would be incorrect to reason that the radiation can be neglected at frequencies where the plasma is relatively transparent. To the extent that $\Gamma \to 1$, the peak of the spectral intensity appears at ever higher frequency (we ignore the breakdown of the Rayleigh-Jeans formulation, which occurs in the x-ray spectral region).

PROBLEMS

1. Suppose that we were to build a deuterium thermonuclear reactor whose sole energy loss is by black-body radiation from its surface. For simplicity, assume uniform density n and temperature T throughout a sphere of radius r.

(*a*) Show that

$$r = \frac{6aT^4}{n_D{}^2\, \overline{\sigma v}_{\mathrm{DD}} U_{\mathrm{DD}}},$$

where U_{DD} is the energy released per fusion.

(*b*) Suppose that $n_D = 10^{28}/\mathrm{m}^3$. Considering that $\overline{\sigma v}_{\mathrm{DD}}$ is a function of T, find the temperature giving the least radius, and the particular radius.

2. A charge q is subjected to a uniform oscillating electric field $E \cos \omega t$ pointed along the z-axis. Compute the radiation from the charge and the angular distribution. Next compute the damping constant of the motion. Finally find the intensity of radiant energy as a function of frequency.

3. For a small transverse acceleration of an electron in the field of a heavy ion, show that the Fourier transform $\dot{v}(\omega)$ of the acceleration $\dot{v}(t)$ is

$$\dot{v}(\omega) = \frac{Ze^2\omega}{4\pi^2\epsilon_0 mv^2} K_1\left(\frac{\omega b}{v}\right),$$

where K_1 is the modified Bessel function. Then show that $\dot{v}(\omega)$ is approximately constant for $\omega < v/b$, and decreases exponentially thereafter.

4. For a simple deuterium plasma in which the tritium is burned but He^3 is not, calculate the initial temperature at which $w_{\mathrm{DD}} + w_{\mathrm{DT}} = w_x$. Note that $T_e \neq T_{\mathrm{D}}$; assume $\ln \Lambda = 20$, and neglect heating of the electrons by the reaction products.

5. We usually assume that the bremsstrahlung power density w_x generated in a thermonuclear plasma escapes freely. Make some reasonable assumptions, and discuss as quantitatively as you can the absorption of this radiation as it might be generated and absorbed in

(*a*) A plasma of controlled thermonuclear interest.
(*b*) The sun.
(*c*) Interstellar space.

Discuss the nature and limitations of your approximations. The temperature in the sun's interior is about 2 kev, and in interstellar space is about 1 ev. Specific gravity of the sun is about 1, and the density in interstellar space is about $10^6/\text{m}^3$.

6. Suppose that the initial density of electrons is n_1 and the initial energy of deuterons is U_{D1}. The density of the electrons after a time τ is n_2, and the energy of the deuterons is U_{D2}, both quantities increasing linearly with time. Take the transfer of energy to the electrons from the deuterons and the bremsstrahlung of the electrons into account. Find the energy of the electrons U_{e2} at the end of the heating cycle. Find an approximate expression valid if the compression and heating are great.

7. Find the relativistic Maxwell-Boltzmann distribution in energy.

8. Calculate the electron temperature in a hydrogenic plasma slab, for which

$$\begin{aligned} &\text{Thickness } L = 0.5 \text{ m} \\ &n_e = n_i = 5 \times 10^{20}/\text{m}^3 \\ &T_i = 100 \text{ kev} \\ &B = 8 \text{ webers/m}^2. \end{aligned}$$

Assume that the mean atomic weight of the ions is 2.5, corresponding to a D-T mixture, that the electrons gain energy through heating from the ions and lose it through radiation. Which is larger, the bremsstrahlung or the cyclotron radiation?

9. The curves of Fig. 11.7 are inapplicable for calculation of cyclotron-radiation loss, if reflecting walls surround the plasma. This fact is true even if the radiation loss calculated that way were reduced by the factor $1 - \Gamma$ (Γ is the reflection coefficient). Explain why these statements are correct and whether such a calculation will underestimate or overestimate the radiation loss.

REFERENCES

1. W. K. H. Panofsky and M. Phillips, *Classical Electricity and Magnetism*, Addison-Wesley Publishing Co., Inc., Reading, Mass. (1955), p. 308.
2. H. A. Kramers, *Phil. Mag.*, **46**, 836 (1923).
3. W. Heitler, *Quantum Theory of Radiation*, Oxford University Press (1944).
4. B. A. Trubnikov, *Soviet Phys.* "*Doklady*," **3**, 136 (1958); [*Dokl. Akad. Nauk. S. S. S. R.*, **118**, No. 5, 913 (1958)].
5. B. A. Trubnikov and A. E. Bazhanova, *Plasma Physics and the Problem of Controlled Thermonuclear Power*, L. A. Leontovich, Editor, English Translation, Vol. III, Pergamon Press, London (1959), pp. 141–175.
6. B. A. Trubnikov and V. S. Kudryavtsev, *Proc. 2nd U.N. Conf. on Peaceful Uses of Atomic Energy*, Vol. 31, United Nations, Geneva (1958), p. 93.

7. D. B. Beard, *Phys. Fluids*, **2**, 379 (1959).
8. W. E. Drummond and M. N. Rosenbluth, *Phys. Fluids*, **3**, 45 (1960).
9. J. L. Hirshfield, D. E. Baldwin, and S. C. Brown, *Phys. Fluids*, **4**, 198 (1961).
10. G. A. Schott, *Electromagnetic Radiation*, Cambridge University Press, Cambridge (1912).
11. J. Schwinger, *Phys. Rev.*, **75**, 1912 (1949).
12. L. Landau and E. Lifshitz, *The Classical Theory of Fields*, M. Hamermesh, Translator, Addison-Wesley Publishing Company, Reading, Mass. (1951).
13. E. Jahnke and F. Emde, *Tables of Functions*, Dover Publications, New York (1945), p. 149.
14. S. Chapman and T. G. Cowling, *The Mathematical Theory of Non-Uniform Gases*, 2nd ed., The University Press, Cambridge (1952), pp. 370–377.
15. G. N. Watson, *A Treatise on the Theory of Bessel Functions*, 2nd ed., The University Press, Cambridge (1944).

12

Plasma stability

The stability of a plasma against deformations is a vital problem. The achievement of controlled thermonuclear power depends principally upon its solution. A cooperative action between one part of a plasma and another may generate such fields and motions that a part of the plasma moves rapidly away from the region where it is supposedly confined. The instabilities referred to here do not relate to the rather slow diffusion of the plasma across lines of induction nor to the gradual loss of plasma through mirrors, as a result of Coulomb scattering. Slow losses of that sort might be tolerable. Catastrophic instabilities obviously would not.

A number of plasma confinement schemes are discussed in this book. These schemes are based on the hydromagnetic and/or particle models. In each case an equilibrium of some sort is found in which the plasma appears to be confined in the desired way. Equilibrium is a necessary but insufficient condition. For all practical schemes of generating controlled thermonuclear energy, it is required that a perturbation away from the equilibrium configuration not grow indefinitely. We examine the matter in this chapter.

The theoretical investigation of plasma stability is unfortunately very difficult. All possible types of instabilities must be examined before a plasma can be said to be stable.

Because of these difficulties, the chapter deals only with basic concepts and is not meant to be complete. An elementary derivation of stability criteria will precede a more formal, general development. These criteria will be applied to a few particular confinement schemes. We conclude that stability theory requires further development.

12.1. Elementary Remarks on Stability

The mathematical development of stability theory is often so involved that we tend to lose sight of the basic nature of the problem. Therefore, it seems advisable to illustrate the problem in the simplest terms.

Consider Fig. 12.1. The ball at point A in Fig. 12a is in stable equilibrium; the ball at point B in Fig. 12b is in unstable equilibrium. These statements can be made in a more sophisticated way, as follows. If no

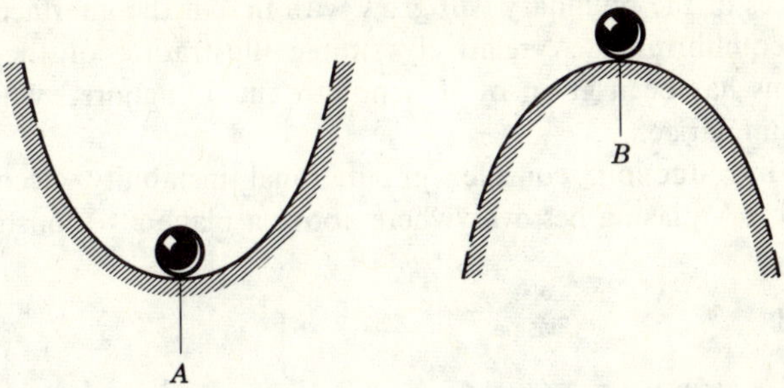

(*a*) Stable (*b*) Unstable

Fig. 12.1 Elementary definition of stable and unstable equilibrium.

energy is added to or subtracted from the system, the total energy (here that of the ball alone) is constant. If the ball at point A moves, its potential energy must increase. Since its kinetic energy cannot decrease, motion is impossible. On the other hand, the ball at point B can increase its kinetic energy by decreasing its potential energy. An infinitesimal displacement will grow, and the ball is no longer confined at its equilibrium position.

The analogy applies quite well to plasma stability; the basic concept is no more difficult. We assume that a static equilibrium of some sort has been found. Then we make a small perturbation and investigate the consequences. Analogously, we may slightly displace the ball at point A, release it, and find whether or not it will oscillate about its equilibrium position. We define the oscillations about point A in Fig. 12.1a to be stable, because the ball remains forever in the vicinity of point A. In the case of a plasma the complexity arises entirely through the multiplicity of types of energy and the multiplicity of possible displacements. Neglect of any one may lead to entirely fallacious conclusions.

Since the total energy of the system is to be preserved in treating these systems, it is essential that we so define its boundary that all interactions are included. For example, loss of part of the plasma may affect the stability of the remainder. In simple analogy, if the balls of Fig. 12.1 are charged dielectric spheres, their stability cannot so readily be determined and may depend upon the nature of other surfaces in the vicinity.

12.2. Simple Derivation of Plasma Surface Instability

Since all plasmas of thermonuclear interest must be confined in a vacuum away from material walls by magnetic fields, stability of the plasma-vacuum interface is of prime importance. We expect that small disturbances of the boundary will grow with time if the interface exists in unstable equilibrium. A relatively simple illustration of the physical mechanisms has been given by Rosenbluth and Longmire,[1] whose work we now summarize.

As an introduction, consider gravitational instability, according to Fig. 12.2*a*. A plasma lies everywhere above a plane z = constant. We

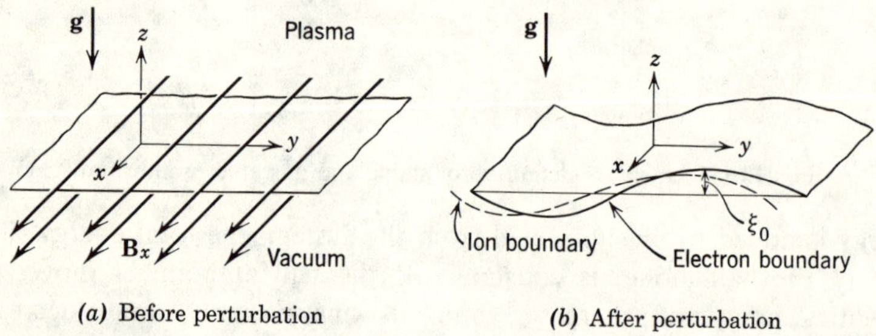

Fig. 12.2. Gravitational instability of a plasma supported by a magnetic induction.

assume that the plasma pressure is low enough that the induction **B** substantially penetrates all space without modification by the presence of the plasma. The gravitational acceleration is **g** in the $(-z)$-direction, and the magnetic induction lies in the x-direction. Let the plasma initially have constant temperature and density, and a sharp boundary at the lower side. In general, the equations governing the plasma motion are nonlinear. However, small departures from equilibrium will satisfy linear equations. Since we are interested principally in the stability or instability of the equilibrium configuration, and not in its large-amplitude behavior, the linearized approximation suffices. For such linear systems, the general perturbation may be expressed as a superposition of sine waves. In the present case, consider the typical surface perturbation

$$\xi \equiv \Delta z = \xi_0 \sin \kappa y \tag{12.1}$$

as in Fig. 12.2*b*, where κ is the wave number. Thus the **B**-lines are still parallel to the surface. We suppose that initially Eq. 12.1 gives the surfaces of both ions and electrons, but this charge neutrality is soon altered. There is a drift due to the gravitational force, which is readily derived by duality. In crossed electric and magnetic fields, a drift appears:

$$\mathbf{v}_E = \mathbf{E} \times \mathbf{B}/B^2, \tag{12.2}$$

which is identical to Eq. 10.13. The electric force here is $q\mathbf{E}$, and could properly be replaced by any conservative force field. Thus the motion of the guiding centers due to gravity is

$$\mathbf{v}_g = m\mathbf{g} \times \mathbf{B}/qB^2, \tag{12.3}$$

which will make the ions drift in the $(-y)$-direction and the electrons in the $+y$-direction. The charge separation shown in Fig. 12.2*b* develops. Since the electron drift is much less than the ion drift, it will be ignored.

If the amplitude of the perturbation is small compared with the wavelength, the time rate of change of surface charge density $\sigma(y)$ resulting from charge separation is

$$\begin{aligned} \frac{\partial \sigma}{\partial t} &= -n_i q_i |\mathbf{v}_{gi}| \kappa \xi_0 \cos \kappa y \\ &= -\frac{n_i m_i g}{B} \kappa \xi_0 \cos \kappa y. \end{aligned} \tag{12.4}$$

In considering the surface charge density, we assume that the charged layer is thin. For small amplitudes ξ_0, we may also assume that the charged layer lies in the original plane of the boundary. The electric field is then easy to compute.

To proceed, it is convenient to demonstrate the effective dielectric constant of a magnetized plasma, by considering a sudden incremental change of the perpendicular component of the electric field appearing in Eq. 12.2. Let us view a particle, both before and after the change, in moving coordinate systems in which the guiding center is at rest. The difference in velocity of the two systems is

$$\Delta \mathbf{v} = \Delta \mathbf{E}_{\mathrm{T}} \times \mathbf{B}/B^2. \tag{12.5}$$

Now if a particle at point $\mathbf{r}$ has velocity $\mathbf{v}$, its guiding center is at the point

$$\mathbf{r}_g = \mathbf{r} + \frac{m}{qB^2} \mathbf{v} \times \mathbf{B}; \tag{12.6}$$

therefore, the instantaneous displacement of the guiding center is

$$\Delta \mathbf{r}_g = \frac{m}{qB^2} (-\Delta \mathbf{v}) \times \mathbf{B} = \frac{m}{qB^2} \Delta \mathbf{E}_{\mathrm{T}}. \tag{12.7}$$

The step is in opposite directions for charges of opposite sign, the ions making the largest contribution to the current in a neutral plasma. We can take account of this induced current in first-order orbit theory by including a polarization current†

$$\mathbf{j}_p = \frac{n(m_e + m_i)}{B^2}\frac{\partial \mathbf{E}_\mathrm{T}}{\partial t} = \frac{\rho_m}{B^2}\frac{\partial \mathbf{E}_\mathrm{T}}{\partial t}. \tag{12.8}$$

We can define a permittivity consistent with Eq. 12.8:

$$\epsilon = \epsilon_0\left(1 + \frac{\rho_m}{B^2\epsilon_0}\right). \tag{12.9}$$

We shall assume merely for simplicity that the plasma density is high, so that $\rho_m \gg B^2\epsilon_0$. It can be shown that the electrostatic field inside a dielectric medium of permittivity ϵ, having a substantially plane surface of charge density

$$\sigma = \sigma_0 \cos \kappa y \tag{12.10}$$

independent of x (see Fig. 12.2*b*), is‡

$$\left.\begin{aligned} E_y &= \frac{\sigma_0}{\epsilon}\sin \kappa y\, e^{-\kappa z}, \\ E_z &= \frac{\sigma_0}{\epsilon}\cos \kappa y\, e^{-\kappa z}. \end{aligned}\right\} \tag{12.11}$$

These electric fields also cause a drift of the plasma, according to Eq. 12.2. The components of this drift velocity are

$$\left.\begin{aligned} V_y &= \frac{\sigma_0}{\epsilon B}\cos \kappa y\, e^{-\kappa z}, \\ V_z &= -\frac{\sigma_0}{\epsilon B}\sin \kappa y\, e^{-\kappa z}. \end{aligned}\right\} \tag{12.12}$$

The velocity field is divergenceless, and therefore it does not change the plasma density inside but merely moves the boundaries. The velocity V_z at the boundary ($z \approx 0$) causes the amplitude of the perturbation to grow according to

$$\frac{d\xi_0}{dt} = -\frac{\sigma_0}{\epsilon B}. \tag{12.13}$$

By comparing Eqs. 12.4 and 12.10 and by combining Eqs. 12.9 and 12.13, there results (if $\rho_m \gg B^2\epsilon_0$)

$$\frac{d^2\xi_0}{dt^2} = g\kappa\xi_0, \tag{12.14}$$

† See Problem 3 of Chap. 5 and Problem 4 of Chap. 10.
‡ Note that $\nabla\cdot\mathbf{D} = 0$ inside, and the boundary conditions are satisfied.

with solutions

$$\xi_0 \propto \exp\left(\pm\sqrt{g\kappa}\, t\right). \tag{12.15}$$

If the sign of g is reversed (plasma fluid put on the bottom), the solutions are oscillatory (stable). Another development of the same problem is given by Kruskal and Schwarzschild.[2]

It is interesting to note that the rate of growth here is exactly the same as in the so-called Taylor instability of a fluid supported under gravity by a second fluid which exerts pressure but is weightless.[2] The charge separation is able to overcome exactly the restraining influence of the magnetic induction, and the charge does not appear in Eqs. 12.14 or 12.15. The exact compensation occurs, however, only in the limit $\epsilon \gg \epsilon_0$. If ϵ is not so large, the proper value Eq. 12.9 must be used. The rate of growth is reduced and is eventually proportional to $\sqrt{n}$ for low-density plasmas.

The essential mechanism of the instability is the charge separation produced by the gravitational force. However, any force perpendicular to **B** which is independent of the sign of the charge would do as well. Suppose now that the plasma is *not* confined by lines of **B** which are straight and uniform but have the general configuration of Fig. 12.3 or Fig. 12.4. From the motions of individual particles we can recognize two such forces.

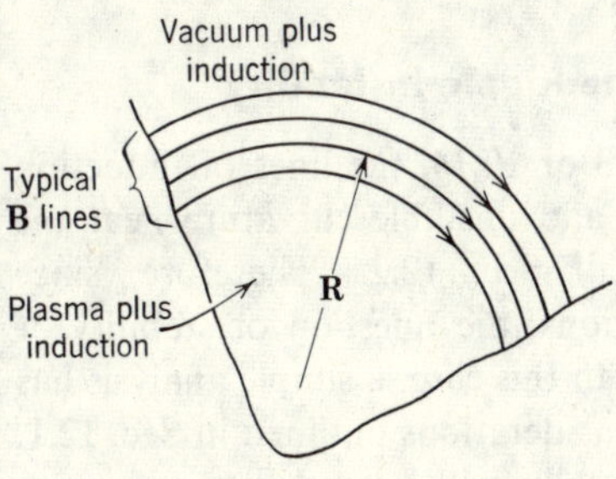

Fig. 12.3. Plasma surrounded by **B** lines concave toward the plasma.

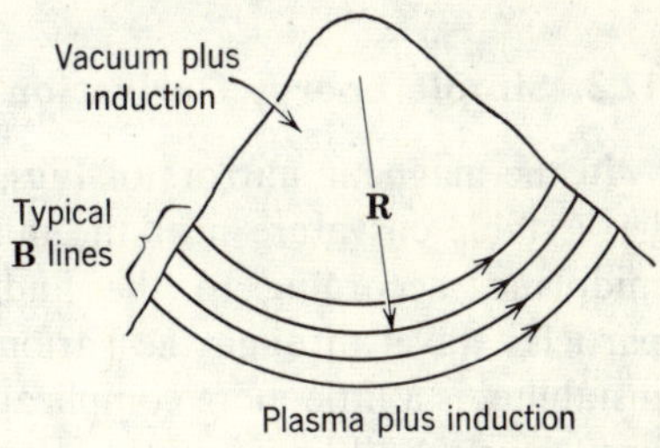

Fig. 12.4. Plasma surrounded by **B** lines convex toward the plasma.

These two forces were discussed in Sec. 10.2. From Eq. 10.15 we find that

$$\mathbf{v}_g = \frac{(U_\mathrm{T} + 2U_\parallel)}{qB^2R^2}(\mathbf{R}\times\mathbf{B}) + \frac{U_\mathrm{T}\mu\mathbf{B}\times(\mathbf{B}\times\mathbf{j})}{qB} + \mathbf{v}_\parallel. \tag{12.16}$$

The terms involving U_T result from the first term of Eq. 10.15 by application of a familiar vector identity to reduce $\nabla(\mathbf{B}\cdot\mathbf{B})$ and by use of Eqs.

5.13 and 5.20 with neglect of displacement current. The term in $U_\parallel$ of the present result arises from the second term of Eq. 10.15, as discussed in Sec. 10.2. The term involving $\mathbf{B} \times \mathbf{j}$ is negligible because the current of one particle is trivial. Thus, we must replace

$$\mathbf{g} \rightarrow \frac{\mathbf{R}}{R^2}(v_\parallel^2 + \tfrac{1}{2}v_b^2). \tag{12.17}$$

If the radius of curvature **R** points out of the plasma, as in Fig. 12.3, it will tend to be locally unstable. In accord with the meaning of this equation and our discussion, we define the radius of curvature **R** to be positive if **R** points out of the plasma. Conversely, the configuration of Fig. 12.4 is locally stable according to the present criterion.

Note that the instability appears as ripples in the plasma surface. The crests and troughs are parallel to the **B** lines; from its appearance, the particular mode is sometimes referred to as a flute instability. It is the principal mode of instability to be discussed, and we shall have occasion to enlarge upon its physical interpretation in subsequent sections.

From Eq. 12.15, we observe that short-wavelength ripples or flutes grow fastest. We suspect, however, that the theory must be modified if the wavelength is very short and that ripples with wavelength shorter than a certain minimum value will not grow. The question is related to our neglect here of the electrostatic energy and other matters; we delay discussion of these points until Sec. 12.8.

12.3. Simple Energy Derivation of the Ripple Instability

In the magnetic mirrors of Figs. 10.13 or 10.14, the lines of induction have stable curvature near the mirrors and unstable curvature near the midplane, according to the findings of Sec. 12.2. Therefore, since particles travel through the various regions, the question of stability or instability is a little more complicated. In this case, a simple analysis has been developed,[1] based on the energy considerations outlined in Sec. 12.1. We present the method, which incidentally will predict the ripple instability of Sec. 12.2, as an introduction to the formalism of Secs. 12.5 to 12.8.

Assume that the plasma pressure p is a scalar, and consider the plasma shown in Fig. 12.5, where we interchange plasma and magnetic induction in regions 1 and 2. Let the system be axially symmetric, and let the plasma pressure be small compared with the magnetic pressure. Then the magnetic induction is nearly identical to that in a vacuum, and any distortion of it increases its energy. Hence, the only dangerous perturbations are those that leave the induction unchanged.

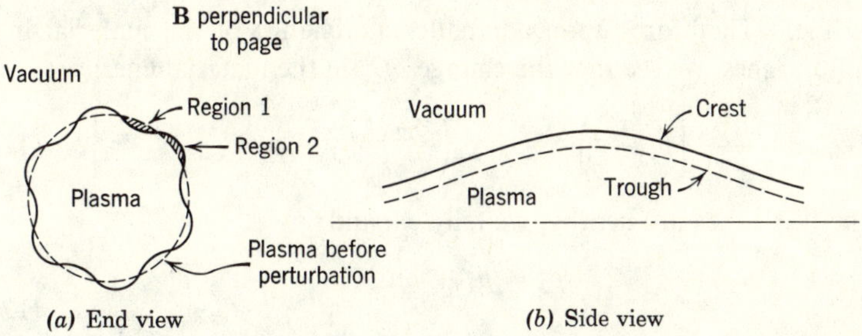

Fig. 12.5. Interchange (*or* ripple, *or* flute) instability.

The flutes are thus constructed so that their surfaces are bounded by lines of unperturbed **B**. The net effect of the flutes is to interchange the field and matter between regions 1 and 2, and we now proceed to calculate the change of potential energy δQ.

The magnetic energy in a flux tube is

$$Q_M = \int d\mathbf{r}\,\frac{B^2}{2\mu_0} = \int dl\, s\,\frac{B^2}{2\mu_0}, \tag{12.18}$$

where l is the length of the tube, s is the cross-sectional area, and the last integral is taken along a complete flux line. But

$$sB = \Phi = \text{flux}, \tag{12.19}$$

which is constant along a flux tube. Thus

$$Q_M = \frac{\Phi^2}{2\mu_0}\int\frac{dl}{s}, \tag{12.20}$$

and the change in magnetic energy on interchange of flux tubes 1 and 2 is

$$\delta Q_M = \frac{1}{2\mu_0}\left[\left(\Phi_1{}^2\int_2\frac{dl}{s} + \Phi_2{}^2\int_1\frac{dl}{s}\right) - \left(\Phi_1{}^2\int_1\frac{dl}{s} + \Phi_2{}^2\int_2\frac{dl}{s}\right)\right]. \tag{12.21}$$

From Eq. 12.21, the magnetic energy is left unaltered if we interchange tubes containing equal amounts of flux, so that $\Phi_1 = \Phi_2$.

The material energy in any small volume $\mathscr{V}$ is calculated from the fact that an energy $nkT\mathscr{V}/2$ exists for each degree of freedom. From Eq. 7.73, we find that

$$Q_p = \frac{nkT\mathscr{V}}{\gamma - 1} = \frac{p\mathscr{V}}{\gamma - 1}, \tag{12.22}$$

where γ is the ratio of specific heats. Furthermore, $p\mathscr{V}^\gamma = \text{const}$ (Eq. 7.19) for the adiabatic motion of a fixed quantity of material. Finally, from the discussion of Sec. 7.2, the scalar pressure p is constant along a flux line. Therefore, upon adiabatic interchange of the material in the two flux tubes, we see that the change δQ_p in the material energy is

$$\delta Q_p = \frac{1}{\gamma - 1}\left[p_1\left(\frac{\mathscr{V}_1}{\mathscr{V}_2}\right)^\gamma \mathscr{V}_2 + p_2\left(\frac{\mathscr{V}_2}{\mathscr{V}_1}\right)^\gamma \mathscr{V}_1 - p_1\mathscr{V}_1 - p_2\mathscr{V}_2\right]. \quad (12.23)$$

If the flux tubes are nearby, we may expand

$$\left.\begin{aligned} p_2 &= p_1 + \delta p, \\ \mathscr{V}_2 &= \mathscr{V}_1 + \delta\mathscr{V}, \end{aligned}\right\} \quad (12.24)$$

and find

$$\delta Q_p = \delta p \delta\mathscr{V} + \gamma p \frac{(\delta\mathscr{V})^2}{\mathscr{V}} = \mathscr{V}^{-\gamma}\delta(p\mathscr{V}^\gamma)\delta\mathscr{V}. \quad (12.25)$$

The condition for stability is that $\delta Q > 0$. In general, p will decrease with distance away from the axis to the plasma edge. Hence near the boundary, $\delta(p\mathscr{V}^\gamma)$ is negative, since as $p \to 0$, $|\delta p/p| > \gamma(\delta\mathscr{V}/\mathscr{V})$. Under these circumstances, the condition for stability becomes $\delta\mathscr{V} < 0$. We now observe that the volume of a flux tube is

$$\mathscr{V} = \int dl\, s = \Phi \int \frac{dl}{B}. \quad (12.26)$$

Since the flux Φ is unchanged, the stability condition is

$$\delta \int \frac{dl}{B} < 0. \quad (12.27)$$

The geometrical significance of Eq. 12.27 can be seen from Fig. 12.6. The integral in Eq. 12.27 is computed over the whole length of a flux line.

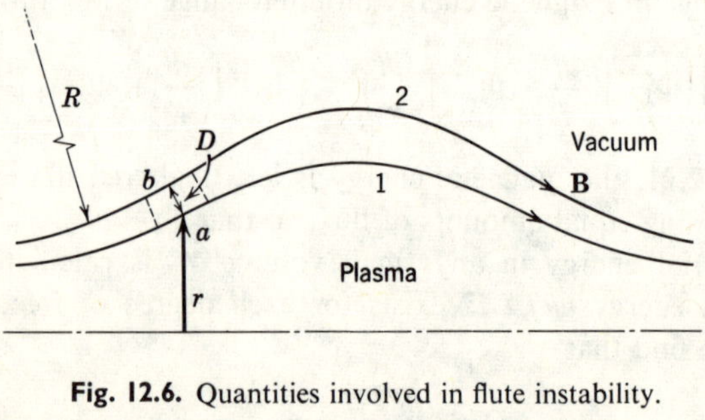

Fig. 12.6. Quantities involved in flute instability.

The difference in the values of $\int dl/B$ along lines 1 and 2 is to be negative according to Eq. 12.27. Consider any point a on line 1 and the directly opposite point b on line 2, joined to point a by the perpendicular distance D.

Since a negligible current flows between the two lines 1 and 2 at any point, $\nabla \times \mathbf{B} = 0$, a scalar magnetic potential $\psi = \int dl\, B$ exists; the points a and b are at the same potential. We can now write Eq. 12.27 as

$$\delta \int \frac{dl}{B} = \oint_2 \frac{dl\, B}{B^2} - \oint_1 \frac{dl\, B}{B^2}, \tag{12.28}$$

where by the symbol $\oint$ we mean that the integrals are taken around complete flux lines, with a cut in line 1 opposite a cut in line 2. The potential is the same at the ends and at all corresponding points a and b. Therefore, the limits of the integrals are identical; if the flux lines are adjacent, the variation $\oint \delta\, dl\, B = 0$; Eq. 12.28 can then be written

$$\delta \int \frac{dl}{B} = \int dl\, B\delta\left(\frac{1}{B^2}\right). \tag{12.29}$$

From the curvature of the lines we see that

$$\frac{\delta B}{B} = -\frac{D}{R}, \tag{12.30}$$

where again R is the radius of curvature, negative if the center of curvature lies outside the plasma. Further, D can be eliminated by using the fact that the flux between the lines is constant along their length; that is,

$$2\pi r D B = \Phi = \text{const}, \tag{12.31}$$

where r is the radius shown in Fig. 12.6. Thus the condition for stability (Eq. 12.27) becomes

$$-\int \frac{dl}{R r B^2} > 0. \tag{12.32}$$

Since, by definition, R is negative at the ends and positive in the middle, the middle section promotes instability. In general, the fact that B is large at the ends will make the middle region dominate.

Note that $dl/R \approx -d\theta$, the angle of the flux line with the z-axis, and that Br^2 is roughly a constant; then Eq. 12.32 may be approximated by

$$\int \frac{d\theta}{B^{3/2}} > 0 \tag{12.33}$$

for stability. Since B is smaller in the region of negative $d\theta$, the configuration is unstable, according to these criteria. Furthermore, note that the criterion of Sec. 12.2 is recovered if the curvature does not change.

In addition to these developments, Rosenbluth and Longmire also demonstrate that a similar criterion exists if the pressure is not a scalar and that the growth rate of large flutes is considerably less than the mean thermal particle speed. We leave the details, which are quite straightforward, to the interested reader.

12.4. Simple Stability Criteria Applied to Confinement Schemes

The term interchange instability is also often applied to the phenomenon just described, and the genesis of the term is evident from this derivation. The **B** lines act like stretched elastic bands (see Sec. 5.6). They tend to contract lengthwise and are kept apart by their mutual repulsion. If now we stretch them around a plasma as in Fig. 12.6, they tend to slip inward, while the plasma leaks out between them. The total energy of the system is thereby decreased by this interchange, and the system is unstable.

According to these criteria, the simple magnetic mirror (Fig. 10.13 or Fig. 10.14) should be unstable. In actual fact, experiments to date have not conclusively demonstrated the instability. Therefore we suspect some inadequacy of the theory, to which we return in Sec. 12.8. Be that as it may, other confinement schemes have been postulated, for which the curvature of **B** lines should promote stability. The simplest of these schemes is a cusped system, found by reversing the current in one of the coils in a conventional mirror. The device is shown schematically in Fig. 12.7. The **B** lines enter (for example) through each coil and leave

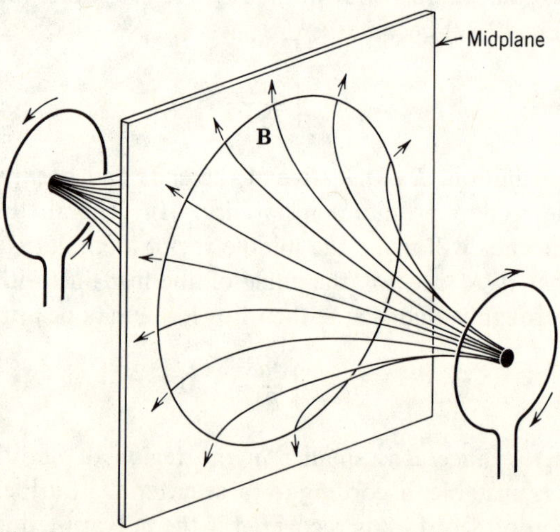

Fig. 12.7. Cusped geometry.

at the midplane. The disadvantage of the system is that the plasma can escape through the large ring cusp at the midplane, as well as along the axis. The confinement properties of the system cannot readily be calculated, and we shall not attempt to do so. We observe in passing, however, that a calculation similar to that for a conventional mirror, based on invariance of the magnetic moment, will not apply here. The reason is that $\mathbf{B} = 0$ at the center of the system: the motion of charged particles in the region of the center cannot be described at all well in terms of perturbed circular gyrations. More nearly, the particles bounce against the surrounding magnetic walls. In principle, the system should be capable of confining a plasma in which the quantity $\beta = nkT/(B^2/2\mu_0) = 1$, according to the normal convention of its definition. In such a case, the induction is excluded entirely from the interior region. A series of such configurations can be laid end to end, resulting in a structure sometimes called a picket fence. We consider cusped systems further in Sec. 15.7.

A more complex system involving linear cusps is shown in Fig. 12.8. Here, six wires are arranged cylindrically, and each carries current in the opposite direction. The figure is included because the configuration is similar to the stabilizing windings of the stellarator. If the cylinder is

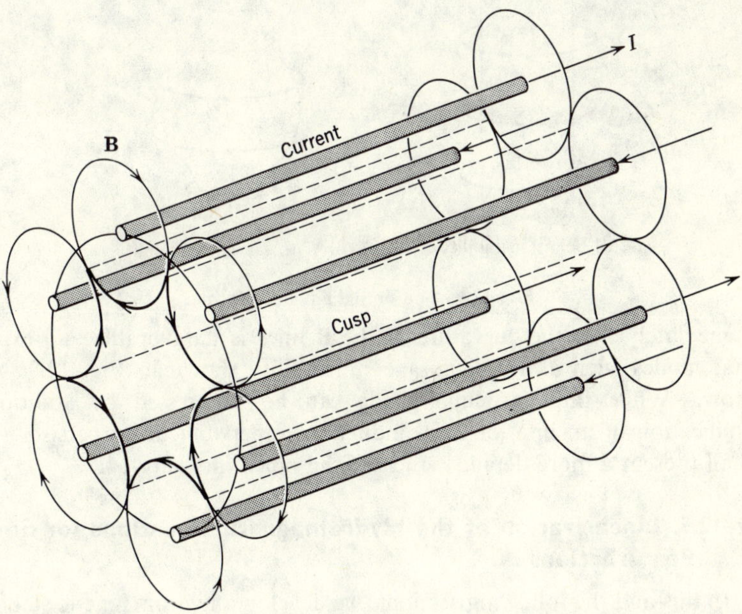

Fig. 12.8. Arrangement of six conductors along a cylinder, with adjacent currents flowing oppositely, forming line cusps.

extended axially, twisted (as a rope), bent into a torus, and provided with a conventional toroidal induction, a stellarator configuration results.

Several other simple applications of the criterion on curvature of **B** lines can be made. In the simple torus, such as Fig. 10.8, the **B** lines near the periphery of the confined plasma have unfavorable curvature. Thus we expect that the plasma will be unstable on that side, a fact demonstrated in Sec. 10.2 by a different analysis.

The simple plasma column carrying a current confined by its own magnetic induction is called a pinch and is shown in Fig. 12.9*a*. We see

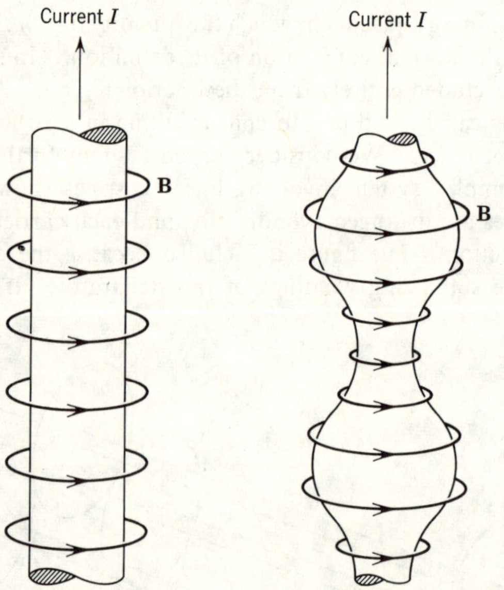

(a) Before instability *(b)* With sausage instability

Fig. 12.9. The simple pinched plasma.

immediately that the curvature of the **B** lines is unfavorable and expect that modes such as the "sausage" instability shown in Fig. 12.9*b* can grow. While this particular mode can be suppressed by additional application of an interior axial induction, other difficulties persist. We shall present a more detailed study of the pinch in Chap. 14.

★ 12.5. Linearization of the Hydromagnetic Equations for Small Perturbations

In this and the following sections, we develop a hydromagnetic stability formulation, following the work of Bernstein, Frieman, Kruskal, and Kulsrud.[3] Generally similar analyses are given by Berkowitz, Grad,

Rubin, and co-workers,[4,5] and by Kadomtsev and Braginsky.[6] Our choice for discussion is somewhat arbitrary. The development is complicated and proceeds con brio through several movements.

As usual, the plasma is assumed to be confined in static equilibrium. Here, however, no other restrictions are placed upon the plasma-field configuration, nor upon the arbitrary first-order displacement except that $\boldsymbol{\xi}$ must satisfy certain boundary conditions.

The plasma has mass density ρ_m and pressure p (Bernstein and co-workers consider also a stress tensor P, but we shall not, because the results are much the same). In addition, fields **E** and **B** exist in the plasma and throughout space. We also assume at this stage that a gravitational potential ϕ also exists, because its inclusion here offers no particular difficulty; we shall in most subsequent developments neglect its effects. All the quantities may be functions of position. Then the total energy can be written as an integral over the entire contributing volume of the sum of the various forms of energy:

$$\begin{aligned}\text{Total energy} &= \int d\mathbf{r}\left(\frac{\rho_m V^2}{2} + \frac{B^2}{2\mu_0} + \frac{\epsilon E^2}{2} + \frac{p}{\gamma - 1} + \rho_m \phi\right),\\ &= U + Q = \text{const},\end{aligned} \tag{12.34}$$

where U is the kinetic energy, and Q is the potential energy. In static equilibrium, $\mathbf{V} = 0$ everywhere; thus $U = 0$. We now perturb the system by displacing the plasma slightly. That is, the plasma at an initial point $\mathbf{r}_0$ is displaced by an amount $\boldsymbol{\xi}(\mathbf{r}_0)$, and such displacements may be made throughout the plasma. As described in Sec. 12.1, the criterion for stability is then that the variation δQ of the potential energy be such that

$$\delta Q > 0 \tag{12.35}$$

for all possible displacements $\boldsymbol{\xi}(\mathbf{r}_0)$.

In passing, we note that the present technique, while not simple, is often analytically simpler than the usual normal-mode technique. This latter technique consists of solving the linearized equations of motion for small perturbations about an equilibrium state. The system is unstable if any solution increases indefinitely in time; if no such solution exists, the system is stable.

The energy-principle technique, on the other hand, depends upon a variational formulation of the equations of motion. Its advantage lies in the fact that if the mere existence of stability or instability is in question (which is our question), we can afford to sacrifice knowledge about the exact eigenfrequencies of the system. Thus the calculation is often simplified and can be made more general.

The case of a plasma with finite conductivity and arbitrary electric fields has not been much investigated. Therefore, we confine our formulations to a simple plasma with infinite conductivity and hence no net charge density.

The equation of motion for a plasma with no net charge density is from Eq. 6.61

$$\rho_m \frac{\partial \mathbf{V}}{\partial t} = -\nabla p + \mathbf{j} \times \mathbf{B} - \rho_m \nabla \phi. \tag{12.36}$$

The mass-conservation condition, Eq. 6.55, is

$$\frac{\partial \rho_m}{\partial t} + \nabla \cdot (\rho_m \mathbf{V}) = 0. \tag{12.37}$$

Ohm's law for the case of infinite conductivity and the further assumptions listed below is

$$\mathbf{E} + \mathbf{V} \times \mathbf{B} = 0 \tag{12.38}$$

from Eq. 6.68. The adiabatic law appropriate to a fixed observer (see Eqs. 6.21 and 7.19) is

$$\left(\frac{\partial}{\partial t} + \mathbf{V} \cdot \nabla\right)(p \rho_m^{-\gamma}) = 0. \tag{12.39}$$

Finally, Maxwell's equations are required in the forms

$$\nabla \times \mathbf{E} = -\frac{\partial \mathbf{B}}{\partial t}, \tag{12.40}$$

$$\nabla \times \mathbf{B} = \mu_0 \mathbf{j}, \tag{12.41}$$

$$\nabla \cdot \mathbf{B} = 0. \tag{12.42}$$

We may readily check from earlier discussions (see Sec. 6.6) that the use of Eqs. 12.36 to 12.42 implies the following further assumptions. (1) The quantities $\mathbf{V}$ and $\mathbf{j}$ are small, so that quadratic terms in those quantities may be neglected. (2) The system is quasi-neutral; that is, the Debye shielding distance is small compared with all the dimensions of the system. In particular, this assumption precludes study of effects due to the plasma sheath. We return to this point later. (3) The ratio m_e/m_i is negligible compared with unity. (4) The kinetic stress tensor is isotropic, an assumption that implies many collisions in some characteristic time. (5) Displacement currents are negligible, which approximation is valid if the speed of Alfvén wave propagation (Eq. 9.49) is small compared with the speed of light. (6) Heat flow is negligible, thus justifying the adiabatic law (Eq. 12.39). (7) All distances are large compared with a Larmor orbit, a point to which we also return later.

Corresponding to these assumptions and equations, there is a set of boundary conditions, either at a fluid-fluid interface or between fluid and vacuum. Let **n** be the outward-pointing unit vector normal to the plasma surface, **K** be the surface current density, and $[\![G]\!]$ the increment in any quantity G across the boundary in the direction **n**. From Maxwell's equations, we find immediately

$$\mathbf{n}\cdot[\![\mathbf{B}]\!] = 0, \tag{12.43}$$

$$\mathbf{n}\times[\![\mathbf{B}]\!] = \mu_0\mathbf{K}. \tag{12.44}$$

Since $[\![\mathbf{B}]\!]$ lies in the surface, we find by taking components of Ohm's law (Eq. 12.38) that the tangential component of $[\![\mathbf{E}]\!]$ is

$$\mathbf{n}\times[\![\mathbf{E}]\!] = (\mathbf{n}\cdot\mathbf{V})[\![\mathbf{B}]\!]. \tag{12.45}$$

At a fluid-fluid interface,

$$\mathbf{n}\cdot[\![\mathbf{V}]\!] = 0; \tag{12.46}$$

This equation is inapplicable at a fluid-vacuum interface, since the mass velocity of a vacuum is without meaning. Finally, the total pressure is continuous, or

$$\left[\!\!\left[p + \frac{B^2}{2\mu_0}\right]\!\!\right] = 0. \tag{12.47}$$

If the boundary is a rigid, perfectly conducting wall, we are further restricted by the condition that there can be no field **E** parallel to the surface, no velocity **V** normal to it, and no motion of the **B** lines in the wall. Thus

$$\mathbf{n}\times\mathbf{E} = 0, \tag{12.48}$$

$$\mathbf{n}\cdot\mathbf{V} = 0, \tag{12.49}$$

$$\mathbf{n}\cdot\frac{\partial\mathbf{B}}{\partial t} = 0. \tag{12.50}$$

We further specify that any surface of discontinuity carries no sheet mass. In that case, the lines of magnetic induction must lie in the interface. Were this not so, refraction of the lines would cause infinite acceleration owing to the unbalanced tangential forces, as may be seen from the stress tensor. Therefore we require

$$\mathbf{n}\cdot\mathbf{B} = 0. \tag{12.51}$$

In a discussion of the change of energy with respect to departures from an equilibrium, the static form of the equation of motion (Eq. 12.36),

Ampère's law (Eq. 12.41), and the divergenceless character of the induction (Eq. 12.42) are required. These equations in static equilibrium reduce to

$$\nabla p_0 - \mathbf{j}_0 \times \mathbf{B}_0 + \rho_{m0} \nabla_0 \phi_0 = 0, \tag{12.52}$$

$$\nabla \times \mathbf{B}_0 = \mu_0 \mathbf{j}_0, \tag{12.53}$$

$$\nabla \cdot \mathbf{B}_0 = 0, \tag{12.54}$$

where the subscript zero denotes the fact that these quantities relate to the equilibrium configuration.

The motion of the fluid will be described by following an element. Such a description is called Lagrangian. Since we consider departures from the equilibrium configuration, the initial position $\mathbf{r}_0$ of a fluid element and its later position $\mathbf{r}$ at time t are significant. Let

$$\boldsymbol{\xi}(\mathbf{r}_0, t) = \mathbf{r} - \mathbf{r}_0, \tag{12.55}$$

where $\boldsymbol{\xi}(\mathbf{r}_0, t)$ is the displacement at time t of the fluid from equilibrium. Clearly this displacement is a function of the original equilibrium position $\mathbf{r}_0$ of the element. Since we are investigating the perturbation of systems that are in static equilibrium at time $t = 0$, we must have

$$\mathbf{V} = \frac{d\boldsymbol{\xi}}{dt} = \dot{\boldsymbol{\xi}}, \tag{12.56}$$

$$\boldsymbol{\xi}(\mathbf{r}_0, 0) = 0, \tag{12.57}$$

$$\dot{\boldsymbol{\xi}}(\mathbf{r}_0, 0) = 0. \tag{12.58}$$

All the differential equations and their boundary conditions written so far depend upon the position $\mathbf{r}$ of an element of the plasma at time t. We must examine the stability of the equilibrium configuration (described by $\mathbf{r}_0$) and must therefore recast our equations in terms of the equilibrium variable $\mathbf{r}_0$ and the hypothesized displacement $\boldsymbol{\xi}$. Derivatives with respect to $\mathbf{r}$ can be easily related to derivatives with respect to $\mathbf{r}_0$ as follows: for any quantity G

$$\frac{\partial G}{\partial x_\alpha} = \sum_\beta \frac{\partial x_{0\beta}}{\partial x_\alpha} \frac{\partial G}{\partial x_{0\beta}}, \tag{12.59}$$

where x_α is the α component of $\mathbf{r}$, and $x_{0\beta}$ is the β component of $\mathbf{r}_0$. This expression defines the α component of ∇G. The quantity $\partial G/\partial x_{0\beta}$ in Eq. 12.59 is the β component of $\nabla_0 G$, where ∇_0 denotes the gradient with respect to $\mathbf{r}_0$. Likewise, the quantity $\partial x_{0\beta}/\partial x_\alpha$ may be regarded as the $\alpha\beta$ component of $\nabla \mathbf{r}_0$. An expression such as $\nabla \mathbf{r}_0$, involving two vectors or vector operators placed in juxtaposition, is called a dyadic.† Various

† A dyadic is a tensor of rank 2, cf. Sec. 5.2.

operations may be performed on a dyadic, but our development will concern only the scalar product of a dyadic with a vector. Of course, the ordering of the factors must be specified, since $\mathbf{AB}\cdot\mathbf{C}$ is generally different from $\mathbf{C}\cdot\mathbf{AB}$. Any dyadic K is completely specified by its components $\mathsf{K}_{\alpha\beta}$, two indices being required. For example, we may define the unit dyadic I by the matrix

$$\mathsf{I} = \begin{bmatrix} 1 & 0 & 0 \\ 0 & 1 & 0 \\ 0 & 0 & 1 \end{bmatrix}. \tag{12.60}$$

In accordance with these discussions, we now apply the operator ∇ to Eq. 12.55 and obtain

$$\nabla \mathbf{r}_0 = \nabla_r \mathbf{r} - \nabla_r \boldsymbol{\xi}, \tag{12.61}$$

or in component form

$$\frac{\partial x_{\beta 0}}{\partial x_\alpha} = \frac{\partial x_\beta}{\partial x_\alpha} - \frac{\partial \xi_\beta}{\partial x_\alpha}, \tag{12.62}$$

whence

$$\nabla_r \mathbf{r}_0 = \mathsf{I} - \nabla_r \boldsymbol{\xi}. \tag{12.63}$$

Since $\boldsymbol{\xi}$ is a perturbation and we may replace $\nabla_r \boldsymbol{\xi}$ by $\nabla_0 \boldsymbol{\xi}$, we find to first order

$$\nabla = [\mathsf{I} - \nabla_0(\boldsymbol{\xi})]\cdot\nabla_0 = \nabla_0 - (\nabla_0 \boldsymbol{\xi})\cdot\nabla_0. \tag{12.64}$$

The investigation of stability requires the equation of motion for the displacement $\boldsymbol{\xi}$. For small displacements, which suffice for the examination of stability, the linearized approximation to the equation of motion (Eq. 12.36) is adequate. To this end, we must first linearize all dependent variables appearing in this equation and express them in terms of $\mathbf{r}_0$ and $\boldsymbol{\xi}$.

Consider first Faraday's law (Eq. 12.40). Substitute successively Eq. 12.56 into Ohm's law (Eq. 12.38), and substitute the result into Faraday's law. With the application of Eq. 6.21 for the total derivative with respect to time, we find

$$\left(\frac{d}{dt} - \frac{d\boldsymbol{\xi}}{dt}\cdot\nabla\right)\mathbf{B} = \nabla \times \left(\frac{d\boldsymbol{\xi}}{dt} \times \mathbf{B}\right). \tag{12.65}$$

The term $d\mathbf{B}/dt$ is the one desired from which the perturbation in $\mathbf{B}$ is to be obtained. The equation can be integrated directly with $\mathbf{B}$ set equal to $\mathbf{B}_0$ as a first approximation in the higher-order terms. To first order,

$$\mathbf{B} = \mathbf{B}_0 + \mathbf{G} + (\boldsymbol{\xi}\cdot\nabla_0)\mathbf{B}_0, \tag{12.66}$$

where

$$\mathbf{G} = \boldsymbol{\nabla}_0 \times (\boldsymbol{\xi} \times \mathbf{B}_0). \tag{12.67}$$

Next, Ampère's law (Eq. 12.41) becomes

$$\begin{aligned}\mu_0 \mathbf{j} &= [\boldsymbol{\nabla}_0 - (\boldsymbol{\nabla}_0 \boldsymbol{\xi})\cdot\boldsymbol{\nabla}_0] \times [\mathbf{B}_0 + \mathbf{G} + (\boldsymbol{\xi}\cdot\boldsymbol{\nabla}_0)\mathbf{B}_0] \\ &= \mu_0 \mathbf{j}_0 - [(\boldsymbol{\nabla}_0 \boldsymbol{\xi})\cdot\boldsymbol{\nabla}_0] \times \mathbf{B}_0 + \boldsymbol{\nabla}_0 \times \mathbf{G} + \boldsymbol{\nabla}_0 \times [(\boldsymbol{\xi}\cdot\boldsymbol{\nabla}_0)\mathbf{B}_0]\end{aligned} \tag{12.68}$$

to first order in $\boldsymbol{\xi}$. This last expression can be reduced by expansion of $\boldsymbol{\nabla}_0 \times [(\boldsymbol{\xi}\cdot\boldsymbol{\nabla}_0)\mathbf{B}_0]$. The quantity $(\boldsymbol{\xi}\cdot\boldsymbol{\nabla}_0)$ is a scalar, and we can write the identity

$$\boldsymbol{\nabla}_0 \times [(\boldsymbol{\xi}\cdot\boldsymbol{\nabla}_0)\mathbf{B}_0] = (\boldsymbol{\xi}\cdot\boldsymbol{\nabla}_0)(\boldsymbol{\nabla}_0 \times \mathbf{B}_0) + [\boldsymbol{\nabla}_0(\boldsymbol{\xi}\cdot\boldsymbol{\nabla}_0)] \times \mathbf{B}_0. \tag{12.69}$$

In applying this identity, we must bear in mind that the first $\boldsymbol{\nabla}$ operator in the last term does *not* act on $\mathbf{B}_0$. The quantity appearing there in brackets is

$$\sum_\beta \frac{\partial \xi_\beta}{\partial x_{0\alpha}} \frac{\partial}{\partial x_{0\beta}} = (\boldsymbol{\nabla}_0 \boldsymbol{\xi})\cdot\boldsymbol{\nabla}_0. \tag{12.70}$$

Insertion of Eq. 12.70 into Eq. 12.69, and Eq. 12.69 into Eq. 12.68, and use of Ampère's law (Eq. 12.41) yield

$$\mathbf{j} = \mathbf{j}_0 + (\boldsymbol{\xi}\cdot\boldsymbol{\nabla}_0)\mathbf{j}_0 + \boldsymbol{\nabla}_0 \times \mathbf{G}/\mu_0. \tag{12.71}$$

The density ρ_m is similarly expressed in terms of ρ_{m0} and $\boldsymbol{\xi}$ by means of the conservation condition (Eq. 12.37) rewritten as (see Eq. 6.21 and note that we follow the fluid)

$$\frac{\partial \rho_m}{\partial t} + \rho_m \boldsymbol{\nabla}\cdot\left(\frac{d\boldsymbol{\xi}}{dt}\right) = 0; \tag{12.72}$$

thus to first order

$$\rho_m = \rho_{m0} - \rho_{m0}(\boldsymbol{\nabla}_0\cdot\boldsymbol{\xi}). \tag{12.73}$$

The total derivative in the adiabatic law (Eq. 12.39) is displayed explicitly there. After performing the total differentiation, applying the new conservation condition (Eq. 12.72), and integrating, we find to first order

$$p = p_0 - \gamma p_0(\boldsymbol{\nabla}_0\cdot\boldsymbol{\xi}). \tag{12.74}$$

Finally,

$$\phi = \phi_0 + (\boldsymbol{\xi}\cdot\boldsymbol{\nabla}_0)\phi_0 \tag{12.75}$$

corresponding to a simple Taylor expansion.

The equation of motion for $\boldsymbol{\xi}$ may now be derived from Eq. 12.36 and the preceding results. Direct substitution yields

$$\begin{aligned}\rho_{m0}\ddot{\boldsymbol{\xi}} = &-[\boldsymbol{\nabla}_0 - (\boldsymbol{\nabla}_0\boldsymbol{\xi})\cdot\boldsymbol{\nabla}_0][p_0 - \gamma p_0(\boldsymbol{\nabla}\cdot\boldsymbol{\xi})] \\ &+ [\mathbf{j}_0 + (\boldsymbol{\xi}\cdot\boldsymbol{\nabla}_0)\mathbf{j}_0 + (\boldsymbol{\nabla}_0 \times \mathbf{G}/\mu_0)] \times [\mathbf{B}_0 + \mathbf{G} + (\boldsymbol{\xi}\cdot\boldsymbol{\nabla}_0)\mathbf{B}_0] \\ &- [\rho_{m0} - \rho_{m0}(\boldsymbol{\nabla}\cdot\boldsymbol{\xi})][\boldsymbol{\nabla}_0 - (\boldsymbol{\nabla}_0\boldsymbol{\xi})\cdot\boldsymbol{\nabla}_0][\phi_0 + \boldsymbol{\xi}\cdot\boldsymbol{\nabla}_0\phi_0].\end{aligned} \tag{12.76}$$

Equation 12.76 will now be reduced. The identity

$$\mathbf{j}_0 \times (\boldsymbol{\xi}\cdot\boldsymbol{\nabla}_0)\mathbf{B}_0 - \mathbf{B}_0 \times (\boldsymbol{\xi}\cdot\boldsymbol{\nabla}_0)\mathbf{j}_0 = (\boldsymbol{\xi}\cdot\boldsymbol{\nabla}_0)(\mathbf{j}_0 \times \mathbf{B}_0) \tag{12.77}$$

will simplify our result. The quantity $\boldsymbol{\xi}\cdot\boldsymbol{\nabla}_0(\mathbf{j}_0 \times \mathbf{B}_0)$ may be computed from the equilibrium force law (Eq. 12.52). The terms appearing in Eq. 12.76 may be further reduced by the following four identities:

$$\boldsymbol{\nabla}(\boldsymbol{\xi}\cdot\boldsymbol{\nabla}p) = \boldsymbol{\nabla}\boldsymbol{\xi}\cdot\boldsymbol{\nabla}p + \boldsymbol{\xi}\cdot\boldsymbol{\nabla}\boldsymbol{\nabla}p, \tag{12.78}$$

$$\boldsymbol{\xi}\cdot\boldsymbol{\nabla}(\rho_m\boldsymbol{\nabla}\phi) = (\boldsymbol{\xi}\cdot\boldsymbol{\nabla}\rho_m)\boldsymbol{\nabla}\phi + \rho_m\boldsymbol{\xi}\cdot\boldsymbol{\nabla}\boldsymbol{\nabla}\phi, \tag{12.79}$$

$$[\boldsymbol{\nabla}\cdot(\rho_m\boldsymbol{\xi})]\boldsymbol{\nabla}\phi = \rho_m(\boldsymbol{\nabla}\cdot\boldsymbol{\xi})\boldsymbol{\nabla}\phi + (\boldsymbol{\xi}\cdot\boldsymbol{\nabla}\rho_m)\boldsymbol{\nabla}\phi, \tag{12.80}$$

$$\boldsymbol{\nabla}(\boldsymbol{\xi}\cdot\boldsymbol{\nabla}\phi) = (\boldsymbol{\nabla}\boldsymbol{\xi})\cdot\boldsymbol{\nabla}\phi + \boldsymbol{\xi}\cdot\boldsymbol{\nabla}\boldsymbol{\nabla}\phi. \tag{12.81}$$

After these operations, Eq. 12.76 for the motion may be expressed in the form

$$\rho_{m0}\ddot{\boldsymbol{\xi}} = \mathbf{F}(\boldsymbol{\xi}), \tag{12.82}$$

where

$$\mathbf{F}(\boldsymbol{\xi}) = \boldsymbol{\nabla}_0[\gamma p_0(\boldsymbol{\nabla}_0\cdot\boldsymbol{\xi}) + (\boldsymbol{\xi}\cdot\boldsymbol{\nabla}_0)p_0] + \mathbf{j}_0 \times \mathbf{G} - \frac{1}{\mu_0}\mathbf{B}_0 \times (\boldsymbol{\nabla}_0 \times \mathbf{G}) + [\boldsymbol{\nabla}_0\cdot(\rho_{m0}\boldsymbol{\xi})]\boldsymbol{\nabla}\phi_0. \tag{12.83}$$

Equation 12.83 is the form desired. Note that the first-order equations for $\mathbf{B}$, $\mathbf{j}$, ρ, p, ϕ, and $\ddot{\boldsymbol{\xi}}$ all depend on $\boldsymbol{\xi}$ but not on $\dot{\boldsymbol{\xi}}$. This property follows from the conservative nature of the system. Equation 12.82 with appropriate initial and boundary conditions determines $\boldsymbol{\xi}$. Equations 12.66, 12.68, 12.73, 12.74, and 12.75 then determine the perturbed field quantities.

The boundary conditions (Eqs. 12.43 to 12.47) at the plasma-vacuum interface must now be treated in the same manner as the preceding differential equations. The position variable $\mathbf{r}$ is to be expressed in terms of the initial equilibrium position of the plasma and the departure from this position. Denote by $\mathbf{A}$ the disturbance of the vector potential in the vacuum region resulting from a perturbation of the plasma away from its equilibrium position. All other quantities in the vacuum will be denoted by circumflexes over them. For example,

$$\hat{\mathbf{B}} = \boldsymbol{\nabla} \times \mathbf{A} + \hat{\mathbf{B}}_0 \tag{12.84}$$

by definition. Furthermore, from Eqs. 5.45 and following and in view of the fact that the charge density $\rho = 0$ in the vacuum,

$$\hat{\mathbf{E}} = -\frac{\partial \mathbf{A}}{\partial t} + \hat{\mathbf{E}}_0. \tag{12.85}$$

We shall choose the gauge condition (see Sec. 5.4) so that the scalar potential for the equilibrium configuration vanishes. A seemingly simple condition on **A** is that

$$\nabla \times \nabla \times \mathbf{A} = 0, \tag{12.86}$$

since the current density is always zero in the vacuum. However, we shall return to this point later. It is convenient now to denote by $\hat{\mathbf{n}}$ the normal unit vector leading out of the vacuum either into the plasma or a surrounding wall. Note that at the plasma-vacuum interface only $\hat{\mathbf{n}} = -\mathbf{n}$. From the boundary condition on the electric intensity (Eq. 12.48) and the relation of the electric intensity to the perturbation vector potential (Ec 12.85), we find that

$$\hat{\mathbf{n}} \times \frac{\partial \mathbf{A}}{\partial t} = 0 \tag{12.87}$$

at a rigid perfectly conducting wall bounding the vacuum. Since at time $t = 0$, $\partial \mathbf{A}/\partial t = 0$ at this boundary, our first boundary condition is that at all times

$$\hat{\mathbf{n}} \times \mathbf{A} = 0 \tag{12.88}$$

at the conducting vacuum wall. Furthermore, at a perfectly conducting rigid wall bounding a fluid, Eq. 12.49 for the normal component of the material velocity becomes

$$\mathbf{n} \cdot \boldsymbol{\xi} = 0. \tag{12.89}$$

The boundary condition (Eq. 12.47) on the total pressure is reduced as follows. Since the quantity $p + (B^2/2\mu_0)$ is always continuous across the boundary, we may equate only the changes in the quantities inside and outside the interface. Thus,

$$(p - p_0) + \frac{(B^2 - B_0^2)}{2\mu_0} = \frac{(\hat{B}^2 - \hat{B}_0^2)}{2\mu_0}. \tag{12.90}$$

The change $p - p_0$ in the particle pressure is given by Eq. 12.74, and the change $\mathbf{B} - \mathbf{B}_0$ in induction inside the plasma is given by Eq. 12.66. The change $\hat{\mathbf{B}} - \hat{\mathbf{B}}_0$ in the induction outside the plasma is given by Eq. 12.84, but we must also take into account the change in position of the boundary with time. The induction **B** must be expanded in a Taylor series. With these insertions, Eq. 12.90 gives the third boundary condition

$$-\mu_0 \gamma p_0 (\nabla_0 \cdot \boldsymbol{\xi}) + \mathbf{B}_0 \cdot [\mathbf{G} + (\boldsymbol{\xi} \cdot \nabla_0)\mathbf{B}_0] = \hat{\mathbf{B}}_0 \cdot [\nabla \times \mathbf{A} + (\boldsymbol{\xi} \cdot \nabla)\hat{\mathbf{B}}_0] \tag{12.91}$$

to first order in small quantities.

The boundary condition (Eq. 12.45) on the mass velocity is re-expressed by use of Eqs. 12.56 and 12.85; we find for the gauge condition specified

$$\mathbf{n}_0 \times \left(-\frac{\partial \mathbf{A}}{\partial t} - \mathbf{E}\right) = \left(\mathbf{n}_0 \cdot \frac{d\boldsymbol{\xi}}{dt}\right)(\hat{\mathbf{B}} - \mathbf{B}). \tag{12.92}$$

From Ohm's law (Eq. 12.38) we may eliminate **E**. By expanding the resulting triple cross product, by using Eq. 12.56 and the boundary condition (Eq. 12.51) on **B**, we find that Eq. 12.92 becomes

$$\mathbf{n}_0 \times \frac{\partial \mathbf{A}}{\partial t} = -\left(\mathbf{n}_0 \cdot \frac{d\boldsymbol{\xi}}{dt}\right)\hat{\mathbf{B}}. \tag{12.93}$$

This equation can be integrated immediately to give in first order the final boundary condition

$$\mathbf{n}_0 \times \mathbf{A} = -(\mathbf{n}_0 \cdot \boldsymbol{\xi})\hat{\mathbf{B}}. \tag{12.94}$$

Note that our present result reduces to the previous boundary condition (Eq. 12.88) on the vector potential at a rigid boundary.

With these results, linearization of the hydromagnetic equations has been completed.

★ 12.6. The Energy Principle

Following the discussion of the previous sections, we wish to calculate the change in potential energy δQ of the system under a perturbation $\boldsymbol{\xi}(\mathbf{r}_0)$. If δQ is negative, clearly the system must be unstable. We shall state the result and then derive it by two different methods. The incremental potential energy δQ is given by

$$\delta Q = -\tfrac{1}{2} \int dr_0 \, \boldsymbol{\xi} \cdot \mathbf{F}(\boldsymbol{\xi}), \tag{12.95}$$

where $\mathbf{F}(\boldsymbol{\xi})$ is given by Eqs. 12.82 and 12.83. Since the integrand of Eq. 12.95 is a second-order quantity, the integral could in principle be taken over either the initial or perturbed volume; choosing the initial volume, as has been done, is simpler.

The first derivation is a simple heuristic one. From Eq. 12.82, **F** is the force per unit volume of fluid and is a first-order linear quantity in $\boldsymbol{\xi}$. The work done on a unit volume in making the total displacement $\boldsymbol{\xi}$ is $\boldsymbol{\xi} \cdot \mathbf{F}(\boldsymbol{\xi})/2$ for a linear system. Since the total energy is conserved, this work appears at the expense of the potential energy. Equation 12.95 follows immediately. The analogy with a spring whose constant C is positive (as with real springs) or negative is obvious. If $\boldsymbol{\xi} \cdot \mathbf{F}(\boldsymbol{\xi})$ is negative, the force opposes the motion, and the system is locally stable. Incidentally, it is improper to demand that $\boldsymbol{\xi} \cdot \mathbf{F}(\boldsymbol{\xi})$ itself be everywhere negative for stability. It is only the integral over the system that is significant because a positive contribution at one place might necessitate a more negative contribution somewhere else.

A more mathematical derivation of the incremental potential energy (Eq. 12.95) is now presented. We begin the proof by considering the total energy of the system. The kinetic energy U will be a function of the velocity $\dot{\boldsymbol{\xi}}(\mathbf{r}_0, t)$ of each element of the system:

$$U(\dot{\xi}^2, t) = \tfrac{1}{2} \int dr_0 \, \rho_{m0}[\dot{\boldsymbol{\xi}}(\mathbf{r}_0, t)]^2. \tag{12.96}$$

Because the velocity of each element changes with time, the kinetic energy will also change with time as the displacements develop. Similarly, the change δQ in the potential energy about the equilibrium value will be quadratic in the displacement $\boldsymbol{\xi}(\mathbf{r}_0, t)$ if the change is sufficiently small. The potential energy will also depend upon time. Conservation of energy requires that

$$U(\dot{\xi}^2) + \delta Q(\xi^2) = 0. \tag{12.97}$$

The rate of change of kinetic energy is

$$\dot{U} = \int dr_0 \, \rho_{m0} \dot{\boldsymbol{\xi}} \cdot \ddot{\boldsymbol{\xi}} \tag{12.98}$$

as follows from the definition (Eq. 12.96) of the kinetic energy. The equation of motion (Eq. 12.82) and the conservation of energy (Eq. 12.97) then imply that

$$\begin{aligned} \dot{U} &= \int dr_0 \, \dot{\boldsymbol{\xi}} \cdot \mathbf{F}(\xi), \\ &= -\delta \dot{Q}(\boldsymbol{\xi} \cdot \boldsymbol{\xi}), \\ &= -\delta Q(\dot{\boldsymbol{\xi}} \cdot \boldsymbol{\xi}) - \delta Q(\boldsymbol{\xi} \cdot \dot{\boldsymbol{\xi}}). \end{aligned} \tag{12.99}$$

The displacements $\boldsymbol{\xi}$ contemplated above are entirely arbitrary functions of $\mathbf{r}_0$ and time t. Furthermore, the displacement $\boldsymbol{\xi}$ and velocity $\dot{\boldsymbol{\xi}}$ of an element are independent variables; nevertheless, they satisfy the same boundary conditions. Thus, $\dot{\boldsymbol{\xi}}$ may itself be regarded as some arbitrary displacement $\boldsymbol{\zeta}$. Now $\delta Q(\dot{\boldsymbol{\xi}}, \boldsymbol{\xi})$ is a symmetric function of the variables $\dot{\boldsymbol{\xi}}$ and $\boldsymbol{\xi}$. Consequently, by Eq. 12.99, the condition

$$\int dr_0 \, \boldsymbol{\zeta} \cdot \mathbf{F}(\boldsymbol{\xi}) = \int dr_0 \, \boldsymbol{\xi} \cdot \mathbf{F}(\boldsymbol{\zeta}) \tag{12.100}$$

follows. An operator, like $\mathbf{F}$, that satisfies such a relation is said to be self-adjoint.

An operator that is self-adjoint has many useful properties. It is of interest to us now to note that the arbitrary "displacement" $\dot{\boldsymbol{\xi}}$ may be chosen to be $\boldsymbol{\xi}$ itself, since $\dot{\boldsymbol{\xi}}$ and $\boldsymbol{\xi}$ are both perfectly independent and arbitrary. (We can choose both the velocity and displacement of a particle independently by choosing the force function and initial conditions appropriately.) From the condition of self-adjointness (Eq. 12.100), we see that the relation in Eq. 12.95 follows. Our proof is completed.

To evaluate the change in potential energy (Eq. 12.95), Eq. 12.83 must be inserted. The resulting expression requires some reduction before it can be interpreted. Vector identities and several of the boundary conditions will enable us to get a simpler expression for the change δQ. After substituting **F** from Eq. 12.83 into Eq. 12.95, the result can be reduced in the following way. The terms involving the pressure are

$$-\tfrac{1}{2}\int d\mathbf{r}_0\,\boldsymbol{\xi}\cdot\{\nabla_0[\gamma p_0(\nabla_0\cdot\boldsymbol{\xi}) + (\boldsymbol{\xi}\cdot\nabla_0)p_0]\}$$
$$= -\tfrac{1}{2}\int ds_0\,\mathbf{n}_0\cdot\boldsymbol{\xi}(\gamma p_0\,\nabla_0\cdot\boldsymbol{\xi} + \boldsymbol{\xi}\cdot\nabla_0 p_0)$$
$$+ \tfrac{1}{2}\int d\mathbf{r}_0[\gamma p_0(\nabla_0\cdot\boldsymbol{\xi})^2 + (\nabla_0\cdot\boldsymbol{\xi})(\boldsymbol{\xi}\cdot\nabla_0 p_0)], \quad (12.101)$$

where $\int ds_0$ represents the surface integral over the initial configuration. This equation follows by the familiar vector relation concerning the divergence of the product of a vector and a scalar and from Gauss's theorem.

As a result of substituting **F** from Eq. 12.83 into Eq. 12.95 for the change in potential energy, a term involving $\boldsymbol{\xi}\cdot\mathbf{B}_0\times(\nabla_0\times\mathbf{G})$ appears. This term is reduced by transposing the factors in the triple scalar product and by use of the vector identity for the divergence of a cross product and the definition (Eq. 12.67) of **G**:

$$\boldsymbol{\xi}\cdot\mathbf{B}_0\times(\nabla_0\times\mathbf{G}) = \nabla_0\cdot[\mathbf{G}\times(\boldsymbol{\xi}\times\mathbf{B}_0)] + G^2. \quad (12.102)$$

From Gauss's theorem and the boundary condition (Eq. 12.51) on the induction, we see that

$$\int d\mathbf{r}_0\,\boldsymbol{\xi}\cdot\mathbf{B}_0\times(\nabla_0\times\mathbf{G}) = \int d\mathbf{r}_0\,G^2 + \int ds_0\,\mathbf{n}_0\cdot(\mathbf{B}_0\cdot\mathbf{G}\boldsymbol{\xi} - \mathbf{G}\cdot\boldsymbol{\xi}\mathbf{B}_0). \quad (12.103)$$

With the help of these results, we find that δQ is given by

$$\delta Q = \delta Q_F + \tfrac{1}{2}\int ds_0\,\mathbf{n}_0\cdot\boldsymbol{\xi}\left[\frac{\mathbf{B}_0\cdot\mathbf{G}}{\mu_0} - \gamma p_0\,\nabla_0\cdot\boldsymbol{\xi} - \boldsymbol{\xi}\cdot\nabla_0 p_0\right], \quad (12.104)$$

where

$$\delta Q_F = \tfrac{1}{2}\int_{\text{fluid}} d\mathbf{r}_0\left[\frac{G^2}{\mu_0} - \mathbf{j}_0\cdot\mathbf{G}\times\boldsymbol{\xi} + \gamma p_0(\nabla_0\cdot\boldsymbol{\xi})^2 + (\nabla_0\cdot\boldsymbol{\xi})(\boldsymbol{\xi}\cdot\nabla_0 p_0) - (\boldsymbol{\xi}\cdot\nabla_0\phi_0)\nabla_0\cdot(\rho_{m0}\boldsymbol{\xi})\right] \quad (12.105)$$

contains all terms representing volume integrals over the space occupied by the fluid in its equilibrium configuration.

The term $-\frac{1}{2}\gamma p_0 \nabla_0 \cdot \boldsymbol{\xi}$ in the surface integral may be further reduced by means of the boundary condition (Eq. 12.91) and the identity

$$\mathbf{B}_0 \cdot (\boldsymbol{\xi} \cdot \nabla_0)\mathbf{B}_0 \equiv \boldsymbol{\xi} \cdot \nabla_0 \frac{B_0^2}{2}; \tag{12.106}$$

therefore,

$$-\tfrac{1}{2}\gamma p_0 \nabla_0 \cdot \boldsymbol{\xi} = -\tfrac{1}{2}\frac{\mathbf{B}_0 \cdot \mathbf{G}}{\mu_0} - \tfrac{1}{4}\boldsymbol{\xi} \cdot \nabla_0 B_0^2 + \frac{\hat{\mathbf{B}}_0 \cdot \nabla_0 \times \mathbf{A}}{2\mu_0} + \frac{\boldsymbol{\xi} \cdot \nabla_0 B_0^2}{4\mu_0}. \tag{12.107}$$

Substitute Eq. 12.107 into Eq. 12.104, and find

$$\delta Q = \delta Q_F + \tfrac{1}{2}\int ds_0\, \mathbf{n}_0 \cdot \boldsymbol{\xi}\left[-\boldsymbol{\xi} \cdot \nabla_0\left(p_0 + \frac{B_0^2}{2\mu_0}\right) + \boldsymbol{\xi} \cdot \nabla_0\left(\frac{\hat{\mathbf{B}}_0^2}{2\mu_0}\right) + \frac{\hat{\mathbf{B}}_0 \cdot \nabla_0 \times \mathbf{A}}{\mu_0}\right]. \tag{12.108}$$

The term involving $\hat{\mathbf{B}}_0 \cdot \nabla_0 \times \mathbf{A}$ in Eq. 12.108 can be simplified by the boundary condition (Eq. 12.94), simple vector identities, Gauss's theorem, and Ampère's law (Eq. 12.86) for a vacuum:

$$\tfrac{1}{2}\int ds_0\, (\mathbf{n}_0 \cdot \boldsymbol{\xi})\frac{\hat{\mathbf{B}}_0}{\mu_0} \cdot \nabla_0 \times \mathbf{A} = \tfrac{1}{2}\int d\mathbf{r}_0 \frac{(\nabla_0 \times \mathbf{A})^2}{\mu_0}. \tag{12.109}$$

With the aim of further reducing the two gradient terms of the integrand of Eq. 12.108, we note that

$$\nabla_0 \times \nabla_0\left(p_0 + \frac{B_0^2}{2\mu_0}\right) = 0 \tag{12.110}$$

if $p_0 + (B_0^2/2\mu_0)$ is differentiable at least twice. By Stokes's theorem, therefore,

$$\oint d\boldsymbol{\ell} \cdot \nabla_0\left(p_0 + \frac{B_0^2}{2\mu_0}\right) = 0. \tag{12.111}$$

The path of integration is chosen to be a tiny rectangle that is much longer than wide, as shown in Fig. 12.10. Over such a small circuit the integrand changes negligibly. Consequently,

$$\left[\!\left[\mathbf{n}_0 \times \nabla_0\left(p_0 + \frac{B_0^2}{2\mu_0}\right)\right]\!\right] = 0. \tag{12.112}$$

As our final simplifying point, let the displacement vector $\boldsymbol{\xi}$ be resolved into the component along the unit vector $\mathbf{n}_0$ normal to the plasma and the component perpendicular to $\mathbf{n}_0$:

$$\boldsymbol{\xi} = (\mathbf{n}_0 \cdot \boldsymbol{\xi})\mathbf{n}_0 + (\mathbf{n}_0 \times \boldsymbol{\xi}) \times \mathbf{n}_0. \tag{12.113}$$

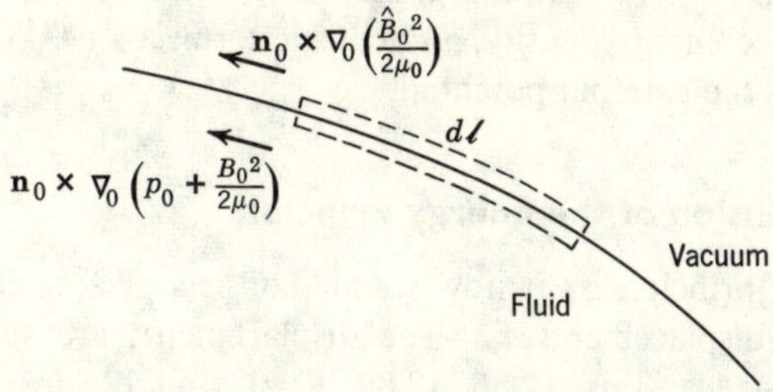

Fig. 12.10. Integration of $\mathbf{n}_0 \times \nabla_0(p_0 + B_0^2/2\mu_0)$ around a path on both sides of a fluid-vacuum interface.

This identity is applied to the vector $\boldsymbol{\xi}$ within the brackets of the integrand in Eq. 12.108. The result is simplified by Eqs. 12.109 and 12.112 and yields finally

$$\delta Q = \delta Q_F + \delta Q_S + \delta Q_V, \tag{12.114}$$

where the surface integral is

$$\delta Q_s = \tfrac{1}{2} \int_{\text{plasma}} ds_0 (\mathbf{n}_0 \cdot \boldsymbol{\xi})^2 \, \mathbf{n}_0 \cdot \left[\!\left[\nabla_0 \left(p_0 + \frac{B_0^2}{2\mu_0} \right) \right]\!\right], \tag{12.115}$$

the vacuum integral is

$$\delta Q_V = \frac{1}{2\mu_0} \int_{\text{vacuum}} d\mathbf{r}_0 (\nabla_0 \times \mathbf{A})^2, \tag{12.116}$$

and δQ_F is given by Eq. 12.105. The three terms comprise integrations over the plasma, the plasma-vacuum interface, and the vacuum bounding the plasma.

A partial physical interpretation can be given to the terms. Consider the surface integral, Eq. 12.115; it is the net work done in displacing the surface. The quantity $\mathbf{n} \cdot [\![\nabla_0(p_0 + B_0^2/2\mu_0)]\!]$ is the effective "spring constant" C of the surface for normal displacement. The quantity $(\mathbf{n}_0 \cdot \boldsymbol{\xi})$ is the normal displacement, and $(\mathbf{n}_0 \cdot \boldsymbol{\xi})^2 C/2$ is the work done. The vacuum term merely represents the increment of total magnetic energy ascribable to the pressure of the vacuum. The quantity δQ_V is *not* the change of magnetic energy *in* the vacuum. This latter quantity is a first-order entity, while $|\nabla_0 \times \mathbf{A}|^2$ is second order. The difference arises from the fact that the trapping of flux requires that a reduction (say) of the size of the vacuum region increases the magnetic energy there, and correspondingly decreases the magnetic energy inside the plasma. The difference is second order, and Eq. 12.116 is one of the contributions.

The fluid term (Eq. 12.105) comprises, in succession, two terms relating to the magnetic energy density, an acoustic term, a "$p\mathscr{V}$" term, and the work done by the external potential.

★ 12.7. Extension of the Energy Principle

The energy principle as it now stands consists of a calculation of δQ arising from displacements $\boldsymbol{\xi}$. The displacements are subject to four conditions, listed as Eqs. 12.86, 12.89, 12.91, and 12.94. Ampère's law (Eq. 12.86) and the condition (Eq. 12.91) that arose from $[\![p + B^2/2\mu_0]\!] = 0$ cause particular mathematical difficulty in that only a restricted class of $\boldsymbol{\xi}$'s is permitted. It is not easy to select this class from among all possible $\boldsymbol{\xi}$'s satisfying only the boundary conditions, Eqs. 12.89 and 12.94.

It is possible to ignore these two conditions, Eqs. 12.86 and 12.91, with qualifications on the latter. In most physical arguments, restrictive conditions cannot be neglected with impunity; the present idealized case is peculiar in this respect. The qualification will be presented after the formal derivations.

We consider first the condition, Eq. 12.86, and inquire into the consequences of a perturbation so chosen that Ampère's law (Eq. 12.86) would not be zero. It is easy to show, and we leave the proof to Problem 3, that the minimum energy requires Eq. 12.86 to be satisfied. The physical interpretation is simple. If $\nabla \times \nabla \times \mathbf{A} \neq 0$, a current flows in the vacuum region. Suppose that some conductors, initially carrying zero current, were disposed about the vacuum region. A current will not start to flow spontaneously in the conductors; we must do work to establish a current. Therefore if $\mathbf{A}$ does not satisfy this equation, another solution $\mathbf{A}'$ can be chosen to satisfy it and certainly decrease δQ_V thereby.

From the immediately preceding discussion, it is clear that any perturbation, even if it does not satisfy Ampère's law (Eq. 12.86), that produces a negative change of potential energy predicts instability. A physically realizable perturbation will produce a change of potential energy even more negative.

Avoidance of the boundary condition, Eq. 12.91, is not so simple and depends upon the physical fact that δQ for certain nonlinear perturbations is independent of the boundary condition. We present a physical argument, based on the original nonlinear equations, Eqs. 12.36 to 12.42. Suppose that a displacement $\boldsymbol{\xi}$ is made, giving the solid pressure–distance curve of Fig. 12.11, that satisfies neither the linearized boundary condition, Eq. 12.91, nor the nonlinear condition $[\![p + (B^2/2\mu_0)]\!] = 0$. Having made the arbitrary displacement $\boldsymbol{\xi}$, we could bring material from somewhere inside the plasma up to the interface, in order to equate the pressures on the two sides, as shown by the transition a of Fig. 12.11. Work must be involved in the process of moving the matter against the pressure. Because this additional material is properly disposed, the nonlinear boundary condition, Eq. 12.47, would be satisfied.

Now let us instead add the required material in a thinner layer at the surface, that is, in a sheath of thickness $\mathfrak{d}$. As $\mathfrak{d} \to 0$, the amount of material moved, and consequently the work done in moving it to satisfy the boundary conditions, becomes negligible. Therefore, we conclude that the matching of the nonlinear boundary condition, Eq. 12.47,

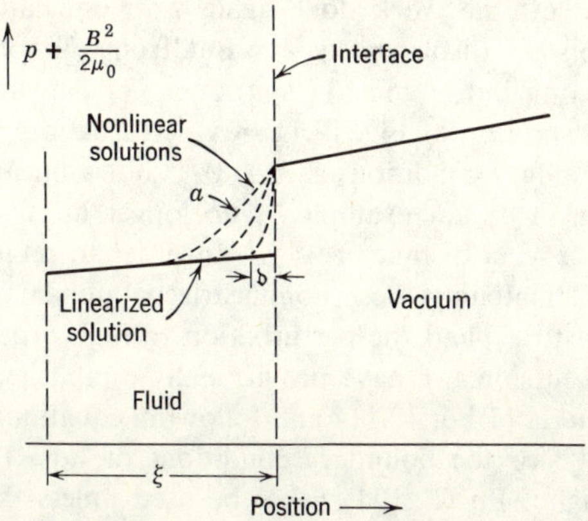

Fig. 12.11. Unbalanced pressure at a fluid-vacuum interface and movement of material to match the boundary condition.

requires in principle no energy. In other words, we have shown that for every linear perturbation $\boldsymbol{\xi}$, whether the boundary condition, Eq. 12.91, is satisfied or not, there exists a nonlinear perturbation which does satisfy the exact nonlinear boundary condition and for which the energy is the same. In conclusion, Eq. 12.114 may be applied to arbitrary displacements satisfying only the boundary conditions, Eqs. 12.89 and 12.94; the condition in Eq. 12.91 may be ignored, provided we are concerned only with the energy of the system.

We emphasize that the present conclusion cannot be properly obtained from the linearized boundary condition and show the fallacy of attempting to do so. Following Bernstein and his co-workers,[3] we make an arbitrary displacement $\boldsymbol{\xi}$ that does not satisfy the linearized boundary condition, Eq. 12.91, and a closely related displacement $\boldsymbol{\xi}'$ that does satisfy Eq. 12.91. Let

$$\boldsymbol{\xi}' = \boldsymbol{\xi} + \mathfrak{d}\mathbf{s}, \tag{12.117}$$

where $\mathfrak{s}$ is a finite vector along $\nabla_0 p_0$, which becomes zero at a distance $\mathfrak{d}$ away from the interface. It can be shown in a straightforward manner that the inclusion of $\mathfrak{d}\mathfrak{s}$ leads in first order to an additional surface-energy term, related to Eq. 12.115 by

$$\delta Q_s' = \delta Q_s + \tfrac{1}{2} \int ds_0\, (\mathbf{n}_0 \cdot \boldsymbol{\xi})(\nabla_0 \cdot \mathfrak{d}\mathfrak{s})\left(\gamma p_0 + \frac{B_0^2}{\mu_0}\right). \tag{12.118}$$

The last term is in general not zero. Since $\mathfrak{d}\mathfrak{s}$ vanishes in a distance $\mathfrak{d}$, we have

$$\nabla_0 \cdot \mathfrak{d}\mathfrak{s} \approx |\mathfrak{s}|. \tag{12.119}$$

The term represents the work done against an unbalanced pressure in making the arbitrary displacement $\boldsymbol{\xi}$. But from our previous physical discussion, the contribution must be spurious. The apparent discrepancy is resolved by inspection of Fig. 12.11. A perturbation as shown, which matches the boundary conditions as $\mathfrak{d} \to 0$, is not a linear one. That is, the perturbation in p_0, for example, is no longer linear in $\boldsymbol{\xi}$, because p changes rapidly over a distance $\mathfrak{d} \ll \xi$. In order to retain this behavior and the hope of matching these nonlinearized boundary conditions, we would have had to expand the perturbation to high order in ξ.

From this discussion, we have provisionally established the important fact that the criteria of Eq. 12.114 and following equations can be applied with regard only for the boundary conditions of Eqs. 12.89 and 12.94. On the other hand, Eq. 12.104 *cannot* be used unless the conditions in Eqs. 12.86 and 12.91 are satisfied as well. Since it is easier to neglect boundary conditions than to apply them, we have seemingly extended the validity of the energy principle, which was the subject of this section. However, a physical difficulty still appears to remain. The assumption has been made that the fluid is a perfect hydromagnetic medium, even within small distances $\mathfrak{d}$. Surely the hydromagnetic model fails in a plasma surface region whose thickness is of the order of the Larmor orbit r_{bi} of the ions. Thus if the plasma dimension is L and if a hierarchy of inequalities $L \gg \mathfrak{d} \gg r_b$ cannot be defined, the stability criteria developed here cannot properly be applied. We shall return to this important point at the end of the next section, where it will be shown that an even more stringent hierarchy of inequalities must exist.

★ 12.8. Stability of a Plasma with No Internal Induction

The developments of Secs. 12.2 to 12.4 have led us to believe that a plasma confined by **B** lines concave toward the plasma, as in Fig. 12.3, will be unstable. Conversely, a plasma whose **B** lines have opposite curvature, as in Fig. 12.4, will be stable. The stability formulations of

Secs. 12.5 to 12.7 also predict this phenomenon. We demonstrate this fact for the special case of a plasma in which the pressure p is constant up to the boundary and is zero outside and in which the induction vanishes inside. At the same time, we shall discuss some limitations of the method and the possible stabilizing effect of the plasma sheath.

To apply the method of the preceding sections, we neglect the gravitational effects and choose a divergence-free displacement $\boldsymbol{\xi}$ for insertion in Eq. 12.114 for the purposes of illustration; the total plasma volume will then remain fixed. Under the present circumstances, all terms in the fluid contribution δQ_F are zero. With regard to the surface term, Eq. 12.115, there are no gradients inside, but in general $\nabla \hat{B}_0{}^2 \neq 0$ outside. We drop the circumflex for convenience and find that

$$\delta Q = \int_{\text{vacuum}} d\mathbf{r}_0 \frac{|\nabla_0 \times \mathbf{A}|^2}{2\mu_0} - \int ds_0\, (\hat{\mathbf{n}}_0 \cdot \boldsymbol{\xi})^2 \hat{\mathbf{n}}_0 \cdot \nabla_0 \left(\frac{B_0{}^2}{4\mu_0}\right), \tag{12.120}$$

where again $\hat{\mathbf{n}}_0$ is the normal to the interface pointing toward the plasma. From Eq. 10.23 and Ampère's law for a vacuum, it follows that

$$\frac{\hat{\mathbf{n}}_0 \cdot \nabla_0 B_0{}^2}{2} = -\frac{\hat{\mathbf{n}}_0 \cdot \mathbf{R} B_0{}^2}{R^2}, \tag{12.121}$$

where $\mathbf{R}$ is the radius of curvature measured to the $\mathbf{B}$ lines as in Fig. 12.3 or Fig. 12.4. The surface term of Eq. 12.120 is negative or positive according to whether $\mathbf{R}$ points away from the plasma or into it. If $\mathbf{R}$ everywhere points into the plasma, δQ is obviously positive, and the system is stable. The cusp shown in Fig. 12.7 is therefore stable, in agreement with the previous findings.

If the radius $\mathbf{R}$ is directed out of the plasma, a simple analysis is required to demonstrate the instability. Consider a small region on the surface where $\mathbf{R}$ is so directed, as shown in Fig. 12.12. Construct a local Cartesian coordinate system about a point on the surface; let the z-axis be normal to the surface and point into the vacuum, and the x-axis lie in the direction of $\mathbf{B}_0$. Choose a displacement $\boldsymbol{\xi}$ so that

$$\xi_z(x, y, 0) = \xi_0 f(x, y) \sin \kappa y. \tag{12.122}$$

The quantity ξ_0 is a constant; $f(x, y)$ is a function of order unity at the origin, which falls to zero in a small distance $a \ll R$ over the surface. The displacement consists of ripples within the region bounded by the periphery a. For reasons to become apparent in a moment, we shall choose

$$\kappa a^2 \gg R. \tag{12.123}$$

The inequality ensures that there are many wavelengths across the

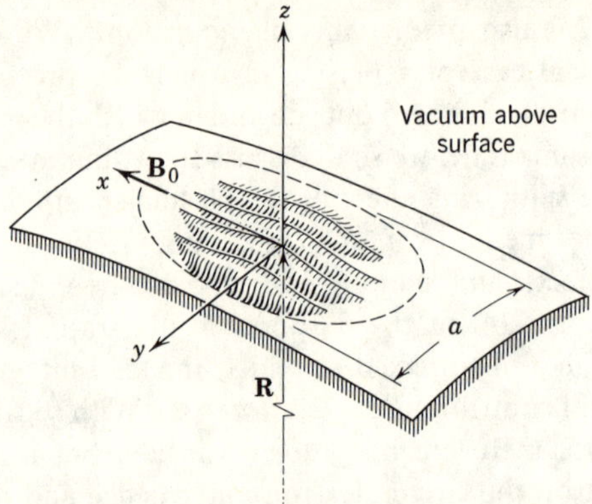

Fig. 12.12. Displacement $\boldsymbol{\xi}$ on plasma surface consisting of ripples within the region a. Radius of curvature of $\mathbf{B}_0$ lines on the unperturbed surface is $\mathbf{R}$.

perturbed region, but for clarity we show only a few in Fig. 12.12. Choose also a trial vector potential

$$\begin{aligned}\mathbf{A}(x, y, z) &= f(x, y)\nabla_0\left[\frac{\xi_0 B_0}{\kappa}\cos \kappa y \exp(-\kappa z)\right] \\ &= f(x, y)\, \nabla_0\Theta. \qquad (12.124)\end{aligned}$$

We readily check that these choices of $\boldsymbol{\xi}$ and $\mathbf{A}$ satisfy the boundary condition, Eq. 12.94, at the surface.

The contributions to Eq. 12.120 can now be evaluated. Since

$$\nabla_0 \times \mathbf{A} = \nabla_0 f \times \nabla_0 \Theta, \qquad (12.125)$$

we find that

$$\begin{aligned}\int \frac{d\mathbf{r}_0\, |\nabla_0 \times \mathbf{A}|^2}{2\mu_0} &\approx \int d\mathbf{r}_0\, |\nabla_0 f|^2 \xi_0^2 \left(\frac{B_0^2}{2\mu_0}\right) \exp(-2\kappa z) \\ &\approx \xi_0^2 \left(\frac{B_0^2}{2\mu_0}\right) \int_0^\infty dz \exp(-2\kappa z) = \frac{\xi_0^2 B_0^2}{4\mu_0 \kappa}. \qquad (12.126)\end{aligned}$$

On the other hand,

$$\delta Q_s = \int \frac{ds_0\, (\hat{\mathbf{n}}_0 \cdot \boldsymbol{\xi})^2 \hat{\mathbf{n}}_0 \cdot \mathbf{R} B_0^2}{2\mu_0 R^2} \approx -\frac{\xi_0^2 B_0^2 a^2}{2\mu_0 R}. \qquad (12.127)$$

Because of the inequality, Eq. 12.123, the vacuum term is negligible compared with the surface term, δQ is given substantially by Eq. 12.127, which is negative, and the system is unstable.

At this point, we have obtained all the conclusions of the simpler theory presented in Secs. 12.2 to 12.4. The instability found here is just the so-called flute or ripple instability described in earlier sections.

To estimate the rate of growth of the instability, choose further the x and y components of the displacement; let

$$\left.\begin{aligned} \xi_x &= 0, \\ \xi_y &= \xi_0 f(x, y) \cos \kappa y \exp(\kappa z), \\ \xi_z &= \xi_0 f(x, y) \sin \kappa y \exp(\kappa z), \end{aligned}\right\} \tag{12.128}$$

the z-component being compatible with the previous choice for it. With this displacement,

$$\nabla_0 \cdot \boldsymbol{\xi} = \xi_0 \left(\frac{\partial f}{\partial y}\right) \cos \kappa y \exp(\kappa z), \tag{12.129}$$

which is smaller than $\partial \xi_y / \partial y$ or $\partial \xi_z / \partial z$. Thus $\nabla_0 \cdot \boldsymbol{\xi} = 0$ is satisfied to order $(\kappa a)^{-1}$.

We now invoke Rayleigh's principle,[7] which states that if the kinetic energy is calculated formally as a function of $\boldsymbol{\xi}$ (not $\dot{\boldsymbol{\xi}}$), then the frequency ω of the system is given by

$$\omega^2 = \frac{\delta Q(\boldsymbol{\xi}, \boldsymbol{\xi})}{U(\boldsymbol{\xi}, \boldsymbol{\xi})}. \tag{12.130}$$

The kinetic energy from Eqs. 12.128 is

$$2U = \int d\mathbf{r}_0 \, \rho_{m0} \xi^2 \approx \frac{\rho_{m0} \xi_0^2 a^2}{2\kappa}. \tag{12.131}$$

Then the frequency from Eqs. 12.127 and 12.131 becomes

$$\omega^2 = \frac{\delta Q}{U} \approx -\left(\frac{2 B_0^2 \kappa}{\mu_0 \rho_{m0} R}\right); \tag{12.132}$$

this frequency increases as the wavelength of the ripples ($\propto 1/\kappa$) decreases, so that short-wavelength instabilities grow the fastest. Substantially the same result is demonstrated in Sec. 12.2.

The question now arises whether the formalism can properly be applied to plasmas of controlled thermonuclear interest. The difficulty will occur in matching the boundary condition, Eq. 12.91, by adjusting material in a thin layer of thickness $\mathfrak{d}$, as described in the previous section. We shall discover that $\mathfrak{d}$ cannot be made arbitrarily small and, consequently, that the energy involved in matching the boundary condition is not negligible.

Let us now inquire into the magnitude of the surface correction. For the plasma with no internal induction, Eq. 12.91 is

$$-\mu_0 \gamma p_0 (\nabla_0 \cdot \boldsymbol{\xi}) = \hat{\mathbf{B}}_0 \cdot \nabla_0 \times \mathbf{A} + \hat{\mathbf{B}}_0 \cdot (\boldsymbol{\xi} \cdot \nabla_0) \hat{\mathbf{B}}_0. \tag{12.133}$$

From Eq. 12.124 for $\mathbf{A}$ and Eq. 12.125 for $\nabla_0 \times \mathbf{A}$, we find (with $\mathbf{B}_0 \equiv \mathbf{B}_x$)

$$\hat{\mathbf{B}}_0 \cdot \nabla_0 \times \mathbf{A} = -\frac{\partial f}{\partial y} \xi_0 B_0^2 \cos \kappa y, \tag{12.134}$$

which is of order of magnitude $\xi_0 B_0^2/a$ at the boundary $z = 0$. On the other hand, the term $\hat{\mathbf{B}}_0\cdot(\boldsymbol{\xi}\cdot\boldsymbol{\nabla}_0)\hat{\mathbf{B}}_0$ is of order of magnitude $\xi_0 B_0^2/R$. Since $a/R \ll 1$, this term is negligible compared with the first term. With insertion of $\boldsymbol{\nabla}_0\cdot\boldsymbol{\xi}$ from Eq. 12.129, Eq. 12.133 is not satisfied. We must postulate a surface correction, which gives a value $(\boldsymbol{\nabla}_0\cdot\boldsymbol{\xi})$ in the fluid of the order of

$$|\boldsymbol{\nabla}_0\cdot\boldsymbol{\xi}| \approx \frac{\xi_0 B_0^2}{\mu_0 a\gamma p_0} = \frac{2\xi_0}{a\gamma}. \tag{12.135}$$

This divergence gives a positive definite contribution to δQ_F, Eq. 12.105. The thickness of the surface pressure correction is $\mathfrak{d}$. Then

$$\begin{aligned}\delta Q_F &\approx \int d\mathbf{r}_0\, \gamma p_0(\boldsymbol{\nabla}_0\cdot\boldsymbol{\xi})^2,\\ &\approx \frac{2\xi_0^2 B_0^2 \mathfrak{d}}{\gamma\mu_0},\end{aligned} \tag{12.136}$$

and this term must be small compared with the surface term of Eq. 12.127 in order that the system be unambiguously unstable. By comparison of Eqs. 12.127 and 12.136, we find

$$\mathfrak{d} \ll \frac{\gamma a^2}{4R}, \tag{12.137}$$

if the present theory is to predict a definite instability. At the same time, we must have

$$\mathfrak{d} \gg r_{bi}, \tag{12.138}$$

the ion Larmor radius, if hydromagnetic theory is to be applied. Finally, we may also impose the reasonable criteria that the wavelength of the ripples be large compared with an ion Larmor radius and that the surface pressure correction be small compared with a wavelength. Collecting all the inequalities from the development of this section, we require for the validity of the theory, that

$$1 \gg \frac{a}{R} \gg \frac{1}{\kappa a} \gg \frac{\mathfrak{d}}{a} \gg \frac{r_{bi}}{a}. \tag{12.139}$$

For plasmas of thermonuclear interest, we expect $R \approx 1$ m and $r_{bi} \approx 10^{-2}$ m. Then in general we cannot satisfy the inequalities very well and conclude that the present theory is inadequate. It seems probable that some of the inequalities of Eq. 12.139 could be removed by judicious adjustment of the theory. It is difficult to speculate upon the predictions of such a modified theory. We see that the physical reason for the difficulty is that hydromagnetic theory fails over distances of the order of a Larmor radius, whose magnitudes are not sufficiently small compared with the plasma dimensions. To put the matter another way, the surface regions must be treated more carefully.

If we consider relatively small experimental plasmas, our statement regarding the inapplicability of the stability criteria can be made *a fortiori*. For example, if the plasma dimension is only a few Larmor radii, the stability criteria seem not to apply at all. Experiments on small mirror systems do not show an unambiguous instability of the sort described. It seems to us that such observations are not at variance with the theory presented here. At the opposite extreme of magnitudes, we conclude that the stability criteria developed here should apply well to very large plasmas, such as the ionosphere or the sun.

It is interesting to speculate on the effect of the electrostatic sheath that must also surround the plasma. Presumably in a device such as a mirror, the high electron-thermalization rate would lead to greater electron than ion loss from the system. Therefore, a space charge builds up in the plasma until the plasma potential becomes comparable with kT_e/e, thus bringing the loss rates into balance. Experience leads us to expect that most of this potential appears in a sheath at the plasma surface. Suppose that the thickness of the sheath is d. Then a surface contribution

$$Q_E \approx \int \frac{ds_0 \epsilon k^2 T^2}{2e^2 d} \tag{12.140}$$

to the potential energy also exists. If we further assume that rippling of the surface does not alter the sheath thickness, then any rippling as shown in Fig. 12.12 leads to a positive definite contribution to δQ, because the surface is stretched. The term Q_E of Eq. 12.140 represents an effective surface tension. It is hard to estimate the magnitude of this term; if d is the Debye length, then the electrostatic energy density in the sheath is about equal to nkT, but the sheath is thin ($\approx 10^{-4}$ m in a thermonuclear plasma).

We can develop a very rough criterion for the effect of the electrostatic sheath in stabilizing the surface by noting that the increase in surface area under the rippling of Fig. 12.12 is about $\xi_0^2\kappa^2/4$ per unit area. Suppose now that the thickness of the sheath does not change as we ripple the surface. Then the increase δQ_E over the area of extent a^2 is

$$\delta Q_E \approx \epsilon \left(\frac{kT}{e}\right)^2 \frac{\xi_0^2 \kappa^2 a^2}{8d} \tag{12.141}$$

by Eq. 12.140. This positive surface contribution is to be compared with the negative surface contribution of Eq. 12.127. After using the approximation $B^2/2\mu_0 = nkT$, we find that the condition for *instability* becomes

$$\frac{\lambda^2\kappa^2 R}{8d} < 1, \tag{12.142}$$

where λ is the Debye length (see Eq. 8.5). We cannot precisely define the sheath thickness d, but we can safely assume that $\lambda < d < r_{bi}$. Therefore the quantity $R/8d \gg 1$ for a plasma of thermonuclear interest. Equation 12.142 then states that the wavelength of growing instabilities must be large compared with the Debye length.

Since the stability criteria developed in these last paragraphs are very qualitative, it would be idle to speculate further upon them. We have seen that the ratios r_{bi}/R and λ/R have appeared in the discussion, which quantities are assumed to be negligibly small in conventional hydromagnetic theory. We therefore conclude our remarks by essaying that this finer structure of the plasma must be included in a more detailed theory before unambiguous answers can be given in practical cases.

PROBLEMS

1. Derive Eq. 12.4.

2. A number of unusual magnetic configurations have been suggested for confining plasmas. Consider a regular octahedron; show that it can be wound along all the edges with a single wire, so that the curvature of the **B** lines inside is favorable for stability. What happens on the major diagonals of the structure? Will a regular tetrahedron or a cube do as well?

3. For the configuration discussed in Secs. 12.6 and 12.7 and for which $\hat{\mathbf{n}} \times \mathbf{A} = 0$ at conducting walls, show that the minimum magnetic energy corresponds to $\nabla \times \nabla \times \mathbf{A} = 0$.

4. Derive Eq. 12.118 in the limit of small $\mathfrak{d}$. Hint: Substitute $\boldsymbol{\xi}'$ in Eq. 12.91 and evaluate the volume and surface integrals.

5. Derive Eq. 12.121.

6. Discuss the difference (if any) between the growth rate of ripple instabilities predicted by the methods of Secs. 12.2 and 12.8.

7. A hydrogen plasma in a small experimental mirror is compressed by rapidly pulsing the mirror. For the compressed plasma, $T \approx 2$ kev, $n_i = n_e = 10^{21}/\text{m}^3$, and its dimension $R \approx 10^{-2}$ m. Discuss the behavior of this plasma with respect to flute instability, using the discussion of Sec. 12.8.

REFERENCES

1. M. N. Rosenbluth and C. L. Longmire, *Ann. Phys.* (*N.Y.*), **1**, 120 (1957).
2. M. D. Kruskal and M. Schwarzschild, *Proc. Roy. Soc.* (*London*), *A*, **223**, 348 (1954).

3. I. B. Bernstein, E. A. Frieman, M. D. Kruskal, and R. M. Kulsrud, *Proc. Roy. Soc.* (*London*), *A*, **244**, 17 (1958).
4. J. Berkowitz, H. Grad, and H. Rubin, *Proc. 2nd U.N. Conf. on Peaceful Uses of Atomic Energy*, Vol. 31, United Nations, Geneva (1958), p. 177.
5. H. Grad and H. Rubin, *Proc. 2nd U.N. Conf. on Peaceful Uses of Atomic Energy*, Vol. 31, United Nations, Geneva (1958), p. 190.
6. B. B. Kadomtsev and S. I. Braginsky, *Proc. 2nd U.N. Conf. on Peaceful Uses of Atomic Energy*, Vol. 32, United Nations, Geneva (1958), p. 233.
7. Lord Rayleigh, *The Theory of Sound* (1878); also 2nd ed., Vol. 1, Dover Reprints, New York (1945), pp. 91–110.

13

Energy balance and material problems

It is obvious that controlled thermonuclear power will not be possible unless the many problems of stable plasma confinement and heating are solved. Therefore, thermonuclear research has so far concentrated, and rightly so, almost exclusively upon these problems. The subject has occupied our thoughts through the preceding chapters. It will again do so in the chapters following this one, where we shall discuss briefly a number of recent experimental developments and enlarge upon some of the principles behind them. But there is another aspect: even if confinement and heating were achieved, the structure of a practical thermonuclear power generator would still not be clear. Little real thought has gone into the energy balance and material problems pertaining to such a hypothetical device. While any such calculations will be highly speculative, some worth-while points can be made. In particular, we shall conclude that a thermonuclear reactor will be large; with somewhat less assurance, we conclude that profitable power production from a fast pulsed system looks extraordinarily difficult of achievement. We conclude also that practical power from a D-D reactor will be very much more difficult to achieve than from a D-T reactor, in spite of the problem of regenerating tritium.

One thing appears clear—the energy of the thermonuclear neutrons must be recovered as heat. This single fact makes necessary the presence of a heat-recovery blanket surrounding the reactor. In addition, if the reactor operates by D-T fusion, the tritium must be regenerated by these neutrons, for no other method seems available for producing tritium in the amount required. The blanket then becomes complicated on this account. Further difficulties arise at the vacuum wall that faces the plasma. The heat flux on it may be large, and special efforts may be required to cool it.

The functions of the blanket must also be compatible with the establishment of the magnetic induction. If the current-carrying coils were normal conductors, they would have to be protected from the neutron flux in order to prevent structural damage and consequent increased resistivity. If such normal conductors were used, the coil power loss would be a significant and perhaps large fraction of the electrical output of the system. Vastly more attractive is the use of superconducting coils for generating static or quasi-static magnetic inductions. Very recent advances in superconducting materials have been unbelievably spectacular. The possibility of reducing the coil power to zero (except for modest refrigeration) will affect the net power behavior of thermonuclear reactors very favorably. This application is only one of many which can be envisaged for exploiting loss-free high magnetic field structures. We shall discuss some of these possibilities in Sec. 13.3.

In Secs. 13.4 to 13.6 we discuss some of the static and dynamic equations that must apply to any steady state or quasi-steady-state controlled thermonuclear reactor. Here, particle and radiation losses and energy transfer among the species are considered. We shall discover that a number of difficulties arise, such as in the calculation of He^4 density in the reacting plasma. At this point, it is much easier to pose questions heuristically than to answer them.

We shall illustrate the problems that arise, partly by general calculations and partly by calculation of specific examples. All of these are meant to indicate the order of magnitude involved, and nothing else. The illustrations will be based almost exclusively on a D-T fusion cycle. However, the D-D power balance, which is much less favorable, will be mentioned.

13.1. The Neutron Blanket

We could develop the subject matter of this chapter in a number of different ways, and our choice is partly arbitrary. The components of any thermonuclear system are interrelated, and the design of each depends upon all the rest. The nature of the neutron-absorbing blanket appears relatively less dependent upon other considerations, and we discuss it first.

The neutron energy can be made available only as heat. All considerations of energy conversion require that the heat be removed at a usefully high temperature, for example 500° C. Since about 80% of the D-T and 35% of the D-D thermonuclear energy appear as kinetic energy of the neutrons, the plasma must be surrounded by a neutron-absorbing blanket. This blanket will bear some resemblance to the moderator in a fission reactor. Neutrons must be moderated in the blanket in order that capture cross sections be large; therefore, the blanket can be made reasonably thin and the volume of exterior coils minimized.

In our discussions, we shall assume that the thermonuclear system is approximately a cylinder, whose length is at our disposal. If the cylinder is bent into a torus, we shall neglect curvature of the axis. A provisional outline of the system is shown in cross section in Fig. 13.1. The coils

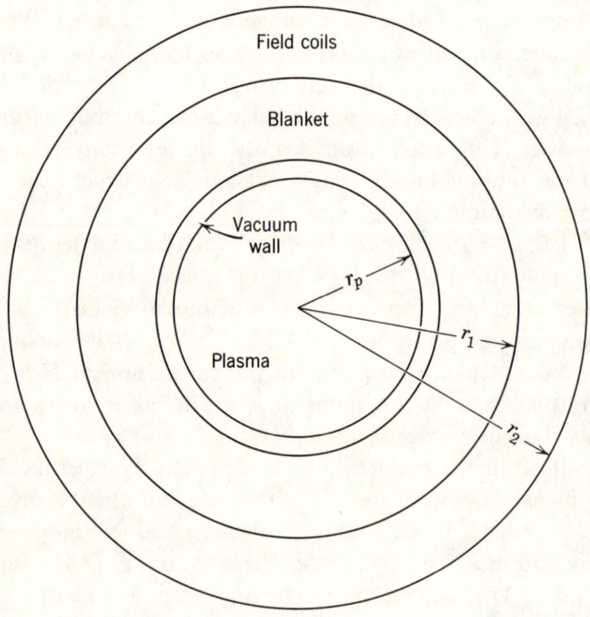

Fig. 13.1. Cross section of cylindrical thermonuclear system.

that produce the confining induction are shown outside the blanket. We shall defend our choice of coil position in subsequent sections; the development here is largely independent of such matters. Further requirements of the vacuum wall will be discussed in Sec. 13.2.

Consider now the blanket for a hypothetical D-T reactor. Since tritium does not occur naturally, it will be necessary to regenerate it somewhere in the system. The only feasible reaction appears to be

$$Li^6 + n \rightarrow T + He^4 + 4.8 \text{ Mev}, \qquad (13.1)$$

whereby the thermonuclear neutrons regenerate the tritium. The cross section of Li^6 for capture of thermal neutrons is 945 barns, which is a large value and is approximately proportional to the reciprocal of the neutron speed. Natural lithium contains about 7.5 atom% Li^6 and appears to be in abundant supply for the purpose. We note also that the total energy released per fusion is substantially increased by this additional reaction in the blanket, and for some later discussion, we define the effective total nuclear heat per reaction as

$$U'_{DT} = U_{DT} + 4.8 \text{ Mev} = 22.4 \text{ Mev}. \tag{13.2}$$

The vacuum wall and blanket will have to contain a certain amount of structural material that absorbs neutrons. We require in addition a significant multiplication of the number of neutrons by some further nuclear reaction. Our over-all requirements for the blanket become the following:

1. Provision for neutron multiplication.
2. Moderation of the neutrons to low energy and their absorption in Li^6.
3. Small parasitic neutron capture.
4. Absorption of γ-rays produced by neutrons in elastic scattering.
5. Extraction of the heat at useful temperature.
6. Recovery of tritium.
7. Stability against structural failure and corrosion.
8. Small electrical loss.
9. If possible, low pressure.

We now discuss these severe requirements.

Because of the high temperature and the gas evolution in the $Li^6(n, \alpha)T$ reaction, it is necessary that the lithium-containing material be fluid. It probably must be recirculated. Pure lithium is undesirable, because the $\mathbf{V} \times \mathbf{B}$ electric field in the moving fluid will cause excessive ohmic loss in the conducting lithium. Of the lithium compounds, we prefer those with strong ionic bonding, rather than less stable covalent bonding. Tentatively, our choice is liquid lithium fluoride, perhaps mixed with sodium or potassium fluoride. A eutectic mixture of some kind is required to lower the melting point to about 400 to 500° C. Fluorine has small neutron absorption, and the resistivity of the molten salts is high ($\approx 10^{-2}$ ohm-m). A small amount of dissociation will be continually taking place in the fluid, because of the nuclear reactions. Lithium metal is slightly soluble in LiF, and a small excess of lithium may suppress any free fluorine. Recent metallurgical developments indicate that these liquid fluoride salts may be contained without excessive corrosion.† The recovery of tritium should not be insuperably difficult; we shall not discuss this chemical problem.

While LiF may be satisfactory for neutron capture and tritium recovery, it may not be the best moderator. The 14.1-Mev neutrons will be inelastically scattered down to an energy of about 2 Mev. Below about 2 Mev, the nuclear energy levels are widely enough separated so that neutrons lose energy principally by elastic scattering in the light elements composing the blanket. Even though the lithium nucleus is light, and hence the fractional energy loss per elastic collision is moderately high, its elastic scattering cross section is very low. Hintermann and Wideroe[1] calculate the blanket thickness required to moderate these neutrons and find about 1.5 m for natural Li or 0.75 m for pure Li^6 (in the latter case, most neutrons are captured at the 255-kev resonance). The effect of the addition of fluorine has not been calculated.

Among other moderating materials, beryllium is efficient, and a mixture of BeF_2 and LiF might be postulated for a homogeneous moderator. Beryllium may also be necessary for neutron multiplication, as discussed in the next paragraphs. Other moderating materials will necessitate heterogeneous systems. Water (H_2O) is an efficient moderator, and a thickness of about 0.2 m would suffice. However, we do not favor its use (or the use of D_2O), principally because of the high pressures involved at the working temperature. Graphite is a good moderator and might be used with the fused salts circulating in thin-walled pipes. Another material of possible interest is zirconium hydride. The hydrogen density in the hydride (the light H atoms are most efficient in slowing the neutrons) is almost that of water. Problems exist in the use of this material at the operating temperature. We cannot choose definitely among these systems at the present time.

So far, we have not provided the additional neutrons to overcome parasitic capture. Calculations to date on fission breeder reactors indicate that it is very difficult to prevent the loss of about 10% of the neutrons to nonuseful capture. A number of reactions are available to provide the make-up, using 14.1-Mev fusion neutrons. An obvious one is direct fast fission of U^{238}, which has a cross section of several barns at this energy. However, we eschew this choice on the grounds that (1) if U^{238} is sufficiently plentiful for the purpose (see Chap. 1), more conventional fission breeder reactors may be easier to construct; (2) the radioactive waste is a serious problem. Incidentally, the fission energy available is about 200 Mev.

† For example, in an alloy known as INOR-8.

The only $(n, 2n)$ reaction with large cross section down to relatively low energy is

$$Be^9 + n \rightarrow 2He^4 + 2n; \qquad (13.3)$$

the cross section is shown in Fig. 13.2. Since only fast neutrons induce the reaction, the beryllium must be located near the vacuum wall ahead of the principal moderating region. This aim could be accomplished by using a mixed Be-Li fluoride, for example, behind the vacuum wall. While other liquid or powder compounds might be used, beryllium metal itself could not, because of the gaseous helium produced in the reaction.

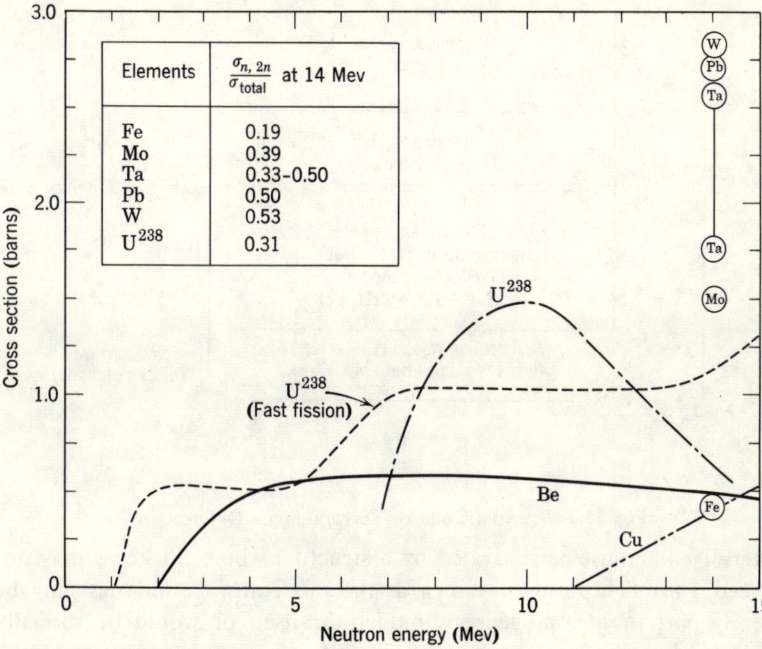

Fig. 13.2. Cross sections for $(n, 2n)$ reactions. Circled elements placed at experimental points.

Another possibility exists, whose attractiveness has not yet been fully evaluated. Many heavy elements (Mo, W, and so on) have $(n, 2n)$ cross sections in excess of 1 barn at 14.1 Mev. Cross sections for a number of materials are shown in Fig. 13.2. For many of these metals, for example tungsten, the cross section for the $(n, 2n)$ reaction at 14.1 Mev is approximately 50% of the total cross section for all processes. Even though the cross section drops precipitously with decreasing energy, a 14.1-Mev neutron stands a good chance of multiplication. Therefore, if the vacuum wall contains an appropriate heavy metal, the neutron multiplication may be provided automatically. In the next section, we shall see some additional reasons for using a refractory metal such as tungsten or molybdenum for the purpose. It is estimated that about 2 cm of such material may provide 10% neutron multiplication. If such a thickness is too great to allow proper cooling, it could be placed in layers.

We summarize our discussion by reference to Fig. 13.3. The dimensions are very approximate and indicate only that the plasma and

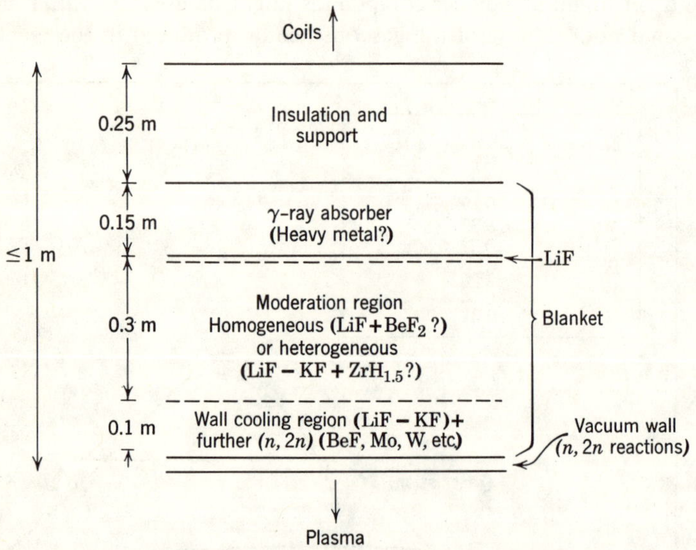

Fig. 13.3. Schematic blanket surrounding a D-T reactor.

exterior coils must be separated by a structure whose thickness may not exceed 1 m. The lithium shown at the moderator boundaries may be merely part of a homogeneous moderator itself or would be specially provided in a heterogeneous assembly. Its purpose is to prevent escape of thermal neutrons. Structural material that must pervade the assembly is not shown.

We propose that the heat be extracted by pumping the fused salts. Low-pressure gases have too little heat capacity; high-pressure systems (including water) do not appear feasible; liquid metals will have large ohmic loss.

Incidentally, we have discovered one of the reasons that the coils must be outside the blanket of a D-T thermonuclear reactor: the neutron economy is important, and we cannot allow a region unfavorable to this economy to intervene between the plasma and the blanket. We shall find additional reasons for our choice in Sec. 13.3.

13.2. The Vacuum Wall

We shall indicate in Secs. 13.5 and 13.6 that a thermonuclear reactor might be profitably designed to operate either at high magnetic induction ($\approx$7 to 10 webers/m^2) or at relatively low induction ($\approx$1.5 webers/m^2). In the latter case, the material and power densities are moderate, and the radiant energy flux on the vacuum wall can readily be dissipated. If the induction is high, the problem is more severe, and the vacuum wall must be designed with particular care. We address ourselves briefly to the problem.

The wall could be damaged by (1) sputtering from its surface by fast plasma ions, (2) the fast neutron flux, (3) failure to remove the radiant heat incident upon it from the plasma. Of these, item 1 is a plasma problem; item 2 must be tolerated in any event; the cross sections, except for the desirable $(n, 2n)$ reaction, are relatively small for 14.1-Mev neutrons. Regarding item 3, the incident flux consists of cyclotron radiation and bremsstrahlung x rays. Both types are absorbed in a small depth of any material that we shall find suitable for a vacuum wall. Therefore, it is proper to assume that the radiant flux is deposited at the surface facing the plasma.

Since the radiant energy should be recoverable as heat and since the moderator behind the wall is at high temperature, the vacuum wall itself should be run at a usefully high temperature. Let us assume 400° C for the back surface in contact with a coolant. Table 13.1 summarizes the properties of a number of materials; the data are taken in part from Craston and co-workers.[2] The second column lists the maximum temperature allowable on the vacuum side, as worked out by Craston for a 10-sec pinch. Since the vapor pressure of a material increases almost exponentially with temperature, the maximum wall temperatures are almost independent of the assumptions made. The third column lists the heat flux obtainable through a 1-cm wall with the front side at T_{max} and the back side at 400° C. Several conclusions can be drawn: (1) electrical insulators are generally unsuitable; (2) if the wall is made of a single material, the refractory metals appear most suitable; (3) if the thermal flux is several kw/cm^2, the wall must be thin ($\leqslant$1 cm), and the coolant behind it must therefore run at low pressure; (4) a sandwich-type wall, arranged for low evaporation on the vacuum side and low corrosion on the coolant side may be most desirable.

In view of the vacuum and structural requirements, we conclude that a maximum radiant flux of about 2 to 3 kw/cm^2 might usefully be tolerated and shall bear this number in mind for later calculations. This heat can probably be removed in the coolant region shown in Fig. 13.3.

Table 13.1. Some properties of materials considered for a thermonuclear vacuum wall

Material	T_{max} for Wall (° C)	Thermal Flux through 1-cm Wall (w/cm²)	Corrosion Resistance against Alkali Metals	Remarks
Be	670	300	good	poisonous
Al	510	240	poor	—
Ti	840	60	good	—
Cu	580	640	fair	—
Mo	1480	1080	excellent	—
W	2500	3500	excellent	—
Alumina	1030	40	good	—
Beryllia	1360	180	?	brittle
Silica	610	4	poor	—
Stainless Steel	900?	100?	excellent	—

13.3. The Confining Field Coils

We turn now to consider losses in the coils and their most advantageous disposition. Figure 13.1 defines the inner coil radius r_1 and the outer radius r_2. The consequences of placing the coils inside the blanket can be calculated by setting $r_1 = r_p$, the plasma radius.

In order to assess the coil problem most accurately, we begin by considering briefly the use of normally conducting coils without refrigeration. Following that exercise, we shall mention the advantage that accrues by suitably refrigerating such coils to reduce their resistance. The concept of refrigeration leads us naturally to consider superconducting systems in which the resistance of the coils is zero. These systems will be found to be quite superior.

The effective resistivity of the coil material is η. This coefficient includes the space factor, that is, the effect of using some coil space for cooling, support, electrical access, and so forth. Efficient coil designs generally allow about 70% of the volume for conductor material. Therefore, η will be about 1.4 times the resistivity of the conductor material, and with this correction we assume the coil to be homogeneous.

The most efficient current density distribution $j(r)$ in a cylindrical coil is readily derived; this derivation is the subject of Problem 8. We express the result in terms of the total circumferential current K per unit axial length of cylinder. This current K is determined by

$$B_z = \mu_0 K \tag{13.4}$$

and is fixed by the requirements of the magnetic pressure. Then

$$j(r) = \frac{K}{r \ln (r_2/r_1)}; \tag{13.5}$$

in terms of the induction, the magnet power is

$$W_m = \frac{2\pi\eta B_z^2}{\mu_0^2 \ln (r_2/r_1)} \quad \text{per unit length.} \tag{13.6}$$

For fixed ratio of coil radii, the magnet power is proportional to the magnetic pressure.

The alternate current distribution $j(r) = \text{const}$ is only marginally less efficient than a $1/r$ distribution if $r_2/r_1 < 2$.

In assessing the virtue of various coil systems, we now require that the magnet power W_m be small compared with the total thermonuclear heat W'_{DT}. For a cylindrical system, the ratio will be

$$\frac{W_m}{W'_{\text{DT}}} = 2\eta B^2/\mu_0^2 n_{\text{D}} n_{\text{T}} \overline{\sigma v}_{\text{DT}} U'_{\text{DT}} r_p^2 \ln (r_2/r_1). \tag{13.7}$$

The equation can be put into a more convenient form for illustrative purposes. We shall consider a D-T reaction, and shall discover later that the most favorable over-all operation proceeds at a 50% D-50% T mixture, at a temperature $T_{\text{D}} \approx T_{\text{T}}$ in the range 30 to 100 kev. Then we may set $\overline{\sigma v}_{\text{DT}} \approx 9 \times 10^{-22}$ m^3/sec. We shall neglect reaction products and the temperature difference between electrons and ions. Equation 13.7 then reduces to

$$\frac{W_m}{W'_{\text{DT}}} = \frac{10^3 \eta T^2}{\beta^2 B^2 r_p^2 \ln (r_2/r_1)}, \tag{13.8}$$

where T is measured in kev, and β is the ratio of total material to magnetic pressure. Note from Eq. 13.8 that for similarly scaled systems the relative magnet loss scales as $B^{-2} r^{-2}$. Large systems with high induction are most efficient, according to this criterion.

The futility of placing the coils inside the blanket can be simply demonstrated from Eq. 13.8. If the coils are inside, they must be run hot so that the energy of neutrons moderated within the coil structure can be recovered. Problem 9 illustrates the calculation. For reasonable choices of η, T, β, and B most favorable for minimizing W_m/W'_{DT}, the coils must be so thick that they ruin the neutron economy, unless the plasma radius is so large that the difference between placing the coils inside and outside the blanket becomes relatively unimportant. We remark in passing that if the values of $\overline{\sigma v}_{\text{DD}}$ and U_{DD} had been used in deriving Eq. 13.8, operation with hot coils would have been *a fortiori* impossible.

If the coils are placed outside the blanket, they could be run at room temperature—the resistivity of copper is about 2.5 times as great at 400° C as it is at room temperature. Thus we already achieve considerable power saving over running the coils hot. Better still, the coils could be refrigerated.

Figure 13.4 shows a perfect (Carnot) refrigeration cycle. Resistance of the magnet coils at the low temperature T_1 is $\mathscr{R}_1$. The ohmic loss $I^2 \mathscr{R}_1$ in the coils at temperature T_1 is to be rejected at a higher temperature

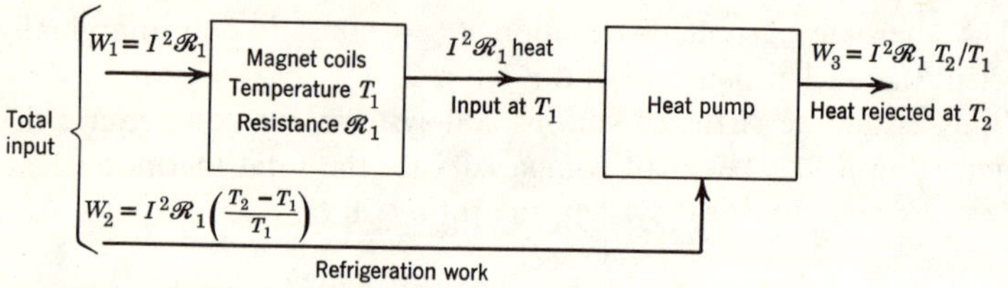

Fig. 13.4. Carnot refrigeration cycle.

$T_2 > T_1$. Let W_2 be the work done in the refrigerator. Then the total energy W_3 to be rejected at T_2 is $W_3 = W_1 + W_2$. From the Carnot cycle efficiency, we find

$$W_3 = T_2 W_1 / T_1. \tag{13.9}$$

The current I must be provided to produce the induction and does not depend upon the temperature. Suppose that the efficiency of the refrigerator is $\mathfrak{h}$. Then the ratio of total magnet powers (including refrigeration) for operation at temperature T_1 or T_2 is

$$\frac{W_m(T_1)}{W_m(T_2)} = \frac{\eta(T_1)T_2}{\eta(T_2)T_1\mathfrak{h}}. \tag{13.10}$$

Consider well-annealed copper, for which $\eta(20° \text{ K})/\eta(300° \text{ K}) \approx 1/700$, as an example. Then for a 30% mechanically efficient system, Eq. 13.10 yields

$$\frac{W_m(20° \text{ K})}{W_m(300° \text{ K})} \approx 0.07. \tag{13.11}$$

A big saving can be effected by refrigeration. The similar use of sodium at 20° K or lower has also been investigated.

We now turn to superconducting coils, which appear by far the most promising for providing large volumes of high induction at low cost. Certain elements and compounds become superconducting below a transition temperature T_c peculiar to the material. More precisely, the transition is not sudden, but the resistance drops toward an immeasurably low value as the temperature is lowered through a small range near T_c. The range of superconducting transition temperatures varies from very near 0 to about 18° K (the latter is for the compound Nb_3Sn, of which more later). Semi-empirical relations involving the average number of valence electrons in the material have been discovered that aid in the prediction of possible superconductivity.

Below the transition temperature T_c, a suitably induced current in the material will persist essentially forever without loss. By the same token, any magnetic induction brought near a superconductor will induce on its surface a current; this current sets up a canceling induction, and none penetrates the material. The concept is consonant with our many previous discussions of diffusion of the magnetic induction: if the conductivity is infinite, the induction is frozen in or out of the material, depending upon the initial conditions, and the material is perfectly diamagnetic at temperatures $T < T_c$.

Until recently, the difficulty with the application of this concept has been that the superconductivity is quenched (the material becomes normal) if the induction exceeds an uninterestingly low value. For many so-called soft superconductors, the critical quenching induction B_c varies with temperature roughly as

$$B_c = B_{c0}\left[1 - \left(\frac{T}{T_c}\right)^2\right]. \tag{13.12}$$

At the transition temperature, the critical induction is zero; the critical induction B_{c0} approached at 0° K has typical values 0.02 to 0.2 weber/m^2 for these soft superconductors. Therefore a superconducting solenoid giving higher induction could not be made with these materials. The physical reason for the behavior according to Eq. 13.12 is not fully understood. In this quenching transition also, the superconductivity is not suddenly destroyed as B exceeds B_c; the resistance increases to normal over a small range of induction.

The recent spectacular advances involve a series of materials—the so-called hard superconductors—that have vastly different properties; B_c decreases as T approaches T_c from below, but Eq. 13.12 is not obeyed. The critical induction is very much higher. This phenomenon, not now clearly understood, also depends upon the magnitude of the current and upon the metallurgical state of the material. Figure 13.5 shows the critical induction for annealed and unannealed samples of niobium[3] at 4.2 and 1.6° K. Note that cold-working can about double the critical induction. Small solenoids at 0.8 weber/m^2 have been built by Autler and Montgomery,[4,5] using niobium. More recently, the ductile alloy Mo_3Re has been developed,[6] giving a critical induction as high as 1.8 webers/m^2

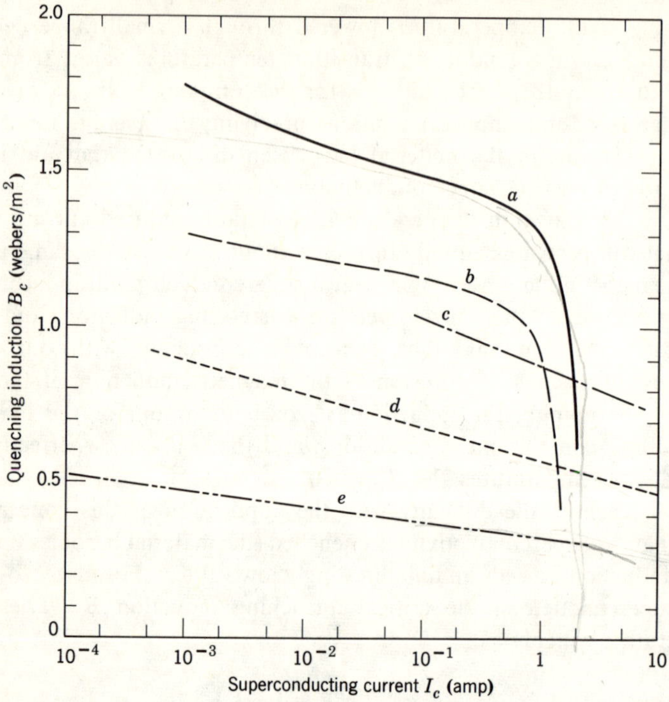

Fig. 13.5. Critical quenching induction B_c versus superconducting current I_c in Mo_3Re and Nb wires. Induction is perpendicular to the wires. (*a*) Mo_3Re, 0.007-cm diam, 1.5° K.[6] (*b*) Mo_3Re, 0.007-cm diam, 4.2° K.[6] (*c*) Nb, 0.01-cm diam, unannealed, 1.6° K.[3] (*d*) Nb, 0.01-cm diam, unannealed, 4.2° K.[3] (*e*) Nb, 0.012-cm diam, annealed, 4.2° K.[3]

at 1.5° K. The current dependence is also shown in Fig. 13.5. For all these materials, the approximate phenomenological behavior $B_c = C_1 - C_2 \ln I_c$, where C_1 and C_2 are constants, is observed at low and moderate currents. Above some value of current (about 1 amp for the Mo_3Re samples in Fig. 13.5), the critical induction falls rapidly.

By far the best material to date is Nb_3Sn, as reported by Kunzler and co-workers.[7] A core of Nb and Sn powders within a Nb tube is drawn to wire and heated at about 1000° C to form Nb_3Sn, which is a very fragile intermetallic compound. Figure 13.6 shows their results to date. Presumably the sintered core material carries all the current at the induction shown, and the niobium is normally conducting. Note that the current density in the core is huge. Even better materials may become available.

This latest development, which promises loss-free inductions in the order of 10 webers/m^2 and perhaps higher, presages a complete

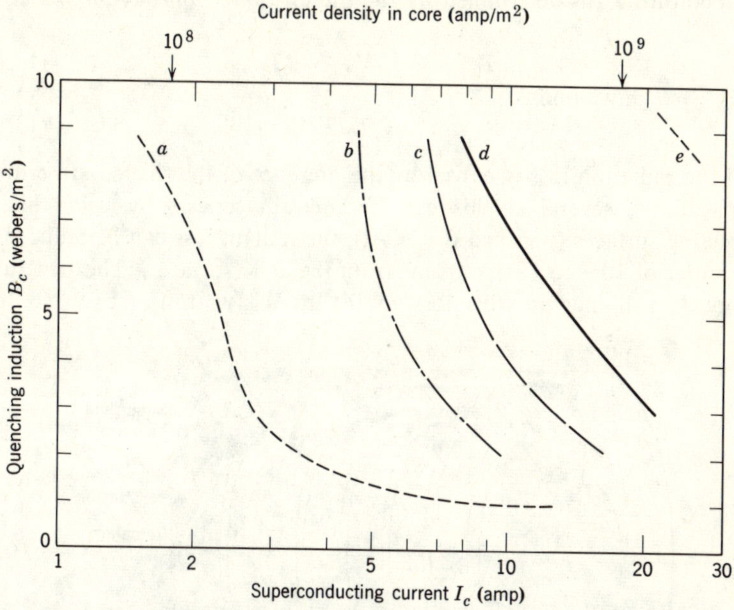

Fig. 13.6. Critical quenching induction B_c versus superconducting current I_c in Nb_3Sn-cored Nb wires.[7] All wires 0.038-cm diam with core $\approx$0.015-cm diam. (*a*) (3Nb + 1.1Sn), 14 hr at 1200° C, 1.5° K. (*b*) (3Nb + 1.0Sn), 16 hr at 1200° C, 4.2° K. (*c*) Same as (*b*), 1.5° K. (*d*) (Nb_3Sn + 0.1Sn), 16 hr at 970° C, 1.5° K. (*e*) (3Nb + 1.1Sn), 16 hr at 970° C, 4.2° K.

revaluation of many energy-conversion schemes, in addition to controlled fusion. It should not be particularly difficult to construct solenoids and other magnetic structures. Figure 13.7 shows a conceptual design of a small superconducting solenoid (0.2-m inner working diameter at 300° K). The superconducting coils are embedded in a strong cylinder and cooled below 2.2° K. At this temperature, a further advantage accrues: helium becomes superfluid; that is, the thermal conductivity of the liquid is essentially infinite, and the viscosity is virtually zero. Thus the small tubes shown at 2.2° K are adequate for cooling. Surrounding the coil cylinder are in succession: a radiation shield, a cylinder at liquid nitrogen temperature (77° K), a second radiation shield, and the enclosing cylinder at room temperature. The whole structure is evacuated to prevent gas conduction loss and air condensation. The radiation shields operate as follows. In equilibrium, the radiant energy emitted by the shield equals the radiant energy absorbed by it. Then for a shield between two surfaces at temperatures T_1 and $T_2 > T_1$, the shield temperature T_{sh} is determined by the Stefan-Boltzmann law; T_{sh} is given by

$$T_2^4 - T_{sh}^4 = T_{sh}^4 - T_1^4. \tag{13.13}$$

If $T_2^4 \gg T_1^4$, we find

$$T_{sh} = T_2/2^{1/4} \tag{13.14}$$

and the radiation loss is halved by the presence of the shield. In critical applications, several shields can be used in series. By using highly reflecting surfaces (polished Cu or Al), the radiant loss can be reduced to the order of 10^{-2} to 10^{-1} watt/m^2 from the 2° K surface. The structural supports will increase this loss to 0.1 to 0.3 watt/m^2. Even so, the

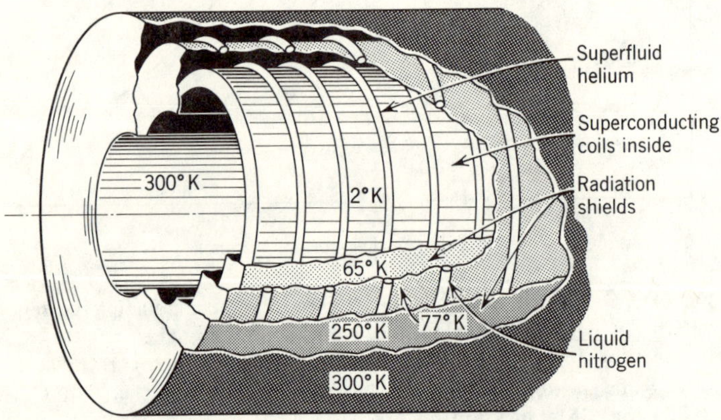

Fig. 13.7. Conceptual design of a superconducting solenoid to operate with superfluid helium. For simplicity, the radiation shields are not shown at the top. The support structure is not shown.

refrigeration power is negligible. With a mechanical efficiency of only 25% in a helium liquifier, the helium refrigeration power will be 60 to 200 watts/m^2 of coil surface.

Further advantages of the system are relatively light weight, low capital cost, better insulation in larger sizes, and smaller over-all size than normally conducting coils at high induction. Because the induction is almost free, hitherto unlikely confinement schemes may be interesting.

At very high induction, the coil-carrying structure must withstand high stress. This requirement is of course also present in any solenoid, whether superconducting or not. One problem peculiar to superconducting solenoids should be recognized. If the superconductivity is quenched anywhere for any reason, the local heating will rapidly spread the quenching throughout the whole system, and the induction will collapse. Large induced voltages will appear on the coils, unless they are somehow protected. The energy of the magnetic field (4×10^7 joules/m^3 at 10 webers/m^2) may damage the coils. One method of protecting the coils against this disaster is to surround them in close proximity by short-circuited turns of a normal conductor massive enough to absorb the energy. For example, in the design of Fig. 13.7, virtually no more magnetic energy would be deposited in the coil structure than that contained inside the innermost radiation shield. For relatively rapid changes in induction, these cylinders could be open-circuit azimuthally, then short-circuited for protection during steady operation. Even if all the cylinders are left short-circuited, the induction can still be changed slowly without much excess dissipation. For example, suppose that an axial induction is applied from an outer solenoidal winding through a hollow cylinder of radius r, shell thickness d, and resistivity η ohm-m. The potential ϕ developed around the cylinder is

$$\phi = \pi r^2 \dot{B}, \tag{13.15}$$

and the power dissipation per unit length in it is

$$\frac{\phi^2}{\mathcal{R}} = \frac{\pi r^3 d}{2\eta} \dot{B}^2. \tag{13.16}$$

The total energy deposited in it while the induction changes during the period 0 to τ is then

$$\text{Energy} = \int_0^\tau dt \frac{\pi r^3 d}{2\eta} \left(\frac{dB}{dt}\right)^2. \tag{13.17}$$

We note that $\int_0^\tau dt(dB/dt) = \Delta B$, and is a fixed constraint on the solution of Eq. 13.17. Therefore if dB/dt is programmed to be small, the energy lost in the structure can be made negligible.

In addition to applications in controlled fusion, superconducting solenoids will be the *sine qua non* for hydromagnetic energy conversion, where also large volumes of high induction are required. In a larger sense—not just in plasma applications—superconducting systems may greatly extend many of our energy conversion concepts. For example, the energy density of 4×10^7 joules/m^3 at 10 webers/m^2 is readily available for pulse applications, and far exceeds the capability of electrostatic capacitors.

★ 13.4. Fractional Burnup and Helium Pressure in a D-T System with Particle Losses

From here to the end of the chapter, we shall attempt to apply some of the principles of the preceding sections and chapters to problems that we have found interesting. We make no claim of completeness for our discussions and offer them partly as provocative examples. Other examples can be found in the literature.[8–10]

The first concerns the fractional burnup of the material in a steady-state D-T system from which particles can escape by scattering. For definiteness, we shall refer to the system as a generalized mirror. Here the plasma escapes from the device at a rate that we shall crudely estimate, and is replenished by high-energy ion injection. We shall find that the build-up of He^4 is alarming. From Sec. 4.10, the rate equations, including now the He^4 nuclei (henceforth called α-particles), are

$$\frac{\partial n_D}{\partial t} = S_D - n_D^2\,\overline{\sigma v}_{DD} - n_D n_T\,\overline{\sigma v}_{DT} - L_D, \tag{13.18}$$

$$\frac{\partial n_T}{\partial t} = S_T + \frac{n_D^2}{4}\,\overline{\sigma v}_{DD} - n_D n_T\,\overline{\sigma v}_{DT} - L_T, \tag{13.19}$$

$$\frac{\partial n_\alpha}{\partial t} = n_D n_T\,\overline{\sigma v}_{DT} - L_\alpha. \tag{13.20}$$

Here the S terms refer to the rate of injection of the fuel, and the L terms refer to the loss rate of each component from the system. The system will be assumed to be in a steady state.

The D-D reaction has negligible influence on our choice of reaction parameters, as is shown in Problem 7. We shall therefore set $n_D = n_T$ in the reactor and ignore both the D-D reaction power and the presence of He^3 and protons.

We now attack the loss term L. The precise application of even our approximate theories would be a toilsome task. Since we are interested only in qualitative and semiquantitative results, we shall make a number of approximations.† Each ion species scatters in angle by collision with other ions; the effect of electron-ion interactions on ion scattering is negligible, as discussed in Sec. 8.3. For angular relaxation of ions of type ζ by ions of type ξ, we can define a rate $(\tau_{\theta\zeta\xi})^{-1}$ for the process. Define also the probability ψ that one chance angular relaxation causes the particle to escape. For example, in a mirror system the factor ψ depends principally upon the mirror ratio and also upon the particular geometry of the system; ψ will be smaller as the mirror ratio is made greater. If ψ does not depend upon the particle species, the loss rate L_ζ of type ζ ions will be

$$L_\zeta = \psi n_\zeta \sum_\xi (1/\tau_{\theta\zeta\xi}). \tag{13.21}$$

This method of loss calculation works well enough for scattering of

† The interested reader is cordially invited to improve upon our approximations.

D and T ions among themselves, because their velocity distributions are approximately Maxwellian, and Eq. 8.23 can be used to calculate the rate. The scattering of D and T ions by the low density of α-particles is relatively small, and we shall ignore it. In addition, we ignore the mass difference between D and T ions, which occurs only in the square root in Eq. 8.23, and assign a mean atomic weight of 2.5. Further, we set $\ln \Lambda = 20$ throughout. Then Eq. 13.21, combined with Eq. 8.23, becomes

$$L_D \approx L_T \approx \frac{\psi n_D(2.8 \times 10^{-17} n_i)}{T_i^{3/2}}, \tag{13.22}$$

where $n_i = n_D + n_T$ is the total density of D and T ions, and T_i is their common temperature.

The fractional burnup f_b of deuterium (or tritium) is defined as

$$f_b = \frac{S_D - L_D}{S_D} = \left(1 + \frac{L_D}{n_D^2\, \overline{\sigma v}_{DT}}\right)^{-1} \tag{13.23}$$

from Eq. 13.18 in the steady state and the equality of n_D and n_T. By inserting Eq. 13.22, we find

$$f_b = \left(1 + \frac{5.6 \times 10^{-17} \psi}{\overline{\sigma v}_{DT}\, T_i^{3/2}}\right)^{-1}. \tag{13.24}$$

We choose also to calculate the energy carried away by the escaping deuterons and tritons, because it can be large and serves well for numerical illustration. Because slow ions scatter more easily, the escaping ion starts with less energy than $3kT_i/2$, but it gains energy corresponding to the plasma potential as it leaves. We ignore these adjustments and assign $3kT_i/2$ per ion. The energy loss rate $w_{LD} + w_{LT}$, because of loss of deuterons and tritons, is from Eq. 13.22

$$w_{LD} + w_{LT} = \frac{2.7 \times 10^{-32} \psi n_D^2}{T_i^{1/2}} \text{ watts/m}^3. \tag{13.25}$$

A more significant quantity is the ratio of $w_{LD} + w_{LT}$ to the total nuclear heat w'_{DT} (including neutron capture energy). For T_i in kev, the ratio is

$$\frac{w_{LD} + w_{LT}}{w'_{DT}} = \frac{7.5 \times 10^{-21} \psi}{\overline{\sigma v}_{DT}\, T_i^{1/2}}. \tag{13.26}$$

We now turn to the α-particles, whose distribution is far from Maxwellian. They are created with 3.5 Mev energy; thus their initial speed far exceeds that of the deuterons and tritons. However, their speed equals that of a 480-ev electron. It will soon become evident that energy transfer from the various ions to the electrons will raise T_e to several kev at least. Then in velocity space, the α-particles lie almost entirely outside the core of D and T ions and inside the cloud of electrons.

The α-particles are born at high energy, interact negligibly among themselves, but lose energy to the D and T ions and electrons, while undergoing scattering out of the system via the D and T interactions. Thus the best method for finding their energy distribution is to set up their rate of flow in energy. The downward flow at the initial energy U_0 (=3.5 Mev) is the rate of production $n_D n_T \overline{\sigma v}_{DT}$. But all these α-particles will not reach some lower energy U, because some will be scattered out. The fraction of the flow remaining is

$$\exp - \int_0^t \frac{dt'}{\tau_{L\alpha}(t')},$$

where t is the time taken to lose energy from U_0 to U, and $1/\tau_{L\alpha}$ is the loss rate at time t' (or corresponding energy U'). The quantity $f_\alpha(U_\alpha)$ is the density of α-particles per unit energy; dU_α/dt is the rate at which the energy of these particles changes. Therefore, in the steady state the downward flow in energy at energy U_α is $f_\alpha(U_\alpha)|dU_\alpha/dt|$. By noting that $dt = dU/(dU/dt)$, we then obtain the distribution function $f_\alpha(U_\alpha)$ of α-particles in the form

$$f_\alpha = \frac{n_D n_T \overline{\sigma v}_{DT}}{|dU_\alpha/dt|} \exp - \int_{U_\alpha}^{U_0} \frac{dU'}{\tau_{L\alpha}(U')|dU/dt|}. \tag{13.27}$$

We shall omit the subscript on U_α where no ambiguity arises.

The loss rate dU/dt arises from two causes. First, the loss to D and T ions can be calculated by dynamical friction, Eq. 8.53; here we assume for simplicity that the ions are all near the origin of velocity space. Second, the loss to electrons can be calculated by Eq. 8.35; here we assume $U_\alpha \gg kT_e$ and $m_e U_\alpha \ll m_\alpha kT_e$. All these various approximations are valid as long as U_α considerably exceeds the plasma energy. The approximation overestimates $|dU_\alpha/dt|$ at low U_α, but f_α is small there, and our final result will be insensitive to conditions at such low energy. Combination of Eqs. 8.53 and 8.35 then gives

$$\frac{dU_\alpha}{dt} = - \frac{Z_\alpha^2 e^4 n_i m_e^{1/2} U_\alpha \ln \Lambda}{3\pi(2\pi)^{1/2}\epsilon_0^2 m_\alpha (kT_e)^{3/2}} \left[1 + \frac{3\sqrt{\pi} m_\alpha^{3/2}(kT_e)^{3/2}}{4 m_r m_e^{1/2} U_\alpha^{3/2}}\right]. \tag{13.28}$$

Here, m_r refers to α-ion interactions, and energies are in joules. The first term in the bracket is the effect of electrons, and the second is the effect of ions.

The loss time $\tau_{L\alpha}$ is calculated from Eq. 8.20 for $\tau_\theta(v)$. But first we shall multiply Eq. 8.20 by the factor 25.8/16, to agree with the correction appearing between Eqs. 8.22 and 8.23, and also by $1/\psi$, to include the reduced probability of escape. Expressed in terms of energy, the result is

$$\tau_{L\alpha} = \frac{25.8\pi\epsilon_0{}^2 m_r{}^2 U_\alpha{}^{3/2}}{\sqrt{2} Z_\alpha{}^2 e^4 m_\alpha{}^{3/2} n_i \psi \ln\Lambda}. \tag{13.29}$$

Integration of Eq. 13.27 is straightforward. Define

$$\left.\begin{aligned} C &= \frac{3\sqrt{\pi} m_\alpha{}^{3/2}(kT_e)^{3/2}}{4 m_r m_e{}^{1/2}}, \\ K &= \frac{8\psi m_\alpha}{25.8 m_r}; \end{aligned}\right\} \tag{13.30}$$

the answer is

$$f_\alpha(U) = \frac{3\pi(2\pi)^{1/2}\epsilon_0{}^2 m_\alpha n_{\mathrm{D}} n_{\mathrm{T}} \overline{\sigma v}_{\mathrm{DT}} (kT_e)^{3/2} U^{(1/2)+K} (U_0 + C)^{2K/3}}{Z_\alpha{}^2 e^4 n_i m_e{}^{1/2} \ln\Lambda\, U_0{}^K (U^{3/2} + C)^{1+(2K/3)}}. \tag{13.31}$$

We display Eq. 13.31 in its general form because it applies equally well for fast protons from the D-D or D-He^3 reaction by relabeling subscripts.

The competition between electrons and ions for the α-particle energy can be seen from Eqs. 13.28 and 13.31. For electrons alone, we neglect the term $C^{2/3}$ compared with U or U_0. Then $dU/dt \propto U$, energy transfer is greatest at high α-particle energy, and $f_\alpha(U) \propto U^{-1}$. For ions alone, we neglect U or U_0 compared with $C^{2/3}$; then $dU/dt \propto U^{-1/2}$, energy transfer is poor at high energy, and $f_\alpha(U) \propto U^{(1/2)+K}$.

We shall be particularly interested in the density n_α and pressure p_α of the α-particles. Integration of Eq. 13.31 is an unwelcome chore; we search for an approximation. For an ion of mean atomic weight 2.5, $m_r = 0.385 m_\alpha$, and $K = 0.805\psi$. We shall find that the only systems of practical interest are those for which $\psi \lesssim 0.1$, whence K is a very small exponent. Thus we set $\psi = 0$ or $K = 0$, an approximation that is equivalent to having the α-particles degrade to very low energy before escaping. From the form of Eq. 13.31, we see that $f_\alpha(U) \to 0$ as $U \to 0$: the processes at low energy will not affect n_α or p_α appreciably. Our answer will be an upper limit to these quantities, because increased α-particle loss will reduce n_α and p_α.

The ratio of n_α to the useful ion density is important. With these approximations, we find

$$\begin{aligned} \frac{n_\alpha}{n(\mathrm{D}+\mathrm{T})} &= \frac{\pi(2\pi)^{1/2}\epsilon_0{}^2 m_\alpha \overline{\sigma v}_{\mathrm{DT}} (kT_e)^{3/2}}{8 e^4 m_e{}^{1/2} \ln\Lambda} \ln\left(1 + \frac{U_0{}^{3/2}}{C}\right) \\ &= 8.2 \times 10^{16}\, \overline{\sigma v}_{\mathrm{DT}}\, T_e{}^{3/2} \ln\left(1 + \frac{U_0{}^{3/2}}{C}\right). \end{aligned} \tag{13.32}$$

The α-particle pressure p_α can be expressed similarly. The distribution is almost isotropic, whence

$$\frac{p_\alpha}{p(\mathrm{D+T})} = \frac{1}{3n_\mathrm{D}kT_i}\int_0^{U_0} dU\, Uf_\alpha(U)$$

$$= \frac{\pi(2\pi)^{1/2}\epsilon_0{}^2 m_\alpha \overline{\sigma v}_{\mathrm{DT}}(kT_e)^{3/2}U_0}{8e^4 m_e{}^{1/2}kT_i \ln\Lambda}\left[1 - \frac{1}{3y}\ln\frac{y-\sqrt{y}+1}{(\sqrt{y}+1)^2} - \frac{2}{y\sqrt{3}}\tan^{-1}\left(\frac{\sqrt{3y}}{2-\sqrt{y}}\right)\right]; \quad (13.33)$$

$$y = \frac{U_0}{C^{2/3}}. \quad (13.34)$$

Note that if T_i and U_0 are expressed in the same units, the numerical coefficients of Eqs. 13.32 and 13.33 are identical.

Figure 13.8 schematizes the energy flow in our plasma. Fast deuterons and tritons are injected; we assume that no energy is needed to inject electrons. If $T_e < T_\mathrm{D} \approx T_\mathrm{T}$, electrons are heated by all ion species. All escaping particles carry off energy; also, the electrons radiate via bremsstrahlung and cyclotron radiation. One-fifth of the fusion energy appears with the α-particles and four-fifths escapes with the neutrons.

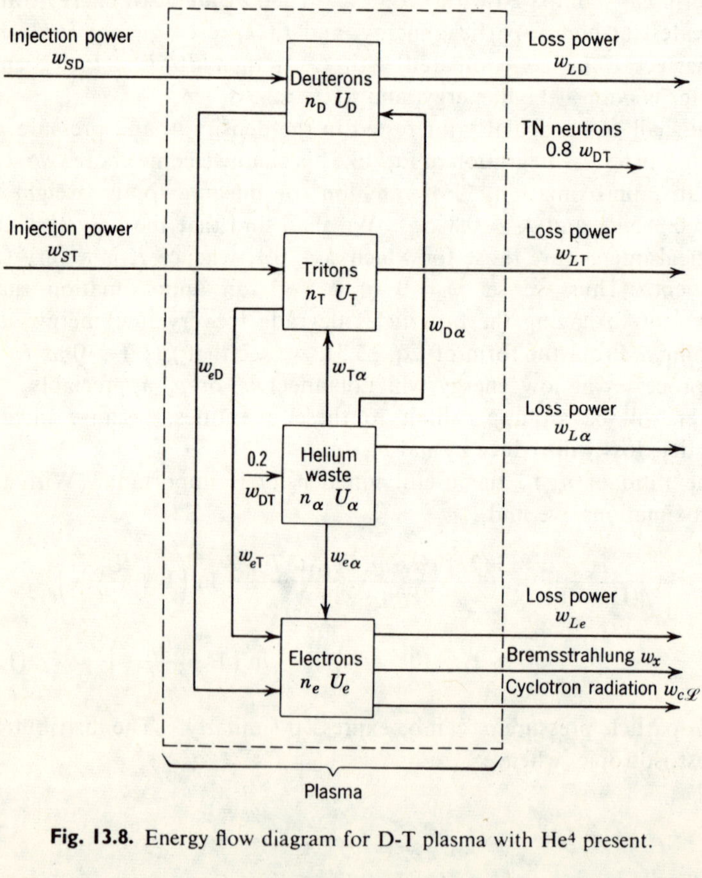

Fig. 13.8. Energy flow diagram for D-T plasma with He[4] present.

Table 13.2 lists the various quantities calculated for five values of T_i. The quantities f_b and $(w_{LD} + w_{LT})/w'_{DT}$ are insensitive to T_e, and are shown for several values of ψ; $n_\alpha/(n_D + n_T)$ and $p_\alpha/(p_D + p_T)$ are insensitive to ψ, and are shown for several values of T_e.

From the table, we make several conclusions. First, the ion-loss power must be made up in large part by a corresponding electric injection power; but w'_{DT} represents heat. Thus we require $(w_{LD} + w_{LT})/w'_{DT} < 0.1$ (let us say); from the table, $\psi \lesssim 0.1$ and $T_i \approx 100$ kev. Here we find justification for the approximations made in computing n_α and p_α. In this regime of parameters, the fractional burnup $f_b \gtrsim 0.1$; note that for lower energy or more lossy systems, f_b is impossibly small.

The α-particle pressure is not large, and n_α itself is very low. We did not calculate equilibrium temperatures T_e determined by energy transfer from D-, T-, and α-particles and radiation losses, but used three arbitrary temperatures in Table 13.2. This scheme is not unrealistic: first, since the cyclotron radiation depends strongly on T_e, T_e is held to $\lesssim$ 50 kev; second, a large fraction of the α-particle energy eventually escapes via transfer to electrons and subsequent radiation. To facilitate this transfer and to keep the electron pressure low, T_e might be deliberately reduced by introducing impurities to enhance the bremsstrahlung.

The α-particles do not have a Maxwellian temperature, but an average energy figured from the ratio $3p_\alpha/2n_\alpha$ is approximately 1.5 Mev.

Table 13.2. Quantities in a D-T mirror as functions of deuteron and electron temperatures T_D and T_e and confinement factor ψ

Quantity	ψ or T_e (kev)	Temperature of Deuterons 10 kev	20 kev	50 kev	120 kev	200 kev
$\overline{\sigma v}_{DT}$(m^3/sec)		1.2×10^{-22}	4.4×10^{-22}	9×10^{-22}	8×10^{-22}	4×10^{-22}
f_b	$\psi = 0.05$	0.0013	0.014	0.10	0.27	0.29
	0.1	0.0007	0.0069	0.053	0.16	0.17
	0.2	0.0003	0.0035	0.027	0.085	0.091
	0.3	0.0002	0.0023	0.018	0.059	0.063
$\dfrac{w_{LD} + w_{LT}}{w'_{DT}}$	$\psi = 0.05$	1.0	0.19	0.059	0.043	0.067
	0.1	2.0	0.38	0.12	0.086	0.13
	0.2	4.0	0.77	0.24	0.17	0.27
	0.3	6.0	1.15	0.36	0.26	0.40
$\dfrac{n_\alpha}{n_D + n_T}$	$T_e = 25$	0.0023	0.0085	0.018	0.016	0.0078
	40	0.0033	0.012	0.025	0.022	0.011
	50	0.0038	0.014	0.028	0.025	0.013
$\dfrac{p_\alpha}{p_D + p_T}$	$T_e = 25$	0.25	0.46	0.37	0.14	0.042
	40	0.39	0.71	0.58	0.21	0.064
	50	0.45	0.84	0.68	0.25	0.076

These remarks conclude our treatment of the α-particles in a mirror. Our intent is to present heuristic examples and orders of magnitude rather than to solve specific cases exactly. To continue by calculating self-consistent electron temperatures, including the α-energy degradation, would be difficult. In the next section, therefore, we shall remove the α-particles magically as they form, in order to obtain a relatively concise insight into some of the remaining phenomena. This condition might be approximated by injecting enough cold electrons or selected impurities to depress T_e and counterbalance the energy transfer from the α-particles.

★ 13.5. Energy Balance in a D-T Lossy Device without Helium

Figure 13.9 shows an energy flow diagram for a D-T system with the helium somehow removed as it forms. By exorcising the helium, we must abandon any attempt to keep the plasma hot through the fusion reaction. We assume as before that the input energy is supplied by injecting energetic deuterons or tritons.

The calculations of the previous section can be usefully applied here if the various terms containing n_α are neglected. We now consider the role of electrons; here, phenomena associated with energy transfer from α-particles are assumed to be balanced out, as discussed above.

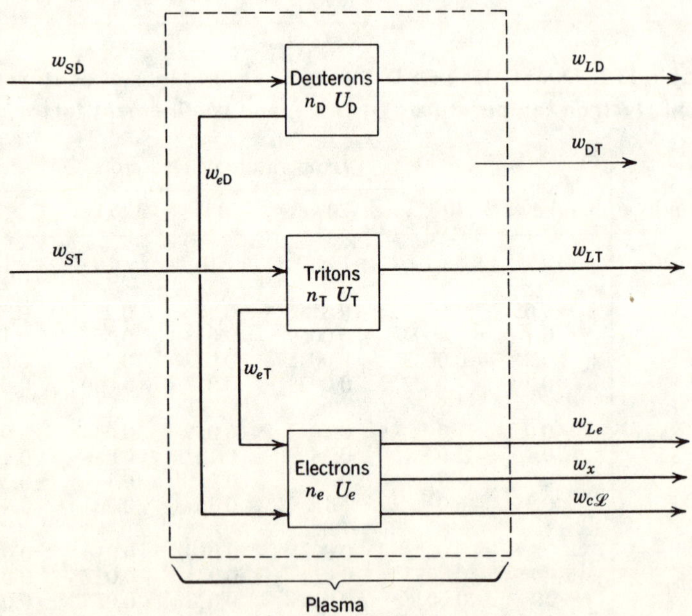

Fig. 13.9. Energy flow diagram for a D-T plasma with purging of He^4.

We set the deuteron and triton temperatures equal to a common ion temperature T_i. Since the net electron heating now arises only by deuterons and tritons, the electron temperature T_e will be less than T_i.

The total energy loss rate by ions to electrons is given by Eq. 8.39 summed over the ion species. In the steady state, this rate equals the total energy loss from the electrons: bremsstrahlung w_x, cyclotron radiation $w_{c\mathscr{L}} = w_c K_{\mathscr{L}}$ (the amount actually escaping), and loss w_{Le}. Thus, in Fig. 13.9

$$w_{eD} + w_{eT} = \frac{3}{2} \sum_{\xi} \frac{n_\xi k T_\xi}{\tau_{u\xi e}} = w_{Le} + w_x + w_{c\mathscr{L}}, \tag{13.35}$$

where the sum is taken over the ion species ξ, and τ is defined in Eq. 8.39. Note that the term $\sqrt{6/\pi}(m_e T_i/m_i T_e)^{3/2}$ in Eq. 8.39 is negligible here.

We assume that electronic and ionic charge losses will be equal and that each electron carries away $3kT_e/2$ energy, all in the interest of simplicity. Then the power density of electron loss is, because one electron leaves for each positive charge escaping,

$$w_{Le} = \frac{3}{2} k T_e (2L_D + 2L_\alpha). \tag{13.36}$$

The bremsstrahlung power density is now taken from Eq. 11.17. The cyclotron radiation power density that actually escapes is given by Eq. 11.83 multiplied by the coefficient $K_{\mathscr{L}}$, as described in Eq. 11.85 and Fig. 11.7. Here, we assume no reflection of the cyclotron radiation at the vacuum wall and will make an approximate correction later. From these equations, the assumption $\ln \Lambda = 20$, and Eq. 13.22 for L_D, we find that Eq. 13.35 can be written term by term as

$$\begin{aligned} &8.0 \times 10^{-34} n_D{}^2 \frac{T_D}{T_e^{3/2}} \left(1 - \frac{T_e}{T_D}\right) \\ &\quad = 3.0 \times 10^{-32}\, n_D{}^2 T_e \left(\frac{\psi}{T_D^{3/2}} + 1.6 \times 10^{16}\, \overline{\sigma v}_{DT}\right) + 1.9 \times 10^{-36}\, n_D{}^2 T_e^{1/2} \\ &\qquad + 1.2 \times 10^{-16}\, n_D T_e B^2 \left(1 + \frac{T_e}{204} + \cdots\right) K_{\mathscr{L}}. \end{aligned} \tag{13.37}$$

In Eq. 13.22 for L_D, we have assumed that the α-particles escape promptly and do not scatter the ions. In the present result, temperatures are in kev and B is webers/m² as usual. The equation can be reduced to

$$\begin{aligned} \frac{T_D}{T_e} - 1 &= 37.5 T_e^{3/2} \left(\frac{\psi}{T_D^{3/2}} + 1.6 \times 10^{16}\, \overline{\sigma v}_{DT}\right) + 2.4 \times 10^{-3} T_e \\ &\quad + 1.25 \times 10^{-4} T_e^{3/2} (T_e + T_D) \left(1 + \frac{T_e}{204} + \cdots\right) K_{\mathscr{L}}/\beta. \end{aligned} \tag{13.38}$$

Comparison of the terms is revealing. The term on the left represents energy transfer to the electrons from the ions, and its magnitude is not far from 1. The second term on the right represents bremsstrahlung. Since $T_e < T_D < 200$ kev, we see that bremsstrahlung has little effect on T_e; thus bremsstrahlung loss from this source is small compared with other electron energy loss mechanisms. The largest loss arises from escaping electrons, represented by the first term on the right side of Eq. 13.38. Furthermore, the term $1.6 \times 10^{16}\,\overline{\sigma v_{DT}}$ represents loss of α-particles and is small compared with the term $\psi/T_D^{3/2}$ that represents loss of the reacting gas. That is, the fractional burnup $f_b \leqslant 0.15$. If cyclotron radiation is neglected, the ratio T_D/T_e can be readily obtained from Eq. 13.38. Figure 13.10 shows (T_D/T_e) as a function of ψ for several deuteron temperatures, neglecting cyclotron radiation.

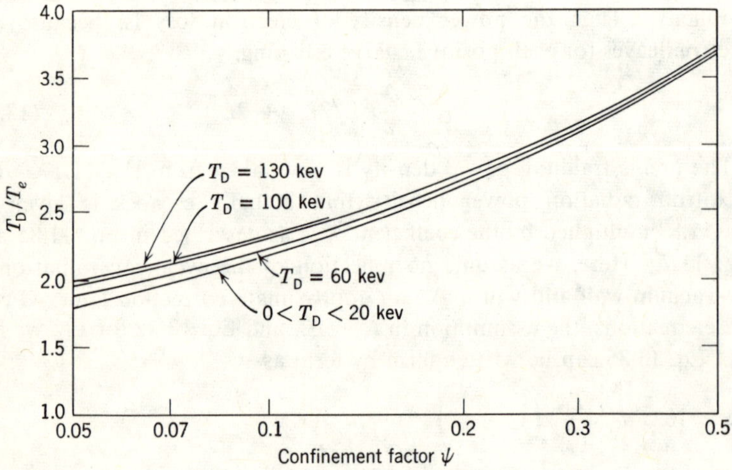

Fig. 13.10. Ratio T_D/T_e of deuteron to electron temperature in a D-T mirror, neglecting cyclotron radiation, as a function of confinement factor ψ.

The cyclotron-radiation term is best worked by example. One case in Fig. 13.10 is $T_D = 100$ kev, $\psi = 0.1$, $T_e = 44.5$ kev. Then the third term on the right is $6.5\ K_{\mathscr{L}}/\beta$. Assuming that $\beta \approx 0.25$, we should have $K_{\mathscr{L}} \ll 0.04$ if cyclotron radiation loss is to be negligible. Note that T_e is reduced by this radiation; of course, this reduction of T_e is actually symptomatic of larger total electronic energy loss, as may be seen from Eq. 8.39. In treating the over-all power balance of a particular reactor, we shall use Fig. 13.10 to compute T_e and check whether the inclusion of w_c makes very much difference.

The total loss

$$w_{LD} + w_{LT} + w_{eD} + w_{eT} = w_{SD} + w_{ST} \quad (13.39)$$

must be made up electrically by the inputs of energy via ions. From Eqs. 13.25 and 8.39, we have an input particle power density

$$w_S = w_{SD} + w_{ST} = \frac{3 \times 10^{-32} n_D^2 \psi}{T_D^{1/2}} \left[1 + \frac{2.67 \times 10^{-2}}{\psi} \left(\frac{T_D}{T_e}\right)^{1/2} \left(\frac{T_D}{T_e} - 1\right)\right], \quad (13.40)$$

where T_D/T_e can be taken from Fig. 13.10 under most circumstances. In the limit of relatively low temperature, the second term in the brackets of Eq. 13.40 is 1.0 (T_e/T_D).

We are now in a position to estimate the energy balance and size of our hypothetical D-T system. Figure 13.11 shows an energy flow diagram.

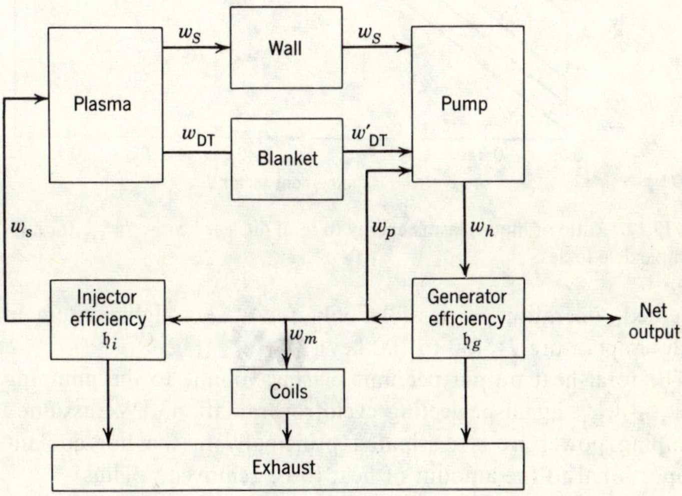

Fig. 13.11. Energy flow in a thermonuclear reactor.

All energy leaving the plasma is assumed to be convertible to useful heat. That total heat output per unit volume is the thermonuclear power density w_{DT} plus the entire injection power density w_S. A figure of merit for the system is the ratio of injection power density to the nuclear heat w'_{DT}, including that of neutron capture from Eq. 13.2. With the aid of Eq. 13.40 and Fig. 13.10, we plot the ratio w_S/w'_{DT} in Fig. 13.12, as a function of confinement factor ψ and deuteron temperature T_D. The nonlinear cyclotron radiation has been neglected. If reasonable efficiencies are

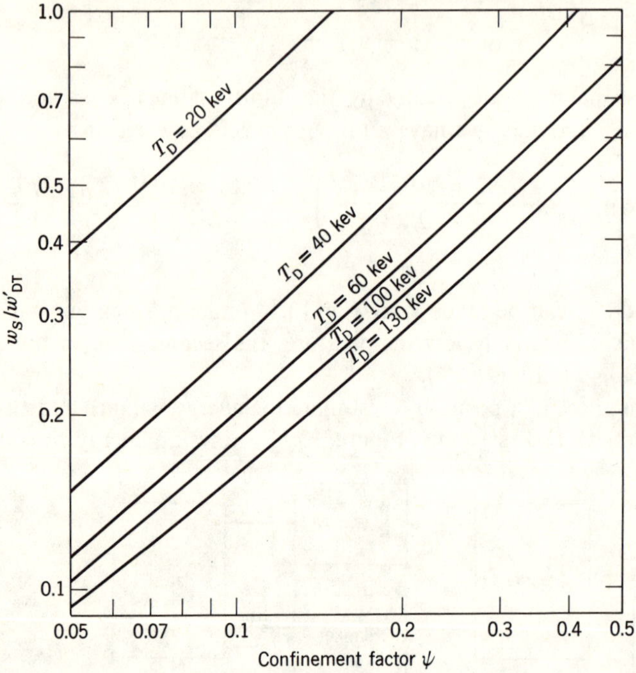

Fig. 13.12. Ratio of input plasma energy to total nuclear heat w_S/w'_{DT} for a D-T system with particle losses.

assumed, operation is possible only for a very tight system and at a high temperature T_D (50 to 100 kev).

The total heat output per unit plasma volume to the pumping system is $w_S + w'_{DT}$, again neglecting cyclotron radiation. We assume that the pumping power w_p is dissipated principally in the hot coolant and is proportional to the amount of heat to be removed. Thus

$$w_p = \mathfrak{h}_p (w_S + w'_{DT}). \tag{13.41}$$

The total heat is converted to electricity with efficiency $\mathfrak{h}_g$, and the fuel is injected electrically with efficiency $\mathfrak{h}_i$. Then, even with neglect of cyclotron radiation and coil losses, we must have

$$(1 + \mathfrak{h}_p)(w_S + w'_{DT})\mathfrak{h}_g - \frac{w_S}{\mathfrak{h}_i} - \mathfrak{h}_p(w_S + w'_{TD}) \geqslant 0 \tag{13.42}$$

for any output at all. For example, if $\mathfrak{h}_p = 0.05$, $\mathfrak{h}_g = 0.40$, and $\mathfrak{h}_i = 0.5$, we find $w_S/w'_{DT} \leqslant 0.23$ for net output at this stage.

These figures force us to extreme measures: a temperature $T_D \approx 100$ kev, a confinement factor $\psi \approx 0.05$, and a very large system, so that cyclotron radiation and other losses are minimized. Suppose that a mirror ratio of 10 suffices. Even if the allowable material stress at the mirrors is as high as 1000 atmospheres (15 webers/m^2), the long central section of the mirror will have $B \approx 1.5$ webers/m^2. The system is a low-pressure one after all. Superconducting coils might be used throughout.

In order to complete the example, we choose $T_D = 100$ kev, $\psi = 0.05$, $B = 1.5$ webers/m^2, except at the mirrors, and $\beta = 0.25$. From Fig. 13.10, $T_e = 50.7$ kev. For the cyclotron radiation, we find easily from Eq. 11.83 that

$$\frac{w_{c\mathscr{L}}}{w'_{DT}} = \frac{2.75 \times 10^{-26} T_D^2 [T_e/T_D][1 + (T_e/204)][1 + T_e/T_D] K_{\mathscr{L}}}{\beta \overline{\sigma v}_{DT}}. \tag{13.43}$$

The ratio $w_{c\mathscr{L}}/w'_{DT}$ must be small. It determines the transparency coefficient $K_{\mathscr{L}}$ and hence the radius of the plasma. In the present example, Eq. 13.43 reduces to

$$\frac{w_{c\mathscr{L}}}{w'_{DT}} \approx 1.2\, K_{\mathscr{L}}; \tag{13.44}$$

if we choose $w_{c\mathscr{L}}/w'_{DT} = 0.02$, $K_{\mathscr{L}} \approx 0.017$. From Fig. 11.7, with $T_e = 50$ kev, the dimensionless depth $\mathscr{L}$ of a plasma slab with $K_{\mathscr{L}} = 0.017$ is $\mathscr{L} \approx 2 \times 10^4$. For a plasma slab in which all ions are at a common temperature T_i, and $n_e = n_i$, the dimensionless distance $\mathscr{L}$ from Eq. 11.78 can conveniently be written

$$\mathscr{L} = 1.5 \times 10^5\, \beta B L/(T_e + T_D), \tag{13.45}$$

where L is the thickness of the slab, and T is measured in kev. With $\beta = 0.25$, $B = 1.5$ webers/m^2, $T_e + T_D = 150$ kev, we find that an equivalent plasma slab ≈ 50 m thick is required if the cyclotron radiation loss is to be desirably small and reflection of this radiation is neglected.

Reflection of the radiation will drastically reduce the size. For a qualitative estimate of this reduction, we refer to Fig. 11.6. There, we see that a reduction in plasma size by a factor of 10 with 90% reflection decreases the radiated power by a factor of about 12. Thus the escaping plasma radiation per unit plasma volume, which is the important quantity, is reduced slightly. We speculate that 95% to 98% reflection is probably attainable and that the plasma size may be correspondingly reduced to 2% to 5% of the above size. If we associate a slab depth L with the plasma radius, the plasma radius will then be 1 to 2 m. Since in this example, $n_D = 4.7 \times 10^{18}$/m^3, $w'_{DT} = 6.3 \times 10^4$ watts/m^3, we find a nuclear heat output of about 0.8 megawatt per meter length for a system of 2-m radius.

We shall not carry this calculation any further, but we note that the system has not been optimized at all. The output would be increased by a factor of about 4 at constant pressure by halving the deuteron temperature at the cost of increased particle scattering and somewhat decreased efficiency. The cyclotron-radiation loss would be much reduced, and a smaller system would suffice. In any event, the system must be long compared with the diameter if radiation losses from the ends are to be negligible. Refrigeration losses for superconducting coils are negligible.

We summarize the discussion by stating the following qualitative conclusions for a system with scattering-out losses, if it could be stabilized and the α-particles removed:

1. The principal difficulty in producing power would be the large loss from the system and the necessity for making up this loss by reinjection.
2. Radiation losses can be made relatively small. As a corollary, the radiation power density on the vacuum wall is small, as can readily be checked from the preceding example.
3. It will be difficult to turn much more than about 15 to 20% of the thermonuclear heat into salable electrical power unless present efficiencies are considerably improved.
4. Superconducting coils must be used if the losses are to be tolerable.
5. A D-D reaction, by analogy, probably cannot be made practical at all.

★ 13.6. Dynamic Equations for a System without Particle Loss

As a final example, we consider a hypothetical system from which the plasma cannot escape. Since the large particle losses of the previous sections do not exist here, we expect that there will be no great problem in obtaining a reasonable power efficiency. However, the α-particles necessarily remain trapped; the fuel must be injected and heated, then exhausted after some optimum burnup.

The densities n_{D}, n_{T}, and n_α change with time as

$$\frac{dn_\alpha}{dt} = -\frac{dn_{\mathrm{D}}}{dt} = -\frac{dn_{\mathrm{T}}}{dt} = n_{\mathrm{D}}n_{\mathrm{T}}\,\overline{\sigma v}_{\mathrm{DT}}, \tag{13.46}$$

$$\frac{dn_e}{dt} = 0. \tag{13.47}$$

We invoke the assistance of Fig. 13.13 to calculate the rate of change of total energy per unit volume of each constituent. A number of approximations must be made if the manipulations are to be kept simple. We begin as usual by neglecting the difference between deuterons and tritons and set the mass number A of deuterons and tritons equal to 2.5. Then from Eqs. 13.35 and 8.39, we calculate the power $w_{e\mathrm{D}} = w_{e\mathrm{T}}$ trans-

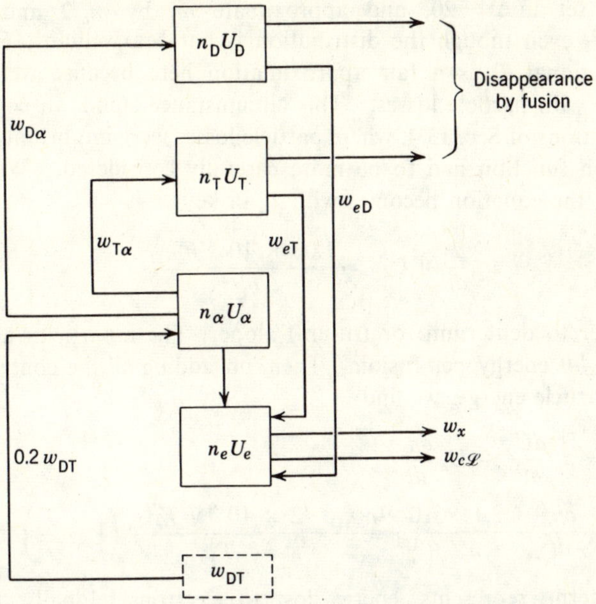

Fig. 13.13. Energy flow diagram of a completely confined system.

ferred to the electrons by deuterons and tritons and the power $w_{e\alpha}$ similarly transferred by the α-particles; we write an energy balance for the electrons. The result is

$$\frac{d}{dt}(n_e U_e) = n_e \frac{dU_e}{dt} = \frac{1.6 \times 10^{-18} n_e n_D U_D}{T_e^{3/2}} \left(1 - \frac{T_e}{T_D}\right)$$
$$+ \frac{2 \times 10^{-18} n_e n_\alpha U_\alpha}{T_e^{3/2}} \left(1 - \frac{T_e}{T_\alpha}\right) - w_x - w_{c\mathscr{L}}. \quad (13.48)$$

Here, the first term on the right represents the transfer from both deuterons and tritons, because we assume (as usual) that $n_D = n_T$ and $T_D \approx T_T$. The second term similarly represents electron heating by α-particles, and the last two terms are the bremsstrahlung and cyclotron-radiation losses.

In this example, where α-particles are retained, we must actually calculate their rate of change of energy. As already outlined in Sec. 13.4, a dynamical friction argument is used. In first approximation, we neglect the speed of the deuterons and tritons compared with that of the α-particles, calculate $m_\alpha v_\alpha (dv_\alpha/dt)$ from Eq. 8.53, and find that the rate of change of energy is given by

$$\frac{d}{dt}(n_\alpha U_\alpha) = -\frac{4e^4 n_D n_\alpha \ln \Lambda}{4\pi\epsilon_0{}^2 m_r} \overline{\left(\frac{1}{v_\alpha}\right)}. \quad (13.49)$$

We now set $\ln \Lambda = 20$, and approximate m_r by $m_\alpha/2$ and $\overline{1/v_\alpha} \approx (m_\alpha/kT_\alpha)^{1/2}$, even though the distribution is not Maxwellian. Our convenient fictional T_α is a fair approximation here because we are not concerned with particle losses. This circumstance stands in contrast to the calculations of Sec. 13.4, where particle losses were important, and the distribution function had to be more carefully considered. With these insertions, the equation becomes with T_α in kev

$$\frac{d}{dt}(n_\alpha U_\alpha) = \frac{1.03 \times 10^{-31} n_\alpha n_{\mathrm{D}}}{T_\alpha^{1/2}} \tag{13.50}$$

for transfer to deuterium (or tritium) alone. The α-particles also gain $0.2U_{\mathrm{DT}}\, dn_\alpha/dt$ energy per fusion. Then, on adding all the contributions to the α-particle energy, we find

$$\begin{aligned}\frac{d}{dt}(n_\alpha U_\alpha) &= n_\alpha \frac{dU_\alpha}{dt} + U_\alpha \frac{dn_\alpha}{dt} \\ &= 0.2U_{\mathrm{DT}} \frac{dn_\alpha}{dt} - \frac{2.1 \times 10^{-31} n_\alpha n_{\mathrm{D}}}{T_\alpha^{1/2}} - \frac{2 \times 10^{-18} n_\alpha n_e U_\alpha}{T_e^{3/2}} \left(1 - \frac{T_e}{T_\alpha}\right). \end{aligned} \tag{13.51}$$

The last term represents energy lost to electrons. Finally, for the deuterons,

$$\begin{aligned}\frac{d}{dt}(n_{\mathrm{D}} U_{\mathrm{D}}) = \frac{1.03 \times 10^{-31} n_\alpha n_{\mathrm{D}}}{T_\alpha^{3/2}} &- U_{\mathrm{D}} n_{\mathrm{D}} n_{\mathrm{T}} \overline{\sigma v}_{\mathrm{DT}} \\ &- \frac{0.8 \times 10^{-18} n_e n_{\mathrm{D}} U_{\mathrm{D}}}{T_e^{3/2}} \left(1 - \frac{T_e}{T_{\mathrm{D}}}\right). \end{aligned} \tag{13.52}$$

Here, the first term on the right side represents energy gain from α-particles, the second represents material lost to fusion, and the last term represents energy lost to the electrons.

Equations 13.48, 13.51, and 13.52 can also be conveniently expressed in terms of rates of change of temperature in kev. Thus we find, upon writing w_x and $w_{c\mathscr{L}}$ explicitly,

$$\begin{aligned} n_e \frac{dT_e}{dt} = \frac{1.6 \times 10^{-18} n_e n_{\mathrm{D}} T_{\mathrm{D}}}{T_e^{3/2}} \left(1 - \frac{T_e}{T_{\mathrm{D}}}\right) &+ \frac{2 \times 10^{-18} n_e n_\alpha T_\alpha}{T_e^{3/2}} \left(1 - \frac{T_e}{T_\alpha}\right) \\ -4 \times 10^{-21} T_e^{1/2} n_e (n_{\mathrm{D}} + 2n_\alpha) &- 0.26 n_e B^2 T_e \left(1 + \frac{T_e}{204} + \cdots\right) K_{\mathscr{L}}; \end{aligned} \tag{13.53}$$

$$\begin{aligned} n_\alpha \frac{dT_\alpha}{dt} = (2350 - T_\alpha) n_{\mathrm{D}} n_{\mathrm{T}} \overline{\sigma v}_{\mathrm{DT}} &- \frac{8.6 \times 10^{-16} n_\alpha n_{\mathrm{D}}}{T_\alpha^{1/2}} \\ - \frac{2 \times 10^{-18} n_e n_\alpha T_\alpha}{T^{3/2}} \left(1 - \frac{T_e}{T_\alpha}\right); \end{aligned} \tag{13.54}$$

and

$$n_D \frac{dT_D}{dt} = \frac{4.3 \times 10^{-16} n_\alpha n_D}{T_\alpha^{1/2}} - \frac{8 \times 10^{-19} n_e n_D T_D}{T_e^{3/2}} \left(1 - \frac{T_e}{T_D}\right). \quad (13.55)$$

Equations 13.46, 13.53, 13.54, and 13.55 are five equations in the five unknowns $n_D(=n_T)$, n_α, T_e, $T_D(=T_T)$, and T_α. The solution of even this simplified set would be a difficult task. Notwithstanding these difficulties, a number of observations can be made, provided that they are made in the correct order.

First, we consider the rate of heating of electrons by deuterons and tritons compared with the burnup rate of the fuel. At early times, when $n_\alpha \ll n_D$, we wish to compare the electron heating rate, which is $(1/\tau_{ueD}) + (1/\tau_{ueT})$ from Eq. 13.53 with the rate $1/\tau_{DT} = n_D\, \overline{\sigma v}_{DT}$. We note that $\overline{\sigma v}_{DT} \leqslant 9 \times 10^{-22}$ m³/sec. Then

$$\left(\frac{1}{\tau_{ueD}} + \frac{1}{\tau_{ueT}}\right) \tau_{DT} \geqslant \frac{1800 T_D}{T_e^{5/2}} \left(1 - \frac{T_e}{T_D}\right). \quad (13.56)$$

We shall conclude from Eq. 13.56 that initially cold electrons will be heated to a temperature not far from their equilibrium value before an appreciable fraction of the ions has reacted. For example, if $T_D = 100$ kev and $T_e = 0$ initially, and if we neglect all losses (in this case radiation only), we shall find that, by conservation of energy, $T_D = 60$ kev and $T_e = 40$ kev some time later. At that time, $[d(\ln T_e)/dt]/[d(\ln n_D)/dt] \approx 4$ from Eq. 13.56. There is always some advantage gained in reducing radiation by injecting the electrons at low energy. But we see that over most of the burnup time the electrons will have been heated by the ions near to an equilibrium temperature. This temperature is determined by energy input from the ions and output by radiation.

Given the fact that the electrons are fairly hot, we now inquire how the α-particles divide their energy loss. The losses to the ions and to the electrons is given by the last two terms of Eq. 13.54. Thus

$$\frac{\alpha\text{-loss to } (D + T)}{\alpha\text{-loss to electrons}} = \frac{210}{(1 - T_e/T_\alpha)} \left(\frac{T_e}{T_\alpha}\right)^{3/2}. \quad (13.57)$$

The smallest value of this ratio occurs for the largest value of T_α, 2350 kev. If we assume the reasonable value $T_e = 50$ kev (see previous examples), the ratio is about 0.7. This ratio becomes 1 for $T_\alpha/T_e = 36$, which is satisfied, for example, by $T_e = 50$ kev, $T_\alpha = 1800$ kev or $T_e = 60$ kev, $T_\alpha = 2160$ kev. In all likelihood, the electron temperature will rise somewhat, and certainly T_α will decrease as the fuel burns. We expect then that more than half the α-particle energy will be transferred to the ions.

The two energy-transfer terms whose ratio appears in Eq. 13.57 are also contained in Eq. 13.51, which describes the rate of change of α-particle pressure. The upper limit of this partial pressure can be investigated by setting $dp_\alpha/dt = 0$. Then from Eq. 13.51, we find

$$\frac{n_\alpha T_\alpha}{n_D T_D} = \frac{p_\alpha}{p_D} = 2.4 \times 10^{-3}\left(\frac{T_\alpha^{3/2}}{T_D}\right)\left[1 - 0.96\,\frac{n_\alpha n_e T_\alpha}{n_D{}^2 T_e^{3/2}}\left(1 - \frac{T_e}{T_\alpha}\right)\right], \quad (13.58)$$

when the α-pressure is a maximum. We do not know exactly when during the cycle this maximum occurs, but Eq. 13.58 allows us to estimate the time. Since $p_\alpha/p_D > 0$, the term in brackets must be positive. We are not concerned with the apparently spurious solution $T_e = T_\alpha$, which would apply only to thermal equilibrium, long after burning is complete. Since we expect T_α to be considerably greater than T_e, we conclude that the maximum α-particle pressure is reached early in the cycle, when $n_\alpha/n_D \ll 1$. This maximum pressure must be less than $2.4 \times 10^{-3}\, T_\alpha^{3/2}/T_D$ from Eq. 13.58. If we set $T_D = 60$ kev, $T_\alpha < 2350$ kev, we find

$$p_\alpha/p_D < 4.5. \quad (13.59)$$

The total pressure of deuterons, tritons, and electrons will be perhaps $3p_D$. Thus the α-particles will raise the total pressure by a factor of 2.5 at most. Furthermore, if the pressure rise is to be as great as this, it must occur very early in the cycle, when the pressure is low and before the fuel has heated itself very much by fusion. Over the remaining part of the cycle, the adverse effects of the α-particle pressure will be less. We conclude therefore that, while the plasma temperature and total pressure will both rise during the burning cycle, the extra effort required to confine the α-particles will be moderate. The result is to be compared with the mirror calculation of Sec. 13.4. The orders of magnitude are similar. Here the α-particles collect from reaction of a single charge of fuel with relatively high burnup f_b; in the previous example, f_b was low, but the throughput of ionic fuel gave a long-time accumulation.

The information on the system obtained up to here may be used to predict further properties of the system. We note from Eq. 13.53 that the rate of transfer of energy to electrons is roughly a linear function of the pressure of each kind of ion. We have also found that the electrons adjust their temperature rapidly compared with the burnup time. Then the radiation from the plasma will be rather more than doubled solely on account of energy transfer from the α-particles.

An important question is that of the reaction sustaining itself without our supplying additional energy. This question is answered by inspection of Eq. 13.55. Let us assume that during the early reaction stage we supply power to keep T_D (and T_T) constant and inquire how long this power must be supplied. We can stop at a time when the two terms on the right of Eq. 13.55 cancel. Since $n_e = 2n_D$, we must add heat until the α-density reaches

$$\frac{n_\alpha}{n_D} \approx 3.7 \times 10^{-3}\left(\frac{T_\alpha}{T_e}\right)^{1/2}\left(\frac{T_D}{T_e} - 1\right). \tag{13.60}$$

The solution for T_α and T_e would involve considerable effort, but Eq. 13.60 has a simple upper bound. We know that $T_\alpha < 2350$ kev; if we set $T_D = 100$ kev, then our previous experience shows that $T_e \approx 50$ kev, so $n_\alpha/n_D \leqslant 2.5\%$ when the reaction is self-sustaining. More likely, $n_\alpha/n_D \approx 1\%$ at that point. In any event, we must supply only a modest amount of energy for a brief period early in the burning cycle. After that, the system runs by itself. By inspection of Eqs. 13.53 and 13.55, we find that this early heating will be less than or equal to the early bremsstrahlung and cyclotron-radiation rates. In a reasonably designed system, these losses should be small.

The system will not be extinguished by the radiation losses under any normal circumstances. The reason is that throughout most of the fusion cycle, the electron temperature is maintained by the α-particles rather than by deuterons and tritons; we have just seen that the deuterons and tritons are actually heated also, thus heating the electrons further. Therefore, the radiation will increase significantly during the cycle. This method is merely a roundabout way of extracting the thermonuclear heat. By this means, the fusion energy appearing in the α-particles is removed by radiation, which of course must eventually take place in a confined system.

In summary, we find that a completely confining system, if it could be developed, appears superior to a mirror. Such a system could be run with superconducting coils. We note that $w_{DT} \propto n^2 \propto B^4$ at fixed T and β (Secs. 4.10 and 7.3). Similarly from Eq. 11.83, $w_c \propto nB^2 \propto B^4$. On the other hand, the cyclotron-radiation escape factor $K_{\mathscr{L}}$ decreases with increasing n/B (Fig. 11.7). Hence $w_{c\mathscr{L}}/w_{DT}$ decreases with increasing induction, other parameters being kept fixed. Alternately, for constant $w_{c\mathscr{L}}/w_{DT}$, the system can be made smaller at higher induction. At an induction of 5 webers/m^2, the plasma might be about 0.5-m diam and have much higher power density than the mirror discussed in Secs. 13.4 and 13.5.

PROBLEMS

1. Calculate the mean free path for capture of neutrons in Li^6 and natural Li if the neutrons are (*a*) thermal, (*b*) at 10 ev.

2. Calculate the approximate thickness of BeF_2 required to give 10% multiplication of 14.1-Mev neutrons, assuming no competing processes.

3. Find the ratio r_2/r_1 of outer and inner coil radii for which a uniform current density gives 10% greater loss than a $1/r$ current density.

4. Calculate the radiation power per m^2 between two plates at 77° K and 4° K, assuming 95% reflectivity of the surfaces.

5. It is desired to support a superconducting coil by foamed polystyrene. Assume the density of the foam to be 0.03 g/cm³, and assume for simplicity that the material is arranged as the walls of cubic cells. Calculate the thermal conductivity of the material, neglecting radiation and assuming that the interstices have been evacuated.

6. Assume that 1 megawatt/m² of γ-rays are generated in the blanket of a thermonuclear reactor because of inelastic scattering of fast neutrons. The distribution of these γ-rays is uniform per energy interval over the range 1 to 2 Mev and is zero outside the range. Assume that the moderator does not absorb the γ-rays, and assume that we place a lead absorber between the moderator and external superconducting coils. The superconducting shell with its support is equivalent to a 10^{-1}-m thickness of iron. Calculate the thickness of lead so that the γ-rays deposit 1 watt/m² in the superconducting structure. Hint: Assume a thickness and use graphical methods.

7. It is desired to maximize the fusion power density of a D-T reactor at constant temperature and constant total hydrogenic density n. Show that $n_D/n_T = 1/(1 - 2K)$, and $w = n^2 \overline{\sigma v}_{DT} U_{DT}/4(1 - K)$, where the parameter $K = \overline{\sigma v}_{DD} U_{DD}/2\overline{\sigma v}_{DT} U_{DT}$ at the maximum.

8. Show that the current distribution for maximum induction per watt throughout a solenoidal coil, inner radius r_1, outer radius r_2, is given by Eq. 13.5. Hint: The power loss per unit axial induction must be constant from all contributions; were this not so, we are wasting power somewhere, and a more efficient current distribution could be found.

9. It is proposed to put field coils inside a thermonuclear reactor blanket and run them at 400° C in order to recover the thermonuclear neutron energy. The resistivity of copper is 5.9×10^{-8} ohm-m, including the space factor. Let $B = 8$ webers/m², $\beta = 0.4$, $T = 40$ kev, and $W_m/W'_{DT} = 0.1$. Find the radial thickness of the coils if the plasma radius is (*a*) 0.5 m; (*b*) 2.0 m.

REFERENCES

1. K. O. Hintermann and R. Wideroe, *Proc. 2nd U.N. Conf. on Peaceful Uses of Atomic Energy*, Vol. 32, United Nations, Geneva (1958), p. 440.
2. J. L. Craston, R. Hancox, A. E. Robson, S. Kaufman, H. T. Miles, A. A. Ware, and J. A. Wesson, *Proc. 2nd U.N. Conf. on Peaceful Uses of Atomic Energy*, Vol. 32, United Nations, Geneva (1958), p. 414.
3. L. J. Donadieu and S. H. Autler, *Research Laboratory of Electronics Quarterly Progress Report*, No. 59, M.I.T., Cambridge, Mass. (October 15, 1960), p. 27.
4. S. H. Autler and D. B. Montgomery, *M.I.T. Lincoln Laboratory Quarterly Progress Report on Solid State Research*, Lexington, Mass. (July 15, 1960), p. 63.
5. S. H. Autler, *Rev. Sci. Instr.*, **31**, 369 (1960).

6. J. E. Kunzler, E. Buehler, F. S. L. Hsu, B. T. Matthias, and C. Wahl, *J. Appl. Phys.*, **32,** 325 (1961).
7. J. E. Kunzler, E. Buehler, F. S. L. Hsu, and J. H. Wernick, *Phys. Rev. Letters*, **6,** 89 (1961).
8. G. Boulegue, P. Chanson, R. Combe, M. Feix, and P. Strasman, *Proc. 2nd U.N. Conf. on Peaceful Uses of Atomic Energy*, Vol. 32, United Nations, Geneva (1958), p. 409.
9. W. N. Huss, *Reaction Kinetics of Deuterium-Tritium Mixtures*, S.M. Thesis, Department of Nuclear Engineering, Massachusetts Institute of Technology (1960).
10. T. H. Jensen, O. Kofoed-Hansen, and C. F. Wandel, *Proc. 2nd U.N. Conf. on Peaceful Uses of Atomic Energy*, Vol. 32, United Nations, Geneva (1958), p. 431.

14

Devices based upon the pinch effect

In this and the final two chapters, we shall describe more precisely some of the specific plasma confinement and heating schemes recently explored. The basic principles of these schemes have been presented in many of the preceding chapters. We shall also describe typical experiments as illustrations both of these principles and of the recent state of the art. No attempt is made to be complete or even to include all important developments. Our objective is rather to give the interested reader some feeling for current thermonuclear research with plasmas and to enable him to pursue further details for himself.

We shall cover pinch-like, mirror-like, and stellarator systems, the first being the subject of the present chapter. Pinch-like systems are defined as those in which the principal confining induction arises from the plasma current itself. Without the plasma current, there is no confining mechanism. In mirror devices and stellarators, on the other hand, a vacuum field is established by external conductors to provide the confining means.

We have already obliquely encountered examples of the simple pinch and shall enlarge upon the elementary concepts.† In principle, the pinch is no more than a current-carrying conductor compressed radially by its girdling magnetic induction. It might appear that, because the plasma is both heated and confined by a magnetic pressure of its own creation,

† Chapter 5, Problems 8, 9, and 10.

these processes will be the least expensive in power and would lead to a high value of $\beta = nkT/(B^2/2\mu_0)$. The idea is partly true; unfortunately the simple pinch is very unstable, as simple illustrations will show. Thus, we are led to supply additional stabilizing fields from external sources at the cost of reducing β. Even more unfortunately, further analysis shows the system to be still unstable, and no certain cure for the standard pinched plasma is now at hand.

The pinched plasma is in essence a pulsed system, and generally it is pulsed rapidly. The apparent instabilities mentioned above, plus our pessimism as shown in Chap. 13 for the use of fast pulsed systems in any practical device, lead us virtually to dismiss the conventional pinch as a means for producing controlled thermonuclear power.

In spite of these criticisms, study of these systems is rewarding. First, we find good examples of experiments illustrating the theory. Second, the sheet or inverse pinch systems, which are derived from the standard pinch discharges, appear to be much more stable than the conventional ones. Third, such systems, whether useful for thermonuclear power or not, are useful experimental devices for studying plasma physics.

The experimental methods used to measure the properties of plasmas are important. Magnetic and electrostatic probes, microwave propagation, and optical spectroscopy are among the techniques used. We shall not devote specific sections to these techniques but shall show their application by example in discussing appropriate experiments. The principles underlying each method have been presented in the basic theory already covered.

14.1. The Simple Pinch

Confinement in the simple cylindrical pinch originates from the $\mathbf{j} \times \mathbf{B}$ force, which arises from the magnetic induction created by the plasma current itself. For this reason, the hydromagnetic formulation of Eqs. 7.3 and following is most appropriate. Figure 14.1 shows a section of a pinch in cylindrical coordinates. We assume that the pinch is straight and that a total current I_0 flows in the $+z$-direction. In order to establish this current physically, electrodes must be placed at some remote positions $\pm z$, but we ignore the end effects. Alternately, the configuration

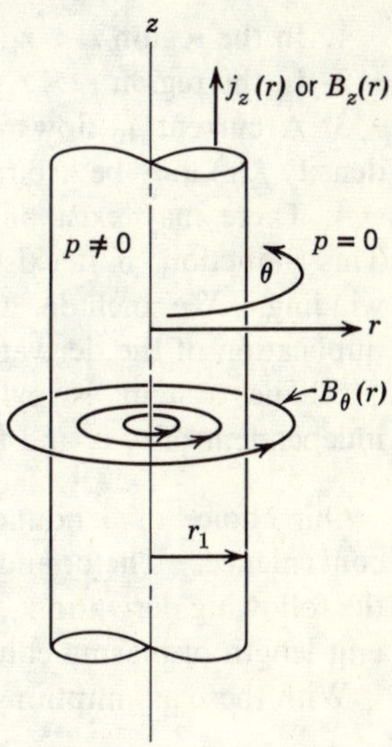

Fig. 14.1. Nomenclature for a simple pinch.

may be that of a torus, as in Fig. 14.2. In that case, the torus is the one-turn secondary of a transformer; a pulse applied to the primary then induces a voltage around the secondary torus and establishes the ring current I_0 in the gas. The early stages of the gas breakdown were described in Sec. 4.7. Analogous to our neglect of end effects in the straight pinch, we assume that the minor radius is small compared with the major radius and neglect the curvature of the pinch column.

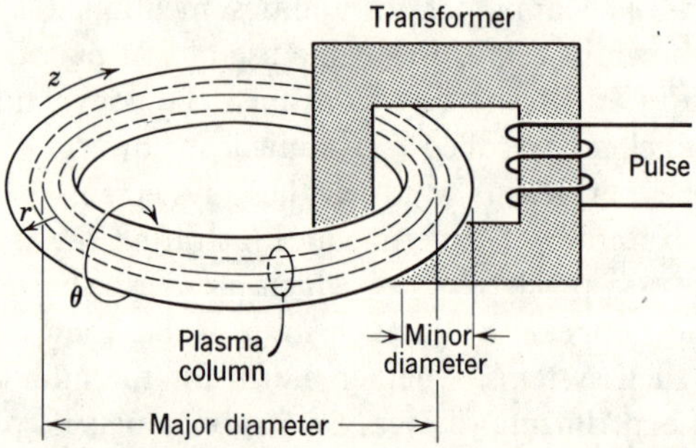

Fig. 14.2. Toroidal pinch system.

In deriving the conditions for a stationary pinched plasma, we make the following reasonable assumptions:

1. In the region $r < r_1$, the plasma pressure p is not zero.
2. In the region $r > r_1$, the plasma pressure is zero.
3. A current I_0 flows in the z-direction in the plasma. The current density $j_z(r)$ may be arbitrary but depends on r only.
4. There may exist an induction $B_z(r)$ and corresponding current j_θ. This induction, if it exists, must be supplied by an external solenoidal winding. We include it here purely for later convenience, to avoid duplication of the derivation when we consider its stabilizing effects.
5. The system is cylindrically symmetric, and all quantities are independent of θ, z, and the time t.

Our choice of a boundary at $r = r_1$ is in fact merely for illustrative convenience. The boundary may be removed to infinity without affecting the following derivation. In that case, we demand only that the mass per unit length of plasma column be finite.

With these assumptions, Eqs. 7.3 and 7.4 are combined to give

$$\frac{dp}{dr} = -j_z B_\theta + j_\theta B_z = -j_z B_\theta - \frac{B_z}{\mu_0}\frac{dB_z}{dr}. \tag{14.1}$$

Also from Ampère's law, we have

$$2\pi r B_\theta = \mu_0 \int_0^r dr' \, 2\pi r' j_z(r') = \mu_0 I_z(r), \tag{14.2}$$

where $I_z(r)$ is the total z-current inside a cylinder at radius r. The two equations give

$$2\pi r \frac{d}{dr}\left(p + \frac{B_z^2}{2\mu_0}\right) = -\mu_0 I_z(r) j_z(r). \tag{14.3}$$

But $2\pi r j_z(r) = dI(r)/dr$, the radial rate of change of enclosed current. Thus j_z is eliminated in Eq. 14.3, and the equation is integrated by parts from the center $r = 0$. We obtain

$$4\pi^2 r^2\left(p + \frac{B_z^2}{2\mu_0}\right)\Bigg|_{r=0}^{r=r_1} - 4\pi \int_0^{r_1} dr \, 2\pi r\left(p + \frac{B_z^2}{2\mu_0}\right) = -\mu_0 I_0^2/2. \tag{14.4}$$

Now the first term is zero at $r = 0$, and $r^2 p$ is zero at r_1 even if $r_1 \to \infty$, because of the remark following assumption 5 above. The induction B_z may or may not be zero at $r = r_1$, depending on the way the plasma was initially established. The remaining integral is a measure of the total energy per unit length being confined.

Let us now set $B_z = 0$, corresponding to no induction imposed by external currents, and put

$$p = n_e k T_e + n_i k T_i, \tag{14.5}$$

corresponding to Maxwellian velocity distributions of electrons and ions; since the plasma is neutral to a good approximation, the average temperature T is

$$T = (T_e + T_i)/2. \tag{14.6}$$

If this average temperature is independent of position, Eq. 14.4 becomes

$$\mu_0 I_0^2/8\pi = NkT, \tag{14.7}$$

where N is the total number of particles per unit length of the column.

The desired result, Eq. 14.7, is independent of the form of $j_z(r)$. For a given temperature and density of charge, the current I_0 given by Eq. 14.7 is just sufficient to provide radial equilibrium of the column. If I_0 is less, the column expands; if it is greater, the column is compressed. This scheme is used both to confine and heat the plasma. At the start of the cycle, an initial relatively cold plasma fills a long tube; high voltage is suddenly applied to its ends from a capacitor bank through low-inductance circuits. A high current flows in excess of that required by Eq. 14.7.

The plasma is compressed to a thin filament and heated by the compression (see Sec. 7.5).

The temperature T then rises, and presumably a pinch is established at the degree of compression satisfying Eq. 14.7. Unfortunately the equilibrium is not stable, as we shall see in the next section.

The result, Eq. 14.7, can be expressed in an alternate interesting way. Almost all the current I_0 will be carried by fast electrons; let their axial drift velocity be v_z. Then since $N = N_e + N_i \approx 2N_e$, we find that Eq. 14.7 can be written

$$(\mu_0 e^2/4\pi m_e)N_e = 4kT/m_e v^2. \tag{14.8}$$

The quantity $\mu_0 e^2/4\pi m_e$ is the classical electron radius, 2.81×10^{-15} m. Roughly speaking, if the drift energy of the electrons is comparable with the mean particle energy, the electron density per unit length of column is about 3.5×10^{14}/m, the inverse of the classical electron radius.

Any number of models of the static pinch can be made up, depending on the form of $j_z(r)$ or $p(r)$. One simple distribution is that corresponding to a constant electron drift speed v_z throughout the plasma. It is called the Bennett distribution[1] after the discoverer of the pinch effect and is treated in Problem 2. Another particularly simple case is that obtained by considering the current to be confined in a thin sheath of thickness $\mathfrak{d}$ at the plasma edge. In this case,

$$p_{\text{inside}} = \mu_0 I_0^2/8\pi^2 r_1^2 = (B^2/2\mu_0)_{r=r_1} = \text{const}, \tag{14.9}$$

since the pressure within the sheath is independent of position. This result is to be expected from Sec. 7.3. The magnetic pressure at the surface balances the plasma pressure inside, and Eq. 7.13 is satisfied out to the edge of the sheath. These parameters are shown in Fig. 14.3.

This model with all the current in a thin surface layer is a good physical approximation in some devices. If the plasma conductivity is high, the magnetic induction cannot penetrate quickly. If the current is suddenly increased, it must flow at the surface.

A dynamical theory has been developed for this case of infinite conductivity.[2] In actuality, fast pinches develop in about 1 μsec; inertial and inductive effects cannot be neglected. Figure 14.4 shows the configuration. The plasma cylinder is confined within conducting walls of radius r_0. Application of a potential at the ends of the cylinder causes the current I to flow in a thin shell at the surface of the plasma cylinder. The instantaneous radius of the plasma is r_s, and the inductance L of the coaxial system per unit length is

$$L = (\mu_0/2\pi)\ln(r_0/r_s) \quad \text{henrys/m}. \tag{14.10}$$

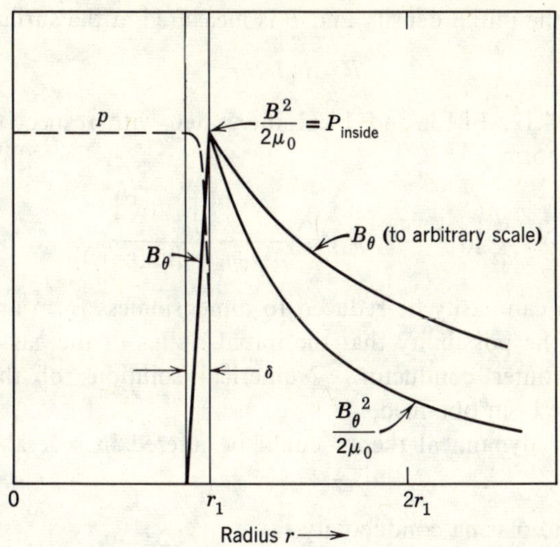

Fig. 14.3. Variation of magnetic field B and pressure p for a simple pinch with current sheet of thickness δ at $r = r_1$.

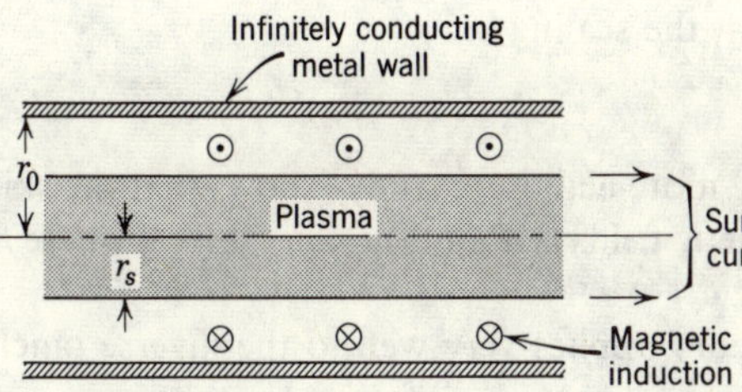

Fig. 14.4. Configuration of a dynamic pinch.

We therefore obtain the electric field $E = -\partial(LI)/\partial t$ applied to the system as

$$E = -\frac{\partial}{\partial t}\left[\frac{\mu_0 I}{2\pi}\ln\left(\frac{r_0}{r_s}\right)\right]. \tag{14.11}$$

The acceleration of the material is now required. Consistent with infinite conductivity, we use a snowplow model in which all material swept up by the imploding surface is compressed into a thin layer at the surface r_s. The material inside r_s is unaffected. We also assume that the material is cold, and that therefore the thin layer exerts negligible tangential stress on itself. Newton's second law is then

$$\frac{d}{dt}\left[\pi\rho_{m0}(r_0^2 - r_s^2)\dot{r}_s\right] = -2\pi r_s\left(\frac{B^2}{2\mu_0}\right), \tag{14.12}$$

where ρ_{m0} is the initial density and B is measured at the surface; but

$$B = \mu_0 I / 2\pi r_s. \tag{14.13}$$

From Eqs. 14.12, 14.13, and 14.11 integrated with respect to time, we find the final form

$$\frac{d}{dt}[(r_0^2 - r_s^2)\dot{r}_s] = -\frac{\left(\int_0^t dt\, E\right)^2}{\mu_0 \rho_{m0} r_s [\ln (r_0/r_s)]^2}. \tag{14.14}$$

The equation can easily be reduced to dimensionless form and modified to allow for the possibility that the initial radius of the gas is less than that of the outer conductor. Numerical solutions of the modified equation have been obtained.[2]

This simple dynamical theory could be altered in several respects to include:

1. The finite plasma conductivity.
2. The reverberation of shocks from the center. This reverberation has been demonstrated both by more elaborate theories and by experiments.
3. The finite Larmor orbits.

From Eq. 14.14, the scaling law,

$$\dot{r}_s \propto (E^2/\rho_m)^{1/4}, \tag{14.15}$$

is apparent. The more elaborate study[2] shows that the skin depth for the electric and magnetic fields is about $(m_e/2n_0\mu_0 e^2)^{1/2}$, where n_0 is the density of electrons.

This simple theory applies very well to the inverse pinch (Sec. 14.9).

14.2. Simple Pinch Instabilities

The simple pinch has a number of serious instabilities. We shall qualitatively describe a few of them and the measures that can be taken to help stabilize the configuration. We shall not dwell long on these elementary corrective measures, because they do not appear sufficient. The more sophisticated analysis of Secs. 14.3 and following will establish this fact and present more severe stability criteria. We note first that the lines of induction confining the pinch shown in Fig. 14.1 (or 14.4) bend around the plasma in the unstable sense shown in Fig. 12.3 and described in Secs. 12.2 and following. Therefore, we expect that instabilities will develop.

An analysis has been performed by Rosenbluth[3] on the stability of the pinch against perturbations of the general form $\mathcal{Re} \exp i(\pm m\theta + \kappa z)$, when θ and z are the coordinates in Fig. 14.1. We shall take up analyses of this sort in Secs. 14.3 and following. All of the modes $m \geqslant 2$ correspond to longitudinal fluting of the pinch column. The mode $m = 1$, $\kappa \neq 0$ corresponds to a radial displacement of the pinch column, the amount varying sinusoidally with z. The appearance is that of a kinked column, whence the name kink instability. The mode $m = 0$, $\kappa \neq 0$ corresponds to an azimuthally symmetric disturbance in which the column tends to neck itself off at one or more axial positions, whence the name sausage instability. These latter two modes have simple physical interpretations.

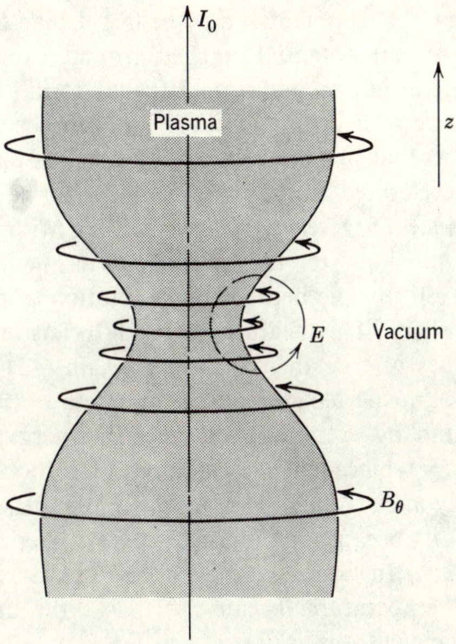

Fig. 14.5. The $m = 0$ (sausage) instability of the pinch.

The sausage instability is shown in Fig. 14.5, which is similar to Fig. 12.9. Since the current I is continuous, the induction at the neck is higher than elsewhere. The magnetic pressure $B_\theta^2/2\mu_0$ is thus greatest at the thinnest part of the plasma and causes the plasma to narrow down further at that part. Material at the neck is squeezed out into the bulges; there is no adequate restoring force.

The $m = 0$ instability induces large electric fields because of the fast inward motion of the B lines at the narrow portions. For one explanation, consider the loops shown in Fig. 14.5. At first the flux through it increases as the plasma surface moves across; then after the surface has passed, the flux remains sensibly constant. An electromotive force is induced which accelerates ions across the gap between the adjacent fat portions of the plasma. A detailed calculation of the effect has been given by Anderson and his co-workers,[4] but a qualitative argument suffices to show that the voltage can be high. Let the radius of the plasma that is in process of narrowing down be r. Then $B_\theta \approx \mu_0 I/2\pi r$. This induction extends over a z-distance of the order of r, so the total flux Φ through a suitable loop is about $\mu_0 Ir/2\pi$. The flux builds up because the plasma is being driven in, and the velocity $\dot{r} = r/\tau$ must be comparable to the ion velocity v_i. Since $\dot{\Phi} \approx \Phi/\tau$, we find a loop voltage $V \approx \mu_0 I v_i/2\pi$. For example, if $I = 2 \times 10^5$ amp, $v_i = 10^5$ m/sec, then $V \approx 4 \times 10^3$ volts. This potential may be larger than that applied to the ends of the tube (or the potential induced around a toroidal pinch).

This local acceleration mechanism is believed to be responsible for the specious appearance of neutrons in deuterium pinches where the average ion energy is so low that surely no true thermonuclear plasma is produced. Deuterons are accelerated toward the cathode in these local high-field regions; they induce D-D reactions by colliding with the deuterons in the nearby fat portions of the plasma. The neutron spectrum is characteristically anisotropic because of the motion of the center of mass of the reacting pair. This feature is used to distinguish these false neutrons from ones of true thermonuclear origin. The phenomenon is also used as a diagnostic technique to show the $m = 0$ instability. The anisotropy is found by examining the neutron energy distribution with directional detectors placed at the anode and cathode. Although the neutrons definitely arise from D-D reactions in the plasma, the process is not of interest to us because it cannot be extrapolated to produce useful amounts of power. In addition, the process confuses the diagnosis of the true plasma temperature, because it masks the appearance of any true thermonuclear neutrons.

The $m = 0$ sausage instability can be cured by application of an axial induction B_z to the system before the gas is ionized. The intense electric field, established shortly after the axial induction, ionizes the plasma and makes it highly conducting. The induction is now trapped within the gas. The outward magnetic pressure of this axial induction is the greatest at those points where the plasma is narrowest. Thus, an axial induction can prevent growth of the sausage instability, but at the cost of decreased compression. In this connection, see Problem 6. Figure 14.6 shows the configuration. The axial induction has already been taken into account in our static force balance equation (Eq. 14.4).

An experiment showing this instability and the effect of adding the axial induction B_z is described briefly in Sec. 14.7.

The other instability having a simple physical interpretation is the $m = 1$ or kink instability. It is shown in Fig. 14.7. The magnetic pressure is higher at the inside of a curve than at the outside. Thus the instability grows and typically turns the pinch column into a corkscrew with pitch angle about 45°. Figure 14.8 shows photographs of this instability growing with time.[5] Here, the pinch is established by induction in a torus, and one section of the metal wall is perforated with small circular windows (hence the pattern). Each photograph pair consists of two views

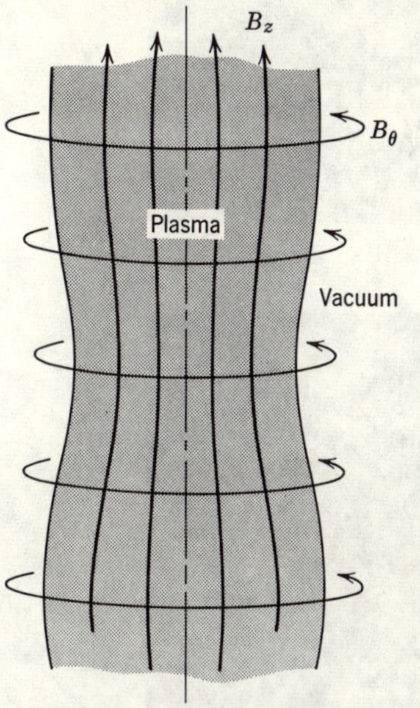

Fig. 14.6. B_z-stabilized pinch.

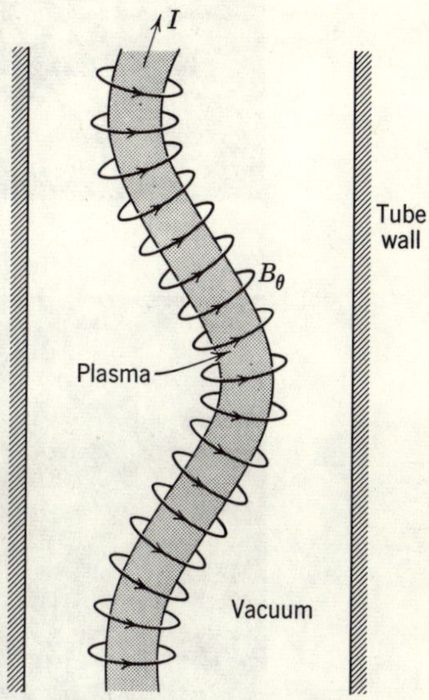

Fig. 14.7. Kink instability.

at 90°. Exposure time was a few microseconds, and the kink growth was typically 2–5 × 10^2 m/sec, increasing both with applied field and with lower pressure.

This $m = 1$ instability can be partially cured by making the walls of metal, so that as the kinked plasma approaches the wall the induction trapped between the plasma and the wall is compressed. If this mechanism is to be effective, the degree of compression of the plasma cannot be very great. A radial reduction of 3 or 4 to 1 seems attainable at best. Again, some stability is achieved but at a cost of limiting the compression. An experiment to observe the effect of maintaining the conducting walls very close to the plasma has been performed by Sawyer and co-workers.[6] Magnetic probes placed in the tube indicate that the

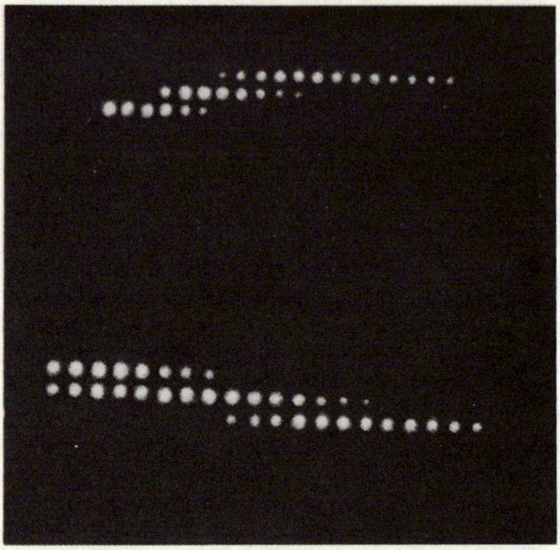

(*a*)

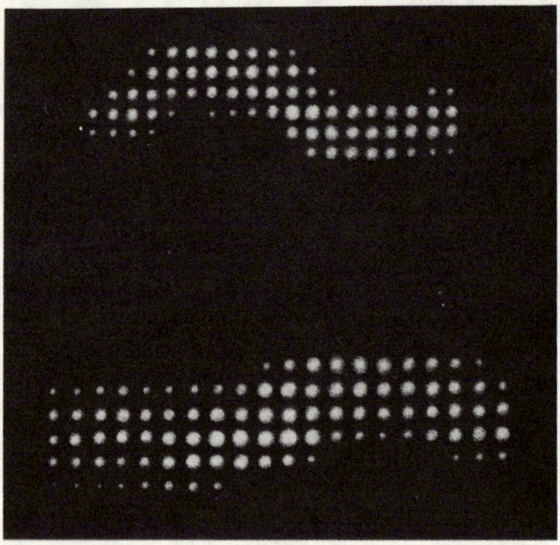

(*b*)

Fig. 14.8. Example of the development of the kink instability in a metal torus with argon gas. (*a*) 10 μsec after initiation of discharge; (*b*) 20 μsec after initiation of discharge. After T. E. Allibone and co-workers.[5]

plasma has the appearance of a straight cylinder from which a corkscrew has been excised. The current rotates screwlike as it runs down the tube.

In addition to the $m = 0$ and $m = 1$ instabilities, a number of other morbid effects may occur. For example, see Fig. 15.4 for end-on photographs of a plasma compressed by a different but related scheme.

Beside stability, another simple criterion must be satisfied if the pinch is to be successfully established: its formation must be fast enough that the plasma does not diffuse across the magnetic induction. While it is easy to state the criterion, its proper evaluation is not generally simple. At early times in the pinch, the electron and ion temperatures are low, and the density gradients extend over many Larmor radii. Therefore if the plasma is highly ionized, the diffusion takes place only by virtue of electron-ion or ion-electron collisions. Like-particle collisions play a negligible role (see Sec. 10.7). We may approximate the diffusion time by the time

$$\tau = L^2/D_\mathsf{T}, \tag{14.16}$$

where L is a characteristic distance, and D_T is the transverse diffusion coefficient relating to particle density gradients. This coefficient is derived directly from $\mathsf{D}'_{\alpha\beta ep}$ (Eq. 7.37) for electrons, or $D'_{\alpha\beta ip}$ (Eq. 7.40) for ions, which equations relate particle currents to pressure gradients of each species. With the insertion of $\mathsf{D}_\mathsf{T} = kT\mathsf{D}'_{\alpha\beta}$ for either particle, Eq. 14.16 for τ becomes

$$\tau_{e,i} = L^2B^2/\eta nkT_{e,i}. \tag{14.17}$$

The ratio of the two τ's is then approximately

$$\tau_i/\tau_e \approx T_e/T_i \tag{14.18}$$

for singly charged ions.

The magnitude of each time τ is proportional to the first power of the particle speed. By insertion of numerical values, we find that if the plasma is at low temperature (for example 1 ev), the diffusion time is in the order of a few milliseconds, if the density is $10^{20}/\mathrm{m}^3$, and $B = 1$ weber/m^2. Such a time is generally long compared with the time of formation of the instabilities described above. Problem 7 illustrates the point. For pinches of thermonuclear temperatures, densities, and sizes, the diffusion takes seconds. However, in this case, the plasma boundary may be sharp, and like-particle diffusion (Sec. 10.7) may predominate.

The diffusion and intermixing of the B_z and B_θ inductions of Fig. 14.6 as the pinch disassembles will further heat the plasma by a considerable amount (see Problem 8).

★ 14.3. Hydromagnetic Stability Criterion for a Linear Pinch

More sophisticated criteria for the stability of a pinch have been developed by Suydam[7] and by Newcomb.[8] Their work is discussed in the following four sections. Each adapts the hydromagnetic formulation of Secs. 12.5 and following to the linear pinch and derives an equation for the change in potential energy of the plasma. The minimum value of the potential energy, hence the least stable configuration, must then be found. To this end, we develop the formalism for minimizing integrals in Sec. 14.4. This formalism is called the Euler-Lagrange equation. In Secs. 14.5 and 14.6 we apply the method. We shall first derive Suydam's criterion in detail, as an example of the methods employed. Newcomb's analysis is much longer and cannot properly be compressed for inclusion here. However, after the discussion of Suydam's work, we shall review part of Newcomb's analysis in Sec. 14.6.

Suydam's criterion relates to the radial pressure gradient, which must be severely limited if flutes are not to grow along the spiraling lines of induction. In this connection, we note that the azimuthal component B_θ of the magnetic induction and the stabilizing axial component B_z will mix at least at the edge of the plasma, giving lines of induction that spiral along the plasma. In this section we derive the integral that is to be minimized to determine the state of lowest energy. The work of this section then merely consists in applying the integral in Eq. 12.105 in some detail to the linear pinch.

We assume the plasma to be bounded by a rigid, perfectly conducting cylinder. Accordingly, we choose cylindrical coordinates, assume that

$$\mathbf{B} = (0, B_\theta, B_z), \tag{14.19}$$

and further that B_θ, B_z, and p (the unperturbed quantities) are functions of r alone. Any physically realizable displacement can be represented as the superposition of Fourier components at least for sufficiently small displacements, since these will be linear. Thus let

$$\boldsymbol{\xi} = \sum_{m,\kappa} \boldsymbol{\xi}_{m,\kappa}(r) \exp\,[i(m\theta + \kappa z)], \tag{14.20}$$

where for reality

and

$$\left.\begin{aligned} \boldsymbol{\xi}_{-m,\kappa} &= \boldsymbol{\xi}^*_{m,\kappa} \\ \boldsymbol{\xi}_{m,-\kappa} &= \boldsymbol{\xi}^*_{m,\kappa} \end{aligned}\right\} \tag{14.21}$$

are required. Here κ is some real number, and m is some integer. Since the displacements are presumably single valued and continuous, the present functions will represent the displacements as accurately as desired.

The plasma boundary is assumed to be fixed. Therefore the stability criterion is that δQ_F must be positive for any displacement whatsoever away from the equilibrium configuration. To apply this criterion, we must evaluate certain functions of $\mathbf{G}$. By Eq. 12.67,

$$\mathbf{G} = \nabla \times (\boldsymbol{\xi} \times \mathbf{B}). \tag{14.22}$$

In the determination of stability it suffices to examine those $\boldsymbol{\xi}$'s giving the least stability. Now consider (see Eq. 12.105)

$$\delta Q_F = \tfrac{1}{2}\int dr_0\,[(G^2/\mu_0) - \mathbf{j}_0\cdot\mathbf{G}\times\boldsymbol{\xi} + \gamma p_0(\nabla_0\cdot\boldsymbol{\xi})^2 + (\nabla_0\cdot\boldsymbol{\xi})(\boldsymbol{\xi}\cdot\nabla p_0)]. \tag{14.23}$$

The last two terms may be written as

$$(\gamma - 1)p(\nabla\cdot\boldsymbol{\xi})^2 + (\nabla\cdot\boldsymbol{\xi})(\nabla\cdot p\boldsymbol{\xi}). \tag{14.24}$$

The first term is positive definite. For small perturbations, log ξ changes more rapidly than log p, so that $\nabla\cdot(p\boldsymbol{\xi})$ has the same sign as $\nabla\cdot\boldsymbol{\xi}$. The terms involving $\nabla\cdot\boldsymbol{\xi}$ thus give a nonnegative contribution to δQ_F, and we shall examine only the worst perturbations, for which $\nabla\cdot\boldsymbol{\xi} = 0$.

We must then evaluate the quantity

$$(G^2/\mu_0) - \mathbf{j}_0\cdot\mathbf{G}\times\boldsymbol{\xi}. \tag{14.25}$$

From Eqs. 14.22 and 14.25 we note that the integrand of Eq. 14.23 will be bilinear in $\boldsymbol{\xi}$ and its first-order derivatives. Derivatives of $\boldsymbol{\xi}$ with respect to any spatial variable will lead to products of exp $i(m\theta + \kappa z)$ with exp $i(m'\theta + \kappa' z)$. Expression 14.25 must be integrated over θ from zero to 2π and over the length of the assembly. Orthogonality of our functions then indicates that there will be a contribution to δQ_F if and only if $m' = -m$, $\kappa' = -\kappa$; that is, only terms composed of complex conjugate factors will contribute. In these terms the exponential dependence integrates to unity with appropriate normalization. If stability occurs in each possible mode, the pinch will be stable.

Since the only terms contributing to the integral in Eq. 14.23 involve the product of a particular mode with its own complex conjugate, we may consider each mode independently of all others. We now work out the term of $\mathbf{j}\cdot\boldsymbol{\xi}\times\mathbf{G}$ that arises from the (m, κ) mode. Since only the (m, κ) mode will be considered, the quantities $\xi_{rm\kappa}$, $\xi_{\theta m\kappa}$, $\xi_{zm\kappa}$, $G_{rm\kappa}$, $G_{\theta m\kappa}$, and $G_{zm\kappa}$ appear in all subsequent equations. For reasons of convenience, however, we suppress the subscripts m and κ in these quantities. Thus only ξ_r, ξ_θ, ξ_z, G_r, G_θ, and G_z appear, but bear in mind that these are not the components of $\boldsymbol{\xi}$ and $\mathbf{G}$ unless $\boldsymbol{\xi}$ is one pure mode. The radial component of current is zero. Therefore, the only term of $\mathbf{j}\cdot\boldsymbol{\xi}\times\mathbf{G}$ that contributes is of the form

$$j_\theta[(\xi_z G_r{}^* - \xi_r{}^* G_z) + cc] + j_z[(\xi_r{}^* G_\theta - \xi_\theta G_r{}^*) + cc], \tag{14.26}$$

where the asterisk denotes the complex conjugate of the appended quantity and cc designates the complex conjugate of the previous term. Likewise, we may work out the term in G^2 arising from thc (m, κ) mode. It is

$$|G_r|^2 + |G_\theta|^2 + |G_z|^2. \tag{14.27}$$

To simplify the present results, we find the three components, $G_j \exp i(m\theta + \kappa z)$, of the (m, κ) mode of $\mathbf{G}$ from Eq. 14.22. The result is

$$\left.\begin{aligned} G_r &= if\xi_r, \\ G_\theta &= i\kappa\alpha - \frac{d}{dr}(\xi_r B_\theta), \\ G_z &= -\left[\frac{im\alpha}{r} + \frac{1}{r}\frac{d}{dr}(r\xi_r B_z)\right], \end{aligned}\right\} \tag{14.28}$$

where

$$\alpha = \xi_\theta B_z - \xi_z B_\theta, \tag{14.29}$$

$$f = \kappa B_z + (m/r)B_\theta; \tag{14.30}$$

α and f stand for $\alpha_{m\kappa}$ and $f_{m\kappa}$. To obtain Eqs. 14.28, we use Eqs. 14.19, 14.20, and 14.22. The term in G^2 arising from the (m, κ) mode can be reduced to

$$|f\xi_r|^2 + \left|\frac{d}{dr}(\xi_r B_\theta)\right|^2 + \left|\frac{1}{r}\frac{d}{dr}(r\xi_r B_z)\right|^2 + \left(\kappa^2 + \frac{m^2}{r^2}\right)\alpha\alpha^*$$
$$+\left\{\alpha^*\left[i\kappa\frac{d}{dr}(\xi_r B_\theta) - \frac{im}{r^2}\frac{d}{dr}(r\xi_r B_z)\right] + cc\right\} \tag{14.31}$$

by means of Eqs. 14.28.

The reduction of Expression 14.26 requires further effort. We now work out the consequences of the divergenceless character of $\boldsymbol{\xi}$. Again, because of the orthogonality, each mode may be considered separately. Thus

$$\nabla\cdot[\boldsymbol{\xi}\exp i(m\theta + \kappa z)]$$
$$= \left[\frac{1}{r}\frac{d}{dr}(r\xi_r) + \frac{im}{r}\xi_\theta + i\kappa\xi_z\right]\exp i(m\theta + \kappa z) = 0. \tag{14.32}$$

The real and imaginary parts of this equation may be separated to find ξ_θ and ξ_z:

$$if\xi_\theta = i\kappa\alpha - \frac{B_\theta}{r}\frac{d}{dr}(r\xi_r), \tag{14.33}$$

$$-if\xi_z = \frac{im\alpha}{r} + \frac{B_z}{r}\frac{d}{dr}(r\xi_r). \tag{14.34}$$

With Eqs. 14.29, 14.33, and 14.34, we may reduce the term of $\mathbf{j}\cdot\boldsymbol{\xi}\times\mathbf{G}$ (Expression 14.26) arising from the (m, κ) mode by eliminating the explicit dependence on ξ_θ and ξ_z. We obtain

$$j_\theta\left\{\xi_r^*\left[\frac{i}{r}\frac{d}{dr}(r\xi_r B_z) + \frac{B_z}{r}\frac{d}{dr}(r\xi_r) + \frac{2im\alpha}{r}\right] + cc\right\}$$
$$+ j_z\left\{\xi_r^*\left[-\frac{d}{dr}(\xi_r B_\theta) - \frac{B_\theta}{r}\frac{d}{dr}(r\xi_r) + 2i\kappa\alpha\right] + cc\right\}. \tag{14.35}$$

Further simplification of our results is effected by eliminating the derivatives of B_z and B_θ with respect to r from Ampère's law, Eq. 5.13:

$$\frac{dB_z}{dr} = -\mu_0 j_\theta, \tag{14.36}$$

$$\frac{dB_\theta}{dr} = \mu_0 j_z - \frac{B_\theta}{r}. \tag{14.37}$$

We are now ready to form the sum of the (m, κ) mode of G^2/μ_0 and the (m, κ) mode of $\mathbf{j}\cdot\boldsymbol{\xi}\times\mathbf{G}$. Equations 14.36 and 14.37 are used to eliminate dB_z/dr and dB_θ/dr from Expressions 14.31 and 14.35. All terms in the coefficient of j_θ cancel; also two terms involving $\alpha\xi^* j_z$ and its complex conjugate cancel. Accordingly, the contribution of the (m, κ) mode to the integrand of Eq. 14.23 is

$$\begin{aligned}
&\frac{f^2|\xi_r|^2}{\mu_0} + \left(\kappa^2 + \frac{m^2}{r^2}\right)\frac{\alpha\alpha^*}{\mu_0} + \frac{1}{\mu_0}\left|B_\theta\frac{d\xi_r}{dr} + \xi_r\left(\mu_0 j_z - \frac{B_\theta}{r}\right)\right|^2 \\
&\quad + \frac{1}{\mu_0}\left|\frac{\xi_r B_z}{r} + B_z\frac{d\xi_r}{dr}\right|^2 + \frac{1}{\mu_0}\left\{\alpha^*\left[i\kappa\left(B_\theta\frac{d\xi_r}{dr} + \xi_r\mu_0 j_z - \frac{\xi_r B_\theta}{r}\right)\right.\right. \\
&\quad \left.\left. - \frac{im}{r^2}\left(\xi_r B_z + rB_z\frac{d\xi_r}{dr}\right)\right] + cc\right\} \\
&\quad + j_z\left\{\xi_r^*\left[-B_\theta\frac{d\xi_r}{dr} - \frac{\xi_r\mu_0 j_z}{2} + i\kappa\alpha\right] + cc\right\}.
\end{aligned} \tag{14.38}$$

It is now merely a matter of much algebra to show that Expression 14.38 reduces to

$$\begin{aligned}
&\frac{1}{\mu_0}\left(\kappa^2 + \frac{m^2}{r^2}\right)\left|\alpha + \frac{i\kappa B_\theta[(d\xi_r/dr) - (\xi_r/r)] - \dfrac{imB_z}{r}[(d\xi_r/dr) + (\xi_r/r)]}{\kappa^2 + (m^2/r^2)}\right|^2 \\
&\quad + \left(\frac{f^2}{\mu_0} - \frac{2j_z B_\theta}{r}\right)|\xi_r|^2 + \frac{B_z^2}{\mu_0}\left|\frac{d\xi_r}{dr} + \frac{\xi_r}{r}\right|^2 + \frac{B_\theta^2}{\mu_0}\left|\frac{d\xi_r}{dr} - \frac{\xi_r}{r}\right|^2 \\
&\quad - \frac{\left|i\kappa B_\theta[(d\xi_r/dr) - (\xi_r/r)] - \dfrac{imB_z}{r}[(d\xi_r/dr) + (\xi_r/r)]\right|^2}{\mu_0[\kappa^2 + (m^2/r^2)]}.
\end{aligned} \tag{14.39}$$

The only term dependent on ξ_θ and ξ_z is the one involving α. The energy change is smallest (that is, most negative or most nearly negative if positive) provided that

$$\alpha = -\frac{i\kappa B_\theta[(d\xi_r/dr) - (\xi_r/r)] - (imB_z/r)[(d\xi_r/dr) + (\xi_r/r)]}{\kappa^2 + (m^2/r^2)}. \tag{14.40}$$

Let us define

$$g = \kappa B_z - (m/r)B_\theta, \tag{14.41}$$

and

$$h = (2\mu_0/r)j_z B_\theta, \tag{14.42}$$

where we note that g stands for $g_{m\kappa}$. From Eqs. 14.30 and 14.41,

$$B_\theta = (f - g)r/2m, \tag{14.43}$$

$$B_z = (f + g)/2\kappa, \tag{14.44}$$

by means of which we eliminate B_θ and B_z from Expression 14.39. When all products are expanded and terms canceled, we find that

$$\frac{\mu_0}{\pi}\,\delta Q_{Fm\kappa} = \int dr\, r\left\{\frac{[rf_{m\kappa}(d\xi_{rm\kappa}/dr) + g_{m\kappa}\xi_{rm\kappa}]^2}{m^2 + \kappa^2 r^2} + (f_{m\kappa}^2 - h)\xi_{rm\kappa}^{\;2}\right\}, \tag{14.45}$$

where the explicit dependence of all variables on m and κ is indicated.

We now revert to our previous notation, where the subscripts are omitted, even on $\delta Q_{Fm\kappa}$.

The displacements $\xi_r(r)$ are now to be so chosen to give δQ the least possible value. If, for such a choice, δQ is negative, then the system is unstable. However, the displacements are restricted to be physically realizable. For this reason

$$\xi_r(0) = 0, \qquad m \neq 0, \tag{14.46}$$

for otherwise the displacement would be multiple valued. In addition, the outside wall at radius R is fixed, so that

$$\xi_r(R) = 0, \qquad \text{for all } m. \tag{14.47}$$

★ 14.4. The Euler-Lagrange Equation

The displacement $\xi(r)$ is to be found that minimizes Eq. 14.45. If the minimum is positive, then this constrained plasma is stable against any perturbation. Accordingly, we must digress for a moment to develop one aspect of a subject known as the calculus of variations.

The integral in question is of the form

$$I(\xi) = \int_a^b dr\, F\!\left(r, \xi, \frac{d\xi}{dr}\right). \tag{14.48}$$

It is a function of the function $\xi(r)$ in that the value of the integral will be different for different functions ξ of r. A function of a function is called a functional; $I(\xi)$ is a functional. Consider two possible neighboring functions as shown in Fig. 14.9. The integral I will have two different values $I(\xi)$ and $I(\xi + \delta\xi)$, where ξ and $\xi + \delta\xi$ represent the two different curves. The difference between the two curves is $\delta\xi$ and is, of course, a function $\delta\xi(r)$ of r. At the point $r = M$, $\delta\xi = QP$ in the figure.

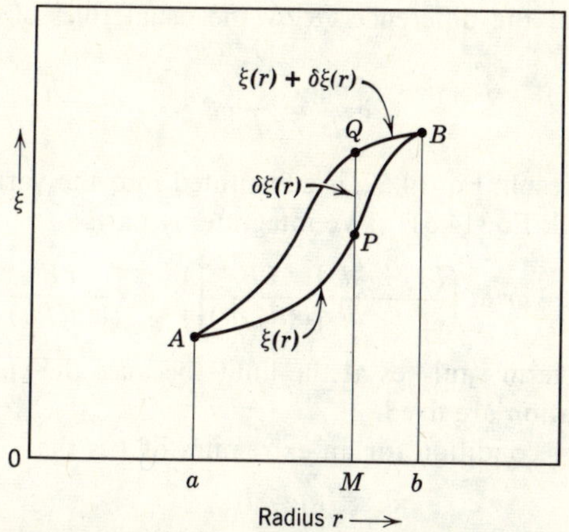

Fig. 14.9. Two curves for the calculation of the variational integral.

Let δI denote the difference between the two integrals $I(\xi)$ and $I(\xi + \delta\xi)$. We wish to evaluate this difference. The difference can be found by finding the difference in the integrands and evaluating the integral over the difference. Thus

$$\begin{aligned}\delta I &= I(\xi + \delta\xi) - I(\xi) \\ &= \int_a^b dr \left\{ F\left[r, \xi + \delta\xi, \frac{d}{dr}(\xi + \delta\xi)\right] - F\left[r, \xi, \frac{d\xi}{dr}\right]\right\}. \quad (14.49)\end{aligned}$$

Now the difference in the integrands can be found by the familiar Taylor expansion:

$$\begin{aligned}&F\left[r, \xi + \delta\xi, \frac{d}{dr}(\xi + \delta\xi)\right] \\ &\quad = F\left[r, \xi, \frac{d\xi}{dr}\right] + \frac{\partial F}{\partial \xi}\delta\xi + \frac{\partial F}{\partial(d\xi/dr)}\delta\left(\frac{d\xi}{dr}\right) + \text{higher-order terms.} \quad (14.50)\end{aligned}$$

In carrying out the above differentiation we must remember that F is a function of ξ and $d\xi/dr$, among other things. Both ξ and $d\xi/dr$ differ on the two paths in general. Let the result of Eq. 14.50 be substituted into the integral δI:

$$\delta I = \int_a^b dr \left[\frac{\partial F}{\partial \xi}\,\delta\xi + \frac{\partial F}{\partial(d\xi/dr)}\,\delta\!\left(\frac{d\xi}{dr}\right)\right]. \tag{14.51}$$

Next, we must consider the expression $\delta(d\xi/dr)$; it is the difference between the derivatives on the two curves. This difference is equal to the derivative of the difference $\delta\xi$ by the usual rules of differentiation; that is,

$$\delta\,\frac{d\xi}{dr} = \frac{d}{dr}\,\delta\xi. \tag{14.52}$$

Our present result, Eq. 14.52, is substituted into the variational integral δI, as it is called, Eq. 14.51. We integrate by parts:

$$\delta I = \int_a^b dr\,\delta\xi\left\{\frac{\partial F}{\partial \xi} - \frac{d}{dr}\left[\frac{\partial F}{\partial(d\xi/dr)}\right]\right\} + \left\{\frac{\partial F}{\partial(d\xi/dr)}\,\delta\xi\right\}\Bigg|_a^b. \tag{14.53}$$

The integrated term vanishes at the limits because the end points of the paths of integration are fixed.

The necessary condition for an extremum of I is that

$$\delta I = 0 \tag{14.54}$$

for any small displacement $\delta\xi$ whatsoever. This criterion can be satisfied only if the integrand of Eq. 14.54 itself is zero, or if

$$\frac{\partial F}{\partial \xi} - \frac{d}{dr}\left[\frac{\partial F}{\partial(d\xi/dr)}\right] = 0. \tag{14.55}$$

In other words, we must have

$$\frac{\partial^2 F}{\partial(d\xi/dr)^2}\,\frac{d^2\xi}{dr^2} + \frac{\partial^2 F}{\partial(d\xi/dr)\,\partial\xi}\,\frac{d\xi}{dr} + \frac{\partial^2 F}{\partial(d\xi/dr)\,\partial r} - \frac{\partial F}{\partial \xi} = 0 \tag{14.56}$$

upon writing out the total derivative. This equation is the so-called Euler-Lagrange equation. It will now be applied to the integrand of Eq. 14.45 to determine the least value of δQ.

The Euler-Lagrange equation is a necessary but not sufficient condition for an extremum. For example, $\delta I = 0$ at horizontal inflection points also; we shall not discuss these special cases. In addition, the extremum may be a maximum or minimum; but we can readily distinguish between them on physical grounds.

★ 14.5. Final Form of Suydam's Stability Criterion

The Euler-Lagrange equation, Eq. 14.56, enables us to compute the condition that the integrand must satisfy such that the energy integral in

Eq. 14.45 be a minimum. According to Eq. 14.56, we must compute the following derivatives:

$$\frac{\partial F}{\partial \xi_r}, \qquad \frac{\partial F}{\partial(d\xi_r/dr)}, \qquad (m^2 + \kappa^2 r^2)\frac{\partial^2 F}{\partial r\, \partial(d\xi_r/dr)}$$

$$\frac{\partial^2 F}{\partial \xi_r\, \partial(d\xi_r/dr)}, \qquad \frac{\partial^2 F}{\partial(d\xi_r/dr)^2}.$$

The various complicated expressions for the derivatives are substituted into the Euler-Lagrange equation, Eq. 14.56. After collecting terms in $d^2\xi_r/dr^2$, $d\xi_r/dr$, and ξ_r and multiplying the resulting equation by $(m^2 + \kappa^2 r^2)/2r^3 f^2$, we find

$$\frac{d^2\xi_r}{dr^2} + \left[\frac{3}{r} - \frac{2r\kappa^2}{(m^2 + \kappa^2 r^2)} + 2\frac{d(\ln f)}{dr}\right]\frac{d\xi_r}{dr}$$

$$+ \left[\frac{2m^2 g}{r^2 f(m^2 + \kappa^2 r^2)} - \frac{g^2}{r^2 f^2} - \frac{(m^2 + \kappa^2 r^2)(f^2 - h)}{r^2 f^2} + \frac{1}{rf^2}\frac{d}{dr}(fg)\right]\xi_r = 0. \tag{14.57}$$

The factor multiplying ξ_r in this complicated result can be simplified somewhat. In equilibrium

$$\frac{dp}{dr} = (\nabla p)_r = (\mathbf{j} \times \mathbf{B})_r = j_\theta B_z - j_z B_\theta. \tag{14.58}$$

Eliminate j_θ by Eq. 14.36, eliminate $j_z B_\theta$ by the definition of h, Eq. 14.42, and substitute Eq. 14.44 for B_z. Then obtain successively

$$\frac{2\mu_0\kappa^2}{rf^2}\frac{dp}{dr} = \frac{\kappa^2 h}{f^2} + \frac{1}{4rf^2}\frac{d}{dr}(f + g)^2; \tag{14.59}$$

$$\frac{2\mu_0\kappa^2}{rf^2}\frac{dp}{dr} = \frac{\kappa^2 h}{f^2} + \frac{m^2}{rf^2}\frac{d}{dr}\left(\frac{B_\theta^2}{r^2}\right) + \frac{1}{rf^2}\frac{d}{dr}(fg); \tag{14.60}$$

$$\frac{2\mu_0\kappa^2}{rf^2}\frac{dp}{dr} = \frac{\kappa^2 h}{f^2} + \frac{m^2 h}{r^2 f^2} - \frac{(f - g)^2}{r^2 f^2} + \frac{1}{rf^2}\frac{d}{dr}(fg). \tag{14.61}$$

Equation 14.60 follows by Eq. 14.43; Eq. 14.61, by use of Eq. 14.37 to eliminate B_θ, by Eq. 14.42 for h, and by Eq. 14.43. We solve Eq. 14.61 for $(1/rf^2)\, d(fg)/dr$ and substitute the result into the reduced Euler-Lagrange equation, Eq. 14.57:

$$\frac{d^2\xi_r}{dr^2} + \left[\frac{3}{r} + 2\frac{d(\ln f)}{dr} - \frac{2\kappa^2 r}{(m^2 + \kappa^2 r)}\right]\frac{d\xi_r}{dr}$$

$$- \left[\left(\kappa^2 + \frac{m^2 - 1}{r^2}\right) + \frac{2\kappa^2 g}{f(m^2 + \kappa^2 r^2)} - \frac{2\mu_0\kappa^2}{rf^2}\frac{dp}{dr}\right]\xi_r = 0. \tag{14.62}$$

The Euler-Lagrange equation, Eq. 14.62, may be used to reduce the integrand of the variational expression, Eq. 14.45, for the energy to a perfect differential. We find that the change of energy in the mode (m, κ) for plasma between r_1 and r_2 is given by

$$\frac{\mu_0 \delta Q}{\pi} = \left. \frac{r^2 f \xi_r (rf(d\xi_r/dr) + g\xi_r)}{m^2 + \kappa^2 r^2} \right|_{r_1}^{r_2}. \tag{14.63}$$

We have only to choose ξ_r in accord with the Euler-Lagrange equation, Eq. 14.62, to find the maximum or minimum of δQ.

In order to choose ξ most pessimistically, we now appeal to physical reasoning. Lines of induction may not cross each other, for if they did the magnetic induction would point in two directions at the point of crossing, a configuration that is clearly contrary to the meaning of a vector field. Lines of induction may, however, be interchanged with each other. The bending of lines of induction requires energy. We would, therefore, expect that modes which interchange lines of induction without bending them would be the least stable. For this reason, we look into such modes in more detail. The lines of induction describe a set of spirals of varying pitch. The pitch is the inverse of the quantity

$$b = B_\theta / r B_z. \tag{14.64}$$

The lines of ξ, on the other hand, describe a set of spirals of constant pitch. At any region or point in space where the two pitches are identical, the lines of induction will not be bent by a displacement of the plasma (the lines are presumably frozen into the plasma). The most unstable region of the plasma is then that at which

$$f = 0. \tag{14.65}$$

Let the radius at this point be a.

Upon referring to the differential equation, Eq. 14.57, we see that the coefficient of the first term is finite at $r = a$, the coefficient of the second term has a simple pole, that is, diverges like $1/(r - a)$, and the coefficient of the third term has a double pole, that is, diverges like $1/(r - a)^2$. The properties of a second-order differential equation having these types of divergences in the coefficients are well known; it is said to have a regular singularity at the point $r = a$. Such a differential equation can be solved by a power-series expansion of the form

$$\xi_r = \sum_{n=0}^{\infty} C_n (r - a_n)^{\nu + n}, \tag{14.66}$$

where C_n are constants to be determined, and ν must be such that

$$\nu^2 + \nu - \frac{2\mu_0}{rB_z^2}\left(\frac{dp}{dr}\right)\left(\frac{b}{db/dr}\right)^2 = 0, \tag{14.67}$$

evaluated at $r = a$. This fact can be seen by substituting our power series directly into the differential equation and by requiring that the solution be nontrivial. There are clearly two roots of Eq. 14.67:

$$\nu_{\pm} = -\tfrac{1}{2} \pm \tfrac{1}{2}\left[1 + \frac{8\mu_0}{rB_z^2}\left(\frac{dp}{dr}\right)\left(\frac{b}{db/dr}\right)^2\right]^{1/2}, \tag{14.68}$$

and two cases arise according to whether these roots are real or complex. Real roots arise when

$$\frac{8\mu_0}{rB_z^2}\left(\frac{dp}{dr}\right)\left(\frac{b}{db/dr}\right)^2 > -1, \tag{14.69}$$

and complex roots arise when

$$\frac{8\mu_0}{rB_z^2}\left(\frac{dp}{dr}\right)\left(\frac{b}{db/dr}\right)^2 < -1. \tag{14.70}$$

Let us study the second case:

$$\nu_{\pm} = -\tfrac{1}{2} \pm i\beta, \tag{14.71}$$

where

$$\beta = +\left[\frac{8\mu_0}{rB_z^2}\left(\frac{dp}{dr}\right)\left(\frac{b}{db/dr}\right)^2 - 1\right]^{1/2}. \tag{14.72}$$

For this case, the solution of the Euler-Lagrange equation is

$$\xi_r = |r - a|^{-1/2} \cos\,[(\beta/2)\ln|r - a| + \varphi][1 + \mathcal{O}(|r - a|^2)], \tag{14.73}$$

where φ is some constant determined by the boundary conditions. We shall not need to find it, however. It is immediately noted that there is a singularity at $r = a$, and that if ξ from Eq. 14.73 is substituted into the integral Eq. 14.45 for δQ, the integral diverges. In general, the Euler-Lagrange formalism cannot be carried past a singularity:† the reader might therefore suspect that all our work has been for nothing. This suspicion disappears when he recalls that the real purpose of the Euler-Lagrange formalism is to delimit the region of search for possible ξ's giving instability. We have obtained in Eq. 14.73 a ξ that certainly represents an extremum but is improper. Our course now consists of investigating any other displacement ξ' which is close to that given by Eq. 14.73, but which *is* proper. Since ξ' no longer represents the completely minimized δQ, we cannot use Eq. 14.63 to evaluate δQ but must use the original integral, Eq. 14.45. We recognize from these statements that any stability criterion found will be a necessary one; it will not be a sufficient one because ξ' will generally not be the worst possible perturbation.

† Newcomb[8] discusses this point exhaustively.

We now complete the analysis along the lines just indicated. An acceptable displacement ξ' to investigate is the following one:

1. ξ' is the solution of Eq. 14.73 in the range $(0 \leqslant r \leqslant a - \mathfrak{d})$, and in the range $(a + \mathfrak{d} \leqslant r \leqslant R)$.
2. The small distance $\mathfrak{d}$ is to be so chosen that the leading term in Eq. 14.73 dominates at $r = a - \mathfrak{d}$.
3. ξ' is constant in the range $(a + \mathfrak{d} < r < a + \mathfrak{d})$.

Equation 14.45 may then be split to four integrals, two of them covering the ranges $(0 \leqslant r \leqslant a - \mathfrak{d})$ and $(a - \mathfrak{d} < r \leqslant a)$, and two more covering the similar range $(a < r \leqslant R)$. We shall look at the behavior of the two integrals only in the range $0 \leqslant r \leqslant a$, those over the range from a to R being exactly analogous. Suydam finds that with this choice of ξ Eq. 14.45 can be integrated in two ranges from 0 to a to give

$$\int_0^a dr\, r\left[\frac{(rf(d\xi_r/dr) + g\xi_\xi')^2}{m^2 + \kappa^2 r^2} + (f^2 - h)(\xi_r')^2\right]$$
$$= \frac{a^3 B_z^2 (db/dr)^2}{4(1 + a^2 b^2)}\left\{\left[1 + \frac{4\mu_0 (dp/dr)}{rB_z^2}\left(\frac{b}{db/dr}\right)^2\right][1 + \cos 2\psi] + \beta \sin 2\psi\right\}; \tag{14.74}$$

here

$$\psi = (\beta/2)\ln \mathfrak{d} + \varphi, \tag{14.75}$$

and the quantities b, db/dr, and dp/dr are to be evaluated at $r = a$. The multiplying coefficient on the right side of Eq. 14.74 is positive definite; its particular value is of no interest to us. Since $\mathfrak{d}$ can be chosen freely, as long as it is small, ψ may take on any value whatsoever. Then the expression in braces in Eq. 14.74 oscillates between the values $+1$ and $-\beta^2$ as ψ varies. As the latter quantity is negative, the integral can therefore be made negative. Similarly, the integral from a to R can be made negative; hence δQ itself can be negative. We find in this way that complex roots of Eq. 14.67 imply instability.

We shall not explore the consequence of real roots but state that in this case also the solution ξ analogous to Eq. 14.73 diverges. If now we choose a substitute ξ' in the same manner as before—the divergent solution except near the singularity and constant across the erstwhile singularity—we find that δQ is positive. Note well, however, that this discovery is not a proof of stability, because ξ' does not represent an extremum, as mentioned a few paragraphs ago.

We have now found a necessary but not sufficient condition for stability against $m \neq 0$ perturbations: there must be no complex roots of Eq.

14.67; that is, Eq. 14.69 must be satisfied. We can express the criterion in the desired final form

$$\frac{2\mu_0}{B_z^2}\left(\frac{dp}{dr}\right) + \frac{r}{4}\left(\frac{1}{b}\frac{db}{dr}\right)^2 \geqslant 0 \tag{14.76}$$

at every point in the plasma.

The criterion, Eq. 14.76, has a number of consequences. First, and somewhat trivially, the inequality is satisfied if the pressure is constant or increases with radius. This fact agrees with our previous notions of stability and will find application to the inverse pinches described in Sec. 14.9. More important to the present discussion is the application of Eq. 14.76 to conventional pinches, where the gas is inside. Since $dp/dr < 0$, the criterion represents a restriction on the radial density gradient. Alternately, if dp/dr is specified, the induction must be programmed so that the rate of change of pitch is sufficiently large. This rate of change of direction of the B lines in successive magnetic surfaces is commonly (though perhaps inappropriately) called magnetic shear. The configuration is shown in Fig. 14.10; in the present context, the radial

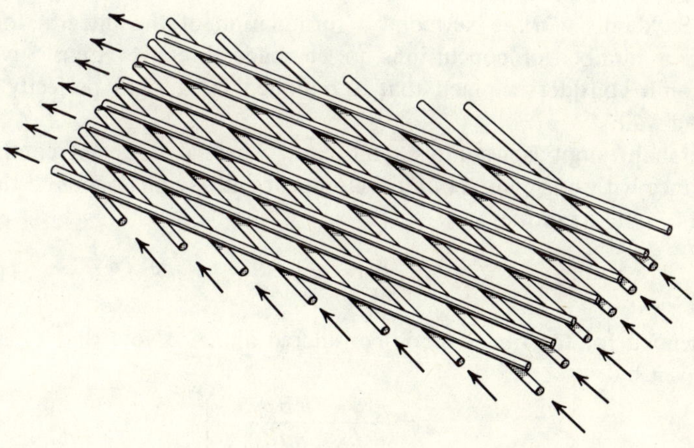

Fig. 14.10. Magnetic surfaces having shear.

direction lies normal to the surfaces. The virtue of the arrangement consists of the fact that magnetic lines cannot be interchanged between layers without distorting them. Such distortion requires energy, whence the configuration with shear proves more stability than one without shear. The physical interpretation of the present criterion is that since $dp/dr < 0$ the potential energy can be decreased by expanding the plasma (Sec. 12.3). If the shear is great enough, an even greater energy increase is required for the concomitant movement of the induction, so the system will be stable. Note that the criterion is not related to any such simple thing as unwinding the shear by plasma currents. We shall meet this concept again in Chap. 16, applied to the stellarator.

According to the criterion of Eq. 14.76 and the associated intuitive concept, we should program the induction surrounding the pinch so that the shear is as large as possible. Seemingly this aim could best be achieved by reversing the induction B_z in the vacuum after the plasma has pinched. However, there appears to be a difficulty in this procedure. The externally imposed B_z presses on the plasma surface and increases the pressure gradient. This pressure gradient instigates the instability according to Eq. 14.76. Thus applying a reversed external B_z may not in fact stabilize a previously unstable pinch.

★ 14.6. Introduction to Newcomb's Stability Analysis

The detailed analysis of the linear pinch by Newcomb[8] cannot properly be condensed for inclusion here. However, we can present a brief summary of the first part of his paper. There are interesting parallels with Suydam's work. Newcomb's formulation of the integral for δQ allows a number of conclusions to be made directly. Like Suydam, Newcomb considers a pinch that is bounded by a rigid perfectly conducting wall.

We shall adopt Newcomb's symbolism, where it does not conflict in substance with our own. In a development entirely analogous to that of Sec. 14.3, the quantities

$$\eta = \boldsymbol{\nabla}\cdot\boldsymbol{\xi} - \frac{1}{r}\frac{d}{dr}(r\xi_r) = \frac{im}{r}\xi_\theta + i\kappa\xi_z \tag{14.77}$$

are defined for the (m, κ) mode considered alone. Note that ξ_θ and ξ_z are given by

$$\left.\begin{aligned} \xi_\theta &= \frac{\kappa r\alpha - irB_\theta\eta}{\kappa rB_z + mB_\theta}, \\ \xi_z &= -\frac{m\alpha + irB_z\eta}{\kappa rB_z + mB_\theta}. \end{aligned}\right\} \tag{14.78}$$

Equation 12.105 for δQ_F is treated much as before. The displacements may have a nonvanishing divergence, however. The current density $\mathbf{j}$ and pressure gradient $\boldsymbol{\nabla}p$ are eliminated by Ampère's law and Eq. 7.3, which are valid in the steady state, the stress being assumed to be isotropic. As previously, only the terms involving $\xi_{m\kappa}\xi_{m'\kappa'}$ that contribute are retained, so that each mode can be individually studied. The resulting expression is minimized with respect to α as before. After a very considerable amount of algebra, Newcomb finds that

$$\delta Q(r_1, r_2; \xi, \eta, \alpha) \propto \int_{r_1}^{r_2} dr\, r\left\{\Lambda\left(\xi_r, \frac{d\xi_r}{dr}\right) + \mu_0\gamma p\left[\eta + \frac{1}{r}\frac{d}{dr}\,\xi_r\right]^2 + \frac{i(\kappa^2r^2 + m^2)}{r^2}\left[\alpha - \alpha_0\left(\xi_r, \frac{d\xi_r}{dr}\right)\right]^2\right\} \quad (14.79)$$

for a particular mode (m, κ), where

$$\Lambda\left(\xi_r, \frac{d\xi_r}{dr}\right) = \frac{1}{\kappa^2r^2 + m^2}\left[(\kappa rB_z + mB_\theta)\frac{d\xi_r}{dr} + (\kappa rB_z - mB_\theta)\frac{\xi_r}{r}\right]^2 + \left[(\kappa rB_z + mB_\theta)^2 - 2B_\theta\frac{d}{dr}(rB_\theta)\right]\frac{\xi_r^2}{r^2}, \quad (14.80)$$

and

$$\alpha_0\left(\xi_r, \frac{d\xi_r}{dr}\right) = \frac{r}{\kappa^2r^2 + m^2}\left[(\kappa rB_\theta - mB_z)\frac{d\xi_r}{dr} - (\kappa rB_\theta + mB_z)\frac{\xi_r}{r}\right]. \quad (14.81)$$

Under the simplest analytic conditions, the lower limit r_1 will be zero, and the upper limit r_2 will be the tube wall. However, consider also the case of the inverse pinch (Sec. 14.9), in which there is no plasma near the axis; in addition, if the integrand contains singularities, the integration may have to be divided into subintervals.

All the terms in Eq. 14.79 are positive definite, with the possible exception of the one in Λ containing $B_\theta\, d(rB_\theta)/dr$. Rosenbluth has pointed out[9] that the pinch is always stable if $|rB_\theta|$ decreases everywhere with increasing radius. This unnecessarily extreme criterion could not be satisfied in a conventional pinch, but it might be satisfied in a hollow pinch with a solid conductor running down the axis (see Sec. 14.9 again). It can be shown easily that if $|rB_\theta|$ decreases everywhere, Suydam's criterion is trivially satisfied (see Problem 10).

Equation 14.79 can obviously be minimized by the choices

$$\left.\begin{aligned} \eta &= -\frac{1}{r}\frac{d}{dr}(r\xi_r), \\ i\alpha &= \alpha_0\left(\xi_r, \frac{d\xi_r}{dr}\right). \end{aligned}\right\} \quad (14.82)$$

Note that this choice of η is equivalent to setting $\nabla\cdot\boldsymbol{\xi} = 0$. Hence δQ is minimized by incompressible displacements, as was proved by a different argument in Sec. 14.3. For these choices, the quantity Λ alone remains in Eq. 14.79. The integral is independent of the ratio of specific heats γ, whence the stability criterion is unchanged by considering incompressible fluids. Equation 14.79 becomes

$$\delta Q(r_1, r_2; \xi) \propto \int_{r_1}^{r_2} dr\, r\Lambda\left(\xi_r, \frac{d\xi_r}{dr}\right), \tag{14.83}$$

which can be integrated by parts to give

$$\delta Q(r_1, r_2; \xi) \propto \int_{r_1}^{r_2} dr\left[C\left(\frac{d\xi_r}{dr}\right)^2 + D\xi^2\right]; \tag{14.84}$$

here

$$C = \frac{r(\kappa r B_z + mB_\theta)^2}{\kappa^2 r^2 + m^2}, \tag{14.85}$$

$$D = \frac{(\kappa r B_z - mB_\theta)^2}{r(\kappa^2 r^2 + m^2)} + \frac{(\kappa r B_z + mB_\theta)^2}{r} - \frac{2B_\theta}{r}\frac{d}{dr}(rB_\theta) - \frac{d}{dr}\left(\frac{\kappa^2 r^2 B_z^2 - m^2 B_\theta^2}{\kappa^2 r^2 + m^2}\right). \tag{14.86}$$

The coefficient C is always positive, but D may have either sign.

By excluding the case $m = 0$, let us compare the effect of displacements with different m, but the same value of $\kappa/m = q$. Then substitute $\kappa = mq$ in Eqs. 14.85 and 14.86 and find that the second term in Eq. 14.86 is the only one depending upon m. This term is proportional to m^2 and is positive definite. Therefore the least stable pinches are those for which $m = 1$, $-\infty < \kappa < \infty$.

If $m = 0$, Eq. 14.84 can be reduced to

$$\delta Q(r_1, r_2; \xi) \propto \delta Q_0(r_1, r_2; \xi) + \kappa^2 \int_{r_1}^{r_2} dr\, rB_z^2\xi^2, \tag{14.87}$$

where

$$\delta Q_0(r_1, r_2; \xi) = \int_{r_1}^{r_2} dr\left[rB_z^2\left(\frac{d\xi_r}{dr}\right)^2 + \left(\frac{B_z^2}{r} + 2\mu_0\frac{dp}{dr}\right)\xi^2\right]. \tag{14.88}$$

The term containing κ is positive definite. Therefore the pinch with $m = 0$ is stable for all κ if it is stable in the limit $\kappa \to 0$.

Newcomb thus is able to state the following general theorem: A linear pinch is stable for all values of m and κ if and only if it is stable for $m = 1$, $-\infty < \kappa < \infty$, and for $m = 0$, $\kappa \to 0$. This theorem considerably simplifies the task of searching for possible instabilities.

With δQ in the form of Eq. 14.84, the Euler-Lagrange formulation is particularly simple, to wit:

$$\frac{d}{dr}\left(C\frac{d\xi_r}{dr}\right) - D\xi_r = 0. \tag{14.89}$$

If there are no singularities in the interval $r_1 < r < r_2$,

$$\delta Q(r_1, r_2; \xi) \propto \int_{r_1}^{r_2} dr \left[C\left(\frac{d\xi_r}{dr}\right)^2 + \frac{d}{dr}\left(C\frac{d\xi_r}{dr}\right)\xi \right]$$

$$= \int_{r_1}^{r_2} dr \frac{d}{dr}\left(C\xi_r \frac{d\xi_r}{dr}\right) = C\xi_r \frac{d\xi_r}{dr}\Big|_{r_1}^{r_2}. \quad (14.90)$$

In particular, if $C\xi_r$ vanishes at r_1 and r_2, and if $d\xi_r/dr$ is finite, the contribution in the range $r_1 < r < r_2$ vanishes.

We shall not carry Newcomb's development further; he derives a number of formal stability criteria and discusses in detail the validity and necessity of computing δQ in well-behaved subintervals in case singularities exist in the integrand. However, apropos of Suydam's criterion, we repeat here a point made by Newcomb. If the current in a pinch is all carried in a thin sheath of thickness $\mathfrak{d}$, the derivatives in Suydam's criterion, Eq. 14.76, are nonzero only over the region $\mathfrak{d}$. Then for small $\mathfrak{d}$, the first term in Eq. 14.76 varies as $\mathfrak{d}^{-1}$, while the second term varies as $\mathfrak{d}^{-2}$ and is positive. Therefore the criterion predicts stability, whereas in fact it is known[3] that this particular configuration is unstable. We thus find further confirmation of our statement in Sec. 14.5 that Suydam's criterion is a necessary but not sufficient one for stability.

The analyses of these four sections, Secs. 14.3 to 14.6, suffice to give the reader a detailed view of typical methods employed to investigate hydromagnetic stability. With this remark, we close our discussion of the problem.

14.7. Specific Conventional Linear Pinch Developments

The literature relating to the development of pinched plasma systems is extensive. In this section we select a few specific, typical matters for discussion as illustrations of the points discussed previously. The interested reader can pursue such items as interest him further by consulting the references listed, some of which are review papers.

As mentioned earlier, a pinch apparatus may be either a straight cylinder or a torus. Much of the work on pinch discharges has been conducted at the Los Alamos Scientific Laboratory. At this laboratory the straight pinch devices are designated by the word Columbus and the toroidal pinch devices are given the name Perhapsatrons for good reasons.

Essentially, one hopes to compress and heat the gas discharge so rapidly in these devices that any instabilities that may be present will not have had time to grow before the thermonuclear reaction has taken place. Accordingly, almost all of the straight pinch assemblies are designed for operation in a microsecond or shorter. They are characterized by very low inductance condensers and by transmission lines. A typical straight pinch configuration is shown schematically in Fig. 14.11.[10]

In this particular device, which is a moderately large one, the total energy stored is 10^5 joules, the maximum current is 8×10^5 amp, and the time to the first current peak is 2.2 μsec. The maximum rate of current rise is thus about 10^{12} amp/sec, as is typical of these large fast devices. Switching such currents is a special problem, and spark gaps are almost invariably used. One vacuum arrangement[11] is shown in Fig. 14.11. Another utilizing a small high-impedance trigger is shown in Fig. 14.12. Here, the gaps are air-filled; the voltage is below that required for breakdown of the two gaps in series, but exceeds that of one gap. The spark plug discharge releases a few photoelectrons from the electrodes, shifts

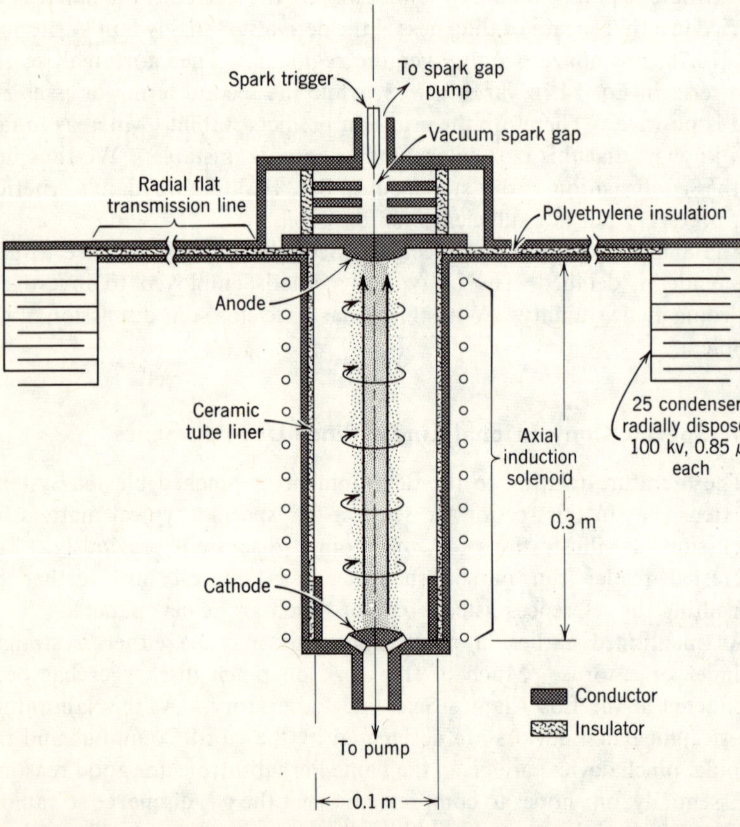

Fig. 14.11. Schematic arrangement of a typical straight pinch experiment (Columbus II). The tube is approximately to scale, but the condensers are actually about 0.4-m diam and 1-m high. After J. L. Tuck.[10]

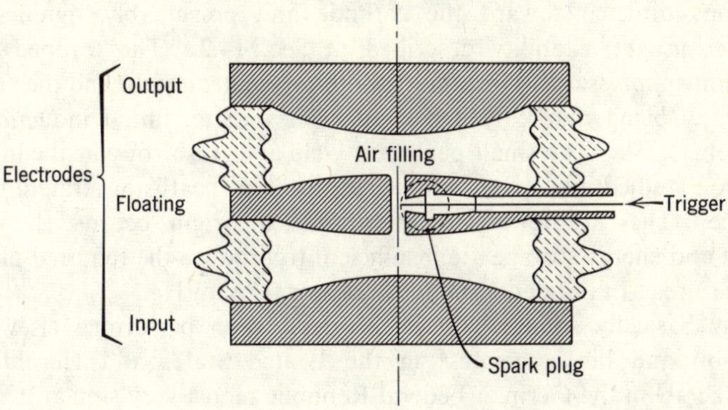

Fig. 14.12. Schematic representation of four-electrode spark gap. After J. L. Tuck.[10]

the potential of the middle electrode, and the whole gap breaks down.

The neutron production from a deuterium pinch experiment[12] in the Columbus II apparatus (Fig. 14.11) is shown in Fig. 14.13. The discharge forms, compresses to a small radius, oscillates in and out a few times, and breaks up in turbulent disorder. At zero axial induction, the yield is 10^8 neutrons per pulse over a period of about 1.5 μsec, with an energy anisotropy of 57.5 kev. This anisotropy is characteristic of

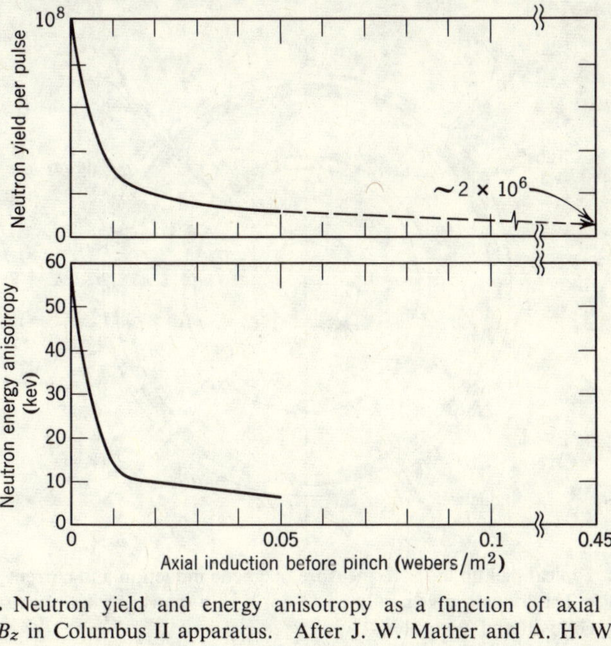

Fig. 14.13. Neutron yield and energy anisotropy as a function of axial magnetic induction B_z in Columbus II apparatus. After J. W. Mather and A. H. Williams.[12]

deuterons directed toward the cathode and presumably signifies the $m = 0$ sausage instability described in Sec. 14.2. The trapped axial induction suppresses this instability, both the neutron yield and the energy anisotropy being markedly decreased even with an initial induction of 10^{-2} webers/m^2. The small persistent yield and anisotropy at the highest induction studied may arise from acceleration in sheaths or straight down the tube. They are not of true thermonuclear origin, because the pinch current and energy storage are insufficient to produce the required plasma temperature. Initial deuteron pressure was 0.2 mm Hg.

Many basically similar experiments have been performed elsewhere. Anderson and his associates[4] in the United States and Herold and co-workers[13] in the German Federal Republic report very similar neutron emission from similar experiments. A large number of linear pinches have been developed in the U.S.S.R.[14–23] Golovin and his associates[23] have performed experiments in a large cylinder 0.23-m diam and 0.8-m length. An axial induction of 2.7-webers/m^2 maximum can be applied from a 5×10^5-joule condenser bank, prior to the pinch formation. The inductance of this system is relatively high. Thus the pinch forms slowly, and a relatively stable pinch is maintained for times up to several hundred μsec. Initial deuterium gas pressure was varied from 0.005 to 5 mm Hg.

Figure 14.14 shows a cross section of the tube used by Golovin and associates, with circular pickup (Rogowski) coils placed inside the tube.

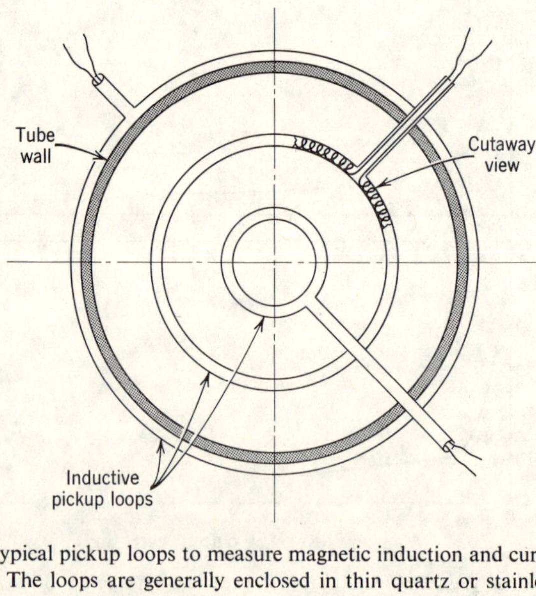

Fig. 14.14. Typical pickup loops to measure magnetic induction and current in a pinch experiment. The loops are generally enclosed in thin quartz or stainless steel tubes. After I. N. Golovin and associates.[23]

Such coils, or variants of them, are very useful diagnostic tools. The voltage induced in each coil is proportional to $\dot{B}$ through it; hence by Maxwell's equations and time integration, the current in the various annular zones can be determined. With this technique, Golovin and his associates determined that with an initial axial induction of 1.5 webers/m^2, over 95% of the current was carried in the innermost zone at times less than 250 μsec. The plasma temperature in this slow pinch is estimated to be 50 ev or less. Corresponding to the initial deuterium filling, the initial ion density may be typically about 10^{22}/m^3, and the final pinched density as high as 10^{24}/m^3. These numbers are representative of many similar experiments;[22] temperatures as high as several hundred electron volts are estimated in some work;[20] current rise rates $> 10^{12}$ amp/sec, and total current $\approx 2 \times 10^6$ amp have been achieved.

Photographs with Kerr cell shutters or rotating mirrors are useful in recording the plasma motion. Figure 14.15 shows the plasma motion and direction of current flow in a developing pinch with maximum current 9×10^5 amp flowing in a cylinder of 0.19-m diam. The mean axial

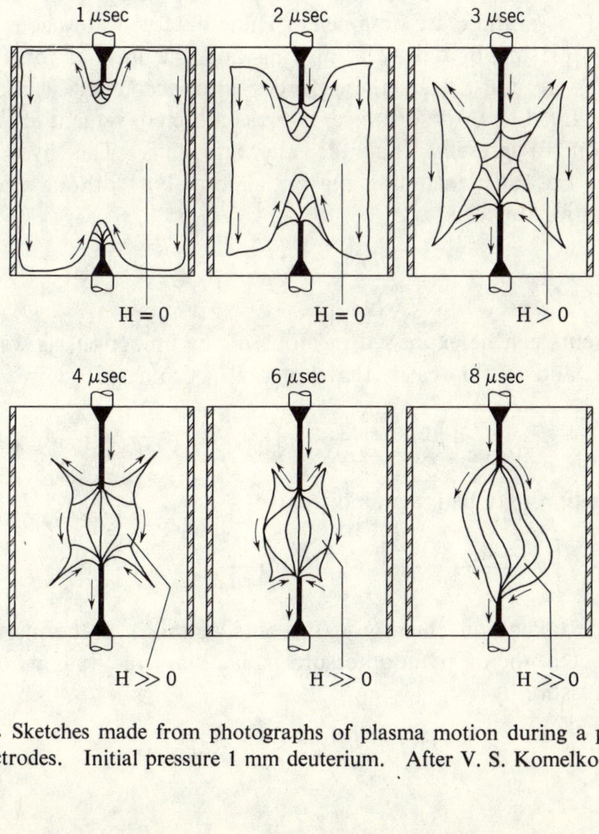

Fig. 14.15. Sketches made from photographs of plasma motion during a pinch with needle electrodes. Initial pressure 1 mm deuterium. After V. S. Komelkov.[21]

speed of the contracting region was about 7×10^4 m/sec, and the radial speed was about 5×10^4 m/sec. Note the evidence of the current flowing in a layer, as in the snowplow model of the pinch, and the evidence for fast shock waves traveling ahead of the plasma boundary. Similar shock waves have been reported elsewhere.[24]

In all these experiments (and the toroidal pinches shortly to be described), spectroscopic analysis indicates the presence of considerable impurities from the tube walls, particularly after the first one or two oscillations of the pinch. The electrodes of the long straight pinch devices appear to contribute less impurities than do the walls. Therefore the use of these more convenient linear discharge tubes is justified.

Very reproducible records of the induction during a pinch have been obtained by Burkhardt and Lovberg[25] in the Columbus S-4 device, which has a discharge tube 0.13-m diam and 0.61-m length. Peak current is 2.5×10^5 amp in 6 μsec; 15,000 joules are stored. Initial deuterium pressure is 0.08 mm Hg. The inductions B_θ and B_z are measured by means of small magnetic pickup probes responsive to dB/dt. The output is integrated and displayed on an oscilloscope. The data of Figs. 14.16*a*, *b*, and *c* are obtained. These data will now be interpreted. Since the plasma boundary is moving and the inertial forces are not negligible, we must take these forces into account. The dynamical equation, Eq. 14.14, would suffice, but it is more convenient in the present illustration to present a simple alternate one. The hydromagnetic equation, Eq. 6.61, including the initial term but without space-charge or gravitational effects, is

$$\rho_m \frac{\partial \mathbf{V}}{\partial t} = -\nabla p + \mathbf{j} \times \mathbf{B}. \tag{14.91}$$

The currents can be expressed in terms of the induction, as was done in Sec. 14.1, and we find easily that Eq. 14.91 becomes

$$\frac{\partial}{\partial r}\left(p + \frac{B_z^2 + B_\theta^2}{2\mu_0}\right) + \frac{B_\theta^2}{r\mu_0} + \rho_m \ddot{r} = 0. \tag{14.92}$$

This equation may be integrated to give

$$\left[p + \frac{B_z^2 + B_\theta^2}{2\mu_0}\right]_{r_1}^{r_2} + \int_{r_1}^{r_2} dr\left(\frac{B_\theta^2}{r\mu_0} + \rho_m \ddot{r}\right) = \text{const}; \tag{14.93}$$

r_2 may be taken conveniently at the outside wall. If the inertial forces are neglected, then a pseudopressure p^* is obtained. It is related to the actual pressure by

$$p(r) = p^*(r) - \int_r^{r_2} dr'\, \rho_m \ddot{r}'. \tag{14.94}$$

In order to evaluate this integral, the density and acceleration must be known. If the quantity $\rho_m \ddot{r}$ is neglected in Eq. 14.93, the pseudopressures p^* of Figs. 14.16*a* and 14.16*b* are obtained. Evidently, accelerations have been neglected, because the material could not reasonably rearrange itself in such short times. It is easy to check that the acceleration must be outward at 1.9 μsec, inward at 2.2 μsec, and approximately zero at 2.6 μsec. If, with Burkhardt and Lovberg, we assume a snowplow model of the pinch, in which all the material is concentrated in a thin current-carrying

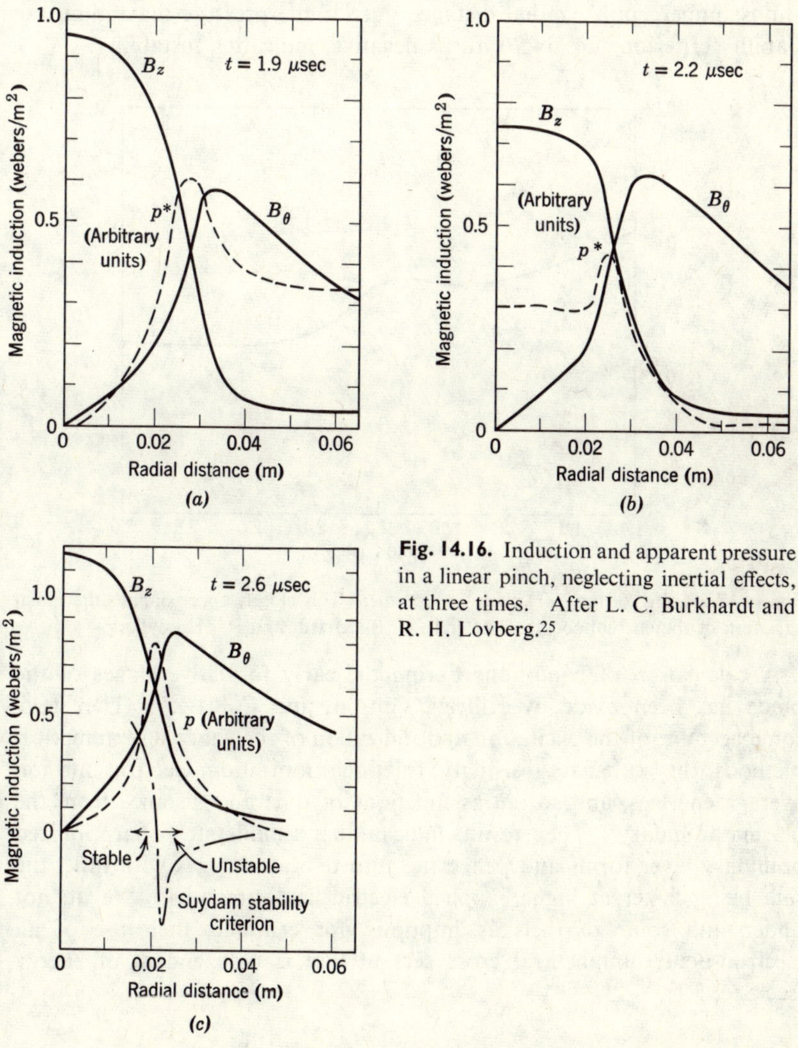

Fig. 14.16. Induction and apparent pressure in a linear pinch, neglecting inertial effects, at three times. After L. C. Burkhardt and R. H. Lovberg.[25]

shell at the position of maximum B_θ, the assumptions regarding the acceleration are borne out. Figure 14.17 shows the position of the B_θ maximum as a function of time. Figure 14.16*c* shows the distributions at a time when the motion was negligible; the pressure distribution indicates that the snowplow model is a reasonable one here. Also shown in this figure is Suydam's stability criterion, Eq. 14.76, calculated after expressing all quantities in terms of B_θ, B_z, and their derivatives. We see that the amount of magnetic shear incidentally provided here is almost negligible, compared with that required for stability. At an almost imperceptible radial distance outside the pressure maximum, the stability criterion, Eq. 14.76, turns negative, indicating instability.

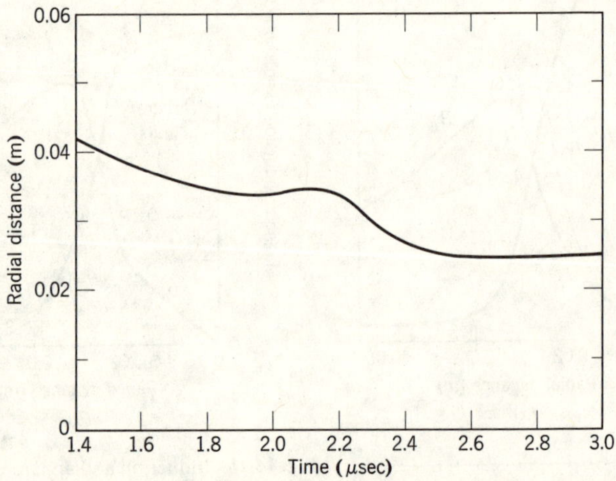

Fig. 14.17. Radius of the $B_\theta(r)$ peak versus time. This closely approximates the radial variation of the current sheath. After L. C. Burkhardt and R. H. Lovberg.[25]

A calculation of conditions during the early formative stages of the pinch has been made by Killeen, Gibson, and Colgate.[26] They take some account of the excitation and ionization of the gas; using numerical methods, they calculate resistivity, fractional ionization, electron and ion average energies, and so on, as functions of time and distance from the plasma boundary. Their results indicate that a moderately sharp ionized boundary layer forms and penetrates into the interior gas with time, the rate being faster at higher applied electric field strengths. We do not understand some of their assumptions, for example, their use of an electron-deuterium neutral cross section that is independent of energy (see Sec. 3.1).

14.8. Specific Conventional Toroidal Pinches

The pinches in toroidal geometry are basically similar to the straight ones, except for the obvious differences that there are no electrodes and that they must be inductively coupled. We shall therefore summarize their operation more briefly.

The U.S. toroidal pinch experiments are described briefly by Tuck.[10] The Perhapsatron S-4 is a small iron-core device of minor diameter 0.07 m, major diameter 0.35 m; the discharge tube is quartz. The arrangement is similar to Fig. 14.2, except that the single-turn primary consists of an aluminum outer torus surrounding the discharge tube, and that there are two iron cores. The pinch is relatively fast for an iron-core device, developing in about 10 μsec. Similar devices, but about twice the size, have been built in France.[27,28] Experiments on metal-wall tori as large as 0.5-m minor diam and 1.25-m major diam are reported from the U.S.S.R.[29] In the United Kingdom, a moderate-sized device called Sceptre III[30–32] and the largest of all toroidal devices, called Zeta, have been operated. We select this last device for brief discussion, because it is representative of many, and its design and operation are well documented.[33–35]

The minor diameter of the torus is 1 m, and the major diameter is about 3.7 m. A simplified schematic diagram is shown in Fig. 14.18. The

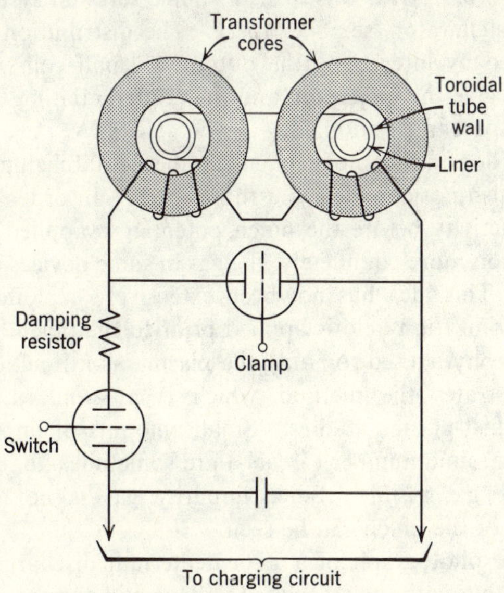

Fig. 14.18. Simplified schematic diagram of Zeta.

damping resistor prevents ringing of the circuit; the clamp, sometimes called a crowbar, shorts the primary at the current maximum; since the primary winding-clamp circuit has low impedance, the primary current remains relatively steady for an extended period, and there is no induced reverse voltage in the secondary discharge tube. The circuit is commonly used both in this application and in sustaining high solenoidal inductions in discharge tubes. Primary current, secondary plasma current (by Rogowski coils), and loop voltage are measured. Small magnetic coils sample the induction inside the plasma, and plots of the same nature as shown in Fig. 14.16 (but on a millisecond time scale) are obtained.

The torus itself has interesting constructional features. For plasma stability, it is metal (2.5-cm aluminum) and should be of one piece. However, such an arrangement would break down at the necessary single gap, because of the induced loop voltage ($\approx$2000 volts). Thus a second segmented torus lies just inside the main wall; the segments are insulated from each other; the annular space between the main wall and liner is evacuated. There are 48 segments; with about 40 volts per insulating gap, arcing is kept tolerably low.

A stabilizing axial induction of approximately 0.016 weber/m^2 is applied to the torus. Initial gas pressure is very low (about 10^{-4} mm deuterium); the plasma current is about 10^5 amp, and the plasma pinches to a radius of 0.2 m. The current channel drifts outward with time, as was shown in Fig. 10.9. It is this drift in simple toroidal systems that is overcome in the stellarator (see Sec. 10.2). The distribution of plasma current is obtained by integrating the output of small coil probes with respect to time to give the induction, and then differentiating this derived induction with respect to position.

The Zeta pinch does not completely entrap the B_z stabilizing induction. The current I_z is also not confined to a thin sheath. In order to increase the plasma conductivity before the pinch potential is applied and hence to trap the induction more completely, the gas in some devices (Columbus II) is preheated. This idea has not been extensively exploited in Zeta, although a rf-preionization field is applied prior to the pinch.

Streak photography is used to study the plasma as a function of time. Figure 14.19 illustrates the method, which is a standard diagnostic technique for all fast-plasma studies. Small amounts of impurity gases of relatively high atomic number (5% N_2) are sometimes added to make the events in the gas visible. Since impurity gas is not completely stripped, the core of the pinch can be seen.

A typical streak photograph of a Zeta deuterium discharge is shown in Fig. 14.20, together with the associated voltage and current waveforms. The pinch breaks up completely in about 3000 μsec; we see that its

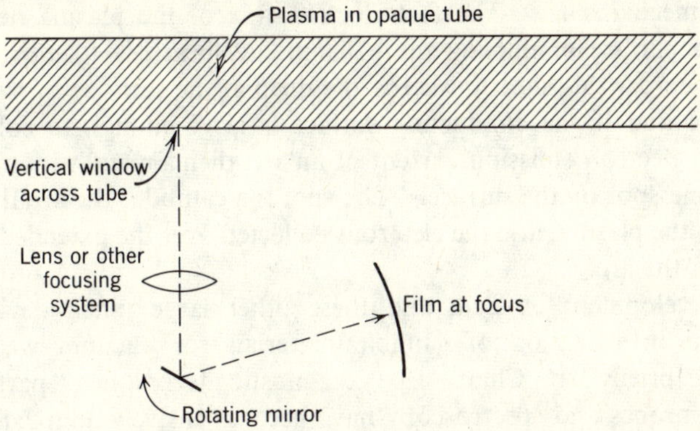

Fig. 14.19. Principle of streak camera.

behavior before that time is anything but simple. There was no addition of nitrogen to this deuterium discharge, so the pinch is not visible. Most of the light comes from impurities outside the core. Spectral lines are frequently used to estimate the temperature of the plasma by a study of the Doppler width. In Zeta, the ions may possibly be as hot as 500 ev.

A number of bright arcs can be seen on the streak photograph of Fig. 14.20. Some of these arcs are unipolar and are sustained by a

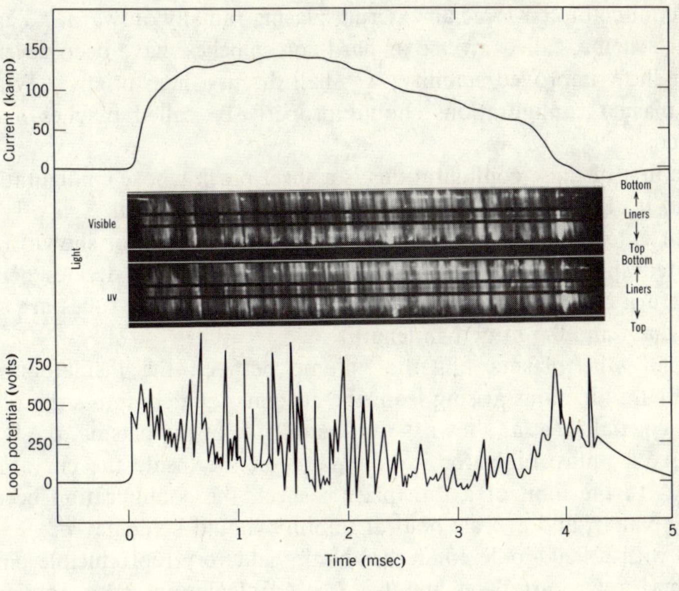

Fig. 14.20. Streak camera photograph of a pinched deuterium plasma in Zeta. Current and voltage waveforms in the plasma are also shown. After E. P. Butt and co-workers.[33]

peculiar mechanism.[36] The fast electron flux of the plasma negatively charges a surface contacting the plasma. A sheath is formed. If the potential across this sheath exceeds the drop at the cathode of a conventional arc and if the conducting surface is large enough to collect the minimum electron emission current of an arc, then a metal vapor arc may form at one spot on the surface. The spot is a cathode; the arc then runs back into the plasma, and the electrons collected over the extended surface return via the arc.

The development of Zeta and these other large pinch devices has stimulated investigation of suitable materials for vacuum walls,[37] as discussed briefly in Chap. 13. Diagnostic techniques, particularly magnetic probes and spectroscopy have been extensively adapted for use in Zeta.[38] The general techniques of magnetic probing and interpretation of the data are described by Lovberg.[39] Some examples of the use of microwaves will be presented in later chapters.

14.9. Sheet and Inverse Pinches

To this point, we have recorded the development of conventional pinches in which the plasma is compressed on its central axis. This method is not the only way in which the forces can be made to act upon the plasma. In particular, the magnetic pressure of the current in a solid axial conductor can force an external plasma radially outward. Variants of this scheme, called inverse or hard-core pinches, have been developed which show improved stability; we shall discuss them briefly. Whether such plasma configurations should properly be called pinches may be debated.

The first of these configurations is a sheet pinch whose configuration is shown in Fig. 14.21. This particular device is called Triax.[40] The current enters the plasma from an annular electrode (not shown) in the insulated space at the upper end of the assembly and divides between two return conductors at the bottom. These particular tubes are 0.1-m outer diam and 0.5-m or 1-m length.

Initially the plasma fills the volume between the insulators. The azimuthal inductions arising from the currents then compress the plasma into a coaxial sheet as shown. On the outside of the plasma, the **B** lines are curved unfavorably for stability, but on the inside the curvature is stable. In the limit of a thin plasma sheet, the configuration becomes locally planar, and at least neutral stability should be achieved. Experiments with a 9000-joule condenser bank yield very reproducible pinches with maximum current at about 3 μsec. Calculations show that under these circumstances, the plasma temperature is uniform at about 15 ev.

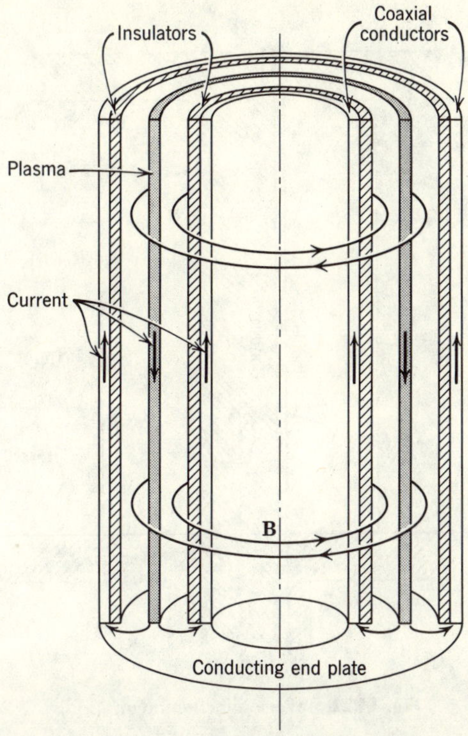

Fig. 14.21. The coaxial Triax pinch.

Experiments with energy input greater than 20,000 joules and initial plasma preheating show small reproducible oscillations of the plasma sheet; here, about 5×10^4 neutrons/pulse are produced in deuterium, probably arising from small instabilities. It is estimated that the plasma temperature in this latter case might possibly be as high as 300 ev because of the joule heating, shock heating, and adiabatic compresssion. A theoretical calculation[41] shows that the plasma in Fig. 14.21 may be unstable with respect to shredding into axial filaments.

The inverse pinch[42] is a more extreme case of the previous example. The configuration is shown in Fig. 14.22; the roles of plasma and return conductor in a simple conventional pinch have been completely reversed. Approximate dimensions of the tube are the following: outer diameter of insulator shown, 1 cm; inner diameter of outer insulator (not shown), 20 cm; electrode spacing, 8 cm. While the direct application of this configuration to thermonuclear power generation is unlikely, there are a number of experimental advantages to the system. First, the pinch surface is visible and highly accessible to probes through the region of

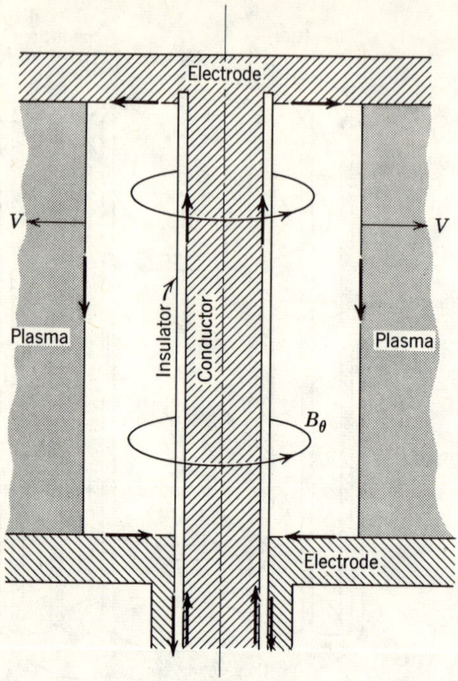

Fig. 14.22. Inverse pinch configuration.[42]

undisturbed gas outside the pinched layer. Second, the expanding boundary is very stable; photographs show the pinch surface to be perfectly straight and without detectable instability.

The outward movement of a deuterium plasma front has been measured by timing the passage of the luminous front across slits placed before the tube. If the pinch behaves according to snowplow theory, Eqs. 14.12 to 14.14 should be applicable, with the distance r_0 now being the inner starting radius. Assume for the moment that the current I has the simple form $I = I_0\omega t$, corresponding to the initial current rise. Define the dimensionless distance

$$y = r/r_0 \tag{14.95}$$

and the dimensionless time

$$\tau = \nu t,$$

where

$$\nu = (\mu_0\omega^2 I_0^2/4\pi^2\rho_{m0}r_0^4)^{1/4}. \tag{14.96}$$

Equation 14.12 then becomes

$$\frac{d}{d\tau}\left[(y^2 - 1)\frac{dy}{d\tau}\right] = \frac{\tau^2}{y}, \tag{14.97}$$

with the boundary conditions $y = 1$ and $dy/d\tau = 0$ at $\tau = 0$. The solution of Eq. 14.97 is shown as the solid curve in Fig. 14.23. More precisely, the current distribution $I = I_0 \sin \omega t$ should be used; in the limit of small ω/ν, the correction is small and readily obtained.[42] The corrected solution is shown by the dashed curve of Fig. 14.23, along with corresponding experimental points measured during slow pulsed experiments. The agreement is remarkably good; other data obtained at different

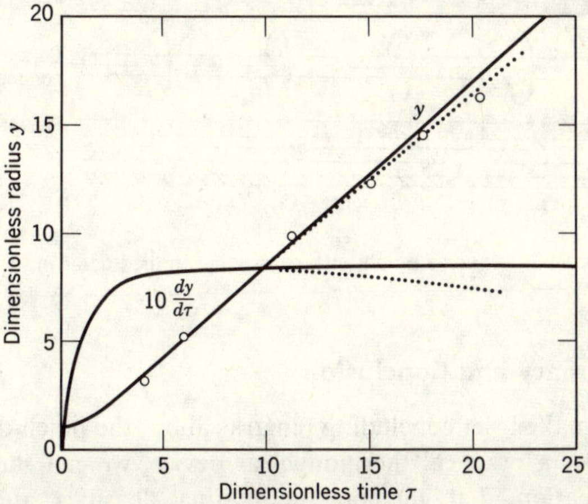

Fig. 14.23. Theoretical dependence of dimensionless radius of inverse pinch $y = r/r_0$ versus dimensionless time $\tau = (\mu_0\omega^2 I_0{}^2/4\pi^2\rho_{m0}r_0{}^2)^{1/4}t = \nu t$. The solid curve shows the case of linearly rising current, and the dotted curve represents the actual dependence $I = I_0 \sin \omega t$, with $\omega/\nu = 0.017$. Experimental points at 1 mm D_2, 5 kv are shown. After O. A. Anderson and co-workers.[42]

pressures and voltages show similar agreement within 10%. The light from the discharge is sharply localized in the pinch front; magnetic probe signals indicate a current-carrying front about 0.9 cm deep. Thus the snowplow model appears moderately well justified in this case.

Experiments with faster pinches show almost as good agreement. Probe measurements in this case indicate that the vacuum behind the plasma front is actually a low-pressure plasma, possibly produced by continual desorption of gas from the inner insulator tube. On the second current half-cycle, some of the original B_θ from the first half-cycle remains trapped in the plasma. Then the reversed B_θ near the central conductor forms a local annular region of rapidly changing induction in the plasma, reminiscent of the Triax configuration of Fig. 14.21.

Figure 14.24 shows the inverse pinch with a superimposed axial B_z induction. It is the precise inverse of the conventional B_z-stabilized pinch. Here, however, the B_z induction may be pulsed to keep the plasma off the outer wall, presumably with at least neutral stability.

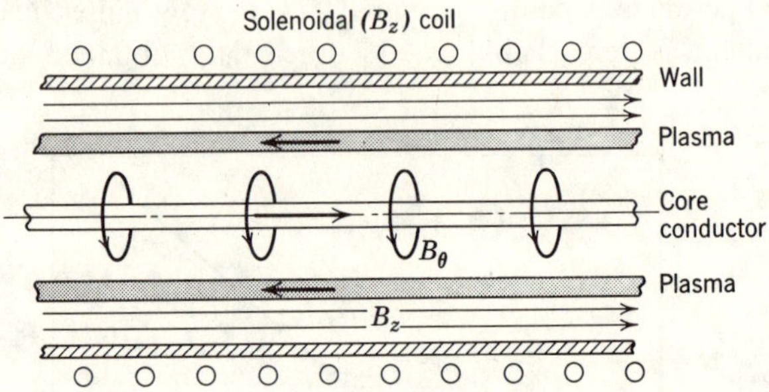

Fig. 14.24. Inverse pinch with externally applied axial induction.

14.10. Summary and Conclusion

We shall make brief concluding remarks about the pinch developments.

In terms of a practical thermonuclear device, we can state with some oversimplification that the present plasma densities are more than adequate, the temperature is about 100 times too low, the volume is several orders of magnitude too small, and the confinement time is about 10^3 times too short. All these shortcomings were, of course, realized even at the time the experiments were designed. For example, the confinement times could not have been very much longer because of limitations on the input energy. A more serious shortcoming is the propensity of the pinch to commit suicide by turbulent instability.

This question of stability, both in velocity and in configuration space, is the central one concerning all proposed thermonuclear systems. The point cannot be overemphasized. With the possible exception of the various inverse pinches, none appears to be particularly stable and therefore suitable for controlled thermonuclear power. The neutrons produced do not seem to be of true thermonuclear origin.

Finally, we refer again to our conclusion of Chap. 13, that fast pulsed systems present very particular difficulties in terms of energy conversion.

In spite of these difficulties, the more nearly stabilized pinches will continue to be useful experimentally, not only because large volumes of hot plasma can conveniently be produced this way, but also because their investigation may suggest more stable configurations.

PROBLEMS

1. Consider an idealized cylindrical static pinch in which the total current I_z flows on the plasma surface at radius r_1. The return current flows on an outer concentric cylinder at radius r_2. There is no stabilizing induction B_z. Compute the ratio

$$\frac{\text{Total particle energy}}{\text{Total magnetic energy}}$$

if the plasma is in equilibrium and the velocities are isotropic inside.

2. If the axial electron drift v_z is constant everywhere in a simple linear static pinch, show that the electron density radial distribution is

$$n_e = \frac{n_0}{[1 + (r/r_0)^2]^2},$$

where

$$\mu_0 e^2 r_0^2 n_0 v_z^2 = 16kT.$$

Check that Eq. 14.7 is satisfied. This distribution is attributed to Bennett.

3. Derive Eq. 14.11 by considering the fields and their rates of change.

4. Derive Eq. 14.12.

5. It is suspected that the $m = 0$ (sausage) instability is generating neutrons in a deuterium pinched plasma by accelerating deuterons locally in the direction of the current and impinging them upon stationary deuterons. What energy resolution is required of neutron detectors placed at the anode and cathode if the mechanism is to be discovered and if the deuterons are accelerated to (*a*) 10 kev, (*b*) 50 kev?

6. Show that for a simple B_z-stabilized pinch with currents carried at the plasma surface, $m = 0$ stability is guaranteed if inside the plasma $B_z = B_\theta/\sqrt{2}$, irrespective of whether or not there is a longitudinal induction B_z outside the plasma.

7. Calculate the diffusion time for a plasma with average energy 10 ev, density $10^{20}/\text{m}^3$, a distance 0.01 m across an induction of 1 weber/m^2. Neglect the fact that the plasma will be compressed by the induction.

8. Prove that if the magnetic induction diffuses across a plasma and propagation may be neglected,

$$\frac{\partial}{\partial t}\int d\mathbf{r}\,(B^2/2\mu_0) = -\int d\mathbf{r}\,(\mathbf{E}\cdot\mathbf{j}),$$

and hence that the intermixing of magnetic inductions in the pinch will heat the plasma.

9. Show that if Eq. 14.69 is satisfied, the pinch discharge is not necessarily unstable.

10. Prove that if $|rB_\theta|$ decreases everywhere with increasing radius in a linear pinch, Suydam's criterion is trivially satisfied. Hint: Re-express Suydam's criterion in terms of $d(\log br^4)/dr$.

11. In a slow pinch, let the minor radius be 0.5 m, axial induction 0.016 weber/m^2, pinch current 10^5 amp, time to current maximum 10^{-3} sec, initial gas pressure 10^{-4} mm D_2. Calculate approximately the equilibrium pinch radius on the assumption that the current flows in a thin layer at the pinch surface. Is the material pressure likely to be important? The parameters are similar to those of Zeta with neglect of the toroidal curvature.

12. The initial design requirements of Zeta were the following: bore of torus = 1 m; maximum resistive emf around axis = 200 volts/m, maintained for 3 msec. Since this device is slow, an ion core transformer will be used. The space factor of the core is 0.6, and transformer steel effectively saturates at 1.6 webers/m^2. Calculate the mean major torus diameter if the core can be magnetically biased.

REFERENCES

1. W. H. Bennett, *Phys. Rev.*, **45**, 89 (1934).
2. M. N. Rosenbluth, *Infinite Conductivity Theory of the Pinch*, Report LA-1850, Los Alamos Scientific Laboratory, Los Alamos, New Mexico (Sept. 1954).
3. M. N. Rosenbluth, *Stability of the Pinch*, Report LA-2030, Los Alamos Scientific Laboratory, Los Alamos, New Mexico (July 1959).
4. O. A. Anderson, W. R. Baker, S. A. Colgate, J. Ise, and R. V. Pyle, "Neutron Production in Linear Deuterium Pinches," *Proc. 3rd International Conf. on Ionization Phenomena in Gases*, Venice (1957), p. 62.
5. T. E. Allibone, D. R. Chick, G. P. Thomson, and A. A. Ware, *Proc. 2nd U.N. Conf. on Peaceful Uses of Atomic Energy*, Vol. 32, United Nations, Geneva (1958), p. 169.
6. G. A. Sawyer, P. L. Scott, and T. F. Stratton, *Phys. Fluids*, **2**, 47 (1959).
7. B. R. Suydam, *Proc. 2nd U.N. Conf. on Peaceful Uses of Atomic Energy*, Vol. 31, United Nations, Geneva (1958), p. 157.
8. W. A. Newcomb, *Ann. Phys.*, **10**, 232 (1960).
9. M. N. Rosenbluth, unpublished communication.
10. J. L. Tuck, *Proc. 2nd U.N. Conf. on Peaceful Uses of Atomic Energy*, Vol. 32, United Nations, Geneva (1958), p. 3.
11. D. C. Hagerman and A. H. Williams, *Rev. Sci. Instr.*, **30**, 182 (1959).
12. J. W. Mather and A. H. Williams, *Proc. 2nd U.N. Conf. on Peaceful Uses of Atomic Energy*, Vol. 32, United Nations, Geneva (1958), p. 26.
13. H. Herold, E. Fünfer, G. Lehner, H. Tuczek, and C. Andelfinger. *Z. Naturforsch.*, **14a**, 323, 329 (1959).
14. Yu. G. Prokhorov and N. V. Filippov, *J. Nuclear Energy*, **4**, No. 2, 203 (1957); *Atomnaya Energ.*, **1**, No. 3, 76 (1956).
15. L. A. Artsimovich, A. M. Andrianov, E. I. Dobrokhotov, I. M. Podgornyi, V. I. Sinitsin, and N. V. Filippov, *J. Nuclear Energy*, **4**, No. 2, 213 (1957); *Atomnaya Energ.*, **1**, No. 3, 84 (1956).
16. S. Yu. Lukyanov and V. I. Sinitsin, *J. Nuclear Energy*, **4**, No. 2, 216 (1957); *Atomnaya Energ.*, **1**, No. 3, 88 (1956).
17. A. L. Bezbatchenko, I. N. Golovin, D. P. Ivanov, V. D. Kirilov, and N. A. Yavlinsky, *J. Nuclear Energy*, **5**, No. 1, 71 (1957); *Atomnaya Energ.*, **1**, No. 5, 26 (1956).
18. V. S. Komelkov, *Zhur. Eksp. i. Teoret. Fiz.*, **35**, 16 (1958).
19. S. I. Braginsky and V. D. Shafranov, *Proc. 2nd U.N. Conf. on Peaceful Uses of Atomic Energy*, Vol. 31, United Nations, Geneva (1958), p. 43.

20. V. S. Komelkov, U. V. Skvortsov, and S. S. Tserevitinov, *Proc. 2nd U.N. Conf. on Peaceful Uses of Atomic Energy*, Vol. 31, United Nations, Geneva (1958), p. 374.
21. V. S. Komelkov, *Proc. 2nd U.N. Conf. on Peaceful Uses of Atomic Energy*, Vol. 31, United Nations, Geneva (1958), p. 382.
22. A. M. Andrianov, O. A. Bazilevskaia, S. I. Braginsky, B. G. Brezhnev, S. Khavaschevski, V. A. Khrabrov, N. G. Kovalski, N. V. Filippov, T. I. Filippova, V. E. Palchikov, I. M. Podgornyi, Yu. G. Prokhorov, and M. M. Sulkovskaya, *Proc. 2nd U.N. Conf. on Peaceful Uses of Atomic Energy*, Vol. 31, United Nations, Geneva (1958), p. 348.
23. I. N. Golovin, D. P. Ivanov, V. D. Kirilov, D. P. Petrov, K. A. Razumova, and N. A. Yavlinsky, *Proc. 2nd U.N. Conf. on Peaceful Uses of Atomic Energy*, Vol. 32, United Nations, Geneva (1958), p. 72.
24. S. L. Leonard and L. O. Heflinger, *Bull. Am. Phys. Soc.*, Ser. II, **5**, 340 (1960).
25. L. C. Burkhardt and R. H. Lovberg, *Proc. 2nd U.N. Conf. on Peaceful Uses of Atomic Energy*, Vol. 32, United Nations, Geneva (1958), p. 29.
26. J. Killeen, G. Gibson, and S. A. Colgate, *Phys. Fluids*, **3**, 387 (1960).
27. R. Aymar, C. Etievant, P. Hubert, A. Samain, B. Taquet, and A. Torossian, *Proc. 2nd U.N. Conf. on Peaceful Uses of Atomic Energy*, Vol. 32, United Nations, Geneva (1958), p. 92.
28. J. Andreoletti, C. Breton, J. Charon, P. Hubert, P. Jourdan, and G. Vendreyes, *Proc. 2nd U.N. Conf. on Peaceful Uses of Atomic Energy*, Vol. 32, United Nations, Geneva (1958), p. 100.
29. G. G. Dolgov-Saveliev, D. P. Ivanov, V. S. Mukhovatov, K. A. Razumova, V. S. Strelkov, M. N. Shepelyev, and N. A. Yavlinsky, *Proc. 2nd U.N. Conf. on Peaceful Uses of Atomic Energy*, Vol. 32, United Nations, Geneva (1958), p. 82.
30. N. L. Allen, T. E. Allibone, D. R. Chick, R. F. Hemmings, T. P. Hughes, S. Kaufman, B. S. Liley, J. G. Mack, H. T. Miles, R. M. Payne, J. E. Read, A. A. Ware, J. A. Wesson, and R. V. Williams, *Nature*, **181**, 222 (1958).
31. T. P. Hughes and S. Kaufman, *Nature*, **183**, 7 (1959).
32. A. A. Ware, *Nature*, **183**, 8 (1959).
33. E. P. Butt, R. Carruthers, J. T. D. Mitchell, R. S. Pease, P. C. Thonemann, M. A. Bird, J. Blears, and E. R. Hartill, *Proc. 2nd U.N. Conf. on Peaceful Uses of Atomic Energy*, Vol. 32, United Nations, Geneva (1958), p. 42.
34. P. C. Thonemann, E. P. Butt, R. Carruthers, A. N. Dellis, D. W. Fray, A. Gibson, G. N. Harding, D. J. Lees, R. W. P. McWhirter, R. S. Pease, S. A. Ramsden, and S. Ward, *Nature*, **181**, 217 (1958).
35. B. Rose, A. E. Taylor, and E. Wood, *Nature*, **181**, 1630 (1958).
36. A. E. Robson and P. C. Thonemann, *Proc. Phys. Soc.* (*London*), **73**, Pt. 3, 508 (1959).
37. J. L. Craston, R. Hancox, A. E. Robson, S. Kaufman, H. T. Miles, A. A. Ware, and J. A. Wesson, *Proc. 2nd U.N. Conf. on Peaceful Uses of Atomic Energy*, Vol. 32, United Nations, Geneva (1958), p. 414.
38. P. C. Thonemann, *Nuovo cimento Suppl.*, **13**, 111 (1959).
39. R. H. Lovberg, *Ann. Phys.*, **8**, 311 (1959).
40. O. A. Anderson, W. R. Baker, J. Ise, Jr., W. B. Kunkel, R. V. Pyle, and J. M. Stone, *Proc. 2nd U.N. Conf. on Peaceful Uses of Atomic Energy*, Vol. 32, United Nations, Geneva (1958), p. 150.
41. H. P. Furth, Paper E-3, 2nd Annual Meeting, Division of Plasma Physics, Gatlinburg, Tenn. (November, 1960).
42. O. A. Anderson, H. P. Furth, J. M. Stone, and R. E. Wright, *Phys. Fluids*, **1**, 489 (1958).

GENERAL REFERENCES

G. H. A. Cole, "The Pinch Effect," *Sci. Progr.*, **47**, 437–458 (1959).

S. Glasstone and R. H. Lovberg, *Controlled Thermonuclear Reactions*, D. Van Nostrand Co., Princeton, N. J., (1960), Chap. 7.

J. L. Tuck, "Controlled Thermonuclear Research at Los Alamos," *Proc. 2nd U.N. Conf. on Peaceful Uses of Atomic Energy*, Vol. 31, United Nations, Geneva (1958), pp. 3–25.

15

Mirror-like devices

The magnetic mirror principle has been applied to a wide variety of plasma confinement and heating devices. We have developed the basic concepts of the simple magnetic mirror in Chap. 10, particularly Sec. 10.6. There it was shown that suitably directed particles can be reflected in a region of increasing induction, that they can be compressed and heated by a time-rising induction or conversely expanded and cooled by a time-decreasing induction. These statements are also in general agreement with the hydromagnetic derivations of Chap. 7. In practice, however, plasma space charge and some further consequences of the Coulomb scattering process will increase the ionic loss rate over that naively expected; we summarize these and a few other general effects in Sec. 15.1.

On the basis of these concepts, we present a number of specific developments. Some of them, such as pulsed-mirror compression experiments and trapping of ion beams in mirrors, are obviously and directly related to mirrors. On the other hand, others, such as the production of the beam to be trapped and the effect of the trapped plasma on the background gas, are related to the mirror only by association; they apply equally well to the systems described in Chaps. 14 and 16. Still others, in which the mirror is part of a more complex configuration, occupy an intermediate relationship.

The first systems to be described are the direct magnetic compression experiments, in which a volume of plasma is heated and compressed by pulsed mirrors. These experiments may be relatively large and slow, or small and fast. They have produced deuteron plasma temperatures of the order of 1 to 3 kev with associated neutron production in the gas. In some of these pulsed experiments, it is proposed to inject the plasma via ion beams, but the injection problem appears formidable.

Following these systems, devices are briefly discussed in which a radial electric field is also applied to the plasma. The $\mathbf{E} \times \mathbf{B}$ azimuthal drift may improve the mirror confinement. The particular devices happen to be pulsed, but in principle need not be.

We devote considerable space to steady-state mirror systems. From the discussion of Chap. 13, we see a number of possible advantages in steady-state (or almost steady-state) systems over rapidly pulsed ones. One of the principal problems with steady-state mirrors is that of producing the plasma inside. We remark on several schemes and develop the concept of burnout of the background residual gas by the trapped plasma. Sections 15.4 and 15.5 describe other problems such a system would have to face in practice.

In addition to these relatively simple configurations, a number of derived ones have been postulated. A simple one is the cusp (Figs. 12.7 and 12.8). A much more complicated arrangement is the Astron, a device that uses a mirror only incidentally to establish a trapped solenoidal current sheet of relativistic electrons; the induction of this current sheet then confines the plasma.

We note here in passing that magnetic mirrors surround us in space. The earth's dipole induction confines high-energy particles (the van Allen belts) for long periods. Similar configurations exist on the surface of the sun, particularly in the vicinity of sunspots. Discussion of the phenomena taking place in these naturally occurring mirrors is beyond the scope of our book.

15.1. Particle Loss from a Mirror

In Secs. 13.4 and 13.5, we considered at some length the energy balance in a lossy confinement system, such as a mirror. We introduced a confinement factor, which was the probability that a particle would escape in one angular relaxation time. We were careful not to relate this quantity in a specific way to the mirror ratio. The actual loss rate depends upon additional considerations that we have now to touch upon.

Most naively, the probability of escape in one random angular relaxation might be expected to be the chance that it scatters into the escape cone in velocity space (see Fig. 10.13). This statement is true. We might then estimate that this fractional probability is to be equated to the fraction of a sphere in velocity space covered by the escape cones, calculated at the midplane. If θ_m is the escape angle at the midplane, as shown in Fig. 10.13, this open fractional area is

$$\frac{\Delta\Omega}{2\pi} = 1 - [1 - (1/R)]^{1/2}, \tag{15.1}$$

where R is the mirror ratio. Equation 15.1 is also the fraction of an isotropic distribution which would be lost from the mirror in the first axial transit (see Problem 8 of Chap. 10). In the limit of a large mirror ratio, $\Delta\Omega/2\pi \rightarrow 1/2R$. This treatment of the escape rate is inadequate.

The first correction to the escape rate involves the potential of the plasma confined inside with respect to the external apparatus. Electrons are much more mobile than the ions in their motion along the field lines, even if the electron energy is considerably less than the ion energy. Thus the electrons escape more readily, until the plasma potential inside becomes about equal to the average electron energy and thereby brings the electron and ion particle currents into balance. This potential tends to accelerate the ions along the axis. Figure 15.1 illustrates the calculation. The ion whose velocity components are $v_{i\parallel}$ and $v_{i\mathrm{T}}$ parallel and transverse, respectively, to **B** is to be accelerated by the axial space-charge electric field so that its new parallel velocity $v'_{i\parallel}$ just brings the particle into the mirror loss cone of angle θ_m. The axial electric field increases the longitudinal ion energy by $m_e v_e^2/2$, so that

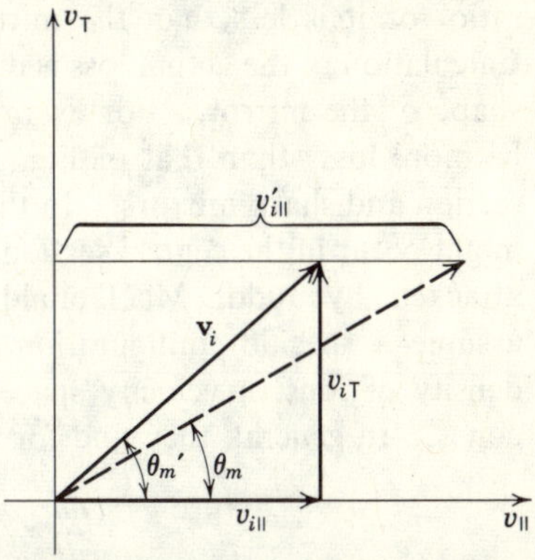

Fig. 15.1. Velocity vectors for an ion accelerated out of a mirror.

$$m_i(v'_{i\parallel})^2 = m_i(v_{i\parallel})^2 + m_e v_e^2. \tag{15.2}$$

The effective escape cone is then defined by the angle θ_m'; the mirror ratio $R = \csc^2 \theta_m'$ (Eq. 10.66) is

$$R = \frac{(v'_{i\parallel})^2 + (v_{i\mathrm{T}})^2}{(v_{i\mathrm{T}})^2} = \frac{v_i^2}{(v_{i\mathrm{T}})^2}\left(1 + \frac{m_e v_e^2}{m_i v_i^2}\right). \tag{15.3}$$

The effective mirror ratio $R' = (v_i/v_{i\mathrm{T}})^2$ is then

$$R' = \frac{R}{1 + (T_e/T_i)}, \tag{15.4}$$

a result that has also been derived by more detailed analysis.[1] Here is evidence further to that presented in Chap. 13 that excessively hot electrons are in general undesirable.

An even more important effect on the escape rate arises from the fact that the particles undergo scattering through all regions of the mirror and not just at the midplane. Examine Fig. 10.13, where a typical particle is reflected at $z = z_h$. If the particle undergoes scattering there, it must return at least to that point and in general will take up a new orbit with increased z_h. Because the axial speed of a particle is least near z_h, the particle has more chance of being scattered there than near the midplane. The matter may be stated in another equivalent way. A trapped particle moving in a mirror sees a magnetic induction whose average value is greater than the value B_0 at the midplane. From the fact that the particle may be scattered anywhere, the effective mirror ratio for it is less than the mirror ratio R calculated at the midplane. Calculation of the actual loss is difficult and depends upon the geometrical shape of the mirror. For example, the configuration of Fig. 10.13 will be more lossy than that of Fig. 10.14, which has a long uniform central section and sharp mirrors. In this latter case, a majority of the collisions might occur in the central section, where $B \approx B_0$. The problem has been attacked by Judd, MacDonald, and Rosenbluth[2] and others. They assume a specific configuration and, as a boundary condition, set the density of ions in velocity space equal to zero at the edges of the loss cones. In general, they find for the loss rate dn/dt

$$\frac{dn}{dt} = -n^2 \overline{Y(v, R)}, \tag{15.5}$$

where $\overline{Y(v, R)}$ is some averaged function of the velocity distribution and the mirror ratio. The n^2 dependence shows the expected behavior for mutual scattering of the particles. From Eq. 15.5, we can define a time constant for their escape:

$$\tau = (n\overline{Y})^{-1}. \tag{15.6}$$

For a parabolic mirror (the most lossy of reasonable configurations) Judd and his associates[2] find for singly charged particles in a mirror with moderately high ratio R

$$\tau \approx \frac{12\pi\epsilon_0{}^2 m^2 \log_{10} R}{ne^4\overline{v^{-1}}\,\overline{v^{-2}}(\ln \Lambda)}. \tag{15.7}$$

For purposes of comparison, we may assume a Maxwellian distribution, set $\overline{v^{-1}} = (2m/\pi kT)^{1/2}$, $\overline{v^{-2}} = (m/kT)$. The result becomes just

$$\tau \approx \tau_\theta \log_{10} R = \tau_\theta/\psi, \tag{15.8}$$

where τ_θ, Eq. 8.23, is the angular relaxation time and ψ is the confinement factor introduced in Sec. 13.4. The factor τ_θ agrees with our intuition regarding the nature of the scattering. The factor $\log_{10} R$ shows that the confinement is very much less than that expected from the naive treatment ending in Eq. 15.1. It must be emphasized that Eq. 15.8 is not applicable to mirrors with long central sections and that the escape rate is reduced from the high value given by this equation.

Because the scattering rate is higher for the slow particles and lower for the fast particles in a velocity distribution, the scattered-out particles will have substantially less than the average energy. Judd and his associates[2] find for their example that the average energy of the escaping particles is about $\frac{1}{3}$ of the average energy inside. This correction is favorable in computing energy balances. For simplicity, this advantage was not included in the examples of Secs. 13.4 and 13.5, because the velocity distributions were not really Maxwellian anyway, and other approximations were made as well.

If a plasma is to be heated by magnetic compression, the compression rate must be high enough so that few particles are lost during the process. Suppose the total number of trapped particles is N and that they are to be compressed by a factor of C in time t_C. Since few are to be lost, we must have

$$\int_0^{t_C} dt \frac{dN}{dt} \ll N. \tag{15.9}$$

But

$$\frac{dN}{dt} \approx -\frac{N\psi}{\tau_\theta} = \frac{Nn(t)\psi}{K[T(t)]^{3/2}} \tag{15.10}$$

from Eqs. 8.23 and 15.8, where K is a constant. The analysis will be approximately correct if the compression (hence the changes in the particle velocity distribution) is isotropic. In that case, from Eq. 7.28, we find that $n/T^{3/2}$ remains constant through the cycle for an adiabatic compression, and Eq. 15.9 becomes

$$\frac{n_0 \psi t_C}{K T_0^{3/2}} \ll 1, \tag{15.11}$$

where n_0 and T_0 are the initial values. For example, for deuterons, $K = 3.2 \times 10^{16}$ in our mks–kev units, for $\ln \Lambda = 20$. Then if $n_0 = 10^{18}/\text{m}^3$, $T_0 = 0.1$ kev, and $\psi = 2$, the compression time τ_C must be much less than 5×10^{-4} sec if few particles are to be lost. On the other hand, if the initial density is much higher and the initial temperature is only a few electron volts, corresponding to small pulsed experiments, the compression must be accomplished in fractions of a microsecond. This simple calculation ignores nonadiabatic effects; if the speed of the magnetic wall exceeds the mean ion speed, a shock wave develops, and the ions are heated nonadiabatically. In fast pulsed experiments, it is this shock wave that causes most of the heating.

15.2. Fast-Pulsed Mirrors

An important class of experiments is that involving the very fast compression of a plasma in a magnetic mirror. Small volumes of plasma at high density and high temperature (1 to 2 kev) have been produced this way. The compressed state lasts for a few microseconds, apparently limited by the experimental parameters. Basically similar schemes have been proposed by a number of authors,[3–7] both in mirror and toroidal configurations.

A basic configuration is shown in Fig. 15.2. The enlarged ends of the single-turn coil make a mirror, because the current runs on the conductor

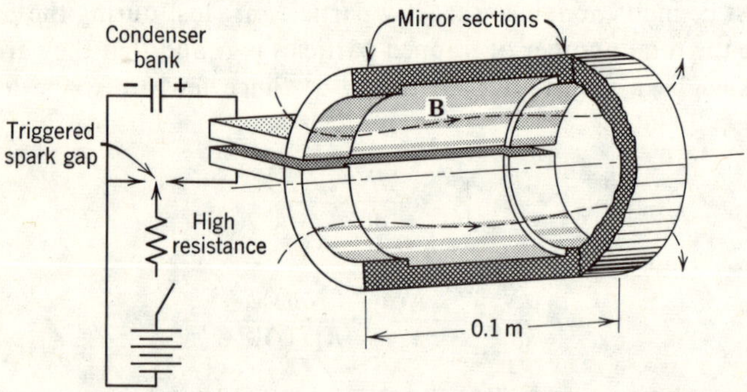

Fig. 15.2. Cutaway view of pulsed mirror configuration (Scylla).[4]

surfaces during the short pulse time. The external circuit is similar to that used for fast pinches: parallel-plate transmission lines, high-voltage condenser banks, and multiple spark-gap switches in parallel with each other to reduce the external inductance. The configuration is sometimes called a collapsed discharge.

We shall summarize some of Kolb's[5] experiments (in some of them the mirror coil is segmented). An insulated discharge tube passes through the coil, similar to that shown in Fig. 15.2. Two alternate sequences of operation are possible. In the first, magnetically driven shock waves from each end of the tube meet in the center; this initial hot plasma is then compressed by pulsing the mirror. In the second mode of operation, the gas is partly ionized by an rf discharge; a preheating condenser bank is discharged through the mirror to complete the ionization; finally, the main condenser bank sets up inward radial and axial shock waves, which heat the plasma and compress it.

In the first sequence, the colliding shock waves produce a plasma with total particle density about $10^{22}/m^3$ and temperature about 100 ev. For protons, the mean thermal speed considerably exceeds the speed of subsequent radial compressions ($\approx 10^5$ m/sec), so that the magnetic compression is adiabatic. The details of the compression can then be worked out relatively easily in the following way. The radius of the coil is r_c, the initial radius of the plasma is r_0, and the radius of the plasma at time t is r_p. The inductance of the coil alone is L_c, and the inductance of the coil with the approximately cylindrical plasma is $L = L_c[1 - (r_p/r_c)^2]$. The circuit equation is then

$$V_L = V_0 - \frac{1}{C}\int_{t_0}^{t} dt\, I = L_c \frac{d}{dt}\left[\left(1 - \frac{r_p^2}{r_c^2}\right)I\right], \tag{15.12}$$

where V_L is the voltage at the coil, and V_0 is the initial voltage. In addition, we have $B^2/2\mu_0 = nkT$ (Eq. 7.14, $\beta = 1$); $B = \mu_0 I/l$, where l is the effective axial length; $T/T_0 = (n/n_0)^{\gamma-1}$, where T_0 and n_0 are the initial values; and $n/n_0 = (r_0/r_p)^2$ for a principally radial compression. These equations yield

$$I = (2l^2 n_0 k T_0/\mu_0)^{1/2}(r_0/r_p)^{\gamma}; \tag{15.13}$$

the first quantity in parentheses is the critical current I_0 required to take the plasma off the walls and start the compression. Substituting Eq. 15.13 into 15.12 yields an equation for the plasma radius r_p as a function of time. Kolb finds the solution of the equations for $\gamma = 2$, and shows that the maximum temperature $T_{\max}$ achieved at the peak of the compression cycle is

$$T_{\max} \approx T_0\left(\frac{r_c}{r_0}\right)\left(\frac{\text{initial energy stored in bank}}{\text{initial energy stored in plasma}}\right)^{1/2}. \tag{15.14}$$

Since the storage bank contains many kilojoules and the plasma initially contains 10 or 20 joules, the maximum temperature can be very high in principle.

In the second sequence—rf preionization, plus an inward shock from the pulsed mirrors—the behavior may be analyzed by an adaptation of the snowplow model of the dynamic pinch (Secs. 14.1 and 14.9). In this connection, note that the imploding magnetic wall consists of B_z lines of induction; the induced current is j_θ; therefore, the configuration is similar to the normal pinch, except that the **B** and **j** lines have each been rotated 90°. For this reason, the configuration is sometimes also called the theta pinch. Figure 15.3 is a drawing made from smear camera photographs of the compression with preionization but no preheating pulse. The first half-cycle (not shown in the figure) of the main pulse completes the ionization; the figure shows the second half-cycle. With a preheating pulse, the phenomena are transferred to the first half-cycle. Most of the

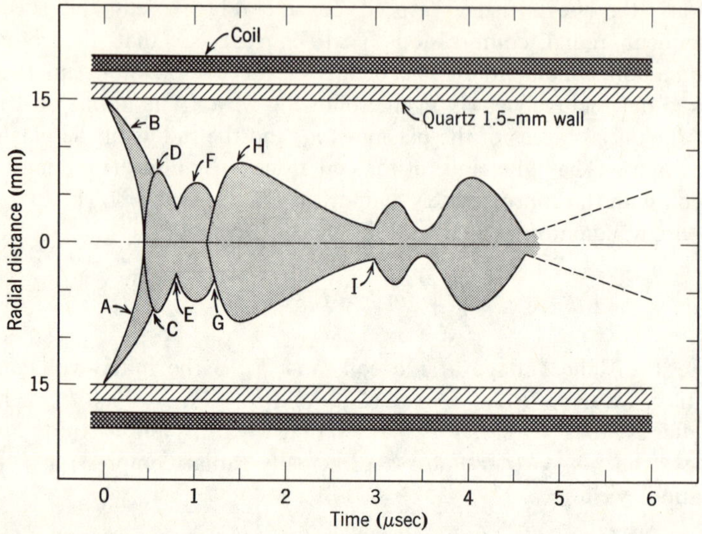

Fig. 15.3. Drawing from smear camera photograph of fast-pulsed mirror compression on the second half-cycle; 4.5×10^4 joules, 0.3 mm D_2, mirror ratio 2:1, $I_{max} = 8 \times 10^5$ amp on the second half-cycle. After Kolb.[5]

behavior shown in the figure is explainable semiquantitatively. At time $t = 0$, the current starts rising on the second half-cycle, and a radial shock wave, labeled A, is formed. The boundary between the plasma and the magnetic induction is at B. The shock wave is reflected from the axis and meets the oncoming plasma boundary at C. The surface is brought to rest at D as the induction increases. This shock bouncing is repeated (E and F), although the interior details are not shown. At G the shock waves from the mirrors collide at the midplane. The collision of these shocks heats the plasma, and it expands against the rising induction to point H. The rising induction then compresses the plasma adiabatically to about one-tenth of its original radius at I. The nature of the radial oscillations beyond point I is not yet understood. Finally, the plasma expands as the current decreases once again. This sequence of events is reproducible in detail.

In this particular experiment, the plasma was compressed by a factor of about 100. In similar experiments at lower pressure, final conditions $n = 10^{23}/m^3$, $T \approx 700$ ev, and volume = 0.7 cm^3 have been achieved. With the entire capacitor bank in use (2.85×10^5 joules), temperatures between 1 and 2 kev are reported. Neutrons are observed in such deuterium experiments; the yield, isotropy, and position of their origin are consistent at first sight with a thermonuclear process. Note that the duration of the maximum compression, perhaps 0.5 μsec, corresponds to about one collisional relaxation time for D^+ ions at the experimental temperature and density. All such matters are not yet finally settled, however, because further experiments[8] indicate that some trapped induction of opposite sign to that of the main external induction is required for neutron production. Thus the neutrons might arise in the region of field mixing. Bremsstrahlung measurements[9] indicate a temperature of about 1 kev, but the bremsstrahlung gives no measure of the high-energy tail of the ion distribution upon which neutron production at these low thermonuclear temperatures would depend.

Maximum available current and induction in these experiments were 8×10^6 amp and 50 webers/m^2. A 2×10^6-joule capacitor storage bank is being constructed in order to allow larger experiments and longer times of maximum compression.

Corroborative results from similar experiments are reported[4] in a device called Scylla. The plasma energy achieved was about 1 kev, as determined by bremsstrahlung[10] and other methods. There was very little energy anisotropy in the neutron energy spectrum. Similarly, the addition of an axial induction inside the plasma antiparallel to that of the main compression produces neutrons during the first half-cycle;[11] this fact indicates again that an acceleration mechanism may in some cases be involved. This suspicion is strengthened by the observation that under these conditions several hundred kev x rays are emitted.[12] These x rays indicate the presence of runaway electrons (Sec. 8.6) produced by the field E_θ during the compression.

Excellent photographs of plasmas produced by these techniques have been taken by Kvartskhava, Kervalidze, and Gvaladze.[13] Figures 15.4*a* to *c* show three of their sequences, each photographed axially with 0.5-μsec exposures, 2 μsec between frames. In this technique, a rotating optical mirror and lens system placed axially at one end of the discharge tube forms successive still images of short duration on different areas of the photographic film. This technique should be very useful for diagnosing the condition of pulsed plasmas.

In each sequence of Fig. 15.4, the plasma forms at the tube wall, then breaks away; small inhomogeneities later lead to instabilities of various sorts. Figure 15.4*a* shows a relatively quiescent compression, but with indication of local luminescent currents extending inward during compression and outward during expansion. Kvartskhava and co-workers associate these streams with runaway electrons. Figure 15.4*b* shows that an originally circular cross section often breaks up into a faceted structure (here pentagonal) that preserves some degree of identity through the entire cycle. If the discharge tube itself is faceted, a corresponding

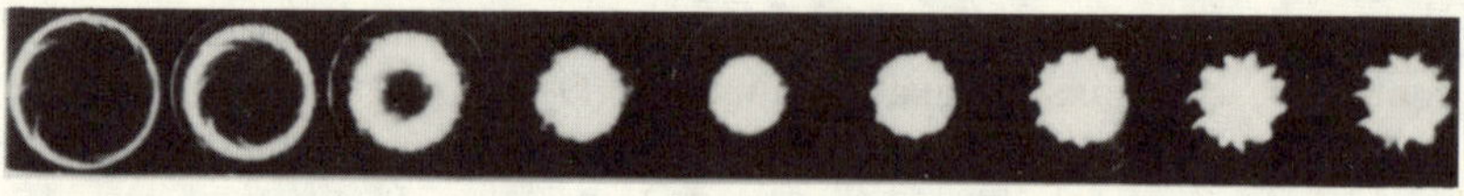

(a)

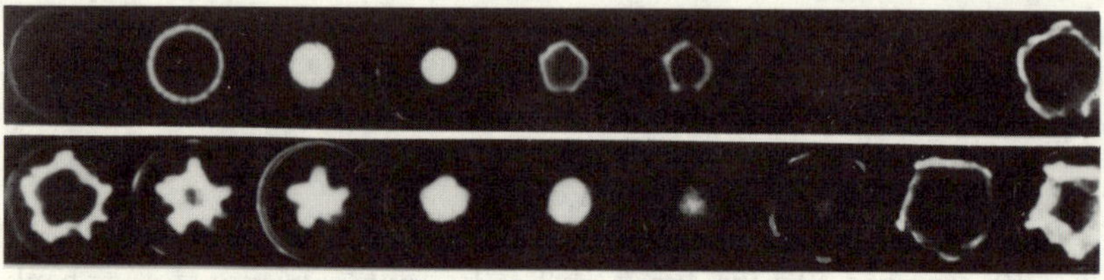

(b)

(c)

Fig. 15.4. Axial photographs of fast magnetically compressed plasmas. (*a*) Round tube, 0.27-m diam by 1-m length, 0.15-m coil length, 10^{-1} mm Kr, 30 kv, 90 μf, $B = 3.5$ webers/m^2 maximum, first half-period. (*b*) Round tube, 0.27-m diam by 1-m length, 0.55-m coil length, 10^{-1} mm He, 30 kv, 60 μf, $B = 1.0$ weber/m^2 maximum, first and second half-periods. (*c*) Octagonal tube, 0.02-m^2 cross-sectional area, 0.5-m length, 0.12-m coil length (round), 4×10^{-1} mm He, 30 kv, 60 μf, $B = 4.5$ webers/m^2 maximum, first half-cycle. After Kvartskhava, Kervalidze, and Gvaladze.[13]

symmetry is generally reflected reproducibly in the plasma. Thus the faceted tubes give more reproducible results than round ones. Figure 15.4*c* illustrates a discharge in an octagonal tube; similar effects are also observed in tubes of square, rectangular, and rhombic cross section. Other photographic sequences taken by Kvartskhava and co-workers show that local plasma density variations produce unequal radial compressions ($\mathbf{j}_\theta \times \mathbf{B}_z$ is approximately independent of θ, but $\nabla \cdot \mathsf{P}$ is not); these radial asymmetries extend axially and approximately uniformly down the tube. They find also that small θ-inhomogeneities early in the cycle can lead to large local radial plasma eruptions that extend out to the wall later in the cycle.

While most of the experiments reported in this system were done in magnetic mirrors, the mirror is not essential. The same phenomena could be produced in a structure where the coil of Fig. 15.2 is bent into a torus. Some of the references cited show this configuration. A betatron accelerating action can be achieved by the application of a steady axial induction to the configuration of Fig. 15.2 and by the later triggering of the switch. This latter induction creates an azimuthal electric intensity, accelerating the plasma. The size and magnitudes of the steady and pulsed inductions must be appropriately chosen. It has been proposed that such a plasma betatron could produce many amperes of current at many millions of electron volts. The stability of such plasmas is not fully understood.

15.3. Slow-Pulsed Mirrors

Some types of devices depend on slowly time-varying mirror systems for heating the plasma or for trapping an injected beam of ions. The compression is almost invariably adiabatic; once the plasma is formed in its final configuration, it is usually intended to be held in a steady state. Such devices have sometimes been called pyrotrons. These slow systems merge into the steady-state ones of Secs. 15.4 and 15.5, and many properties of each class apply to the other. In virtually all of these experiments, the mirror is initially evacuated and a plasma is injected in some manner from a source. This procedure is in contradistinction to the fast pulsed experiments, where the ambient gas is usually ionized to form the plasma. A number of elementary criteria pertaining to straightforward compression in time-dependent mirror systems are given by Post.[14] In addition to the limitation on compression time discussed in Sec. 15.1, the effect of the magnetic mirror shape on the degree of compression is important. In this connection, the action integral (Eq. 10.74) can readily be written

$$[B(0)]^{1/2} \int_{z_2}^{z_1} dz\,[R - R(z)]^{1/2} = \text{const}, \tag{15.15}$$

where $R(z)$ is the mirror ratio at an axial position z, $B(0)$ is the induction at the midplane $z = 0$, and z_1 and z_2 are the turning points of the particle. The equation can readily be recast as an integral over R by expressing $z = z(R)$ for the particular configuration. For example, let the particular dependence on R be inserted and the induction be uniformly increased with time, thus preserving the form of $R(z)$. Then Eq. 15.15 yields the inward motion of the turning points z_h, hence the associated axial compression. Problem 1 illustrates the calculation. This axial compression arising from the principally radial process is easy to understand physically: the particle velocities v_T are increased by the radial compression. Their helix angle θ with respect to the **B** lines increases, and they do not ride so far up on the mirrors. The additional increase in density can be significant for large radial compressions. In the same way, the effect of mirror ratios changing with time may be treated in this adiabatic limit. A number of such simple conclusions derived in Sec. 10.6 are applicable here: radial compression or axial expansion improves the confinement; radial expansion or axial compression worsens the confinement.

In these slow compression experiments, injection of the plasma into the vacuum region is always a problem. The principal schemes proposed are the following:

1. Injection of a plasma cloud from a small pulsed source and trapping of the plasma in transit by the rising induction.

2. Injection of an ion beam from a high-current ion gun placed on the mirror axis and trapping of the ions in individual orbits in the rising induction.

3. Injection of a plasma cloud from a source, as in (1), but of such high density that it distorts the field lines on entry; the cloud disperses inside the mirror. Reassembly is improbable.

The first scheme has been applied to a number of pulsed heating experiments. Figure 15.5 shows the schematic design and magnetic induction

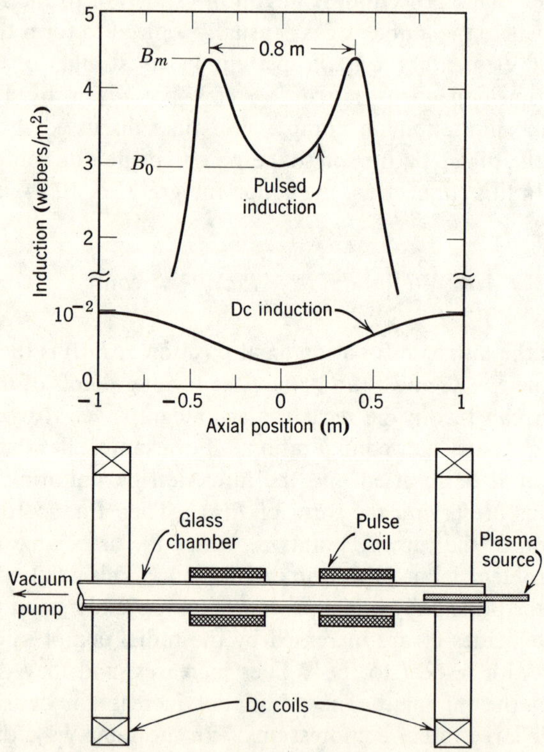

Fig. 15.5. Schematic diagram and magnetic field distributions for a large pulsed mirror. After Coensgen, Ford, and Ellis.[15]

for an experiment by Coensgen, Ford, and Ellis.[15] The plasma source is a stack of alternating titanium and ceramic washers; the front washer and a closing rear disk are deuterium-loaded. A pulsed arc is struck between these two, and an initial plasma density of about $10^{18}/m^3$ with high ion energy (several hundred electron volts) is obtained in the mirror at low dc induction. Some of this plasma is trapped for 50 to 200 μsec by self-collisions near the source. A 10^6-joule condenser bank then compresses the plasma within the pulse coil in about 5×10^{-4} sec. The coils are then clamped; the high induction lasts for about 3×10^{-2} sec.

The radial position of the plasma as a function of time agrees fairly well with the values calculated from the current cycle for radial compression ratios of 16 and 100. The plasma remains confined for several milliseconds.

For the highest compression ratios, estimated to be several hundred radially and 4 or 5 axially, the electron energy becomes high enough to be measured by scintillation techniques. Plastic scintillators are placed axially just outside one of the mirrors; the electron energy spectrum is measured by inserting thin foils ahead of the scintillator. Knowledge of the electron energy degradation as a function of foil thickness then permits calculation of the distribution. For the highest compressions, an approximate electron temperature of 21 kev is found. On the other hand, the ions were estimated to be at a few hundred electron volts at most. The disparity may reflect the fact that only ions of very low energy can be trapped in the initially weak induction. The electron-to-ion energy ratio is preserved through the adiabatic compression. Coensgen, Ford, and Ellis[15] estimate $\beta \approx 0.08$ with some uncertainty.

A three-stage plasma compression experiment has been performed by Coensgen, Cummins, and Sherman.[16] The plasma is formed initially by a source similar to that described above into a large low-induction mirror configuration. One end is then pulsed, forcing the plasma by peristaltic motion axially into a second smaller mirror with higher induction. The process is repeated into a small third mirror, giving an over-all compression of a factor of about 1000. In the experiment, the weak induction (for example 0.007 weber/m^2) in the initial region allowed retention of only the slowest ions. Experiments are now under way using higher induction in the initial region and proportionately less over-all compression. In this manner, the ability of the source to deliver initial ions of several hundred electron volts can be utilized. A plasma with deuteron energy approximately 3 kev has been produced for 1 m sec.[17]

In general, these compressed plasmas have been apparently stable. Referring to the discussion of Chap. 12 and in particular Sec. 12.8, we note that simple theories predict that the configuration should have been unstable. More sophisticated stability theories are required. In actual fact, an instability has been observed,[18] but not of the simple flute type discussed previously. With the large radial compressions used, it is possible to achieve a condition where $p_T^2 > (B^2/2\mu_0)p_{\|}$. A new type of instability then appears in velocity space; this instability has been predicted theoretically.

The second injection scheme already mentioned—beam injection into the rising mirror—is straightforward in principle but would involve enormous ion currents. Suppose that the ion gun is placed just inside the mirror and that the ions are injected normal to the induction. The force on the ion is

$$F = \mathbf{M} \cdot \nabla B = -\frac{m_i v_i^2}{2B}\frac{\partial B}{\partial z}; \tag{15.16}$$

in one gyration time the ion advances a distance

$$\Delta z = \frac{\pi^2 r_b^2}{B}\frac{\partial B}{\partial z} \tag{15.17}$$

toward the midplane. If this displacement permits the ion to clear the end of the gun, it travels to the far mirror and returns. The ion precesses about the axis because of the $\mathbf{B} \times \nabla B^2$ drift (Eq. 10.15), and the configuration might be arranged so that the ions miss the gun on their first return. The ions must then be reflected many times until they have precessed once around the axis. In all this time the induction must rise just enough to move the reflection point for the ion inward a distance Δz. A simple analysis is given by Post,[14] and times of about 10^{-3} sec are available during which the injection must be accomplished. If 10^{20} ions are to be injected, a beam current of about 10^4 amp would be required. This current is discouragingly large.

Similar difficulties are encountered if the gun is placed axially just outside a mirror (see Problem 2).

Beam injection is much more attractive for steady-state mirrors, where there is no time limitation. Here the difficulty arises in trapping the beam; we consider these matters in Secs. 15.4 and 15.5.

The third and last method—forcing aside the **B** lines with a dense plasma cloud—will be discussed later in the section, but mainly in Sec. 15.7.

We turn next to a brief study of two rotating plasma devices; here the induction is steady, but the plasma is pulsed. Figure 15.6 shows a so-called hydromagnetic or plasma capacitor, the general characteristics of which have already been anticipated in the problems.† In this crossed-field configuration, the equilibrium azimuthal velocity is

$$\bar{v}_\theta = [E_r - (m\bar{v}_\theta^2/qr)]/B_z \tag{15.18}$$

† Chapter 6, Problem 6; Chap. 7, Problem 6; Chap. 10, Problem 4.

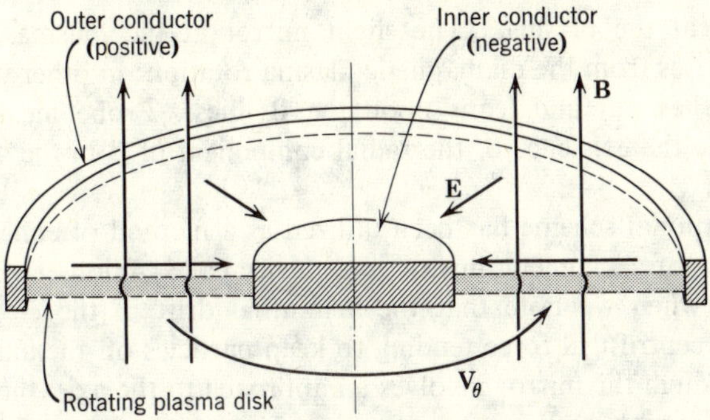

Fig. 15.6. Rotating plasma in crossed **E** and **B** fields.

for either ions or electrons. The second term in the brackets results in slightly different drift velocities for the two types; a net azimuthal current flows. If the plasma resistivity is not zero, an azimuthal electric field also appears, which tends to drive the plasma toward the walls. If this second term is again neglected, the stored energy in the electric field and plasma rotational energy is

$$(\epsilon_0 E^2/2)[1 + (\rho_m/\epsilon_0 B^2)] = K(\epsilon_0 E^2/2), \tag{15.19}$$

where K is the effective dielectric constant. Perhaps the device should properly be called a plasma flywheel, but Eq. 15.19 shows that it behaves electrically as a capacitor with high dielectric constant. The same result is derived in Eq. 12.9 in a different connection. See also Sec. 9.5 and Problem 4 of Chap. 10. The scheme has been proposed as a method of creating a hypersonic plasma wind and of storing energy. A number of analyses have been made of its operation,[19–23] including the effect of collisions.

An experimental rotating plasma device, called Homopolar, has been tested by Anderson and co-workers.[20] The shape of the particular device was that of Fig. 15.6 with outer diameter 0.25 m. Typical operating parameters were 1.5 webers/m^2, 7500 volts, and 0.1-mm gas pressure. Effective dielectric constants between 10^6 and 10^8 were observed, in agreement with the simple theory and corresponding to an effective capacity as large as 75 μf. The device was charged from a condenser bank and could be discharged after several microseconds by shorting it. Rates of change of current up to 5×10^{11} amp/sec were observed. In view of the small size of the device, its short-term energy storage is impressive. Rotational speeds of 3 to 6×10^4 m/sec with argon filling, corresponding to 180 to 720-ev ion rotational energies, were calculated.

Since the azimuthal velocity decreases with radial distance, viscous drag tends to heat the plasma. The slight mirror at the plasma shown in Fig. 15.6 arises from the diamagnetic plasma rotation: in other words, the plasma pushes out and leans upon the **B** lines. Probe measurements demonstrate the existence of the radial component of **B** in the vicinity of the plasma.

This rotational scheme has been utilized as a method of enhancing the confinement of a magnetic mirror in a device called Ixion.[24] The effect is obvious when we note that the azimuthal drift in the electric field produces a centrifugal force tending to keep particles off the axis. Since passage through the mirror involves an approach to the axis, there is thus an effective mirror repulsive force that is independent of sign of the charge and, like the centrifugal force, increases with E^2. For a sufficiently large electric field, particles of arbitrarily high energy can be confined. For small mirror ratios, Boyer and his associates[24] derive the maximum axial kinetic energy that can be contained as

$$U_{\|} = U_{\mathrm{T}}(R_m - 1) + [mv_\theta^2(R_m - 1)/2R_m]. \tag{15.20}$$

There are some difficulties in the scheme. First, for significant additional confinement, the rotational energy must be comparable with or greater than the random kinetic energy, and only the latter is useful in producing thermonuclear reactions. Second, the confinement is poor for the high-energy tail of the ion distribution, which is the most important in determining the thermonuclear reaction rate unless the temperature is very high. Third, the electrons are in practice unaffected (see Eq. 15.20), because the macroscopic v_θ is about the same for ions and electrons. Fourth, the curvature of the **B** lines is increased, thus possibly decreasing the stability; as usual, we have no definite opinion on this last point.

Figure 15.7 shows a simplified view of Ixion with the vacuum walls and access connections omitted. The plasma itself forms the inner conductor. In operation, a plasma gun at the left of the figure injects an ionized plasma through one mirror into the confinement region. This dense plasma is in close contact with cathode rings at each end. A potential is applied between these rings and the outer anode cylinder after the gun has injected the plasma. At each pulse, the valve is opened for about 100 μsec; 400 μsec later, the gas has diffused to the shock coil, which is then energized. The plasma so formed travels into the mirror region, and the radial potential (up to 10^4 volts) is applied. After a further formative delay time of about 200 μsec, the plasma is observed to rotate, and a large charging current is drawn to the device. The measured capacity was about 40 μf; rotational speeds of 4×10^4 m/sec were observed for impurity ions by Doppler shift. The formative delay time appears to be associated with the formation of an anode sheath.

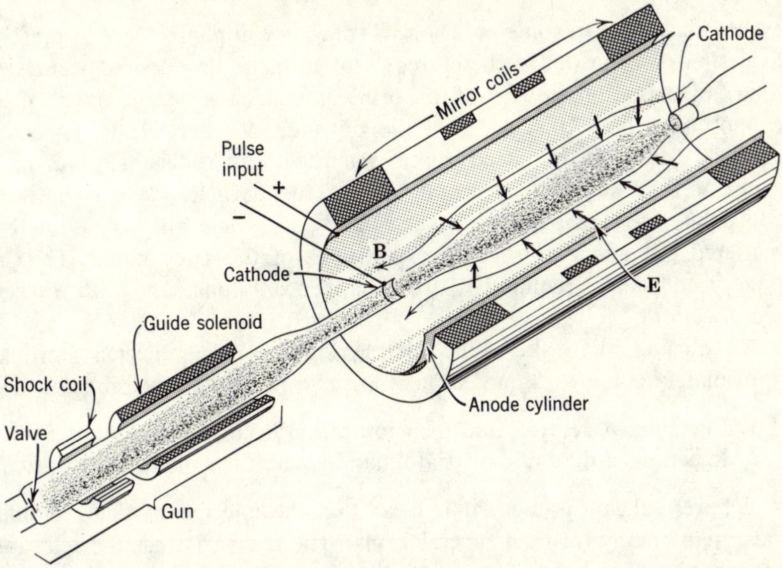

Fig. 15.7. Simplified cutaway view of rotating plasma compression device Ixion. After Boyer and co-workers.[24]

The plasma gun itself will be described in Sec. 15.7. It is of interest because the output plasma energy and density are high enough for it to penetrate a 1-weber/m^2 mirror coil, and therefore to inject plasma into a steady-state confinement device.

15.4. Ion Injection and Gas Burnout in a Steady-State Mirror

In all the plasma assemblies considered in previous sections, the plasma was created by some kind of pulsed operation. In some cases, a low-pressure gas was preionized, then heated by pulsing the mirrors. In other cases, a beam was trapped in a time-rising mirror; in yet others, a plasma was pulsed into a steady-state mirror and made to rotate by a pulsed radial electric field. We shall return to the question of pulsing a plasma into mirror-like systems in Sec. 15.7. In this and the following section, we explore the possibilities available without pulsing. Some advantages of steady-state (or quasi-steady-state) operation for an eventual thermonuclear device have been suggested in Chap. 13.

The problem immediately arises of creating and maintaining the plasma for any such system. It is clear that the injected particles must undergo some sort of real collisions (rather than merely deflection in steady-state fields, however complicated) if they are to be trapped. The proof of the statement is trivial. A trapped particle that suffers no collision will eventually return to some point on its trajectory in phase space as closely as desired. The orbit is, therefore, a closed one to any desired accuracy. Since, by definition, no part of this orbit is outside the system, the particle cannot have been injected from the outside; in other words, particles cannot be trapped in a steady-state system without collisions. Our first problem then suggests a second one: if the particles have significant collisions with the plasma on their first pass, they will very soon be scattered out again by successive collisions of the same kind. Thus a type of collision is required that can happen only once and with a large cross section.

A collision that alters the charge state of the injected particle immediately comes to mind. There are two reasonable possibilities:

1. Injection of neutrals and their ionization inside.
2. Injection of diatomic or triatomic ions and their dissociation inside.

Whichever scheme is used, it is clear that the injected particles should have high energy (at least several kev). The reason is that the relevant cross sections are large; if the desired ions are to be produced deep inside a dense plasma, their speed must be high. As an example, see Chap. 3, Problem 5. Since the ions produced in the plasma will have high energy, further plasma heating may be unnecessary. In this way, we discover the possibility of building up the plasma density, starting at or above the desired final temperature.

The main advantage of injecting neutral particles is that they can cross the magnetic induction and hence in principle enter a plasma deep inside the confining field structure. The disadvantages are:

1. For the energies considered, the particles must first be accelerated as ions, then neutralized again before injection. However, this process can probably be accomplished with 80 to 90% efficiency at useful thermonuclear energies.
2. The path length for trapping is limited to the plasma size.

If we consider molecular ion injection, the advantages and disadvantages are reversed. In particular, the molecular ion orbit must be arranged to enter through the confining induction. The dissociation process at high energy merely decouples the associated neutral atoms without appreciably altering the velocities. Thus the Larmor radius of a trapped D^+ is half that of an injected D_2^+, for example. In spite of this change in Larmor radius, orbits of the trapped D^+ come uncomfortably close to the surrounding structure. There are some compensating advantages. First, the molecular ions may be injected so that their path inside the system is very long; we shall see an example of this point in Sec. 15.5. Second, the possibility exists of injecting a dense ion beam neutralized by electrons. The principle is illustrated in Fig. 15.8; the **V** × **B** electric field polarizes the ion beam, which in this approximation resembles a moving conductor. Particle orbits at the edges are more or less cycloidal, as determined by the induction and the self-electrostatic field. The whole assembly can then move across the induction and into the plasma. We shall not dwell further upon this suggestion because it has never been completely analyzed or yet attempted on any large scale.

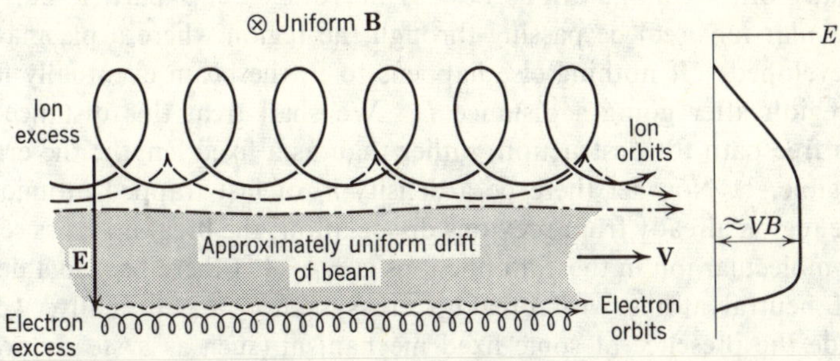

Fig. 15.8. Schematic representation of typical particle orbits and magnitude of electric field in a dense beam of ions and electrons crossing a magnetic induction **B**. The beam is assumed to be thick in the **B**-direction.

A device is presently being constructed by Post[25] in which 20-kev neutral particles will be injected into a mirror and ionized inside—initially on the background gas, then at later times on the plasma itself. The scheme most intensively exploited to date is molecular ion injection, to which we devote the remainder of this section and the next. Much of the analysis to follow can also be applied obviously and directly to neutral particle injection schemes.

The build-up of a plasma by ion injection is closely associated with the presence of background gas in the system. Whatever the pumping speed, some gas remains (for example, $3.5 \times 10^{14}/\text{m}^3$ at 10^{-8} mm Hg). The background gas can serve as the initial breakup mechanism for the input beam. As the high-energy plasma builds up in density, the plasma ions themselves take over the breakup function; the plasma density would grow more or less exponentially, were it not for charge transfer from the trapped ions to the neutral background. This charge transfer (see Sec. 3.2) is an important loss mechanism, especially if the injected beam intensity is small. However, if the energetic ion density is made high enough in some way, the ions will ionize most of the background gas. The charge-transfer loss then drops, and the plasma density rapidly increases. This process of removing the background gas within the plasma volume is called burnout.

It is also possible to provide an additional molecular breakup mechanism inside the mirror. A concentrated arc plasma running along lines of induction through the mirrors to exterior electrodes has been used. Figures 15.13 and 15.14 show such a configuration, which will be described further in Sec. 15.5.

We now develop a simplified theory for the plasma build-up and neutral gas burnout, based on the work of Simon.[26,27] The theory is applicable to many other configurations besides mirrors. Let a particle current I (molecular ions/sec) be passing through the region where a plasma is to be developed. If nothing else happens to it, the beam eventually leaves the region after going a distance L. We shall treat this distance as a mean free path for destruction, rather than as a fixed length; the error is negligible.[27] Now let there be a density n_i of fast trapped atomic ions in the region already from previous dissociation; the breakup cross section for a molecular ion in these atomic ions is σ_{bi}. Let there be also a density n_0 of neutral atoms with breakup cross section σ_{b0}. Finally, we can include the presence of some fixed mechanism (such as a steady arc) by assigning an arbitrary inverse mean free path K to the process. The ion current then is attenuated with distance as

$$\frac{dI}{dx} = -I\left(n_i\sigma_{bi} + n_0\sigma_{b0} + K + \frac{1}{L}\right). \tag{15.21}$$

We now assume for simplicity that all densities are uniform through the plasma. The current and the amount dissociated at some point are then found trivially by integration over all distance. The fraction dissociated is

$$\text{Fraction dissociated} = \frac{(n_i\sigma_{bi} + n_0\sigma_{b0} + K)L}{(n_i\sigma_{bi} + n_0\sigma_{b0} + K)L + 1}. \tag{15.22}$$

We now spread these ions over all the volume $\mathscr{V}$ of the plasma and obtain for the ion density

$$\frac{dn_i}{dt} = \frac{(n_i\sigma_{bi} + n_0\sigma_{b0} + K)LI}{[(n_i\sigma_{bi} + n_0\sigma_{b0} + K)L + 1]\mathscr{V}} - \text{losses}, \tag{15.23}$$

where now and in the following development I is the total beam current.

The ion losses will consist principally of charge transfer to the neutrals n_0 with cross section σ_t and of Coulomb scattering out of the system. The scattering time for a Maxwellian distribution would be the quantity τ_θ (Eq. 8.23). The present distribution is substantially monochromatic. The reaction rate is better approximated by $n_i^2\sigma v_i$, where σ is the appropriate Coulomb cross section calculated for the ion energy. The chance that such a scattering leads to loss is the confinement factor ψ of Sec. 13.4. Equation 15.23 becomes

$$\frac{dn_i}{dt} = \frac{(n_i\sigma_{bi} + n_0\sigma_{b0} + K)LI}{[(n_i\sigma_{bi} + n_0\sigma_{b0} + K)L + 1]\mathscr{V}} - n_0 n_i \sigma_t v_i - n_i^2 \sigma v_i \psi, \quad (15.24)$$

when v_i is the ion speed. We shall assume that the ions are not degraded in energy as they remain trapped. For a calculation that includes the effect of energy degradation, see Problems 5 and 6.

Estimation of the neutral density n_0 inside the plasma in terms of the various parameters of the system is more difficult. The mean neutral density n_0 can be related to the exterior neutral density $\hat{N}_0$ by equating the inward flow of particles to the outward flow plus the interior destruction. The approximate result can be written

$$\hat{N}_0 v_0/4 = (n_0 v_0/4) + n_0 \sigma_d (n_i v_i/4)\bar{l}. \quad (15.25)$$

In Problem 4 a derivation of Eq. 15.25 in two limiting cases is required. Here, $\bar{l}$ is the mean chord length of the plasma, σ_d is the total cross section for destruction of a neutral by a fast ion, and v_0 is the mean speed of the neutral. The destruction occurs chiefly by ionization and to a lesser extent by charge transfer, so that

$$\sigma_d = \sigma_i + \sigma_t. \quad (15.26)$$

Equation 15.25 represents burnout of the gas described earlier in words. Several approximations have been made. First, the principal ionization arises by impact of fast ions; the contribution of electrons, which are heated by the fast ions and which then ionize neutrals, is ignored. This approximation is generally valid. Second, the slow ions produced by ionization are assumed to escape immediately and not contribute to the molecular breakup in Eq. 15.22. This approximation is excellent under all reasonable conditions: if the fast-ion density n_i is high enough to provide a density of slow ions comparable with n_0, the breakup will then proceed mainly on the fast ions n_i. To put the matter another way, breakup on the neutrals and slow ions can be disregarded completely, except for n_i so low that there are negligibly few slow ions.

In discussing the exterior neutral density $\hat{N}_0$, we must account for all sources and sinks of gas. There will in fact be various atomic species present (for example, N_2 from outgassing of the system and H_2 or D_2 from the ion source). We should replace n_0, $\hat{N}_0$, and their functions in Eqs. 15.24 and 15.25 by appropriate sums over the species, using the applicable cross sections; we shall ignore this complication. All sources of gas from the vacuum walls, beam ports, and so forth, which are independent of the beam current or plasma density may be characterized by an influx g (molecules/sec). Next, every fast nucleus (be it contained in an atom, ion, molecule, or whatever) that strikes the wall may eject S neutral molecules from those adsorbed on the walls. The quantity S is an effective desorption coefficient. The probability S may be a function of the precise fate of the particle (for example, the residual beam dumped into a pump, for which S will be small), but we ignore the difference. Against this input, there are the losses (1) the number $\theta \hat{N}_0$ to the pumps, where θ is the pumping speed, and (2) the pumping effect of the plasma itself. Let χ denote the fractional probability that any ion leaving the plasma via the mirror does *not* return as a neutral molecule. Finally, the outgassing rate is given by

$$g = \theta N_0, \tag{15.27}$$

where N_0 is the neutral density when neither beam nor plasma is present.

Each nucleus in the molecular ion beam eventually strikes the wall, except for those that are trapped and disappear out the mirrors by Coulomb scattering. If $\mathscr{V}$ is the plasma volume and $\hat{\mathscr{V}}$ the vacuum volume, the neutral density $\hat{N}_0$ in the vacuum is then given by

$$\hat{\mathscr{V}} \frac{d\hat{N}_0}{dt} = \theta(N_0 - \hat{N}_0) + S(2I - n_i{}^2\sigma v_i \psi \mathscr{V}) - \chi(n_0 n_i \sigma_d v_i + n_i{}^2 \sigma v_i \psi)\mathscr{V}. \tag{15.28}$$

The factor 2 arises because diatomic deuterium molecules are injected, and I is their current.

Three equations, Eqs. 15.24, 15.25, and 15.28, determine the densities as functions of time: Eq. 15.25 is not a differential equation because equilibrium between n_0 and $\hat{N}_0$ is reached rapidly compared with the other times of interest. We see that the set has already been simplified by our failure to distinguish among the ion species and among some of the processes. Further approximations are usually made. The Coulomb scattering terms in Eq. 15.28 generally have negligible effect on the ion density itself. The reason is that the Coulomb scattering is important only when the neutral density has negligible effect upon the kinetics of the system.

Two cases of the coupled equations have been treated by Simon. In the first,[26] the breakup is assumed to occur only by passing the molecules through a fixed dissociating mechanism, such as an arc. The fixed fractional probability is then represented by the fraction $KL/(1 + KL)$ in Eq. 15.24. All effects of ion pumping and variable gas sources are neglected, whence $\hat{N}_0 = N_0$, a fixed density. The steady-state solutions are then characterized by the fast-ion density n_i increasing and the interior neutral density n_0 decreasing monotonically with current. Figure 15.9 shows these densities n_i and n_0.

The interpretation is clear. At low current, the trapped beam is weak and proportional to the input current; the interior neutral density is virtually unaffected by ionizations created by the few fast ions. Fast ions are lost by charge transfer. As the current is increased, the quantity

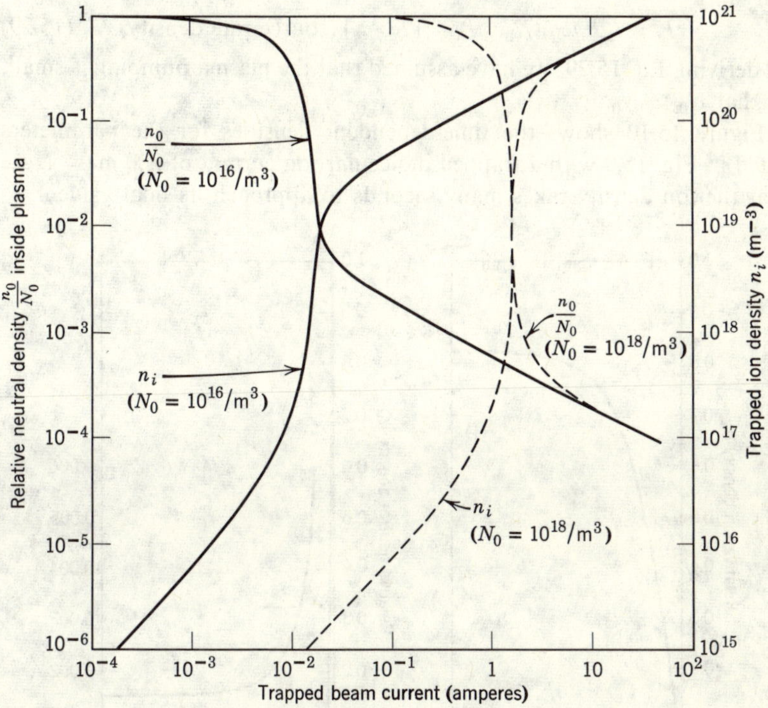

Fig. 15.9. Burnout of neutral gas by a 600-kev molecular ion beam for two background neutral densities N_0. The beam dissociation is assumed to occur by a fixed mechanism (for example, an arc plasma). Here, $v_i = 5.3 \times 10^6$ m/sec (600 kev D_2^+, dissociated to give 300 kev D^+); $\mathscr{V} = 0.047$ m³ (volume of a sphere whose radius is the Larmor diameter at 1 weber/m², 300 kev D^+); $\bar{l}$ = mean chord length of the above sphere; $\sigma_t = 4 \times 10^{-2}$ Å², $\sigma_i = 0.76$ Å²; $v_0 = 1.26 \times 10^3$ m/sec (D_2 mean speed at 300° K); $\psi = 0.3$; $\sigma = 3 \times 10^{-7}$ Å². After Simon.[26]

$\bar{l}v_in_i\sigma_d/v_0$ in Eq. 15.25 becomes comparable with 1, and the interior density n_0 of neutrals decreases by the onset of burnout. Then the charge transfer rate ν_t per ion decreases, and the ion density will increase faster than linearly with current. The process is accentuated with increasing current and density, as illustrated by the steep portions of Fig. 15.9. Finally at high density, Coulomb scattering ($\propto n_i^2$)

predominates, and n_i increases proportional to $\sqrt{I}$. One reasonable criterion for defining a critical trapped current I_c at which burnout has been virtually accomplished is obtained by setting the charge-transfer and Coulomb scattering losses equal in Eq. 15.24. Under these circumstances, $(\bar{l}v_i n_i \sigma_d / v_0) \gg 1$ in Eq. 15.25. Then for this fixed breakup mechanism, we obtain for the critical *trapped* particle current (here *not* the current from the ion gun)

$$I_c = (2\sigma_t/\sigma_d)(N_0 v_0/\bar{l}) = (2\sigma_t/\sigma_d)\,(\text{source rate density}). \qquad (15.29)$$

In deriving Eq. 15.29, we have assumed that the plasma pumping is small, or that $\hat{N}_0 \approx N_0$.

Figure 15.10 shows the time-dependent densities for the parameters listed in Fig. 15.9 with a trapped molecular ion current of 200 ma. Note that the ion density takes many seconds to approach its final value.

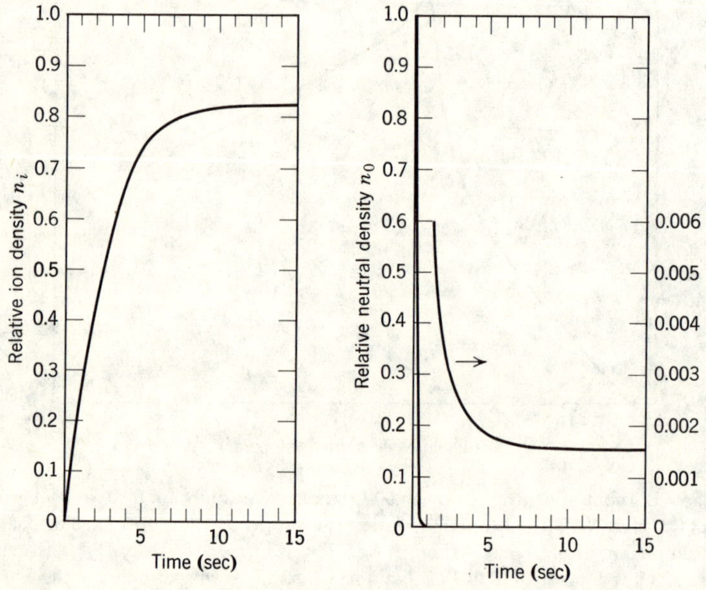

Fig. 15.10. Relative ion and neutral densities as functions of time under burnout conditions for molecular ion injection with fixed breakup mechanism. Pressure $p = 10^{-6}$ mm Hg, D_2^+ energy 600 kev, trapped beam current 200 ma. After Simon.[26]

The other burnout case calculated concerns initial breakup on the residual gas, followed by breakup on the fast plasma ions.[27] No fixed breakup scheme is employed. The dependence of the external gas density on beam current and pumping action of the plasma (phenomena that are represented by the factors S and χ, respectively, of Eq. 15.28) are inserted. Quite different behaviors of n_i vs. I may be obtained, depending upon the circumstances. If the beam gives rise to much gas and the slow ions can return to the system ($S = 1$ and $\chi = 0$, for example), no useful burnout can take place. While the neutral density ratio $n_0/\hat{N}_0$ will decrease with increasing current, this decrease is more than offset by the increase in exterior neutral density $\hat{N}_0$; hence the interior neutral density n_0 itself never decreases.

If the conditions are improved by decreasing S or increasing χ, an essentially different behavior can be obtained, as illustrated in Fig. 15.11. We shall trace out this behavior, starting at low ion density. At very low beam current I and ion density n_i, molecular ion breakup takes place almost exclusively upon residual background gas (Region I of Fig. 15.11): n_i increases linearly with I. Because both the production and loss (by charge transfer) of trapped ions are proportional to gas

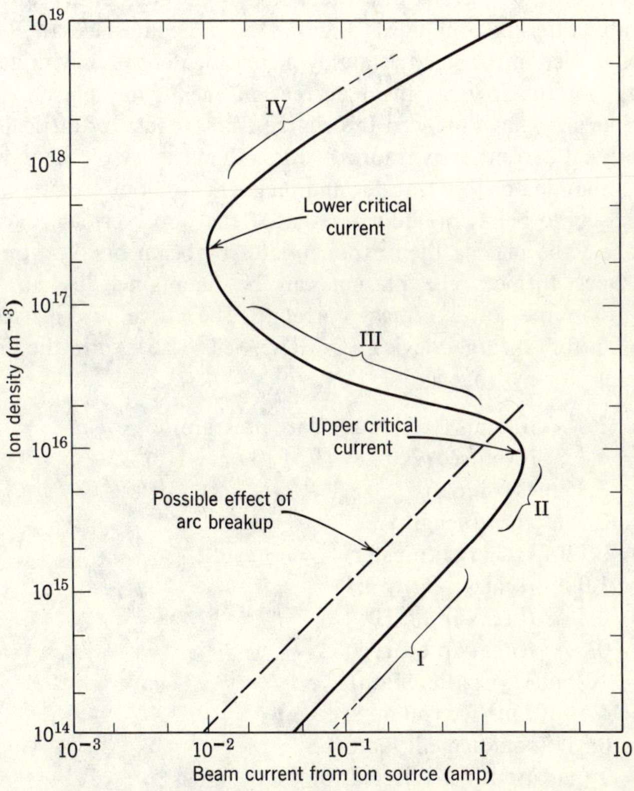

Fig. 15.11. Trapped ion density in a large mirror, showing effect of neutral gas burnout by a 200-kev molecular hydrogen ion beam. Beam dissociation occurs on the background gas and trapped protons. See text for parameters, corresponding approximately to the device OGRA. After Simon.[27]

pressure, the n_i–I curve in Region I is independent of pressure, but depends only upon the experimental configuration, gas and ion species, and ion energy. However, the upper limit of the linear Region I does depend upon pressure, because in Region II, I and n_i are sufficiently large that breakup on the trapped ions has become relatively significant. This additional breakup is not accompanied by any corresponding particle loss (Coulomb scattering is still negligible); therefore the slope of the n_i–I curve increases. Eventually the breakup on the ions alone is sufficient to make up the charge-transfer losses. At this point (the upper critical current) the density n_i is unstable if there is sufficient plasma pumping or burnout: an increase in n_i increases the breakup, the breakup rate, and n_i itself. Simultaneously, the plasma pumping and/or burnout increase and the neutral density decreases, as we enter Region III. If the upper critical current is exceeded, the plasma density grows until Coulomb scattering dominates (Region IV). The intermediate Region III is characterized in the steady state by increasing density n_i with decreasing current. At the lower critical current in the figure, plasma pumping and/or burnout have reduced the charge-transfer losses drastically. The lower critical current, now trapped with high efficiency by the ions, makes up the remaining charge transfer and incipient Coulomb scattering losses.

This S-curve behavior indicates that if the upper critical current can be reached, the plasma then exponentiates by beam breakup on its own ions. Once formed, the plasma can be maintained by any current greater than the lower critical current. The curve of Fig. 15.11 was calculated for a large device (OGRA, Sec. 15.5) with the following parameters for hydrogen:

$N_0 = 3.3 \times 10^{15}/\text{m}^3$ (10^{-7} mm) (base pressure)
$v_i = 4.4 \times 10^6$ m/sec (200 kev H_2^+, 100 kev H^+)
$\mathscr{V} = 2.35\ \text{m}^3$ (volume)
$\bar{l} = 0.5$ m (chord length)
$\sigma_t = 0.12\ \text{A}^2$ (charge transfer)
$\sigma_i = 1.0\ \text{A}^2$ (ionization by H^+)
$\sigma_{bi} = 0.5\ \text{A}^2$ (breakup on H^+)
$\sigma_{b0} = 0.5\ \text{A}^2$ (breakup on H_2^0)
$L = 10^3$ m (H_2^+ path length)
$\theta = 4 \times 10^4\ \text{m}^3$/sec (pump speed)
$\psi = 0.388$ (confinement factor)
$S = 1$ (gas input per injected ion)
$\chi = 1$ (all slow ions pumped out through the mirrors).

An effective Coulomb scattering cross section of $2.6 \times 10^{-6}\ \text{A}^2$ was used in calculating the mirror angular relaxation.

In all these calculations, cross sections from the sources listed in Chap. 3 have been used. The general statement was made there that different measurements of the same cross section do not always agree. That statement is emphatically true in the present case; more favorable values for most of the cross sections used here are reported in the U.S.S.R.[28] The discrepancy, not yet fully resolved, is particularly large for hydrogen. With the U.S.S.R. cross sections, the upper critical current is found at $2 \times 10^{15}/m^3$, 0.04 amp, or a factor 40 less in current. The lower critical current depends in part upon Coulomb scattering and is less drastically dependent on other cross sections.

It may be possible to reduce the effective upper critical current to some extent. The dashed line of Fig. 15.11 shows schematically the effect of dissociation of molecular ions by some fixed mechanism (for example, a suitable arc). To a first approximation, the upper critical current is reduced to the crossover point of the two curves, if the program is arranged so that the density is first built up on the arc, then the arc is extinguished.

The time-dependent behavior of this S-curve type of burnout has not been calculated in detail, but elementary considerations show that the plasma is developed during a period of several seconds. The current in excess of the steady-state values of Fig. 15.11 is not an accurate measure of the build-up rate, because increasing the ion input increases both the gas that must be pumped out and the ion losses. As was stated earlier, such considerations were ignored in the calculations leading to Fig. 15.10.

Another effect not included here is the degradation in energy of the trapped ions as they heat slow ions and electrons. Thus in actuality the temperatures and cross sections will change with time and density. If electrons drift through the plasma and are replaced continually with cold ones, the ion energy loss can be large. This loss has been calculated for some simple cases.[29, 30] Further consideration of the involved computations applicable to actual systems is beyond the scope of this book, but Problems 5 and 6 touch on the question.

In concluding this section, we repeat that these important considerations of charge transfer and neutral gas burnout will be applicable to many other confinement schemes—slow-pulsed systems, the Astron (see Sec. 15.6), cusps (Sec. 15.7), and the stellarator (Chap. 16)—beside the steady-state mirror. However, for the last, neutral gas burnout is the prime objective of present experiments, as we shall see in the next section.

15.5. Steady-State Mirror Devices with Molecular Ion Injection

Several devices have been proposed and constructed for generating a high-energy plasma by molecular ion injection. The most favorable cross sections for plasma formation (principally small charge-transfer cross section σ_t) occur at several hundred kev. The Larmor orbits are large and so are the systems that confine them. Since they are run in the steady state and the gas residue from the beam and other sources must be kept low, the systems are also designed to permit very high pumping speeds.

We introduce this section by illustrating an elegant proposal for focusing an ion beam into the center of a mirror confinement region.[31] Figure 15.12 shows a cross section of a mirror with cylindrical symmetry

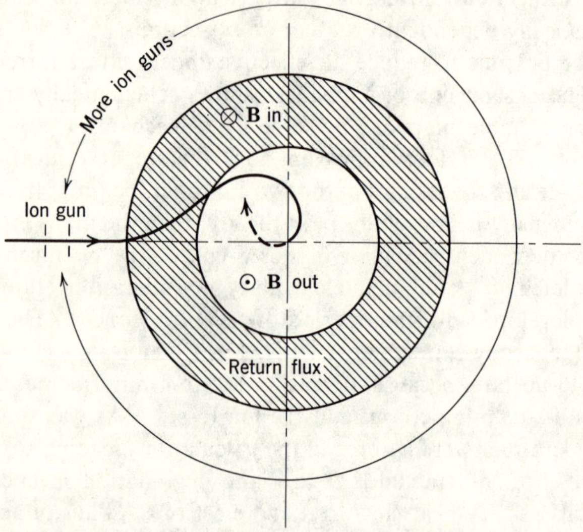

Fig. 15.12. Scheme for radial injection of molecular ions into a magnetic mirror.[31]

at the midplane. The flux in the interior runs one way, and the exterior return flux runs in the opposite direction. The division need not be sharp, but we demand that the induction be substantially zero at the position of the ion guns. The gun is focused directly at the axis. An ion leaving the gun has zero canonical angular momentum (Sec. 10.3). The canonical angular momentum is strictly preserved; therefore the ion crosses the magnetic axis. The ion would then leave by a conjugate path if nothing else happened; but if a number of sources were radially disposed, a dissociating mechanism (for example, an arc) on the axis might produce a plasma of trapped ions. Such a device has not yet been built.

The two principal devices employing molecular ion breakup are the DCX-1 in the U.S. and OGRA in the U.S.S.R.

We shall discuss DCX-1 first; Fig. 15.13 shows a side view, and Fig. 15.14 shows a partial cross section at the midplane. A molecular ion beam (H_2^+ or D_2^+) at 600 kev is focused at an intense arc which runs along lines of induction completely through the system. A small fraction of the beam is dissociated, forming a trapped plasma of protons or deuterons inside. The induction at the midplane is 1 weber/m^2. Total magnet power is 3.7 Mw. The scheme and the figures show a number of interesting features, which we now take up one at a time.[32–35]

The original motivating discovery[33] was that an intense arc can be

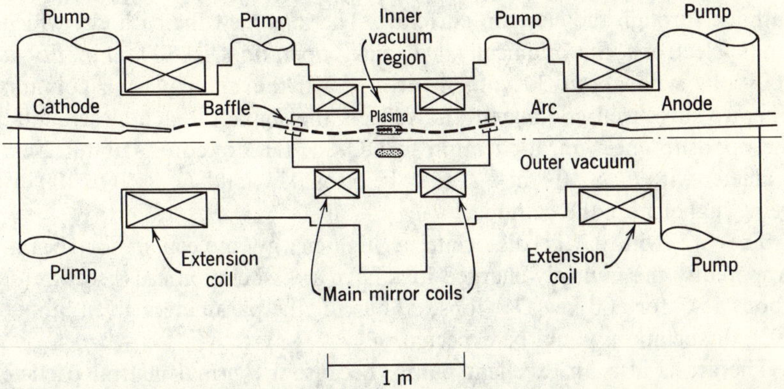

Fig. 15.13. Simplified schematic diagram of molecular-ion injection experiment DCX. The inner vacuum region is pumped by arc ionization and ion flow through the baffles along induction lines.

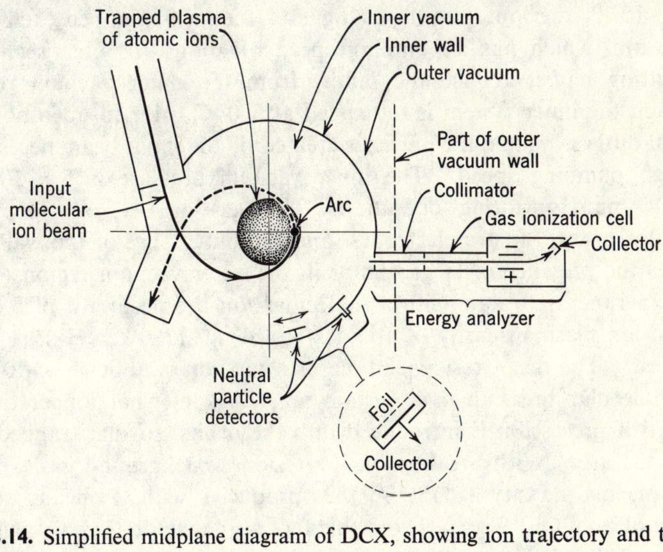

Fig. 15.14. Simplified midplane diagram of DCX, showing ion trajectory and trapped plasma.

run for great distances along induction lines at virtually 100% ionization. In one version, the electrodes are graphite, and the cathode has a small hole bored axially in it. The system is pumped continuously. To start the arc, a puff of gas is let in through the cathode. The gas is ionized by an rf coil, and the main voltage is applied between the electrodes. The arc then strikes, and the gas input and rf excitation are removed. In operation, carbon is vaporized from the anode and ionized progressively to C^+, C^{++}, C^{+++}, and C^{++++} as it drifts down toward the cathode through the electron current. The ions heat the cathode, which emits electrons thermionically. Typical operation is 300 amp dc at 300 volts with about 30 volts/m drop down the main plasma column. The arc stays well collimated and reflects the shape of each electrode at the opposite one. In application to DCX-1, the arc core is about 2-cm diameter with $n_i \approx 10^{20}/\text{m}^3$; T_e may be 5 to 20 ev, and $T_i \approx 10$ to 100 ev, being higher near the cathode.

The arc will partly dissociate a high-energy molecular ion beam. Apparently the multiply charged ions have a dissociation cross section of about 5 A^2 for 600 kev D_2^+ ions. Thus for the parameters given, about 10% dissociation would be expected.

The arc is also an excellent pump, because a thermal neutral particle has little chance of passing through without being ionized. The ions so formed flow along the induction lines to the cathode. The random flux of neutral particles n_0 upon the arc surface is $n_0\bar{v}/4$; if the molecules are N_2 or O_2, the plasma surface has 11 liters/sec-cm^2 effective pump speed. In Fig. 15.13, the inner region of highest vacuum is pumped exclusively by the arc, which has a pumping speed of about 10^4 liters/sec. The close-fitting baffles reduce the inflow from the outer vacuum region. Although the inner system is outgassed at 500° C prior to operation, the residual outgassing and the particles created by the input beam necessitate the high pumping speed. The inner vacuum pressure is 5×10^{-7} to 1×10^{-6} mm Hg during operation. This gas arises from the input molecular beam, the arc electrodes, and the outgassing of the wall, and requires the presence of large pumps in the outer vacuum region.

In experiments to date with a molecular ion beam current of 5 ma, a trapped ion plasma density of about $2 \times 10^{15}/\text{m}^3$ (300 kev H^+) has been produced. The mean residence time of a fast ion is about 3×10^{-3} sec with molecular breakup in a carbon arc. A somewhat longer time is achieved if precession is introduced into the orbits, so that trapped ions escape the arc. With breakup on the residual background gas, a much lower plasma density ($\leq 10^{14}/\text{m}^3$) is produced with a mean particle lifetime of about 10^{-2} sec. From this fact we conclude that a significant fraction of the charge transfer takes place to the ions in the arc.

These facts are shown experimentally by the diagnostic technique of sampling the fast neutrals (H^0) leaving by charge transfer. Figure 15.14 shows a few of the detectors used. The T-shaped type consists of a thin metal window that ionizes a known fraction of the neutrals as they pass through. The charged particles are then measured on an electrode placed behind the window. If one of these detectors is moved azimuthally around the plasma, it is found that at a pressure of 5×10^{-7} mm Hg, about two-thirds of the charge-transfer loss occurs as trapped protons pass through the arc. We can make a simple estimate of the cross section. The arc extends over about 5% of the H^+ orbit; then with an arc density of $10^{20}/m^3$, a proton speed of 7.6×10^6 m/sec, and a residence time of 4.5×10^{-3} sec as a result of the action of this arc, we find a cross section $\sigma_t = 6 \times 10^{-4}$ A^2. Somewhat more refined calculations[36] yield 1×10^{-3} A^2.

The energy distribution of the neutral hydrogen atoms formed by charge transfer is measured by the collimating detector shown in Fig. 15.14. The H^0 atoms are partially reionized in a differentially pumped windowless gas cell and deflected in an electrostatic analyzer. Considerable energy degradation is found, corresponding to cooling of fast ions by dynamical friction to the cool electron cloud.

The burnout theory of Sec. 15.4 can be applied to the device. For the experimental conditions recorded here, burnout is not achieved; beam currents much more than an order of magnitude larger would be required. In addition, energy degradation of trapped ions makes burnout more difficult, because the charge-transfer cross section σ_t increases much more rapidly with decreasing ion energy than does the ionization cross section σ_i. Problems 5 and 6 at the end of the chapter illustrate the calculation of the ion energy distribution function $f(u_i)$.

Beside increasing the input beam current, other improvements are being considered. First, if the arc were completely stripped, no charge transfer to it could take place. For this reason, the properties of energetic hydrogen and deuterium arcs are being investigated. Figure 15.15 shows a composite arrangement. An arc can be run down the lines of induction with gas feed into either the anode or cathode, or both. For 5 cc(STP)D_2 gas feed per second at a potential of 120 volts and a current of 200 amp, an ion density of $10^{20}/m^3$ is typical.[37] Breakup efficiency for fast molecular ions is unfortunately small, $\sigma_{bi} \approx 0.5$ to 1.0 A^2 and is less than that found in carbon arcs. Arcs of this type have been run in DCX-1.

The next improvement concerns reducing the dynamical friction loss to electrons. If the electron temperature T_e is less than about 100 ev, then $\bar{v}_e < \bar{v}_i$ for the fast ions, and the loss rate is almost independent of the electron temperature. However, if this temperature T_e could be raised

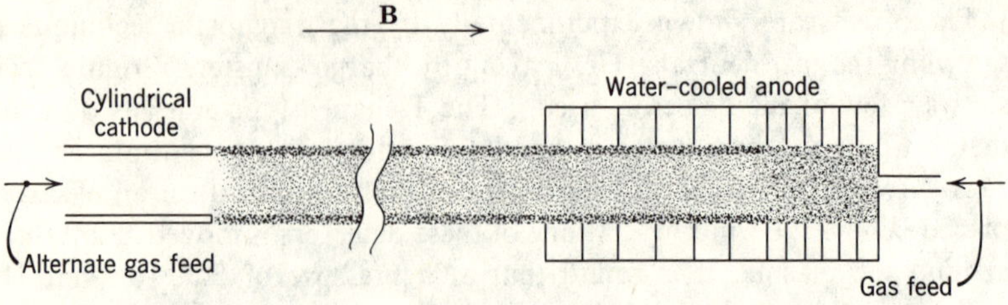

Fig. 15.15. Schematic configuration of an energetic deuterium arc. The plasma distribution is appropriate for anode gas feed; the cathode tip is heated by ion bombardment and emits electrons; the energetic electrons ionize the gas inside the anode. For cathode gas feed, a hollow cathode discharge runs inside the cathode cylinder, and the plasma streams toward the (now passive) anode. The direction of **B** can be reversed. After Mackin and co-workers[35] (p. 47), Mackin and Gibbons,[37] and Luce.[33]

(by rf power at cyclotron resonance, for example), the loss would be substantially reduced (see again Problems 5 and 6). The average electron energy in the deuterium gas arc may be about 100 ev at present. In this respect, dispensing with the arc and its reservoir of cold electrons would be most desirable if appreciable breakup could be achieved some other way.

Finally, the pumping speed may be improved by evaporating metal (usually titanium but sometimes molybdenum) onto large areas inside the various vacuum regions. The clean metal adsorbs chemically active gases (especially H_2 and D_2), so that very high pumping speeds can be achieved.

We now turn briefly to the OGRA device,[28, 38] which is the largest thermonuclear experiment to date; it is shown in Fig. 15.16. The development is proceeding under the direction of I. N. Golovin. Here,

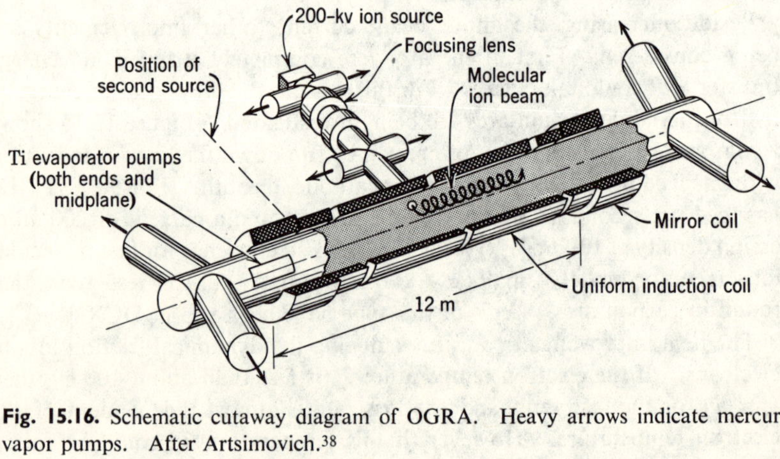

Fig. 15.16. Schematic cutaway diagram of OGRA. Heavy arrows indicate mercury vapor pumps. After Artsimovich.[38]

the 200-kev molecular ions break up on the background gas. The input beam is deflected about 20° out of the midplane as it enters and misses the gun on its first rotation. It then spirals to the mirror; since the induction is inhomogeneous in several regions along the trajectory (as well as at the mirror), the input beam precesses around the magnetic axis. In this way, it misses the gun on its return from the mirror, and spirals back and forth in the mirrors many times. The total path is about 10^3 m. Even at low background gas pressure, appreciable molecular ion breakup can take place over this path length. A typical burnout curve was calculated for this device in Sec. 15.4 (Fig. 15.11). There is no vacuum arc, and the electrons trapped in the plasma will be heated. Titanium evaporator pumps in parallel with mercury diffusion pumps then yield a pressure of 2 to 3×10^{-8} mm Hg. An input H_2^+ beam current of 30 ma produces a 100-kev proton plasma at a density of about $10^{13}/m^3$, which is apparently limited by charge transfer to the background gas and possibly by some nonadiabatic loss through the mirrors. Probe measurements give a positive plasma potential of 1 to 3 kev, indicating an electron temperature of about that magnitude. With ion beam currents an order of magnitude larger, burnout should be closely approached or actually achieved.

The large, hot, low-density plasmas produced to date in DCX-1 and OGRA do not show any definite instability. While the engineering of these large systems is a formidable task (OGRA is about 20 m long over-all), this general approach to producing a plasma of thermonuclear interest appears promising. Intensive work is under way to produce ion sources capable of delivering several hundred ma of molecular ions at several hundred kev.[39] At Oak Ridge National Laboratory (the site of DCX-1), plans are being made for constructing a larger device than DCX-1 and for employing both arc and gas-plasma breakup in ion density ranges where each is most efficient. In this way, the advantages of both types of operation may perhaps be achieved.

★ 15.6. The Astron

The Astron device[40] uses a magnetic mirror incidentally to aid in the formation of toroidal closed field structures. The concept is shown in Fig. 15.17. Pulses of relativistic electrons are to be guided into the mirror. They induce currents in a resistive mesh near the mirrors; their axial motion will be damped, and they will be trapped. One such electron is shown in Fig. 15.17*a*. A sheet of these circulating electrons is then built up to form a so-called E-layer. The circulating sheet current **K** per unit length is to be high enough that the induction in the central region reverses.

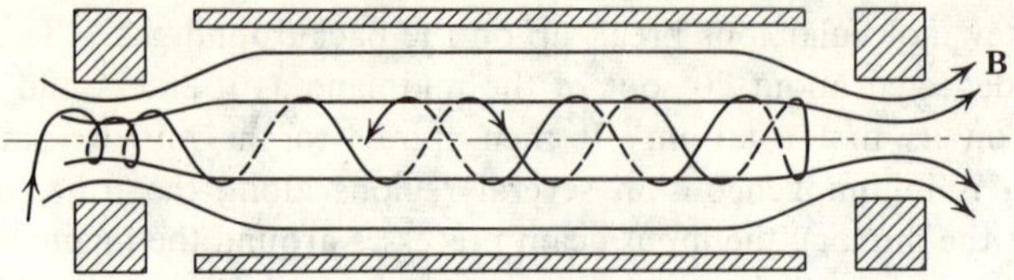

(a)

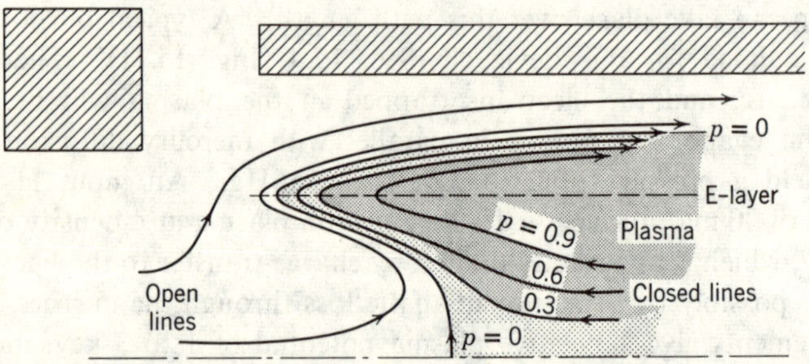

(b)

Fig. 15.17. The Astron, with relativistic E-layer. (*a*) Schematic field configuration, with one relativistic electron spiraling; (*b*) with E-layer established, one quadrant sketched. After Christofilos.[40]

The electron density required of the E-layer is easily calculated. The layer creates an induction

$$B = \mu_0 K. \tag{15.30}$$

The electrons have a density N_e per unit length of cylinder and a speed c close to that of light. Consequently,

$$K = N_e ec/2\pi r_E, \tag{15.31}$$

where r_E is the E-layer radius. If the induction created in this way just cancels that of the solenoid,

$$\mu_0 N_e ec/2\pi r_E = B_{\text{ext}} \tag{15.32}$$

or

$$N_e = 2\pi r_E\, B_{\text{ext}}/\mu_0 ec = (2\pi m_e/\mu_0 e^2)(U/U_0). \tag{15.33}$$

Here, m_e is the electron rest mass, U is the total relativistic energy, and U_0 is the rest energy (0.51 Mev). In deriving Eq. 15.33, we have used the usual cyclotron relation between r_E, B, and m_e. From Eq. 14.8, dealing with the simple pinch, we recognize the same sort of criterion as was derived there. Here, electrons with mean spacing of twice the classical electron radius along the cylinder just cancel the external induction. Motion parallel to the induction has been neglected. It is necessary that the electrons also have considerable axial energy in order that the self-induction created by the E-layer itself does not cause it to collapse axially. Further increase in the electron density in the layer over the critical value given Eq. 15.33 reverses the induction inside and establishes the pattern shown in Fig. 15.17*b*.

It is interesting to note the similarity between the Astron (Fig. 15.17*b*) and the simple torus (Fig. 7.1). In fact, the pattern of closed **B** lines of the Astron is topologically equivalent to the simple torus. This equivalence is the classical problem of turning an inner tube inside out through its valve stem. As an entr'acte, we show the manipulations in Fig. 15.18,

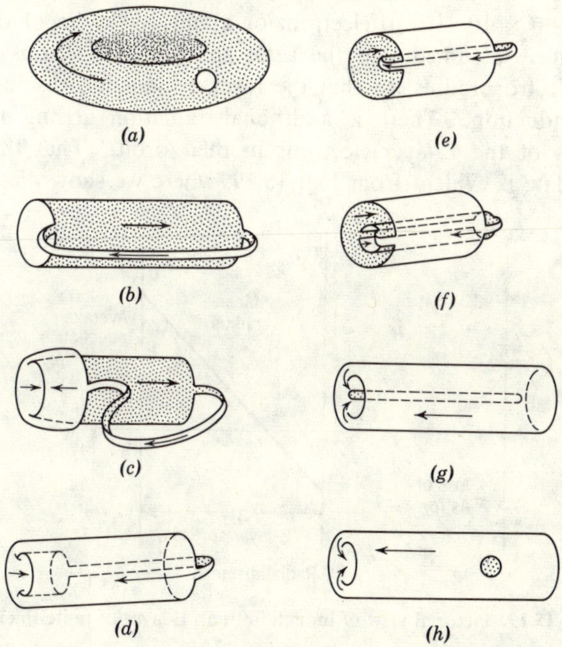

Fig. 15.18. Topological deformation of a simple solenoidal torus to give a second torus in the Astron configuration.

where for clarity the necessary connecting hole has been freely stretched.

All lines of induction inside a limiting line $p = 0$ close on themselves and can in principle confine a plasma in a hydromagnetic sense described in Secs. 7.2 and 7.3. We note that the **B** lines at the ends have the wrong curvature for stability. However, analysis is complicated by the fact that the plasma region is not substantially current free. We do not essay an answer to this question.

Whatever the eventual outcome of stability studies, the static solution can be expressed hydromagnetically in terms of isobaric magnetic surfaces. All plasma in the open induction lines is assumed to escape rapidly out the mirrors, so that the $p = 0$ surface effectively describes the edge of the plasma. The lines $p = 0.3$, 0.6, and 0.9 describe the pressure in terms of the maximum plasma pressure attainable at the midplane in the region of the E-layer. A plasma is to be built up by injecting neutral gas, which is ionized and heated by the E-layer electrons. It appears to us that the light plasma electrons may be heated selectively by this mechanism with resulting increase in the radiation losses.

The E-layer itself will radiate copiously. Suppose that $r_E = 1$ m and $U = 50$ Mev. From either Eq. 11.25 or Eq. 11.79, we find a radiation power of 4.6×10^{-12} watt/electron, or about 160 kw/m of the E-layer. Here we have assumed that the axial spacing of electrons equals the classical electron radius, so that the reverse induction inside equals the original induction. There is additional radiation arising from small oscillations of the E-layer electrons in their orbit. That the electrons must oscillate is evident from Fig. 15.19, where we show pictorial detail

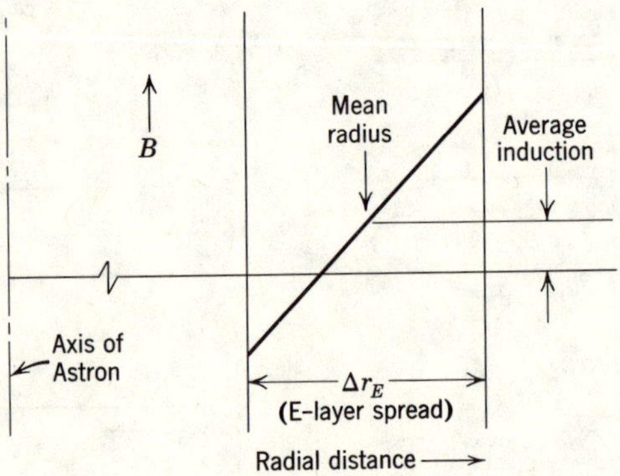

Fig. 15.19. Pictorial view of induction in an E-layer of finite thickness.

of the induction through an E-layer of finite thickness. If the electrons are to travel in the macroscopic orbit, they must run on the average in a small induction whose direction is that outside the layer. But they also oscillate about their mean radius, and the E-layer is therefore spread out over a thickness nominally designated Δr_E. It can be shown that the radiation arising from this motion is almost equal to that ascribable to the macroscopic rotation.† It is calculated that these radiation losses are small compared with the rate of energy transfer by dynamical friction

† These oscillations are similar to those in a betatron. Their frequency is determined by the so-called n-value, where $n = (r/B)(dB/dr)$. Here, n is large.

from the E-layer to a fully developed plasma (see Fig. 15.17*b*). Radiation from the end sections, where the electrons must be sharply reflected in the large B_r, may be locally severe.

In experiments to date, an electron gun capable of producing a pulsed current in excess of 200 amp at 5 Mev is being constructed. The output beam will be used to attempt the establishment of a prototype E-layer.

★ 15.7. Cusps and Related Topics

The instability of the mirror predicted in Chap. 12 has not yet been shown experimentally. This apparent enigma may arise from inadequacy of the theory or from the limited scale of the experiments. Since the enigma is unresolved, it is prudent meanwhile to investigate more stable systems. We have shown, using imperfect theory, that if a plasma excludes the induction completely, the bounding free plasma surface is stable if and only if its center of curvature lies everywhere outside the plasma. There is an immediate corollary: such a surface cannot be smooth, but must contain cusps where the curvature is sharply reversed. Two such configurations were shown in Figs. 12.7 and 12.8.

Cusped systems have been intensively analyzed by Berkowitz, Grad, Rubin, and their associates,[41–43] and also by Kadomtsev and Braginsky.[44] We shall not attempt to analyze their configurations but shall content ourselves with showing a few of them in addition to those already shown in Chap. 12. Following this pictorial representation, we shall discuss an experimental method that might be used to inject plasma into a cusp. The same scheme also works for mirrors.

Our temporary premise is that in the cusp the plasma is dense enough to exclude the confining induction completely. Determination of the plasma configuration then becomes a sort of boundary-value problem. We must solve $\nabla \times \mathbf{B} = 0$ and $\nabla \cdot \mathbf{B} = 0$ under the following conditions:

1. $\mathbf{B}$ reduces to appropriate values near all fixed exterior conductors.
2. $\mathbf{n} \cdot \mathbf{B} = 0$ at the plasma surface.
3. $B^2/2\mu_0 = p =$ constant at the plasma surface.

The problem is then to determine the free plasma surface. Figure 15.20*a* shows the simplest two-dimensional cusped system. Four straight wires carry oppositely directed currents in adjacent quadrants. Since the plasma boundary lies parallel to $\mathbf{B}$, we can, if we wish, replace it by a perfect conductor. The problem then becomes that illustrated by Fig. 15.20*b* of determining the boundary for which $|B| =$ constant over the range $(0 < x < d,\ 0 < y < d)$. Such problems are solved by conformal mapping, an activity that we do not choose to pursue. If the

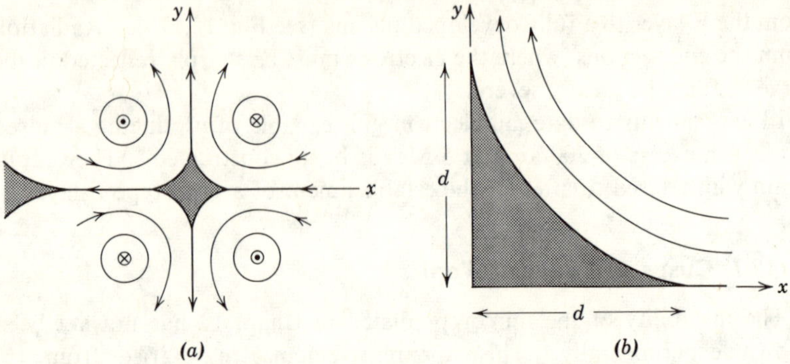

Fig. 15.20. (*a*) Line cusp formed by four wires carrying oppositely directed currents and confining a field-free plasma. (*b*) One quadrant of surface.

distance d is much less than half of the coil spacing, the plasma boundary can be shown to be a hypocycloid:

$$x^{2/3} + y^{2/3} = d^{2/3}. \tag{15.34}$$

Note that the surface $|B|$ = constant with the plasma confined does *not* correspond to the surface $|B|$ = constant with no plasma present. In that latter case, $B_x = -Cx$ and $B_y = Cy$ in the vicinity of the origin; thus $|B|$ = constant without plasma is a circle. The plasma modifies the induction.

If the plasma volume of Fig. 15.20 is not small, the boundary is determined by a much more complicated equation, but it still resembles the inverted quadrants of a circle. The limiting size of the plasma that can be confined corresponds roughly to the cusps lying between adjacent wires, as might be expected.

Once a simple or even trivial configuration has been found, others may be found by conformal mapping or by performing simple operations on simple systems. For example, Fig. 15.20*a* shows one segment of the start of an exterior region, which can be defined by inverting the interior region about a circle passing through the conductors. Physically, the induction could be used to keep the plasma outside a given region, just as well as inside it. We have encountered such a phenomenon in the inverse pinch, Sec. 14.9. The simple ring cusp system of Fig. 12.7 is obtained by rotating the two-dimensional shape of Fig. 15.20*a* about an axis through two opposite cusps. Simple cusps can be combined into a shape sometimes called a picket fence, shown in Fig. 15.21.

Although a cusp system has some resemblance to a conventional mirror, the plasma losses are governed by quite different considerations if the induction is excluded from the interior. Motion of the charged particles is not adiabatic: the trajectory of a particle approaching a cusp might resemble that of Fig. 15.22. The loss rate is determined by such complicated factors as the transient penetration of the particle into the outside region and the space-charge electric field set up by the different penetrations of ions and electrons. This penetration is greater for high-energy particles, whence the cusp confines a high-energy plasma less well than a low-energy plasma. Berkowitz and his associates[41] estimate a confinement time

$$\tau = 10^{-2}(R^2B/T) \tag{15.35}$$

for the simple ring cusp of Fig. 12.7, where R is the radius, B is the induction, and T is the temperature in kev. For $T = 10^{-2}$ kev, $B = 1$ weber/m^2, and $R = 0.1$ m, $\tau = 10^{-2}$ sec, which is satisfactory for a laboratory experiment. However, if $R = 1$ m, $B = 10$ webers/m^2, and $T = 50$ kev, $\tau = 0.002$ sec, which is rather short. Coulomb scattering plays little or no role in this escape mechanism. The numbers quoted here are believed to be very uncertain because of the great difficulty of analysis.

If the plasma does not exclude the induction completely, the analysis is even more complicated. The possibility now exists for some mirror-type reflection to take place as a particle approaches a region that was previously the cusp. However, if the particle passes through an interior point where B is small, the motion is no longer even approximately adiabatic.

Most of the interesting properties of the cusped systems are manifest at high β; the problem arises of filling the cusp rapidly with plasma. One method would be to run a cold plasma into the region, then pulse the cusp. Since the volume of the compressed plasma would be small, much of the available space would be wasted.

Among other possibilities, injecting a high-speed plasma blob or stream axially into the system may be feasible. Two distinct cases arise. In the first, the material does not exclude the induction, but is dense

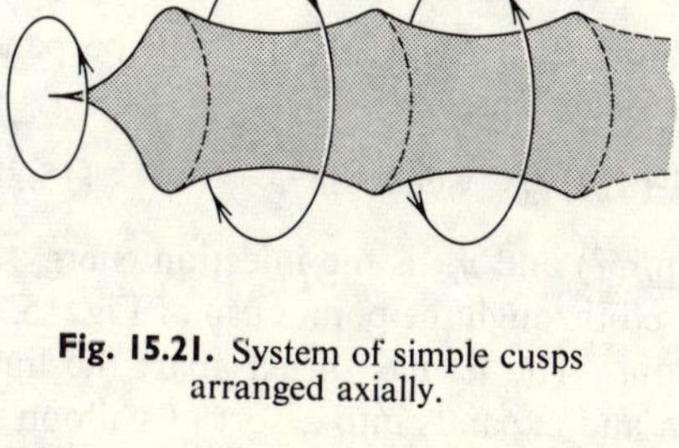

Fig. 15.21. System of simple cusps arranged axially.

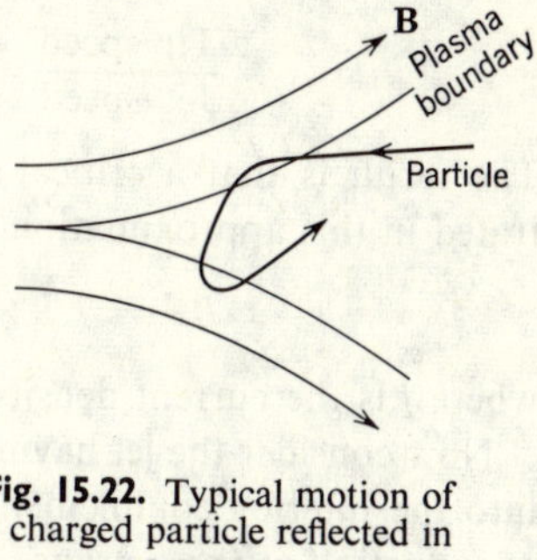

Fig. 15.22. Typical motion of a charged particle reflected in a cusp.

enough to be polarized as shown in Fig. 15.8. It then penetrates to the interior; some of it may disassemble.

In the second case, the material has high density and macroscopic velocity, and pushes the induction aside. An analysis is given by Tuck,[45] which we summarize. Let a plasma jet with density ρ_{m0}, macroscopic velocity V_0, and negligible internal energy be incident upon a region of magnetic induction. The stagnation pressure $\rho_{m0}V_0^2$ of the jet exceeds $B^2/2\mu_0$; therefore the jet penetrates the induction, as shown in Fig. 15.23. The region of induction is large, whence **B** lines do not pile up. Assume

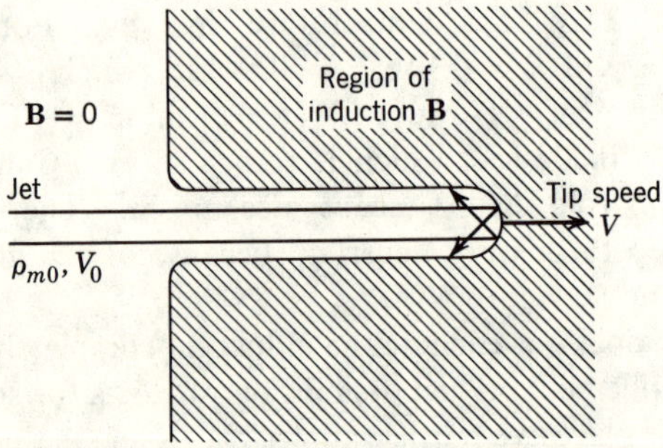

Fig. 15.23. Plasma jet piercing a region of magnetic induction.

that the process of reflection off the curved end of the cavity can be approximated by uniform (cosine-law) scattering from a plane end. Let the tip speed be V. The relative speed of impact of the jet with the moving tip is $V_0 - V$; for cosine scattering, the mean backward particle speed relative to the wall will be $(V_0 - V)/2$.

The axial pressure on the end of the tip is the rate of change of axial momentum. For constant tip speed, the pressure balance reads

$$B^2/2\mu_0 = 3\rho_{m0}(V_0 - V)^2/2, \tag{15.36}$$

or

$$\frac{\text{Tip speed}}{\text{Jet speed}} = \frac{V}{V_0} = 1 - \left(\frac{B^2}{3\mu_0\rho_{m0}V_0^2}\right)^{1/2}. \tag{15.37}$$

The result is that a critical induction $B_c = (3\mu_0\rho_{m0}V_0^2)^{1/2}$ can be penetrated in this approximation. For deuterons,

$$B_c^2 = 7.7 \times 10^{-10}(u_{ev})^{1/2}j \quad \text{weber}^2/\text{m}^4, \tag{15.38}$$

where j is the current density (amp/m^2) and u_{ev} is the injection energy.

Now consider the jet having passed through the point cusp of Fig. 15.24 into the interior confinement region. The jet has thrust apart the lines of induction at the entrance region and expands into a sort of balloon in

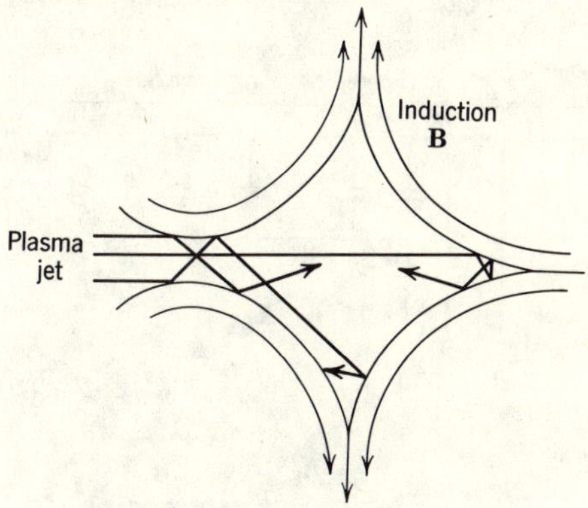

Fig. 15.24. Plasma jet entering a cusp.

the interior. It is statistically less likely that the jet will re-form and pierce the opposite cusp. Thus the plasma is trapped, particularly if the jet is turned off and the entrance closes. Since we know that particles cannot be trapped in a steady-state induction without collisions, we can ask where the collisions took place. The collisions were in fact the sharp deflections of the particles on the magnetic wall. While the jet is entering, the wall has a hole; before the particles can find it again, the hole closes. Thus we are discussing a highly time-dependent case, and not a steady-state one insofar as injection is concerned. We can view the trapping mechanism as one in which the collisions of the incoming jet with the magnetic surface were concentrated in space. Once the particles are scattered on their way in, they are unlikely to reverse the process and concentrate at one point on the magnetic wall to force their way out.

This injection scheme, while most suitable for cusped systems, might be feasible also for injection into mirrors.

Plasma guns have been developed[46–48] that can produce a plasma jet capable of piercing a 1-weber/m^2 induction. One of them[46] is the gun used to generate the central plasma for the Ixion device, Sec. 15.3. Figure 15.25 shows two schemes employed. In both cases, a quick-acting valve is opened by a blow on the valve stem for about 100 μsec, and admits about 1 cc of gas at atmospheric pressure. In one case, a shock coil ionizes the gas after it has had time to enter the barrel, and a pulsed traveling wave accelerates the plasma down the tube. In the other case, the gas is admitted to a coaxial cylinder, and a 15-kv pulse voltage is then applied between inner and outer conductors; the gas is ionized, and the $\mathbf{j} \times \mathbf{B}$ force accelerates the plasma down the tube. In this latter configuration, about 5×10^{19} protons can be accelerated to a velocity of about 1.5×10^5 m/sec (120 ev). About 40% of the input energy is

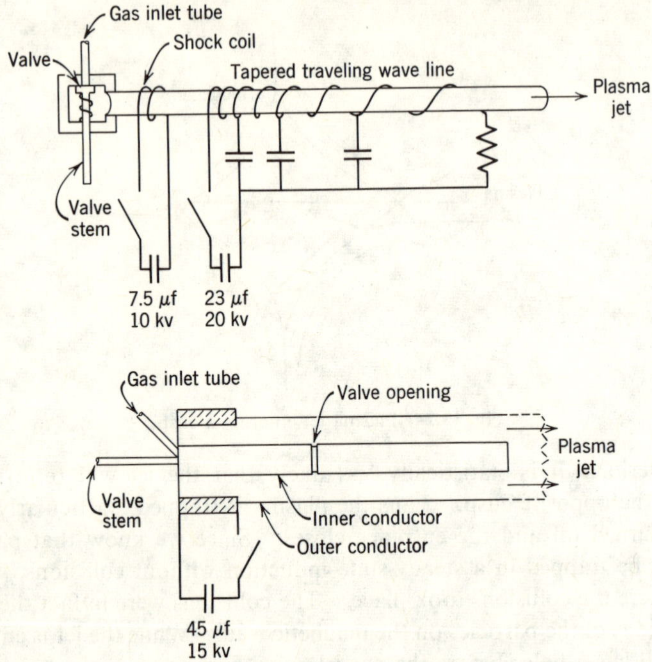

Fig. 15.25. Two types of plasma gun, shown schematically. After Marshall.[46, 47]

delivered to the plasma. In later modifications[48] of the coaxial gun, mean deuteron velocities of 5×10^5 m/sec (≈ 2.5 kev) have been achieved.

Experiments performed to date indicate that if the jet is too small, it will not penetrate into a cusp (or mirror); if it is of intermediate strength, it penetrates and does not leave through the opposite end of the confinement system. If the jet is too strong, it travels completely through. Confinement times are short (≈ 50 μsec). There is some evidence that the plasma in the coaxial gun tends to lose azimuthal symmetry and pinch into a small region at one side of the tube.[49]

15.8. Summary and Conclusions

We have presented at least in outline most of the methods of utilizing mirrors or their adaptations to produce a relatively quiescent hot plasma. It is our opinion that the experiments show more promise of achieving our goal of controlled thermonuclear reactions than did the pinch experiments. If pressed to express a real probability, however, we can only answer that it depends upon the probability that the mirror is stable, or that the cusp is not too leaky, and so forth, all of which questions are yet unanswered. Experiments now being planned and built should provide considerably more insight into these matters.

It is possible to summarize the present thermonuclear experiments in mirrors in terms of the orders of magnitude by which they fail to produce a practical thermonuclear plasma. To have some basis for comparison, let us take a volume 10 m^3, temperature 100 kev, density 10^{21}/m^3, and confinement time 1 second as our criterion. Table 15.1 lists some current experimental approaches and the orders of magnitude they have yet to go. We recognize that one parameter can often be traded for another (for example, volume for density), and we freely admit an uncertainty of an order of magnitude. The sum of all the exponents, in experiments where interesting data exist, is in all cases about 8 to 10. Three years ago (1957), the sum would have been about 12. We close this chapter by speculating that if the sum of exponents can be reduced to 5 or less, some fairly firm predictions regarding our aspirations can be made.

Table 15.1. Approximate orders of magnitude to go in mirror experiments to achieve a practical thermonuclear plasma

Experimental Type	Too Small by	Too Cold by	Too Low Density by	Too Brief by
Fast Shock	10^6	10–10^2	10^{-2}	10^5
Slow-Compression Mirrors	10^4	10–10^2	1–10^2	10^3
Molecular-Ion Mirror Injection	10–10^3	1	10^5–10^7	10^2*
Astron	....	unknown	....	
Cusps	....	unknown	....	

*Based on present lifetime of trapped ions.

PROBLEMS

1. Consider a magnetic mirror in which the mirror ratio $R = 1 + (z/\lambda)^2$, where λ is a constant. A plasma is to be compressed (or expanded) adiabatically by increasing the induction uniformly. If the radial compression ratio is C_r, show that there is an associated axial compression

$$C_z = C_r^{1/4}.$$

2. It is proposed to inject particles into a slowly pulsed magnetic mirror by injecting them through the mirror from outside. The ions barely pass through the mirror, making an angle $(\pi/2) - \delta$ with the **B** lines as they pass through. For small angles δ, show that the mirror induction B_m must rise at the rate

$$\frac{dB_m}{dt} \geqslant \frac{B_m \delta^2}{\tau}$$

in a symmetric system, if only the mirrors are increased with time, and τ is the one-way transit time.

3. The operation of a pulsed mirror circuit is to be analyzed. Each coil consists of 9 turns of $\frac{3}{16}$-in. by $1\frac{3}{4}$-in. copper strips, wound edgewise into a single-layer solenoid. Length of each coil is 5 cm, which allows for insulation between turns, and mean radius is 13 cm. The coils are to be driven in series by the circuit shown. The condenser can be charged to 40 kv, and discharged through

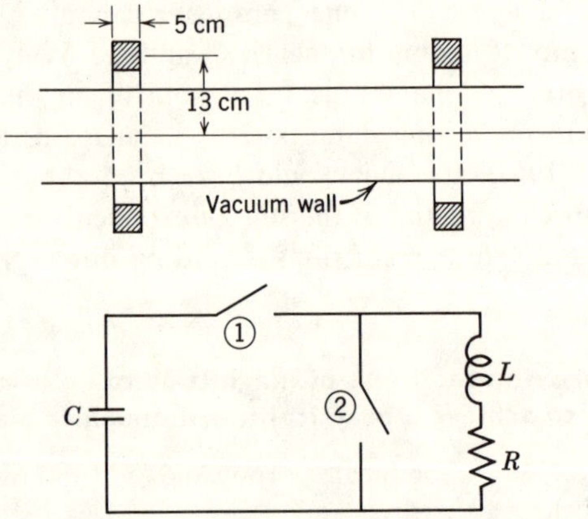

ignitrons (1). When the current is a maximum, it is clamped by ignitrons (2). Assume that the mutual inductance between coils is negligible, and neglect the resistance and inductance of all elements except the coils. Calculate:

(*a*) The total resistance of the coils.
(*b*) The total inductance of the coils.
(*c*) The current required to give 5-webers/m^2 peak at the center of each mirror.
(*d*) The capacity C required to give the peak current.
(*e*) The time after closing ignitrons (1) when the peak induction is reached.
(*f*) The decay rate of the current after the clamp (2) is fired.
(*g*) The energy stored initially.
(*h*) The energy left in C at the instant of clamping.
(*i*) The energy dissipated in resistive losses up to the instant of clamping.

4. Derive Eq. 15.25 approximately in two limiting cases:

(*a*) The plasma is optically very thin.
(*b*) The plasma is optically very thick.

5. We wish to calculate the ion energy distribution function $f(u)$, where u is the ion energy in kev, in a magnetic mirror fed by molecular ion.injection. The assumptions are the following:

(1) A mean molecular ion flux Φ particles/unit area-sec passes through the region. Breakup occurs on trapped ions of density n_i with cross section σ_{bi} and on other species n_j with cross section σ_{bj} for each species.

(2) A trapped ion at energy u is lost by charge transfer to various species n_h with cross section $\sigma_{th}(u)$ for each species.

(3) Trapped ions lose energy by dynamical friction to the electron cloud and to a cloud of slow ions (an arc for example).

(4) $T_e \ll T_i$ (or $\bar{u}$).

(5) Trapped ions that avoid charge transfer and cool to low energy are removed by Coulomb scattering out the mirrors at a negligibly low energy.

(6) Burnout is not yet achieved, and mirror losses at high energy are negligible.

If the electrons are Maxwellian at temperature T_e (kev), show that

(*a*) Energy loss to slow ions is negligible compared with energy loss to the electrons.

(*b*) The energy distribution function $f(u)$ is, with energy loss to slow ions neglected,

$$f(u) = \frac{C_1\Phi\left(\sigma_{bi}n_i + \sum_j \sigma_{bj}n_j\right)u^{1/2}}{n_e}\left[1 + \frac{3\sqrt{\pi}}{4}\left(\frac{m_iT_e}{m_eu}\right)^{3/2}\right]$$

$$\times \exp - \int_u^{u_0} \frac{C_2}{n_e}\, du\, u \sum_h \sigma_{th}n_h\left[1 + \frac{3\sqrt{\pi}}{4}\left(\frac{m_iT_e}{m_eu}\right)^{3/2}\right],$$

where u_0 is the injection energy, and C_1 and C_2 are constants.

(*c*) For protons, with energies in kev, $C_1 = 4.75 \times 10^{12}$, $C_2 = 2.1 \times 10^{18}$.

(*d*) The ion density will increase without limit (until scattering out the mirrors takes over) if

$$\int_{u_l}^{u_0} dx(u)\left[\frac{\Phi\sigma_{bi}}{\sum v\sigma_{th}(u)}\right]\exp\left[-\int_u^{u_0} dx(u)\right] \geq 1,$$

where $dx(u)$ is the quantity under the integral in part (*b*), v is the ion speed, and u_l is some negligibly low energy.

(*e*) The criterion (*d*) for exponentiation of the plasma is more severe than the simple criterion

$$\frac{\Phi\sigma_{bi}}{\sum \sigma_{th}(u_0)v_0n_h} = 1.$$

This problem illustrates the effect of the dynamical friction, and parts (*d*) and (*e*) are modifications of a simple plasma growth derivable from Sec. 15.4.[50]

6. From the results of Problem 5, assume $T_e = 10^{-2}$ kev, corresponding to an arc. Assume reasonable values for the average electron density $\bar{n}_e$ and charge-transfer cross sections $\sigma_t(u)$ from Chap. 3. Compute $f(u)$ and compare it with the distribution $f(u) \propto u^{1/2}$. To what physical conditions does this latter distribution correspond?

7. If controlled thermonuclear power is a failure, we can turn to fanciful interplanetary engines.

The plasma in volume A is somehow kept hot and replenished: it escapes preferentially through the weak mirror.

(*a*) Assume that the plasma actually escapes from the magnetic field. Then for fixed plasma density in A, what is the dependence of the thrust on plasma temperature T in the volume A? Assume $nkT/(B^2/2\mu_0)$ in the region A is suitably less than 1 for all useful temperatures.

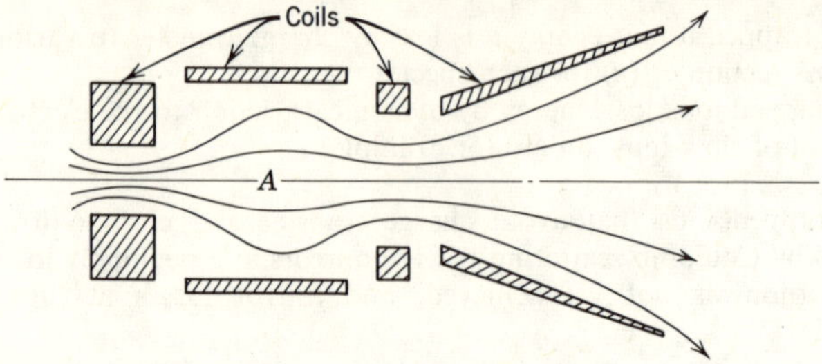

(*b*) Suppose that in *A*, volume = 1 m^3, ion density = $10^{20}/m^3$, T = 1 kev. Compute the order of magnitude of thrust to be expected for a proton plasma.

(*c*) How can the plasma be made to escape from the lines of induction and thereby provide any thrust at all?

(*d*) Point out any similarity in the flow pattern to sub- and supersonic flow as found in a conventional nozzle.

REFERENCES

1. A. N. Kaufman, *Controlled Thermonuclear Reactions, U.S. Atomic Energy Commission Report TID 7536* (Part 2), Technical Information Service Extension, Oak Ridge, Tenn. (1957).
2. D. Judd, W. M. MacDonald, and M. N. Rosenbluth, "End Leakage Losses from the Mirror Machine," *U.S. Atomic Energy Commission Report WASH-289*, Conference on Controlled Thermonuclear Reactions, Berkeley, Calif. (Feb. 1955), p. 158.
3. S. A. Colgate and R. E. Wright, *Proc. 2nd U.N. Conf. on Peaceful Uses of Atomic Energy*, Vol. 32, United Nations, Geneva (1958), p. 145.
4. W. C. Elmore, E. M. Little, and W. E. Quinn, *Proc. 2nd U.N. Conf. on Peaceful Uses of Atomic Energy*, Vol. 32, United Nations, Geneva (1958), p. 337.
5. A. C. Kolb, *Proc. 2nd U.N. Conf. on Peaceful Uses of Atomic Energy*, Vol. 31, United Nations, Geneva (1958), p. 328; *Phys. Rev.*, **112**, 291 (1958).
6. G. Miyamoto, T. Kihara, G. Iwata, S. Mori, T. Ohkawa, and M. Yoshikawa, *Proc. 2nd U.N. Conf. on Peaceful Uses of Atomic Energy*, Vol. 32, United Nations, Geneva (1958), p. 308.
7. S. M. Osovetz, Y. F. Nasedkin, E. I. Pavlov, Y. F. Petrov, and N. I. Schedrin, *Proc. 2nd U.N. Conf. on Peaceful Uses of Atomic Energy*, Vol. 32, United Nations, Geneva (1958), p. 311.
8. A. C. Kolb, C. B. Dobbie, and H. R. Griem, *Phys. Rev. Letters*, **3**, 5 (1959).
9. H. R. Griem, A. C. Kolb, and W. R. Faust, *Phys. Rev. Letters*, **2**, 281 (1959).
10. K. Boyer, E. M. Little, W. E. Quinn, G. A. Sawyer, and T. F. Stratton, *Phys. Rev. Letters*, **2**, 279 (1959).
11. W. E. Quinn, E. M. Little, and F. L. Ribe, *Bull. Am. Phys. Soc.*, Ser. II, **5**, 341 (1960).
12. E. M. Little and W. E. Quinn, *Bull. Am. Phys. Soc.*, Ser. II, **5**, 341 (1960).
13. I. F. Kvartskhava, K. N. Kervalidze, and Yu. S. Gvaladze, *Zhur. Tekh. Fiz.*, U.S.S.R., **30**, 1321 (1960).
14. R. F. Post, *Proc. 2nd U.N. Conf. on Peaceful Uses of Atomic Energy*, Vol. 32, United Nations, Geneva (1958), p. 245.

15. F. H. Coensgen, F. C. Ford, and R. E. Ellis, *Proc. 2nd U.N. Conf. on Peaceful Uses of Atomic Energy*, Vol. 32, United Nations, Geneva (1958), p. 266.
16. F. H. Coensgen, W. F. Cummins, and A. E. Sherman, *Phys. Fluids*, **2**, 350 (1959).
17. F. H. Coensgen, W. F. Cummins, W. E. Nexsen, Jr., and A. E. Sherman, Paper F-5 2nd Annual Meeting, Division of Plasma Physics, Gatlinburg, Tenn. (November 1960).
18. W. A. Perkins and R. F. Post, *Bull. Am. Phys. Soc.*, Ser II, **5**, 353 (1960).
19. H. C. Early and W. D. Dow, *Phys. Rev.*, **79**, 186 (1950).
20. O. A. Anderson, W. R. Baker, A. Bratenahl, H. P. Furth, J. Ise, Jr., W. B. Kunkel, and J. M. Stone, *Proc. 2nd U.N. Conf. on Peaceful Uses of Atomic Energy*, Vol. 32, United Nations, Geneva (1958), p. 155.
21. G. V. Gordeev and A. I. Gubanov, *Zhur. Tekn. Fiz.*, **28**, 2046 (1958).
22. W. O. Schumann, *Z. Naturforsch.*, **13a**, 888 (1958).
23. B. Bonnevier and B. Lehnert, *Arkiv. Fysik*, **16**, 231 (1959).
24. K. Boyer, J. E. Hammel, C. L. Longmire, D. Nagle, F. L. Ribe, and W. B. Riesenfeld, *Proc. 2nd U.N. Conf. on Peaceful Uses of Atomic Energy*, Vol. 31, United Nations, Geneva (1958), p. 319.
25. R. F. Post, *Ann. Rev. Nuclear Sci.*, **9**, 428 (1959).
26. A. Simon, *Phys. Fluids*, **1**, 495 (1958); **2**, 336 (1959).
27. A. Simon, "Critical Current for Burnout in an OGRA-type Device," *ORNL Report 2831*, Office of Technical Services, Dept. of Commerce, Washington 25, D. C.
28. I. V. Kurchatov, *J. Nuclear Energy*, **8**, 168 (1958), and *Progress in Nuclear Energy, Plasma Physics, and Thermonuclear Research*, Vol. 1, Pergamon Press, London (1959). Page 42 gives a summary of U.S.S.R. cross-section data at thermonuclear energies.
29. C. F. Barnett, P. R. Bell, J. S. Luce, E. D. Shipley, and A. Simon, *Proc. 2nd U.N. Conf. on Peaceful Uses of Atomic Energy*, Vol. 31, United Nations, Geneva (1958), p. 298.
30. A. Simon and M. Rankin, *ORNL Report 2354*; T. K. Fowler and A. Simon, *ORNL Report 2552* (USAEC, Technical Information Extension, Oak Ridge, Tennessee).
31. F. Prévot, P. Hubert, and C. Gourdon, *Compt. rend.*, **249**, 997 (1959).
32. C. F. Barnett, P. R. Bell, J. S. Luce, E. D. Shipley, and A. Simon, *Proc. 2nd U.N. Conf. on Peaceful Uses of Atomic Energy*, Vol. 31, United Nations, Geneva (1958), p. 298.
33. J. S. Luce, *Proc. 2nd U.N. Conf. on Peaceful Uses of Atomic Energy*, Vol. 31, United Nations, Geneva (1958), p. 305.
34. *Report ORNL 2802, Thermonuclear Project Semiannual Report for Period ending 31 July 1959*, Office of Technical Services, Washington 25, D. C.
35. *Report ORNL 2926, Thermonuclear Project Semiannual Report for Period ending 31 January 1960*, Office of Technical Services, Washington 25, D. C.
36. Reference 33, p. 10.
37. R. J. Mackin, Jr., and R. A. Gibbons, *Bull. Am. Phys. Soc.*, Ser. II, **5**, 370 (1960).
38. L. A. Artsimovich, *Proc. 2nd U.N. Conf. on Peaceful Uses of Atomic Energy*, Vol. 31, United Nations, Geneva (1958), p. 6.
39. Reference 34, p. 60.
40. N. C. Christofilos, *Proc. 2nd U.N. Conf. on Peaceful Uses of Atomic Energy*, Vol. 32, United Nations, Geneva (1958), p. 279.

41. J. Berkowitz, K. O. Friedrichs, H. Goertzel, H. Grad, J. Killeen, and E. Rubin, *Proc. 2nd U.N. Conf. on Peaceful Uses of Atomic Energy*, Vol. 31, United Nations, Geneva (1958), p. 171.
42. J. Berkowitz, H. Grad, and H. Rubin, *Proc. 2nd U.N. Conf. on Peaceful Uses of Atomic Energy*, Vol. 31, United Nations, Geneva (1958), p. 177.
43. H. Grad and H. Rubin, *Proc. 2nd U.N. Conf. on Peaceful Uses of Atomic Energy*, Vol. 31, United Nations, Geneva (1958), p. 190.
44. B. B. Kadomtsev and S. I. Braginsky, *Proc. 2nd U.N. Conf. on Peaceful Uses of Atomic Energy*, Vol. 32, United Nations, Geneva (1958), p. 233.
45. J. L. Tuck, *Phys. Rev. Letters*, **3**, 313 (1959).
46. J. Marshall, *Proc. 2nd U.N. Conf. on Peaceful Uses of Atomic Energy*, Vol. 31, United Nations, Geneva (1958), p. 341.
47. J. Marshall, *Phys. Fluids*, **3**, 134 (1960).
48. J. E. Osher and D. C. Hagerman, *Bull. Am. Phys. Soc.*, Ser. II, **5**, 351 (1960).
49. L. C. Burkhardt and R. H. Lovberg, *Bull. Am. Phys. Soc.*, Ser. II, **5**, 350 (1960).
50. *Report ORNL 3011*, *Thermonuclear Project Semiannual Report for Period ending 31 July, 1960*, Office of Technical Services, Washington 25, D. C., p. 66.

GENERAL REFERENCE

S. Glasstone and R. H. Lovberg, *Controlled Thermonuclear Reactions*, D. Van Nostrand Co., Princeton, N.J. (1960), Chaps. 9–11.

16

The stellarator

The third and final class of devices to be discussed is the stellarator. Stellarators belong topologically to the class of tori. However, we have seen in Sec. 10.2 and elsewhere that charges will drift in the direction of $\mathbf{B} \times \nabla B^2$ across a magnetic induction having a gradient. A toroidal induction is necessarily stronger on the inner edge and weaker on the outer edge. There is always a gradient that leads to a drift of the particles toward the edges lying in the plane of the torus, as was shown in Fig. 10.8. An electric field is created, since particles of opposite charge drift toward opposite edges. The electric field inhibits any further drift toward these edges, but with the magnetic field it induces a drift of charges of both signs to the outer periphery of the torus, according to Eq. 10.13.

In the stellarator the geometry and fields are designed to overcome the problem of cross drifts and to take advantage of the endless geometry of the toroidal configuration. The essential idea is so to twist the magnetic induction lines, along which the particles travel to a good approximation, that the cross drift in one part of the reaction tube cancels that in another part of the tube. The concept was illustrated in Fig. 10.10 and discussed in the adjoining paragraphs.

The physical picture that we have just painted refers to a single-particle model of the plasma. Alternately, we have developed a hydromagnetic (fluid) model of the plasma in Sec. 7.2. It was shown there that the generalized equilibrium configuration may resemble a torus with isobaric magnetic surfaces. Note particularly Figs. 7.1, 7.2, and the associated discussion. In treatments of the stellarator, both the single-particle model and the hydromagnetic model of the toroidal plasma have their areas of convenient application; we assume that the reader is quite familiar with the material of the sections cited, so repetition is unnecessary here.

It will turn out that there are other physical interpretations of a non-degenerate torus, such as a stellarator, which is arranged to give enhanced stability. We shall find that the configuration of modern stellarators can be interpreted in terms of a multiple-cusp system, and shall point out this feature when it arises in the discussion.

From this preamble, we suspect that a proper analysis of the stellarator configuration would be very complex, as is indeed the case. For the purposes of the present book, discussion of the stellarator must therefore be limited to some of the more qualitative aspects, in contrast to the more detailed quantitative discussions of the pinch and mirror configurations.

The fact that lines of induction do not close on themselves in a stellarator (see again Figs. 7.1 and 7.2) leads us to examine the behavior of lines of induction as they are followed around the torus. This subject is that of rotational transforms; it is discussed in the next section to elucidate the method of cross-drift cancellation. There are in actuality two distinct methods of producing a rotation of the lines: bending the torus into a figure-eight or adding helical windings to a simple torus. The latter method has substantial advantages over the former.

We shall discuss simple drifts of particles in a stellarator, recognizing, however, that a single-particle model is inadequate to describe the behavior of the system.

A number of constructional features of the stellarator are of interest, for example, a device called the divertor, which skims off contaminated plasma near the tube walls and selectively guides it out of the system.

Since the stellarator is a quasi-steady-state device, the usual questions can immediately be raised concerning the method of heating the plasma. The initial heating is accomplished by an induced axial electric field threading the torus; this field leads to ohmic heating. There are disadvantages in this scheme, and we are then led to consider various forms of resonance methods of plasma heating. Of these, ion cyclotron resonance appears most promising. A number of relevant experiments will be described.

The stellarator program differs from the pinch and mirror programs in that almost all research has been done at Princeton University under Project Matterhorn. This circumstance arises principally because the original concept was developed by Lyman Spitzer, Jr., at Princeton. In our description of the stellarator, especially in Sec. 16.1, we shall follow fairly closely a discussion given by Spitzer.[1]

16.1. Rotational Transforms and Magnetic Surfaces

The discussion of this section is a direct logical extension of Sec. 7.2, wherein we now investigate in more detail the behavior of lines of induction as they generate a magnetic surface. We are led to the subject of rotational transforms; in this section we shall describe only the geometrical configuration, and in Sec. 16.2 we shall consider its virtue in confining a plasma.

To this end, consider the cross section of a generalized torus, as shown in Fig. 16.1. In particular, we shall trace out the path of one line of

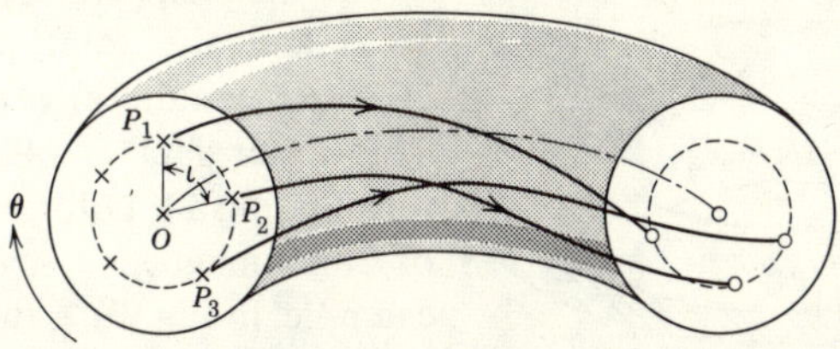

Fig. 16.1. A generalized torus with lines of induction.

induction as it threads the torus. Some lines hit a wall, but one is selected that does not. For the moment, we do not inquire into the method by which the line is created.

Consider first a simple degenerate torus. A line of induction in such a device will close on itself. In other words, a line of induction passing through the point P_1 will return to P_1 upon being followed once completely around the torus. If either the lines are twisted or the torus is twisted, then the line will not in general return to the point P_1 but will pass through some other point P_2. The next time around, the line might pass through a point P_3, all providing it does not first hit a wall. The configuration can be arranged so that most of the lines of induction will not hit a wall; for the moment only those areas of the cross section are considered for which the various points P_1, P_2, P_3, and so forth, do exist. The transformation of the set of all possible points P_1 into the new set of points P_2 is called the magnetic transform of the cross-section plane.

There are interesting and important properties of such transforms. Point P_1 is transformed into P_2, then into P_3, and so forth, on successive circuits of the torus. The set of points transforms into itself. If the density of points P_1 is taken proportional to the induction, the point density must be the same function of position before the transform as after. Such a transformation is called measure preserving. Suppose that the transform is rotational, that is, the transformation of all points P_1 into P_2 involves a rotation in the same direction. In Fig. 16.1 the transformation at the left cut is clockwise. Then all the lines rotate about a magnetic axis O, and this magnetic axis transforms into itself. Each line (except the magnetic axis) generates a magnetic surface. To the approximation that the plasma pressure p is a scalar, p is constant over a magnetic surface. In general, the transformed points will never duplicate themselves exactly but will generate a smooth curve, as shown by the dashed closed curves in Fig. 16.1. These words are merely another way of expressing the intuitive conclusion that the isobaric surfaces will be smooth almost everywhere. Such matters were discussed at length in Sec. 7.2.

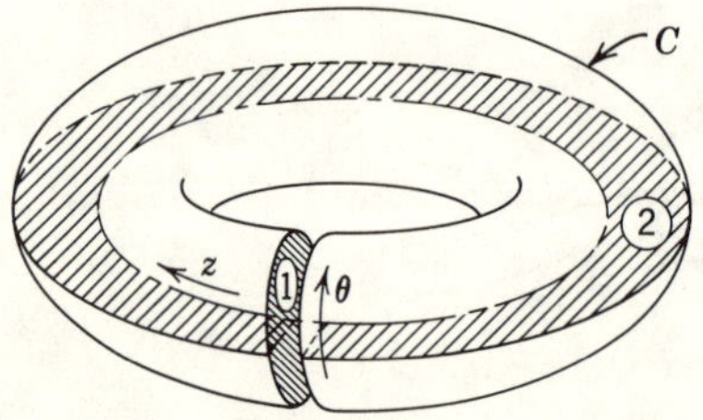

Fig. 16.2. Magnetic surface C with two topologically different cuts made in the interior region.

The magnetic transform preserves areas and the total magnetic flux enclosed within any closed curve. This latter fact is of assistance in formulating a topological description of the magnetic structure.[2] For example, in Fig. 16.2, two orthogonal cuts can be made as shown. For a given magnetic surface C, the total flux passing through each of the two surfaces does not depend on where the cuts are made. That is to say, the fluxes passing through all cross sections topologically similar to cut 1 are the same, and the fluxes through all cross sections topologically similar to cut 2 (linking the magnetic axis and the periphery) are the same. Therefore either of the two fluxes becomes an appropriate independent variable with which to describe the configuration. In a simple degenerate torus, the total magnetic flux passing through any cut 2 is zero.

It is useful to introduce the rotational angle ι. It is the angle between point P_1 and P_2 measured from the magnetic axis O in Fig. 16.1.

There are a number of ways in which a rotational transform can be produced. The earliest consisted of twisting half of a complete torus through almost 180°, as in Fig. 16.3. That a rotational transform results can be seen from this figure. The segment KL of the stellarator lies in one plane, and the segment MN lies in a second plane, parallel to the first. These segments are joined together by the end-cap segments LM and NK, neither of which lies in either plane. The magnetic axis is denoted by the center line. Let us follow a line of induction through one traverse of the stellarator. In Fig. 16.3b it passes inward through the cross section K at point 1. In going along the segment KL the line crosses cross section L at point 2, there being no rotation of the line up to this point. In the

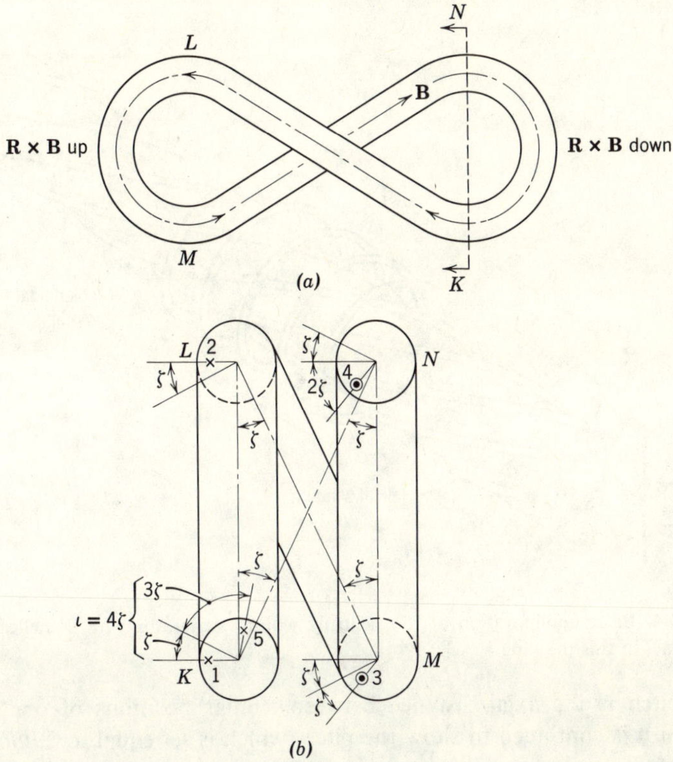

Fig. 16.3. Figure-eight stellarator. (*a*) Plan view; (*b*) cross-sectional view (enlarged) at *KN*.

segment *LM* the line rotates around the magnetic axis by the angle 2ζ, as may be seen from the figure, and comes out through cross section *M* at point 3. In going from *M* to *N* there is no rotation, the line crossing cross section *N* at the point 4. Finally in going from *N* to *K* there is an additional rotation through the angle 2ζ, bringing the line back to the original cross section *K*. The line passes inward through the point 5. The total rotation of the line is thus

$$\iota = 4\zeta. \tag{16.1}$$

A second way to produce the rotational transform is to add helical cusps to a simple torus with solenoidal windings. It will be shown in Sec. 16.2 that this scheme has confining properties superior to that of a figure-eight. Figure 12.8 shows the genesis of the configuration, and Fig. 16.4 shows it wound helically down a cylinder. In general, there are $2l$ conductors so arranged; both figures show six wires, so $l = 3$.

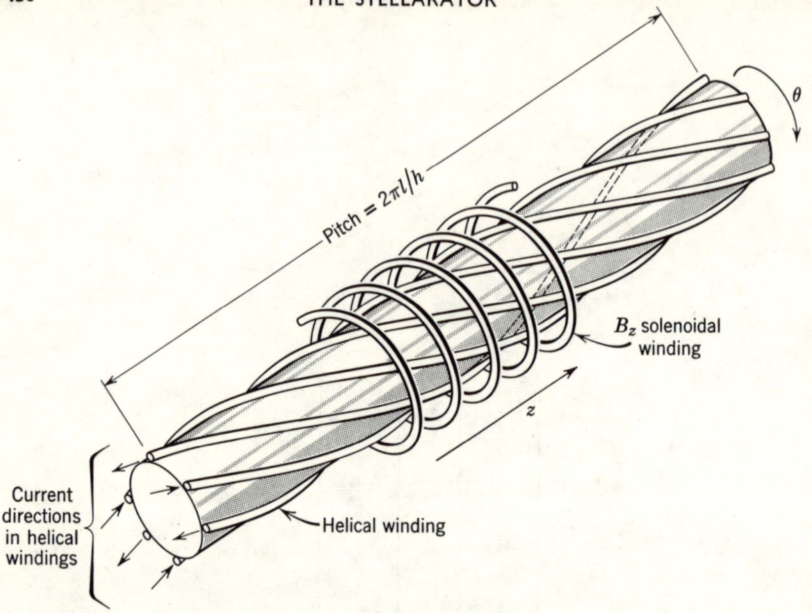

Fig. 16.4. Basic configuration of the helically wound stellarator with $2l$ helical conductors. In this drawing $l = 3$.

The pitch is the axial distance between similar positions of one wire. One turn is continued to show the pitch, which is set equal to $2\pi l/h$, thus defining h.

Certain properties of the helical induction can be found by inspection. It must repeat in the axial distance $2\pi/h$, the distance between conductors carrying current in a particular direction. Likewise, the pattern at a given z repeats in azimuth every $2\pi/l$ radians, since there are $2l$ evenly spaced conductors around the circumference. The induction must then be a cyclic function of $(l\theta - hz)$. Most generally, the induction must be expandable in the orthogonal Fourier series

$$\left.\begin{aligned} B_r &= \sum_g f_r(g, r) \exp ig(l\theta - hz), \\ B_\theta &= \sum_g f_\theta(g, r) \exp ig(l\theta - hz), \\ B_z &= \sum_g f_z(g, r) \exp ig(l\theta - hz), \end{aligned}\right\} \tag{16.2}$$

where g is any integer. The conditions $\nabla \cdot \mathbf{B} = 0$, $\nabla \times \mathbf{B} = 0$, and the boundary conditions determine the functions $f(g, r)$. On the axis, $B_r = B_\theta = 0$ by symmetry. If the pitch is large ($hr \ll 1$), it can readily be shown that the induction from the helical windings is

$$\left.\begin{aligned} B_r &= \sum_g A_g r^{gl-1} \sin g(l\theta - hz), \\ B_\theta &= \sum_g A_g r^{gl-1} \cos g(l\theta - hz) \end{aligned}\right\} \tag{16.3}$$

near the axis, and that B_z is of smaller magnitude. Of these harmonics, the fundamental $g = 1$ dominates; thus

$$\left.\begin{aligned} B_r &= A r^{l-1} \sin (l\theta - hz), \\ B_\theta &= A r^{l-1} \cos (l\theta - hz) \end{aligned}\right\} \tag{16.4}$$

near the axis, where A is a constant characterizing the magnitude of the induction. Problems 2, 3, and 4 illustrate the calculation and show that the functions f are in actuality modified Bessel functions.

Detailed calculation of the induction has been carried out by Johnson, Oberman, Kulsrud, and Frieman.[3] They use a slightly different technique than the one we have outlined here. They helically perturb the boundary of a perfectly conducting cylinder and solve for the perturbation induction inside to second order. They then calculate the rotational transform ι; we shall not follow this complicated calculation but present a simple discussion outlined by Spitzer.[1]

Let us now add the solenoidal winding, which produces a constant induction B_{z0}, choose any line of induction, and follow its r and θ coordinates as z increases. The coordinates are related by

$$\frac{dr}{B_r} = \frac{r\,d\theta}{B_\theta} = \frac{dz}{B_z}. \tag{16.5}$$

If we apply Eq. 16.5 to Eqs. 16.4, we find that in first order each line of induction is twisted into a small spiral, as shown in Fig. 16.5. The rotational transform ι appears in second order for $l \geqslant 2$ as follows. Because r^{l-1} increases with r in Eqs. 16.4, the magnitude of B_θ is larger on the outside of the spiral (where $B_\theta > 0$) than on the inside (where $B_\theta < 0$). As a result, the positive values of $d\theta/dz$ calculated for insertion

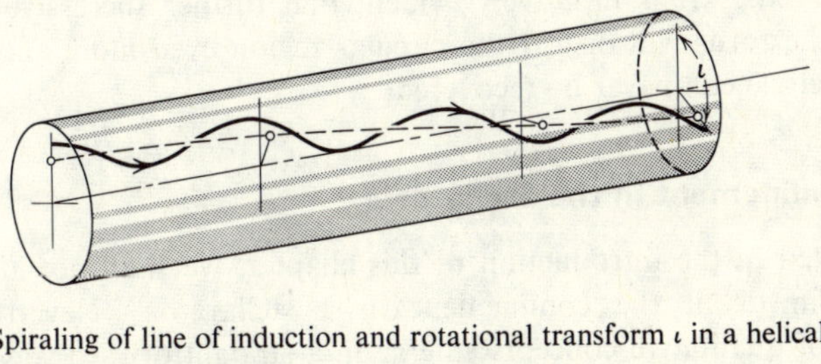

Fig. 16.5. Spiraling of line of induction and rotational transform ι in a helically wound stellarator.

into Eq. 16.5 more than offset the negative ones; therefore the line twists about the magnetic axis. The transform angle ι_h for the angular transformation in the distance $2\pi/h$ (twice the axial wire spacing) is given by Johnson and co-workers[3] as

$$\iota_h = (\pi A^2 r^{2l-4}/h^2 B_{z0}^2)\,[2(l-1) + h^2r^2 + \mathcal{O}(h^4r^4)]. \tag{16.6}$$

The term h^2r^2 is small and may be ignored except for $l = 1$. Note that if $l = 1$ or 2, ι_h is independent of r in this approximation.

The most interesting and useful cases of Eq. 16.6 are those for which $l \geqslant 3$; that is to say, there are six or more helical conductors. In those cases, ι_h increases with r: the magnetic structure has radial shear in the sense described in Sec. 14.5 and shown in Fig. 14.11. Figure 16.6 attempts to show the effect for $l = 3$. The z-axis has been highly compressed, and the incidental spiraling illustrated in Fig. 16.5 has been

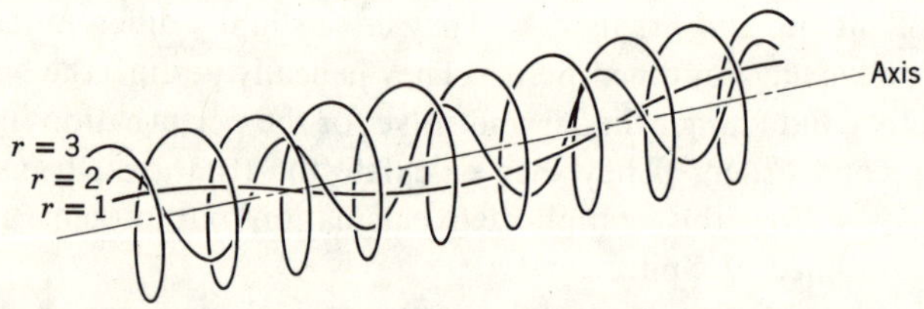

Fig. 16.6. Sheared fields in a stellarator for $l = 3$. The z-axis has been highly compressed, and the shear is much less in practice.

suppressed for clarity. Each line continued around the stellarator axis indefinitely generates a magnetic surface, as described earlier. The induction in each nested surface is topologically different from that in its neighbor. Each line at different radius, followed once around the torus, makes a different (usually nonintegral) number of revolutions around the magnetic axis.

Figure 16.7 shows the effect of a rotational transform on a small circular region for $l = 3$. The figure might represent the effect of following lines of induction once around the torus in the direction of the magnetic axis. It is intuitively evident that further successive passages around the torus will distort the circular region even more and that the original circle can never be recovered.

16.2. Confinement in the Stellarator

As stated in the introduction to this chapter, we shall not attempt to analyze in detail the confinement in a stellarator. Nevertheless, a number of qualitative concepts emerge in a straightforward way.

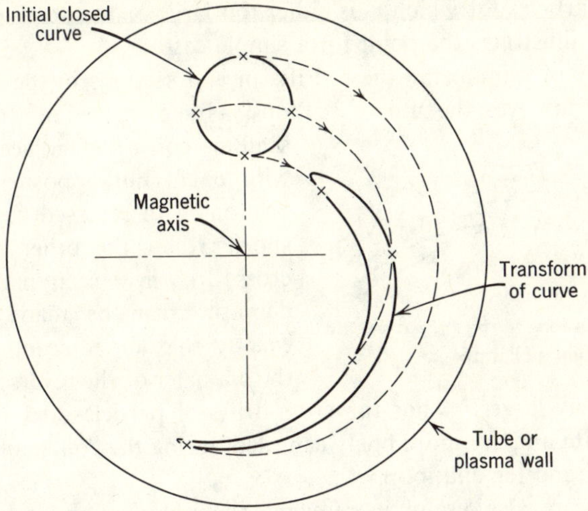

Fig. 16.7. Example of a rotational transform proportional to r^2 (that is, $l = 3$) applied to a small circular region.

To a good approximation the guiding centers of isolated single charges will follow the lines of induction around a stellarator. Thus the lines of Fig. 16.1 could represent equally well (in this approximation) the trajectories of particles in the system. We are then able to define a particle transform analogous to the magnetic transform, to discover that it is also measure preserving, and so on. If we lay the stellarator flat on a table so that a top and bottom are defined, we see that lines of induction travel from the top to the bottom, and vice versa, as they are followed around the device. Similarly, charged particles move from top to bottom and back again as they follow the lines. The electric field arising from the $\mathbf{B} \times \nabla B^2$ drift, and charge separation is canceled by motion of charges along lines of induction. Herein lies one virtue of the configuration.

The figure-eight and the helically wound stellarators have distinctly different drift-cancelation properties, when the two systems are examined in detail. In Fig. 16.3*a*, a charged particle that drifts up in the left loop will drift down in the right loop. A drift cancelation is effected over one complete transit of the particle through the tube. There is a difficulty with this cancelation: each end loop is locally a simple torus, and if the drift is too severe, a particle can be lost before it reaches the opposite loop. The drifts depend both on the ratio $v_{\|}/v_{\mathrm{T}}$ and the magnitudes themselves (see Sec. 10.2). We conclude that a sort of loss cone in velocity space exists in the figure-eight stellarator, analogous to that in a mirror. There are important differences, however. In the stellarator, this loss cone is not so simply defined; depending on the speed of the charges, particles for which $v_{\|}$ is either too large or too small can be lost. Problem 1 illustrates the point for a simple case.

One way of minimizing these drifts in end sections in the figure-eight stellarator involves the use of scallops, as shown in Fig. 16.8. These scallops consist of adjacent sections with equal but opposite curvature. Sections with reversed curvature are shorter than the other ones. The cross drift in a section of reversed curvature can be made to cancel exactly that in adjacent sections if the diameter of the reverse sections is appropriately larger. Thus the cross drifts of particles are divided into small quanta and are more finely canceled during the course of a charged particle around the end loop.

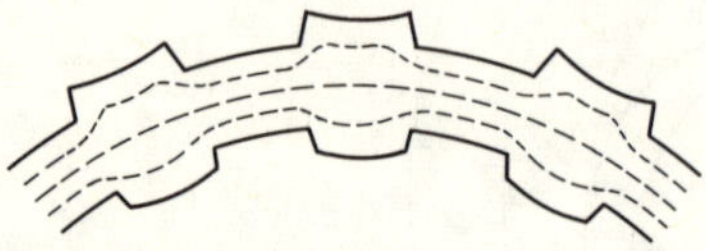

Fig. 16.8. Scallops in the curved section of a figure-eight stellarator.

Scallops are also useful in reducing the polarization of the plasma, which results from the drift due to ∇B^2 in the end sections. The concomitant current between end sections produces an induction across the tube in the straight sections. This additional induction distorts the original configuration of **B** lines, which were designed to confine the plasma.

While single-particle drifts have been suppressed in the figure-eight stellarator, we recognize that the local stability has not been improved. On the other hand, the helically wound configuration can add positive stability, beside canceling drifts continuously all around the tube. This stability question, as applied to the stellarator, has been intensively studied.[2-5] It is these theoretical studies that we do not attempt to develop here. It is evident from Fig. 12.8 that a cusp system of the sort incorporated into the helically wound stellarator will provide stability, because the induction in the cusp has its radius of curvature outside the plasma. Second, for the case $l \geqslant 3$ (six or more helical conductors), the induction has radial shear, as described in the previous section. We recognize from the discussion of Sec. 14.5 that this shear contributes to the stability. In general, the most advantageous configuration is $l = 3$. The reason for this fact is that the helical windings produce an induction proportional to r^{l-1}. We wish to create an induction that penetrates as close as possible to the axis; on the other hand, we also require that the induction have shear. Therefore, since the $l = 1$ and $l = 2$ modes have no shear in first order (see Eq. 16.6), the $l = 3$ mode is the best.

The theoretical analyses state that if a simple hydromagnetic model of the plasma is valid, the stellarator should be stable for β less than about 0.1 or 0.2, depending upon the strength of the helical induction. In

Sec. 16.5 we shall note experimental results that indicate an instability exists if the electrons are highly non-Maxwellian.

One important criterion for stability can be derived intuitively as follows. If there is an axial current around the stellarator, it produces an induction B_θ. This induction coupled with the main solenoidal induction B_z produces a rotational transform of its own. Intuition suggests that if this rotation unwinds the transform impressed externally, the stabilizing effect will be lost. In addition, at any given cross section of the torus, multiples of 2π in the rotational transform are indistinguishable from zero. Thus we expect that, if plasma currents increase or decrease the impressed rotational transform ι taken once around the torus to any multiple of 2π, stability will be lost. In order to calculate the critical axial current I_z, consider it to be concentrated entirely in a thin shell at the plasma radius r_p. We then find the azimuthal induction $\hat{B}_\theta$ just outside the plasma to be

$$\hat{B}_\theta = \mu_0 I_z / 2\pi r_p. \tag{16.7}$$

This induction, coupled with the solenoidal induction B_z just outside the plasma, will produce an angular transform

$$\hat{B}_\theta L / \hat{B}_z r_p = \mu_0 I_z L / 2\pi r_p{}^2 \hat{B}_z \tag{16.8}$$

in an axial distance L. Now let the distance L represent the distance around the torus in the direction of the magnetic axis; also let ι be the rotational transform once around the torus, applied from the external windings or configuration. The criterion for stability then becomes simply

$$\iota + 2\pi K \geqslant \mu_0 I_z L / 2\pi r_p{}^2 \hat{B}_z, \tag{16.9}$$

where K is that integer which gives the left-hand side of Relation 16.9 its smallest value.

Relation 16.9 is called the Kruskal limit, and has been derived much more rigorously by Kruskal.[5] He shows that if the limiting current is exceeded the plasma column should exhibit a long wavelength kink ($m = 1$) instability. Kruskal also shows that the presence of conducting tube walls should not suppress this instability.

We can generalize this notion of unwinding the magnetic transform to consider the likely currents for growth of higher-mode instabilities. For example, the $m = 2$ instability corresponds to the originally circular plasma cross section becoming oval-shaped. Evidently, if the total rotational transform around the system becomes any integral multiple of π, the ovals match up. Similarly, the $m = 3$ instability corresponds to a trefoil cross section; a total rotational transform of any multiple of $2\pi/3$ allows the trefoil to match around the axis. Problem 5 is a simple application of these concepts to a figure-eight stellarator and shows the currents near which one might expect instabilities of a given type to appear.[6]

For a helically wound stellarator, the transform angle ι is a function of radius, so the axial current can effectively unwind the rotational transform only at discrete radii. Whether instabilities grow in the vicinity of such radii is a more complicated matter. An analysis conceptually similar to that of Secs. 14.3 to 14.6 for the pinch could be applied to investigate the stabilizing effect of this magnetic shear. We shall not pursue this analysis further.

16.3. The Divertor and Other Constructional Features of Stellarators

The influx of cold gas impurities from the stellarator walls could severely limit heating of the plasma and even result in a plasma consisting principally of the impurities. We have dwelt on the problem in many previous sections. Ultra-high vacuum techniques are used to minimize this problem. Stainless steel tube sections are built of solid stock or by heliarc welding. This technique is also used for construction of many other devices described in earlier chapters. Demountable tubes are assembled with gold O-ring seals. The tube is outgassed at 450° C by heating it with alternating current and is pumped down to about 10^{-10} mm Hg. A serious impurity influx persists despite these precautions.

A device called a divertor has been designed to alleviate this problem.[7] It is shown in Fig. 16.9. Since the impurities come from the wall, the plasma near the wall is the most contaminated. The divertor is designed to skim off this plasma. A tungsten aperture limiter could skim the surface of the plasma, but then removal of the neutralized gas would be difficult. In the divertor the outer field lines are led off so that the peripheral plasma is guided into a large space which can be pumped easily. The structure is rotationally symmetric about the cross-sectional plane. The divertor has several other advantages beside the one of removing impurity ions. It prevents hot ions that diffuse to the neighborhood of a wall from reaching it and knocking out ions or neutrals from it; that is, the divertor reduces sputtering. Further, in reducing the sputtering it reduces the heat flux to the walls, thus making the wall problem easier. The energy of the particles of the plasma can be dissipated against a wall at the plane of symmetry inside the divertor. This surface can be cooled easily.

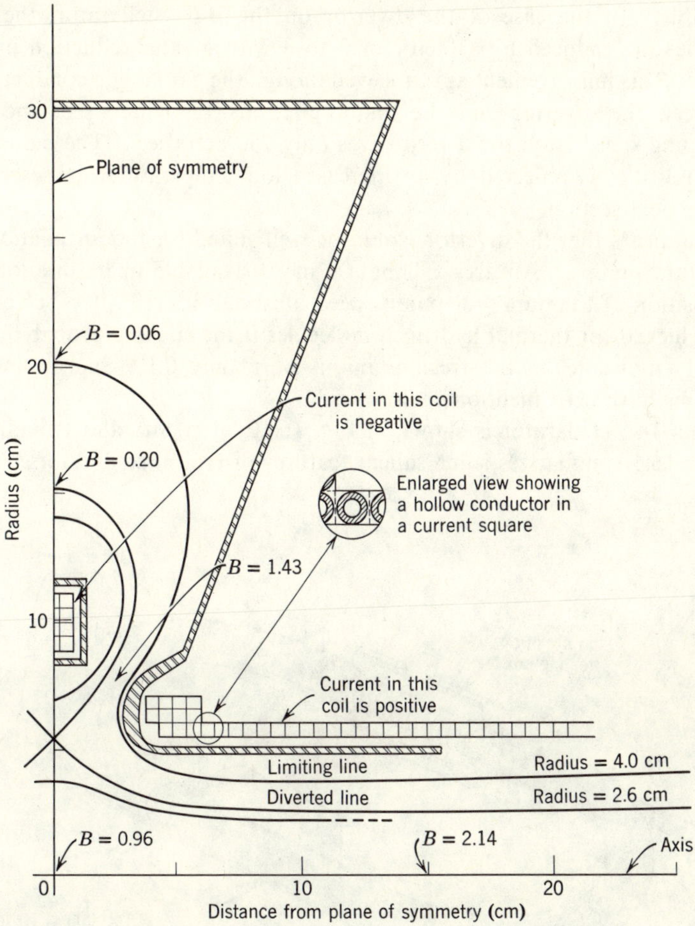

Fig. 16.9. Quarter-section view of the divertor used in the B-65 stellarator. The relative induction is shown at several positions (in units of 10^{-4} weber/m^2-amp). The plasma will distort the **B** lines and the associated isobaric surfaces. After Burnett and co-workers.[7]

The diversion is caused by a reverse solenoidal current. The outer lines of induction in the stellarator tube make a loop into the divertor and around the reversed current. If a disk is put inside the divertor at the plane of symmetry, for example, any ions traveling along the diverted lines of induction will strike this disk and be neutralized. They cannot readily get back into the main tube because of the rather narrow opening between the coils carrying the reversed current and the main tube wall. The neutral impurity atoms are pumped out of the divertor as rapidly as possible. In the case of the divertor on the B-65 stellarator, the impurities are reduced by a factor of 2 to 3 with greater reduction in the core. This improvement appears even though the gas kinetic conductance between the divertor and the main tube was 1700 liters/sec, and the pumping speed from the divertor was only 750 liters/sec. The reduction in impurities is reflected by an increased ion temperature, as described in the next section.

It appears that the divertor would be well suited for the installation of titanium pumps. An area of about 1 m^2 is available in its interior for deposition of titanium. A pump speed of about 4×10^4 liters/sec could be achieved for thermal hydrogen molecules if the sticking probability of an H_2 molecule on the fresh titanium were only 0.1. Such pumping systems have been mentioned elsewhere.

The B-65 stellarator is shown in Fig. 16.10, after Stix and Palladino.[8] Table 16.1 summarizes some salient features of typical stellarators. The

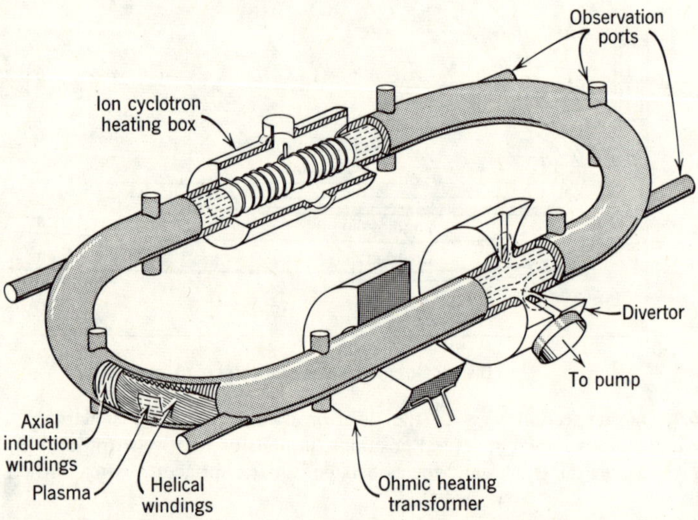

Fig. 16.10. The B-65 stellarator. After Stix and Palladino.[8]

smallest are type A, and intermediate ones are type B. The model C stellarator, in which it is hoped to produce a plasma of real thermonuclear interest, is scheduled to go into operation soon. We shall discuss the plasma heating schemes and associated experiments in the following sections. Some speculation has taken place on the design of a model D stellarator for full-scale power production.

Table 16.1. Features of typical stellarators†

Stellarator Model	Length of Magnetic Axis (m)	Tube Diameter (cm)	Configuration	Divertor	Baking	Induction (webers/m²)	Duration of Induction	Heating Method	T_e (ev)	T_i (ev)	Plasma Density (m^{-3})
A-2 (Etude)	2.6	5.1	Racetrack	No	No	0.8	Steady state	Ohmic	25	?	10^{19}
B-1	4.5	5.1	Figure-eight	No	Yes	3.0	12 msec	Rf, ohmic	< 80	< 5	$\approx 10^{20}$
B-3	6.5	5.1	Figure-eight	Under test	Yes	5.0	12 msec	Rf, ohmic	20	3–20	$\approx 10^{20}$
B-65	5.2	10.2	Racetrack	Yes	No	2.0	6 msec	Ion cyclotron resonance	< 50	120 ?	10^{18}–10^{19}
								Ohmic	> 80	80	
B-66	7.4	12	Racetrack	No	Yes	3.5	12 msec	Ion cyclotron resonance	50	100–400 ?	10^{18}–10^{19}
								Ohmic	?	?	
C	12.2	20.2	Racetrack	Scheduled for later	Yes	5.5	1 sec	Rf, ohmic, magnetic pumping	—	—	—

† Data kindly supplied by M. B. Gottlieb, Project Matterhorn, Princeton, N.J.

16.4. DC Pulse Experiments

Most of the stellarator experiments reported to date have involved the use of a unidirectional electric field pulse induced around the tube. This electric field generates a circulating current around the loop that heats the resistive plasma. The process is called ohmic heating. In this section and the next, we shall discuss a number of these ohmic heating experiments and their interpretation.

An ohmic heating transformer is shown on the B-65 stellarator in Fig. 16.10. Two transformer cores are used[9] on the B-1 stellarator, Fig. 16.11. The arrangement is virtually identical with that used on any

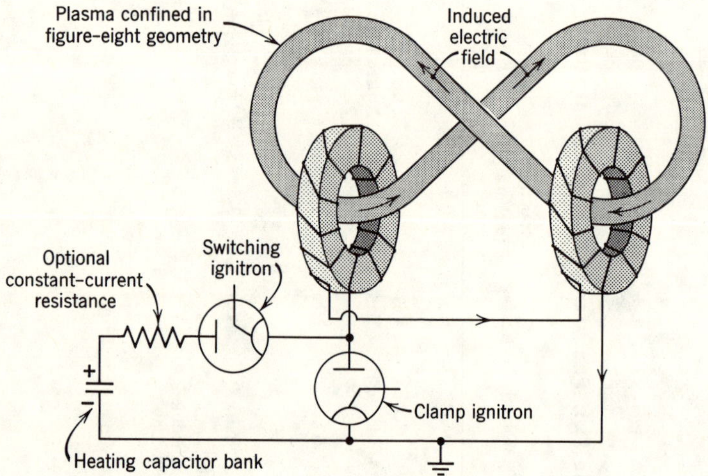

Fig. 16.11. Ohmic heating apparatus and simplified circuitry used on the B-1 stellarator. After Coor and co-workers.[9]

toroidal pinch, Fig. 14.2 and Sec. 14.8. There is an important difference, however. In the pinch, the induced electric field and resulting current are sufficiently high to pinch the plasma and heat it by radial compression. In the stellarator, the field and current are kept much lower, so that none of these effects takes place. Plasma heating arises strictly from the resistive nature of the medium.

A clamp ignitron can be used to terminate the induced electric field at any given time. Insertion of resistance in the primary circuit leads to an approximately constant primary current after the initial transient rise. The plasma forms the secondary; if it is resistive, a constant primary current will induce a constant secondary current. An insulating gap (not shown) in the tube wall is required to prevent short-circuiting the induced loop voltage.

Figure 16.12 shows schematically the instrumentation of the B-1 stellarator. We have described a number of the diagnostic techniques shown in the figure in previous sections and shall describe the others in subsequent paragraphs. Not shown in either figure are the main solenoidal windings that produce the axial induction B_z. The B-1 stellarator has no helical stabilizing windings; the B-3 stellarator, a similar device, has a set of $l = 3$ windings, in addition to the figure-eight twist.

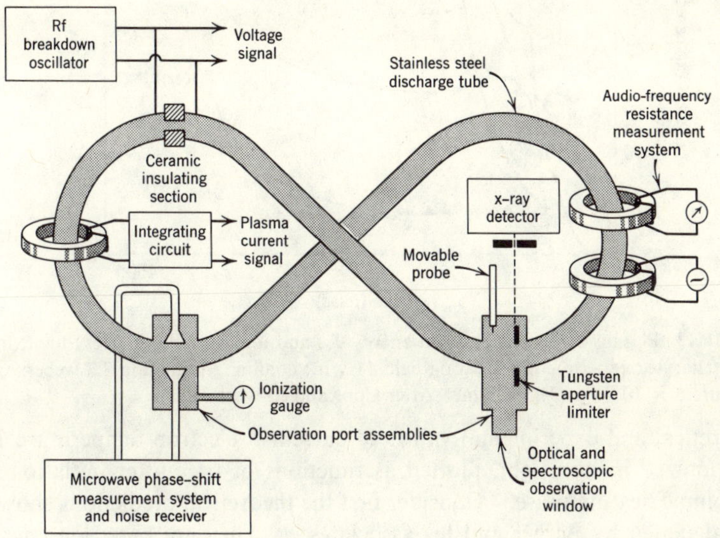

Fig. 16.12. Instrumentation of B-1 stellarator. After Coor and co-workers.[9]

Typical pulsed operation proceeds as follows. The main B_z induction ($\leqslant 3$ webers/m^2 in the B-1 stellarator) is applied from a 10^6-joule capacitor storage bank; the current in the windings rises to its maximum value in about 10^{-2} sec and is clamped there. The induction is then essentially constant over the subsequent experimental time of interest. The rf oscillator (Fig. 16.12) is then turned on for a few milliseconds and ionizes about 10% of the gas. Just after the radio frequency is turned off, the main ohmic heating field is applied; it lasts 1 or 2 milliseconds. Observations of interest are made during and after this latter period.

We shall now summarize a series of experiments, performed principally with the B-1 stellarator, and associated theoretical calculations.

Figure 16.13, taken from the work of Coor and his co-workers,[9] shows

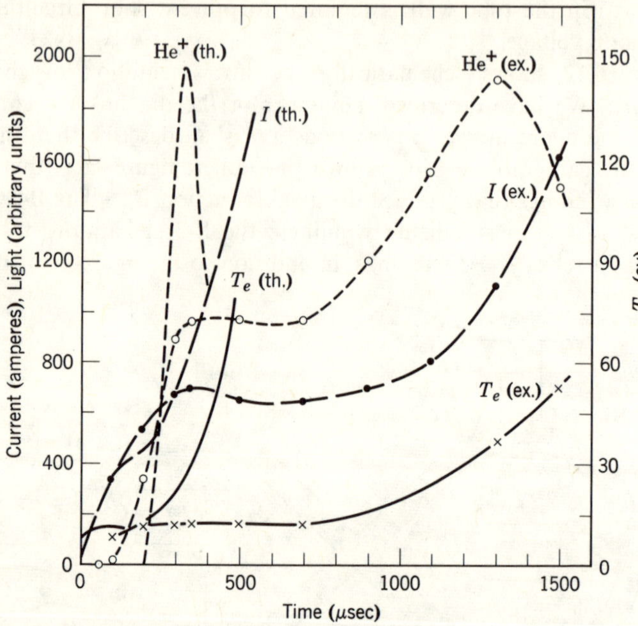

Fig. 16.13. Plasma current I, electron energy T_e, and ionized helium (He^+) light, in a B-1 stellarator experiment. Heating field 11 v/m; confining induction 2.7 webers/m^2; pressure 8×10^{-4} mm Hg helium. After Coor and co-workers.[9]

theoretical and experimental values of current I, electron temperature T_e, and ionized helium light, plotted as functions of time after initiation of the ohmic heating pulse. Consider first the theoretical predictions shown, as calculated by Berger and his associates.[10] They are based on energy and number balance for all types of particles present (electrons, He^0, He^+, and He^{++}). For example, the fast electrons take most of the input power from the heating field. This power goes into electron heating, excitation and ionization of He^0 and He^+, heating of the ions, and bremsstrahlung. Appropriate equations are developed for each type of particle, and the resulting coupled set is solved numerically.

According to the theory, the electron temperature T_e initially rises to about 10 ev in a few microseconds, then remains relatively constant while the He^0 is ionized once. Also during this time, the current levels off if the electric field is not too large; the field recorded in this example is moderately large, but a vestigial plateau can be seen in I(th.). After this dwell time, He^+ becomes excited and eventually ionized; the He^+ light rises and falls accordingly. Finally, the current and electron temperature rise monotonically in the fully ionized plasma (after about 300 μsec in Fig. 16.13).

The theory and experiment are seen to be in fair agreement at short times (up to 200 μsec in Fig. 16.13). At later times, all processes appear to be delayed. For example, if the duration of the current plateau is denoted by τ_1, the theory predicts

$$\tau_1 \approx Cp/EI_1, \tag{16.10}$$

where C is a calculated constant. It is observed by Bernstein and Kranz[11] that the current plateau has the dependence of Eq. 16.10, but that the experimentally determined value of C is much too large, typically by a factor of 5. There are several possible explanations for this delay. The principal one concerns the fact that fresh gas enters the plasma column during the pulse. This gas comes both from the volume between the plasma column and the wall, and from the wall itself. This gas also becomes ionized, and its effect has not been included in the theory. Second, it will be pointed out in subsequent paragraphs that plasma is continually lost from the column in spite of the supposed confinement. If this is the case, the interchange loss of hot for cold plasma from the gas influx will delay formation of a fully ionized plasma. Finally, the gas influx consists in large part of impurity atoms; the additional radiation loss from these heavy impurities (oxygen, carbon, and so on) may be large.

The electric field couples principally to the mobile electrons. During the time shown in Fig. 16.13, the ions remain relatively cold (a few ev); they would be heated by Coulomb collisions with the fast electrons. This energy transfer is inefficient, and the ions are expected to be heated slowly. Ion temperatures are determined experimentally, for example, by time-resolved measurements of the Doppler broadening of spectral lines. Burnett and co-workers[7] report the effectiveness of the divertor on ion temperatures in the B-65 stellarator. They find that He^+ temperatures are raised from 40 without the divertor to 60 ev with it, and O^{4+} temperatures of 130 ev have been observed. It is expected that when the impurity influx is reduced and the divertor pumping is increased, these temperatures will be further improved.

We now turn to experimental observations at larger times than those shown in Fig. 16.13. Figure 16.14 shows the plasma current in the B-1 stellarator as a function of time. At low heating fields, a small stable plasma current flows. At intermediate fields, the current plateau described in the previous paragraphs is seen (500 amp on curve C). Thereafter, the current increases rapidly to about 2000 amp, whereupon the discharge becomes unstable, as may be seen from the oscillations in the figure. Both the plasma current and loop voltage around the stellarator then show violent fluctuations. This current is the Kruskal limit, described in Sec. 16.2; it corresponds to unwinding of the rotational transform by the plasma current. Experiments to detect the instability

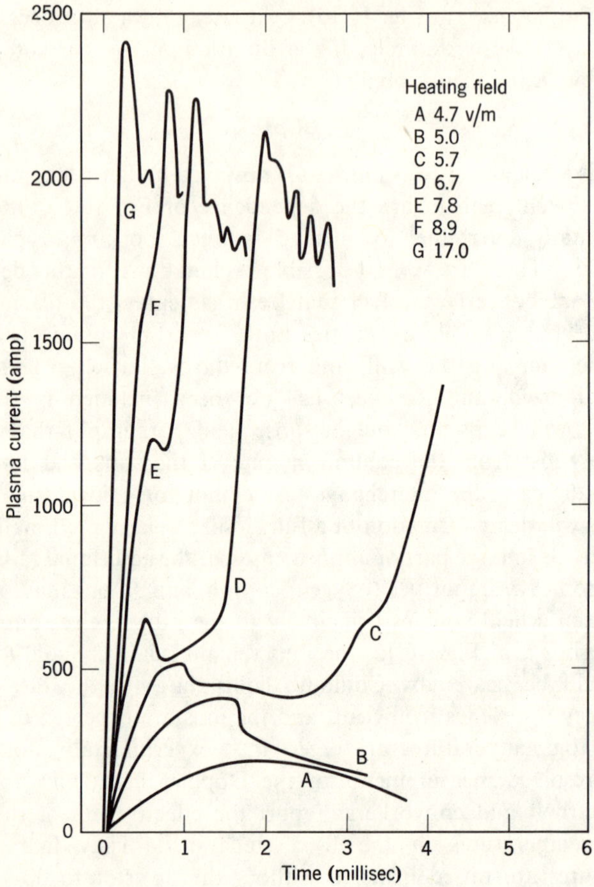

Fig. 16.14. Current in helium discharge for various heating fields. $B_z = 2.7$ webers/m². After Coor and co-workers.[9]

optically are reported by Bernstein, Kranz, and Tenney.[12] By placing narrow slits in the observation ports, they find that at the Kruskal limit the discharge column moves laterally as a whole. They also find some evidence of higher mode instabilities, as described at the end of Sec. 16.2 and in Problem 5, in terms of fast fluctuations in emitted He^0 light as various critical currents are reached. For constant-current operation (see Fig. 16.11), the plasma current can be adjusted to remain safely below the critical $m = 1$ limit during the entire pulse.

16.5. Electron Density, Loss, and Runaway

The dc pulse experiments of the previous section exhibit several additional features of great importance, namely, a mysterious loss of the plasma and the appearance of runaway electrons. The phenomena occur at the later times shown in Fig. 16.14. We shall now discuss these aspects of the experiments and some possible interpretations.

The electron density is measured principally by microwave techniques. A pair of horns is shown in Fig. 16.12; the electric field vector of the propagated wave is usually parallel to the induction B_z, although other orientations can be utilized. It was shown in Secs. 9.2 and 9.3 that the phase velocity of such an electromagnetic wave across the plasma is a function of the electron density and exceeds the velocity of light. Therefore a comparison of the phase shift of the signal propagated across the plasma with the phase shift over a standard reference path yields the electron density. If the plasma frequency ω_p exceeds the applied frequency ω, transmission virtually ceases. This circumstance limits the electron density that can be measured, but allows estimates of electron density also to be made by measurement of the frequency at which transmission ceases. Both these methods are widely applied as plasma diagnostic techniques. The highest frequency reasonably available at present for such purposes is about 100,000 Mc/sec (3-mm waves). The corresponding critical density is $10^{20}/\text{m}^3$. Similar techniques at infrared frequencies are presently being developed. In all these experiments, due account must be taken of diffraction around the plasma column, reflection from vacuum windows, and so forth.

The microwave system can also be used to measure energy radiated from the plasma column. If $\omega_p/\omega \gg 1$, the plasma is essentially opaque and radiates as a black body, as shown in Chap. 11. Measurement of the absolute intensity of radiation and correction for reflection at the plasma surface and windows yields a typical temperature corresponding to the plasma surface. If $\omega_p/\omega \ll 1$, the plasma is relatively transparent, and the radiation per unit plasma area in a frequency range $\Delta\omega$ will be roughly proportional to $nkT\,\Delta\omega$ integrated across the depth of the plasma. The cyclotron radiation itself has not yet been exploited very much as a diagnostic technique.

In the stellarator, the runaway electrons are detected as they strike tungsten limiting apertures placed in the tube and produce x rays. Such a limiter is shown in Fig. 16.12.

With these techniques and others previously described, typical data of the sort shown in Fig. 16.15 are obtained.[9,13,14] The curves sketched do not refer to any one particular experiment but are representative of many. The important things to notice are the decreasing electron density, even as the current remains high, and the x-ray emission. This decrease in electron density corresponds to loss of the plasma. The phenomenon, colloquially called pumpout, is discussed by Ellis and co-workers,[14] who

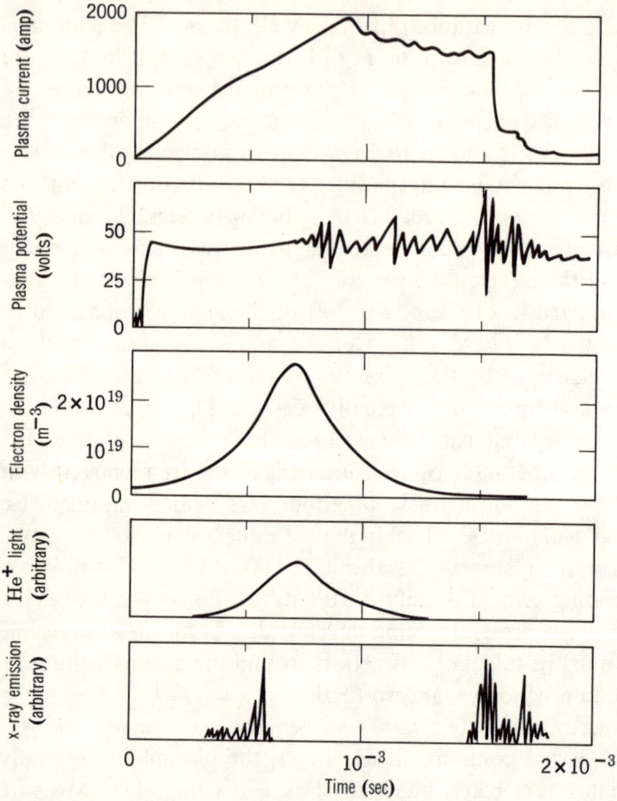

Fig. 16.15. Typical data obtained for operation of the B-1 or B-3 stellarator at high heating field and current exceeding the Kruskal limit.

show that the density decreases exponentially with time. Following their development, we can make the ad hoc assumptions that (1) the rate of electron loss is proportional to the electron density; (2) fresh material is returned from the walls at a rate proportional both to the rate of arrival of electrons there and to the surface density of adsorbed atoms; (3) the surface density of adsorbed atoms is proportional to the pressure. Then we can write

$$\frac{dn_e}{dt} = -C_0 n_e + C_0 C_1 n_e p, \tag{16.11}$$

where the physical process represented by the constant C_0 is left unexplained. If Eq. 16.11 is valid, the electron density should decrease at low gas pressure and remain constant or even increase at high pressure.

This behavior is found experimentally; Fig. 16.16 shows the loss rate as a function of pressure for a hydrogen experiment in the B-3 stellarator.

Extrapolation to zero pressure gives confinement times of the order of 10^{-4} sec in the absence of feedback from the walls. This confinement time is found to vary approximately as $B_z^{1/2}$ and to be somewhat greater for B_z and E_z parallel rather than antiparallel. Very significantly, the confinement time is independent of whether or not the helical stabilizing windings on the B-3 figure-eight stellarator are energized. This loss of plasma therefore does not appear to arise from any known hydromagnetic instability. Recent experiments[15] indicate that this loss also proceeds during the initial ionization phases of the discharge, as well as at the late times so apparent in Fig. 16.15.

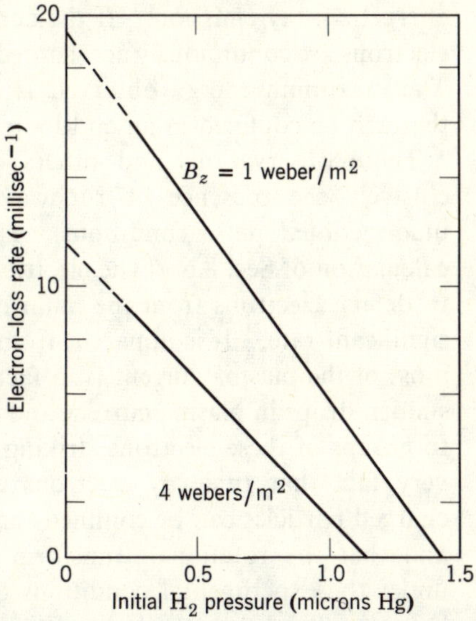

Fig. 16.16. Observed loss rate of electron density during ohmic heating as a function of initial hydrogen pressure. The confining induction and heating electric field are antiparallel. After Ellis and coworkers.[14]

Detailed examination of the energetic x rays sheds light upon some aspects of this mysterious matter and suggests one possible mechanism for generating the instability. The energy spectrum of the x rays has been measured and related to electron energies required to produce them. Observations regarding the maximum electron energy can be crudely summarized by the statement that it equals or exceeds the energy corresponding to unimpeded acceleration over the time during which no instability exists. For example, in the idealized Fig. 16.15, x rays are observed up to a time of 0.6 msec. The maximum x-ray energy measured corresponds to continual electron acceleration from the time at which the heating pulse was applied. The x rays stop at 0.6 msec, corresponding to the onset of the $m = 1$ instability. At that time, this maximum electron energy typically might be about 1.5 Mev. The cessation of x rays during the time the current exceeds the Kruskal limit arises from the fact that electrons are not confined long enough to run away in velocity. As the current at longer times again drops below the Kruskal limit, the energetic x rays reappear. Under some experimental conditions (constant-current operation or abrupt termination of the heating pulse), the plasma current decays in sudden steps, each step being accompanied by copious energetic x-ray emission. If the current is kept below the Kruskal limit, electrons are continuously accelerated during the entire period of operation. The maximum energy observed is 3 Mev, which is the highest energy that can be confined in an end loop of the stellarator.

These observations and others of similar nature demonstrate conclusively the presence of runaway electrons in the stellarator under unidirectional pulse conditions. The conclusion is in accord with the calculation of Sec. 8.6. The electric field in these experiments is sufficient to detach electrons from the main body of the velocity distribution at a significant rate. It is apparent from Fig. 16.15 that at least at late times most of the plasma current is in fact carried by runaway electrons. The sudden drops in plasma current and associated x-ray emission correspond to groups of these electrons striking the tungsten limiting aperture. The very fact that runaway electrons exist establishes the fact that some charged particles can be confined for a relatively long time in a stellarator and that the rotational transform concept has merit. Unfortunately, under the experimental conditions described, runaway electrons may be better confined than the bulk of the plasma.

We may speculate that the two phenomena of electron runaway and plasma loss are related. The runaway electrons should cause an oscillation of the plasma. That some cooperative mechanism exists is shown by the appearance of some electrons with energy in excess of the simple theoretical maximum $eE^2t^2/2m_e$ (ev). Many theoretical analyses have been made of the instabilities that can develop when the electron velocity distribution becomes highly anisotropic. We shall not take up this important subject of instabilities in velocity space. However, we note the fact that a core of electrons near the origin of velocity space and a second isolated group at high velocity appear in a moving coordinate system as two interpenetrating streams. Provided the drift speed exceeds the random speed of each group by an appreciable margin, the velocity distributions can be shown to be unstable. Indeed, this principle is efficiently applied in the design of some electron-tube oscillators. These oscillations will couple to the ions. The calculation of this coupling is difficult and often proceeds by numerical methods. Among the many references on this subject,[16,17] we mention one by Bunemann,[18] which deals directly with this electron-ion coupling at a fairly sophisticated level.

A final understanding of the perplexing and difficult problems of pumpout and runaway in the stellarator is not at hand. Considerably more experimental and theoretical development is required. Operation of the model C stellarator, with its larger size, longer pulse length, and generally improved characteristics, should contribute to this resolution.

We are now in a position to make some general statements about the efficacy of the ohmic heating technique. Its principal advantage is that it is simple and fairly efficient at low plasma energy. Against these factors, we find

1. The electrons are heated selectively, and transfer of electron energy to the ions is inefficient.

2. The Coulomb cross section is proportional to v^{-4}, and the plasma resistivity is proportional to $T^{-3/2}$. From the first of these facts, we discover the likelihood of generating runaway electrons. From the second (related) fact, we discover that large plasma currents are needed for reasonable power input, except at low temperature. For example, if a power input of 10 Mw/m^3 is desired at T = 10 kev, the current is about 10^8 amp/m^2.

3. If the high currents required for reasonable input power are present, the confining effects of the magnetic transform are vitiated.

For these reasons, it appears that ohmic heating may be suitable for achieving temperatures of the order of a hundred electron volts or a few hundred electron volts at most. To achieve higher temperatures and in particular to heat the ions selectively, we must use different techniques. These techniques are the subject of our next section.

16.6. Plasma Heating by Oscillating Electromagnetic Fields

We have just seen that the application of a steady unidirectional electric field to a plasma and its resistive heating is unsuitable for raising it to thermonuclear temperatures. Some of the disadvantages of the schemes outlined in the previous two sections might be overcome by such stratagems as applying a low-frequency (a few tens of kilocycles/sec) axial electric field, so that electrons would not have time enough to run away. However, other difficulties remain, and we shall pass on to other possible schemes employing oscillating electromagnetic fields. Detailed discussion of their merits would involve us in the subject of propagation through anisotropic media. We shall present some basic physical concepts and leave the interested reader to consult the references for the quantitative details. One class of these plasma heating schemes (related to ion cyclotron resonance) appears particularly promising for thermonuclear applications, and is being actively developed for use with the stellarator. It is discussed in more detail in Sec. 16.7.

We can immediately conclude from the work of Chap. 9 that merely aiming rf power of frequency ω at a plasma is an inefficient way to heat it. The rf power does not penetrate the plasma appreciably if $\omega < \omega_p$ in the absence of a magnetic induction. To put the matter another way, the plasma is a highly reactive medium, and the energy absorption is negligible except under special circumstances. For this and other reasons, the problem of heating a hot plasma efficiently is a complicated one.

The general problem of coupling oscillating fields to the plasma has been discussed by Berger and his associates.[19] For any generalized heating process, it is necessary to consider not only the applied angular frequency ω but also the other frequencies (or inversely, the characteristic times) associated with the plasma and its configuration. It is through the processes represented by these frequencies or times that the energy is actually communicated to the plasma. Besides the period $2\pi/\omega = \tau_f$ of the applied field, we can distinguish three other times of possible interest:

1. The period τ_θ for deflection of a charge through a large angle.
2. The transit time τ_{tr} of a charge through the system.
3. The cyclotron period $\tau_b = 2\pi/\omega_{bi}$ of the ions (it is the ions that are to be heated).

In terms of these times, we distinguish a number of possible heating mechanisms characterized as follows:

1. Collisional heating: $\tau_b \ll \tau_\theta \approx \tau_f \ll \tau_{tr}$.
2. Transit-time heating: $\tau_b \ll \tau_{tr} \approx \tau_f \ll \tau_\theta$.
3. Acoustic heating: $\tau_b \ll \tau_\theta \ll \tau_f \approx \tau_{tr}$.
4. Ion cyclotron heating: $\tau_f \approx \tau_b \ll \tau_{tr} \ll \tau_\theta$.

These schemes are distinguished one from the other by the particle densities, geometric size, and applied frequency. We shall discuss the first three methods briefly and the last one in more detail.

A simple example of collisional heating has already been presented in Chap. 7, Problem 13, to which the reader is referred. In that problem, a plasma confined in a solenoid by an induction B_z is to be heated by pulsing the induction between two limiting values B_1 and B_2 and leaving the induction at each value for a time long compared with the collision time τ_θ. For an average energy $\overline{U}_0$ at any given time, the average energy $\overline{U}_0'$ one cycle later is

$$\overline{U}_0' = \overline{U}_0 \left[1 + \frac{2B_2}{9B_1}\left(1 - \frac{B_1}{B_2}\right)^2\right], \tag{16.12}$$

which is the answer to that problem.

This method of heating just described is sometimes called magnetic pumping. While the energy increase per period is a maximum under the conditions described, the energy increase per unit time is maximized if we increase the frequency and do not wait for complete thermalization between half-cycles. An approximate set of differential equations can be written for the effect.

During compression or expansion, U_T/B is constant except for the effect of collisions, since the process is to be carried out adiabatically. This heating increases U_T, and the effect of collisions is to transfer energy to the parallel component of the motion at a rate proportional to the collision frequency and the difference in energy between the equilibrium state and the actual state. Since in equilibrium $U_\parallel = U_T/2$, we find easily

$$\frac{dU_T}{dt} = \frac{U_T}{B}\frac{dB}{dt} + \nu_c\left(U_\parallel - \frac{U_T}{2}\right), \tag{16.13}$$

where $\nu_c = 1/\tau_\theta$ is the frequency for large-angle scattering. Similarly

$$\frac{dU_\parallel}{dt} = -\nu_c\left(U_\parallel - \frac{U_T}{2}\right). \tag{16.14}$$

It has been assumed that $\tau_{tr} \gg \tau_\theta$ so that changes in ν_c or B resulting from axial motion may be neglected.

It is instructive to consider the case where the magnetic field is sinusoidally modulated:

$$B = B_0(1 + \delta \cos \omega t), \tag{16.15}$$

where δ is a small quantity. The term $U_\parallel$ in Eq. 16.13 may be eliminated by using Eq. 16.14. The resulting differential equation has periodically varying coefficients. Berger and co-workers[19] show that the average rate of increase of energy under these circumstances is

$$\frac{dU}{dt} = \frac{\delta^2}{6}\,\omega^2\nu_c\left(\frac{9\nu_c^2}{4} + \omega^2\right)^{-1} U. \tag{16.16}$$

Since $\nu_c \propto U^{-3/2}$ for Coulomb collisions, the rate of energy gain is proportional to $U^{-1/2}$ if $\omega \gg \nu_c$ and is proportional to $U^{5/2}\omega^2$ if $\omega \ll \nu_c$. If losses in tube walls, coils, and the like, are negligible, which they are not, then the rate of gain of energy is higher at higher frequencies. However, if the frequency is too high, only the surface layers of the plasma are compressed and expanded, and only the plasma surface is heated. In other words, high-frequency fields, arbitrarily applied, will not penetrate the plasma.

Transit-time heating, Case 2 above, can be visualized with the aid of Fig. 16.17. A uniform induction B_0 in the z-direction is modulated at frequency ω over a distance of the order a; we can characterize this induction by

$$B = B_0[1 + \delta f(z) \cos(\omega t - \psi)], \tag{16.17}$$

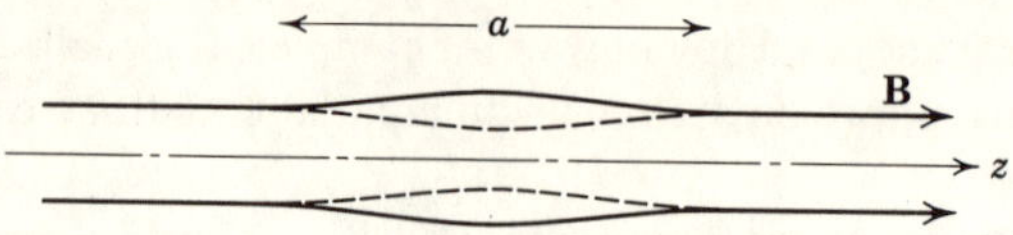

Fig. 16.17. Uniform induction with alternating induction superimposed over a heating region of length a.

where δ represents the magnitude of the perturbation, $f(z)$ is unity at the midplane and drops to zero at the distance a, and ψ is the phase angle specifying the time of entry of an ion into the perturbed region. We assume that an ion passes through the heating region sufficiently fast that collisions during the transit time are negligible, but that the transit is slow enough for the adiabatic approximation to hold. Under these circumstances, the energy U_T of the ion is unaltered by the transit. For the parallel motion, we have

$$m_i \ddot{z} = -M \frac{\partial B}{\partial z}, \qquad (16.18)$$

where M is the (constant) magnetic moment of the particle. Equations 16.17 and 16.18 yield immediately

$$\ddot{z} = -\frac{MB_0\delta}{m_i} \frac{df(z)}{dz} \cos(\omega t - \psi). \qquad (16.19)$$

The z-motion is that of a particle in a time-varying potential well. The heating of selected particles can be seen by considering a particle that enters the heating region in Fig. 16.17 when the induction is shown by the solid lines and leaves at a later time when the induction is shown dotted. Such a particle is accelerated axially both as it enters and as it leaves. Ions with unfavorable phase will be decelerated. Berger and co-workers[19] average over initial phases, assume $f(z) = \exp(-z^2/a^2)$, and find the average increase in energy per transit is

$$\overline{\Delta U} = (\pi/m)(MB_0\delta a\omega/2v_{\|}^2)[(\omega^2a^2/v_{\|}^2) - 2] \exp(-a^2\omega^2/2v_{\|}^2). \qquad (16.20)$$

Equation 16.20 shows that particles whose transit time is somewhat greater than a quarter of the induction period are heated, while those for which the converse is true are cooled. By suitably choosing the frequency, energy can be transferred to the plasma; the velocity distribution will not be preserved, and a Maxwellian distribution will be split into separate fast and slow components. Conversion of this energy parallel to the induction into random energy must take place slowly via collisions.

The heating scheme just outlined is also known as parametric energy conversion; it has also been proposed for use in hydromagnetic energy conversion. In that application, a plasma jet flows axially down the system. Depending upon the frequency ω and half wavelength a of the system, the plasma jet either absorbs energy and is accelerated (for propulsion) or gives up energy to the magnetic pump and is decelerated (for electric power generation). Alternately, a traveling wave of induction can be set up in the z-direction. If the wave speed exceeds the plasma speed, the plasma is accelerated; if the converse is true, the plasma is decelerated and delivers energy to the waves.

We shall not dwell upon the topic of acoustic heating, Case 3 above, because the inequalities cannot reasonably be satisfied in any plasma of controlled thermonuclear interest. The same statement can be made less forcefully both for collisional and ohmic heating.

The last scheme, employing ion cyclotron resonance, appears to be the most promising, and has been extensively studied both theoretically and experimentally. This subject alone is discussed qualitatively in the next section.

16.7. Ion Cyclotron Waves and Heating

The possibility of heating a plasma by the use of frequencies at or near ion cyclotron resonance has been intensively analyzed.[8,19,20–25] These analyses generally extend the earlier work of Alfvén[26] and Åström[27] on hydromagnetic wave propagation and parallel developments pertaining to the ionosphere.

The simplest example of ion cyclotron heating has been discussed in Sec. 9.3. At ion cyclotron resonance, an ion gyrating in phase with a circularly polarized electric field gains energy indefinitely. The configuration is shown again in Fig. 16.18 for such an ion, where the steady induction B_z is perturbed by a small oscillation at the ion cyclotron frequency. Ions gyrating out of phase are decelerated; however, they can lose only whatever energy they have, and then they too start in phase. The heating process is very rapid, so that an ion can be heated to high energy in a short time.

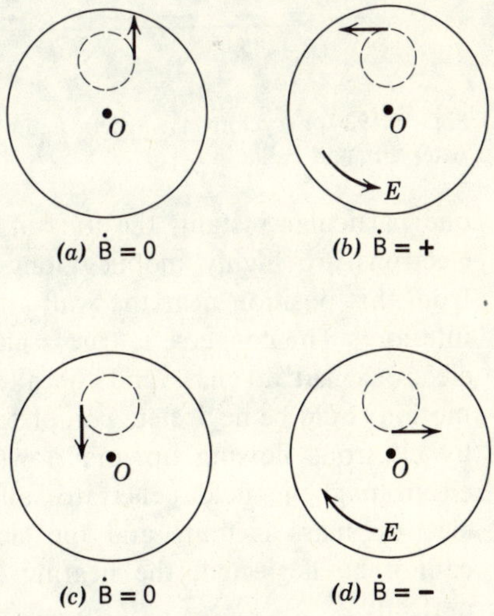

(a) $\dot{B} = 0$ (b) $\dot{B} = +$ (c) $\dot{B} = 0$ (d) $\dot{B} = -$

Fig. 16.18. Particle orbit around its guiding center at phase intervals of $\pi/2$ in phase with ion cyclotron heating field. Steady B_z is directed into the paper.

While the simple cyclotron heating outlined above applies well to isolated charges, it fails in a dense plasma. The reason can be seen from Fig. 16.18. Since the ions are in phase, they are all far from the axis at one time (Fig. 16.18*b*) and close to the axis at another time (Fig. 16.18*d*). On the other hand, the electrons have small Larmor orbits and are tied more strongly to the lines. Thus a radial current is set up, leading to a radial oscillating space-charge electric field. This field suppresses the radial current and hence also the heating. In a sense, the plasma has shielded itself from the applied field.

Stix and Palladino[8,21] overcome this difficulty by reversing the phase periodically along the plasma tube, as shown in Fig. 16.19. Thus at

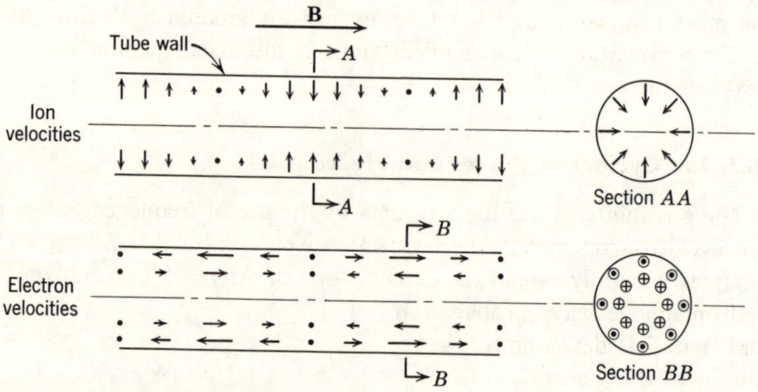

Fig. 16.19. Ion cyclotron wave in a uniform-density plasma in an axial induction. After Stix and Palladino.[21]

one particular instant, the ions at section *AA* are all moving in. The electrons are highly mobile along the *B* lines, and flow axially away from this position near the wall and toward this position in the plasma interior. The converse is true a half wavelength down the tube. If the electrons had no mass and the alternating induction produced by their motion could be neglected, complete charge neutrality would be preserved by electrons flowing up and down lines of induction. Therefore the electromagnetic field penetrates and may heat the plasma. Since the electron mass is finite and the electron currents shown in Fig. 16.19 cannot be neglected, the heating field does not penetrate the plasma perfectly.

The configuration of Fig. 16.19 is in fact that of a standing wave or two waves traveling in opposite directions down the tube. We recognize that the problem of plasma heating has now become the problem of the generation, transmission, and absorption of plasma waves. That such waves can propagate in a magnetized plasma was shown in Sec. 9.4 and Fig. 9.5. In the present application, we are concerned with the region $\omega_b^2/\omega^2 \gg 1$, $\omega_p^2/\omega^2 \gg 1$, or the upper right-hand region of Fig. 9.5. Here, however, the propagation is much more complicated than shown in Fig. 9.5 for simple propagation along the B lines, partly because the electric field vector here is not transverse to the direction of propagation. Analyses of these waves is given by Stix[22–24] and others.[25,28] As stated in the introduction to this chapter and section, we shall not attempt to develop the theory.

Experiments to observe these ion cyclotron waves in the B-65 stellarator have been performed by Stix and Palladino.[8,29] Figure 16.10 shows the B-65 stellarator; we now direct our attention to the ion cyclotron heating box. A cross section of it is shown in Fig. 16.20. The radio-frequency

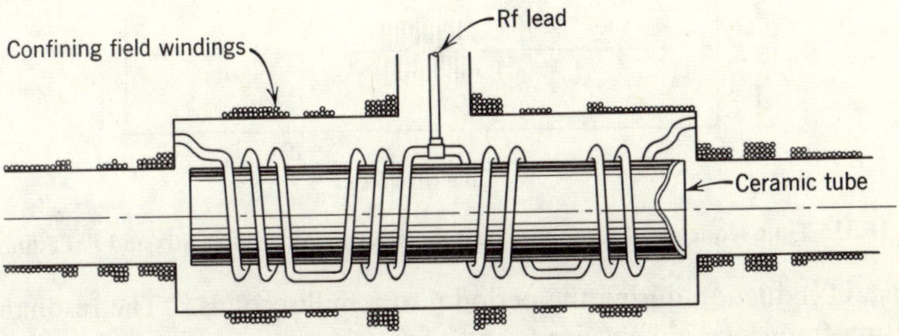

Fig. 16.20. Coil and shield box for ion cyclotron resonance experiments on B-65 stellarator. After Stix and Palladino.[8]

coil modulates the steady induction to produce the wave pattern of Fig. 16.19. Before summarizing the experiments, we shall summarize the results of the theoretical calculation. Stix shows that the dispersion relation for these waves in a cold plasma of uniform density is given by

$$\omega_{bi}^2 = \omega^2[1 + 5 \times 10^{-19}(Z_i^2 n_i \lambda_p^2/A_i)]. \tag{16.21}$$

Here, Z_i and A_i are the ionic charge and mass numbers, respectively, λ_p is the wavelength in the plasma, and n_i is the ion density, each in mks units. From Eq. 16.21, λ_p is real only if $|\omega_b/\omega| > 1$, or if $B > m_i\omega/q_i$. Thus if the applied frequency is kept fixed and the induction is varied, ion cyclotron waves are expected only for inductions exceeding the appropriate critical value.

The time sequence for the first series of experiments is shown in Fig. 16.21. The confining induction, including the helical windings, is pulsed from a capacitor bank (6×10^5 joules maximum) to give a sensibly

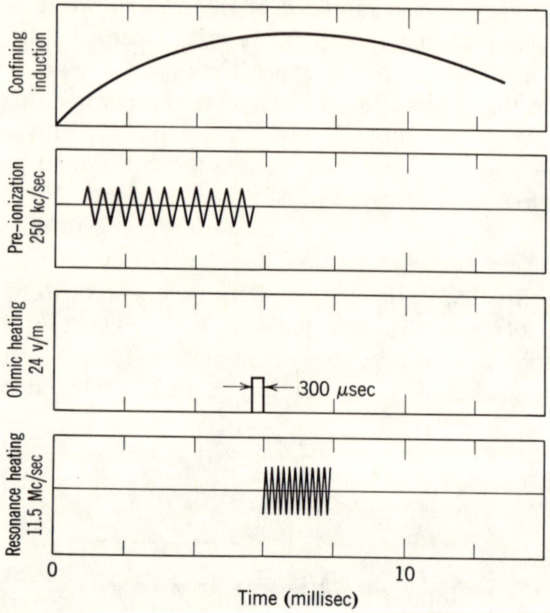

Fig. 16.21. Time sequence of resonance heating experiment. After Stix and Palladino.[21]

constant induction during the period 6 to 8 milliseconds. The resonance heating frequency is kept fixed, and the various resonances are explored by repeating the experiment with different magnitudes of the confining induction. In these experiments, absorption of energy is detected by measuring the loading on the heating coil and comparing this loading with that produced by placing fixed resistors across the coil.

Typical plasma loading is shown in Fig. 16.22 for experiments with deuterium gas. The resonance corresponding to H^+ probably arises from hydrogen gas desorbed from the tube walls during the experiment and is of no interest, except to remind us of the plasma purity problem. For D^+, two close resonances are seen. The first of these, very close to the single-particle cyclotron resonance, is interpreted as power absorption by individual D^+ ions outside the main plasma column. Some neutrons are observed at the D^+ cyclotron resonance. The second resonance at high induction (1.58 webers/m^2 at 300 watts and 1.75 webers/m^2 at 2×10^5 watts) is thought to represent the generation of true ion cyclotron waves that penetrate the plasma. This conclusion is supported by the further observations that

1. Deuterium and impurity spectra in the plasma core are enhanced at the second resonance, but not at the D^+ cyclotron frequency.

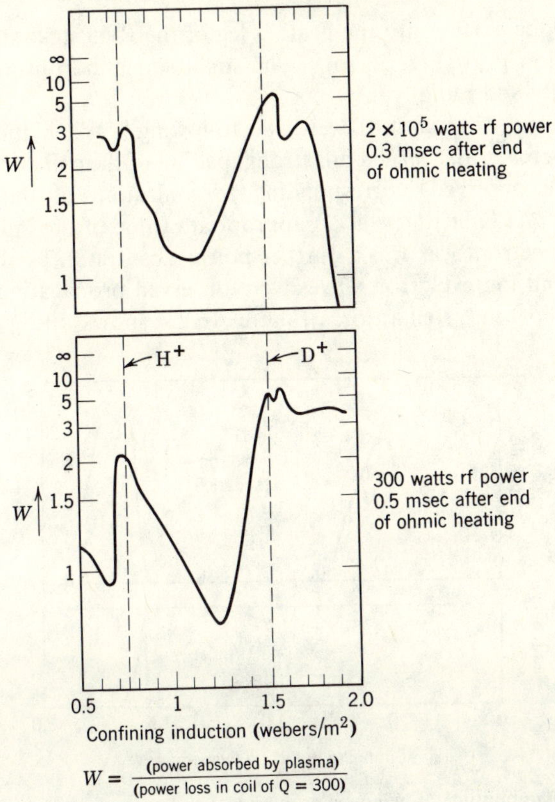

Fig. 16.22. Plasma loading versus confining induction, for low and high cyclotron resonance heating power levels. Frequency 11.51 Mc/sec; $p = 1.1$ microns deuterium; the dashed vertical lines indicate the single-particle resonance inductions for H^+ and D^+. After Stix and Palladino.[21]

2. The fact that no neutrons are observed at the second resonance indicates general heating rather than individual particle heating.

3. The steady induction required to obtain the second resonance increases with rf power; if the plasma density increases with heating power, this shift is in qualitative agreement with predictions from Eq. 16.21.

The ion temperature, measured at the second resonance by Doppler shift, is about 50 ev in these experiments.

We now turn to a second experiment by Stix and Palladino[29] on the B-65 stellarator. There, the helical windings were not energized, and no ohmic heating pulse was applied. Thus the plasma was confined in a simple toroidal induction and was not fully ionized at the start of the ion cyclotron heating pulse. Power level was 2×10^5 watts at 11.5 Mc/sec, as before. The ion cyclotron waves were observed by inserting a rf magnetic probe axially into the heating leg of the B-65 stellarator through an observation port (Fig. 16.10). The small single-loop probe could be moved axially and radially.

The principal observations were the following. When the cyclotron heating power was first turned on, strong loading of the coil and deuterium spectra were observed, corresponding to ionization of the remaining neutral gas. At later times, at an appropriate value of the induction, the deuterium spectrum was weak, further power absorption by the coil was measured, and ion cyclotron waves were observed propagating down the heating leg of the stellarator. Figure 16.23 shows the peak probe

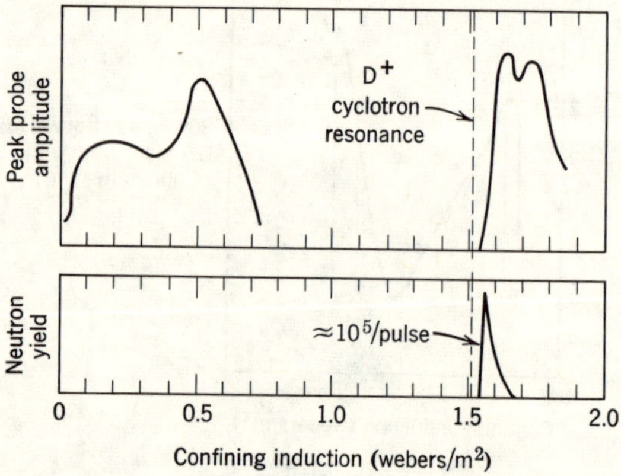

Fig. 16.23. Peak amplitude of magnetic probe signal and neutron yield versus confining induction for a cyclotron wave experiment in the B-65 stellarator. Probe placed on axis, 0.5 m from midplane of induction coil. Frequency 11.5 Mc/sec; initial pressure 1 micron deuterium. After Stix and Palladino.[29]

amplitude during a pulse (the probe was 0.5 m from the center of the induction coil) and the neutron yield, both as functions of the confining induction.

Several qualitative features are evident in Fig. 16.23. First, there is evidence of a wave propagating for values of the induction less than 0.7 weber/m^2. The nature of this wave has not been precisely determined, and no more will be said concerning it at the present time. Second, there is a resonance and associated neutron production close to ion cyclotron resonance, and a further resonance at higher induction. The fact that neutrons are observed for ω_b slightly greater than ω_{bi} may be explainable in terms of spatial nonuniformity of the induction. Third, no propagation exists for $\omega_b < \omega$, in agreement with Eq. 16.21. The probe signal is almost independent of radial position in the tube, showing that the wave does indeed penetrate the plasma.

In principle, the wavelength of these cyclotron waves should be that of the induction coil shown in Fig. 16.20, or 0.228 m. From Eq. 16.21 and the measured values of ω_b and ω, Stix determines the ion density to be about $9 \times 10^{18}/\text{m}^3$. Phase-shift measurements on the probe signal itself, taken as a function of axial distance, yield a wavelength 0.374 m. If this value is inserted in Eq. 16.21, the ion density is $3.3 \times 10^{18}/\text{m}^3$. No microwave measurements were made of the electron density, but similar experiments carried out on the B-66 stellarator gave rise to measured electron densities of 2 to $10 \times 10^{18}/\text{m}^3$. Note that these ion densities correspond to loss of 90 to 95% of the initial deuterium filling. This fact may represent an instability. As an alternate mechanism, Stix and Palladino point out that if the dissociation of D_2 by electron impact is much more likely than ionization, the 2.2 ev deuterium atoms may move quickly to the walls. In regard to this point, see Fig. 3.11, which shows the electronic states of H_2 and H_2^+.

Another similar contributing mechanism is the reaction $D_2 + e \rightarrow D_2^+ + 2e$ followed by $D_2^+ + e \rightarrow D^+ + D^0 + e$, which sequence is much more likely at low electron energy than the direct reaction $D_2 + e \rightarrow 2D^+ + 3e$. By this mechanism half the deuterons are lost. These mechanisms do not seem to be related to the pumpout phenomenon described in Sec. 16.5, which latter appears in the plasma after the ionization processes described here are completed.

The ion temperature in these latter experiments was not directly measured by Doppler broadening or otherwise. The axial attenuation distance of the ion cyclotron waves was measured to be 0.3 m. If this attenuation arises from Coulomb collisions of deuterons with a 20% C^+ impurity, the ion temperature is found to be 0.5 ev. The assumption both of the presence of this impurity and of the thermalization mechanism could be the subjects of considerable discussion. The conclusions regarding ion temperature are therefore highly tentative. Note further that these experiments were performed in a racetrack without rotational transform, so that plasma confinement may have been very poor.

In concluding this phenomenological discussion of ion cyclotron waves, we note that many other effects which we have not yet even mentioned will and do occur. As a wave penetrates into a plasma of spatially increasing density, the wave encounters various resonances and cutoffs, the nature of which depend upon ω_b^2/ω^2, ω_p^2/ω^2, and the directions of $\mathbf{B}$, ∇n, and the propagation vector $\boldsymbol{\kappa}$. Therefore sheaths may form on a plasma heated by ion cyclotron waves. Additionally, various hybrid resonances appear between electrons and ions at $\omega = (\omega_{bi}\omega_{be})^{1/2}$ and between different ion species, if they exist in the plasma. Finally, even if the waves are generated in the plasma, their ordered energy must be converted into random energy if the plasma is to be heated. This disordering may take place either by infrequent Coulomb collisions, by ions of different axial drift speed mixing the phases of the waves and producing a nonthermal randomization, by dissipation of the waves at regions of reduced induction (where the wavelength goes to zero, as shown in Eq. 16.21), or by some other oscillation mechanism yet not understood. For further discussion of these matters, we refer the interested reader to the references already cited.

16.8. Summary and Conclusion

If hydromagnetic theory is to be trusted, the stellarator must be the most stable of all controlled thermonuclear devices that have been extensively developed to date. The magnetic induction with rotational transform has been shown to improve the confinement of individual particles and low-density plasmas very substantially over that obtainable in a simple torus. Against these advantages, we find that the plasmas produced in stellarators to date have very modest energy (about 50 ev) and are in fact confined for rather short times (about 100 μsec). Many of the difficulties presently encountered may arise from contamination and lack of sufficient pumping, and from instabilities generated by dc ohmic heating pulses. However, it is also possible that these present difficulties are symptomatic of more serious shortcomings, the nature of which we cannot yet even suspect. Perhaps experiments on the large model C stellarator will shed some light on these matters.

If asked to give an opinion regarding the outcome of the stellarator program, we could only answer as we did at the end of Chap. 15 for the mirror program. It is too early to guess: if the problems are solved, the device should be very successful.

In a larger sense, we can with some profit look over the whole thermonuclear plasma program and plasma developments in general, to find some current trends. Evidently, the most important problems are those of stable confinement and heating. If theorists are ingenious enough, hydromagnetic stability theory will be replaced by theories based upon ensemble of single particles, their mutual interactions, and their correlations in space and velocity. Such studies, which we have hardly even mentioned in this book, are under way, but it is too soon to predict their outcome. Ideally, such theories should be complete in the sense that stable and unstable oscillations both in space and velocity appear naturally. If that is the case, confinement and heating can be analyzed from a more unified and erudite point of view. Presumably, one superior thermonuclear concept could emerge from the analysis.

At a somewhat more phenomenological level, we have seen examples of this *anschluss* of ideas. For instance, magnetic shear in the pinch and in the stellarator serve much the same purpose. Ion cyclotron heating may be applied to mirrors just as well as to stellarators. More of these valuable generalized concepts will surely emerge in the future.

In conclusion, let us not in a welter of details lose sight of the original motivation for controlled thermonuclear research. Success in this field would bring great benefit to mankind.

ΑΝΑΓΚΗ ΣΤΗΝΑΙ

Aristotle

PROBLEMS

1. In a simple figure-eight stellarator without scallops (Fig. 16.3), we assume that

(*a*) The angle ζ is small.
(*b*) Each end section can be approximated by a semicircle.
(*c*) A particle is barely confined if its orbit just misses the tube walls.

Show that no single particle is confined for which $v_{\|}$ lies outside the limiting real values:

$$v_{\|} = \frac{r\omega_b - v_T}{\pi} \pm \left[\left(\frac{r\omega_b - v_T}{\pi}\right)^2 - \frac{v_T^2}{4}\right]^{1/2},$$

where r is the tube radius.

2. For the helically wound stellarator, show that for the (gl, gh) harmonic of the induction, the quantity $f_\theta(g, r)$, Eqs. 16.2, is given by

$$\frac{d}{dr}\left[r\frac{d}{dr}(rf_\theta)\right] - g^2l^2\left(1 + \frac{h^2r^2}{l^2}\right)f_\theta = 0;$$

the solution of this equation is a Bessel function.

3. Show that $|f_z/f_\theta| = hr/l$ in Eqs. 16.2.

4. Derive Eqs. 16.3 for the approximations stated there.

5. In the model B-1 stellarator (a figure-eight device), it is found that the Kruskal limit for kink instability is reached with an axial circulating current of 1500 amp in one direction (forward) or 1700 amp in the reverse direction.

(*a*) Calculate the rotation transform.
(*b*) Show that further instabilities might be expected near the following currents:

Mode	Forward Current (amp)	Reverse Current (amp)
$m = 2$	1500	100
		1700
$m = 3$	430	630
	1500	1700
$m = 4$	700	100
	1500	900
		1700
$m = 5$	220	420
	860	1060
	1500	1700

REFERENCES

1. L. Spitzer, Jr., *Phys. Fluids*, **1**, 253 (1958).
2. M. D. Kruskal and R. M. Kulsrud, *Phys. Fluids*, **1**, 265 (1958).
3. J. L. Johnson, C. R. Oberman, R. M. Kulsrud, and E. A. Frieman, *Phys. Fluids*, **1**, 281 (1958).
4. M. D. Kruskal and C. R. Oberman, *Phys. Fluids*, **1**, 275 (1958).
5. M. D. Kruskal, J. L. Johnson, M. B. Gottlieb, and L. M. Goldman, *Phys. Fluids*, **1**, 421 (1958); M. D. Kruskal, U.S. Atomic Energy Commission Report No. NYO-6045 (PM-S-12) (1954).
6. W. Bernstein, A. Z. Kranz, and F. Tenney, *Phys. Fluids*, **2**, 713 (1959).
7. C. R. Burnett, D. J. Grove, W. R. Palladino, T. H. Stix, and K. E. Wakefield, *Phys. Fluids*, **1**, 438 (1958).
8. T. H. Stix and W. R. Palladino, *Phys. Fluids*, **1**, 446 (1958).
9. T. Coor, S. P. Cunningham, R. A. Ellis, M. A. Heald, and A. Z. Kranz, *Phys. Fluids*, **1**, 411 (1958).
10. J. M. Berger, I. B. Bernstein, E. A. Frieman, and R. M. Kulsrud, *Phys. Fluids*, **1**, 297 (1958); J. M. Berger and L. Goldman, U.S. Atomic Energy Commission Report No. NYO-7311 (PM-S-21) (1956) (unpublished).
11. W. Bernstein and A. Z. Kranz, *Phys. Fluids*, **2**, 57 (1959).
12. W. Bernstein, A. Z. Kranz, and F. Tenney, *Phys. Fluids*, **2**, 713 (1959).
13. W. Bernstein, F. F. Chen, M. A. Heald, and A. Z. Kranz, *Phys. Fluids*, **1**, 430 (1958).
14. R. A. Ellis, Jr., L. P. Goldberg, and J. G. Gorman, *Phys. Fluids*, **3**, 468 (1960).
15. R. A. Ellis, Jr., J. G. Gorman, and W. Stodiek, *Bull. Am. Phys. Soc.*, Ser. II, **5**, 341 (1960).
16. R. W. Gould, "The Dynamics of Electron Beams," in *Plasma Dynamics*, F. H. Clauser (Ed.), Addison-Wesley Publishing Co., Inc., Reading, Mass. (1960), pp. 78 ff.
17. D. T. Swift-Hook, *Phys. Rev.*, **118**, 1 (1960).
18. O. Bunemann, *Phys. Rev.*, **115**, 503 (1959).

19. J. M. Berger, W. A. Newcomb, J. M. Dawson, E. A. Frieman, R. M. Kulsrud, and A. Lenard, *Phys. Fluids*, **1**, 301 (1958).
20. B. N. Gershman, *Zhur. Eksp. i Teoret. Fiz.*, **24**, 453 (1953).
21. T. H. Stix and W. R. Palladino, *Proc. 2nd U.N. Conf. on Peaceful Uses of Atomic Energy*, Vol. 31, United Nations, Geneva (1958), p. 282.
22. T. H. Stix, *Phys. Rev.*, **106**, 1146 (1957).
23. T. H. Stix, *Phys. Fluids*, **1**, 308 (1958).
24. T. H. Stix, *Phys. Fluids*, **3**, 19 (1960).
25. S. J. Buchsbaum, *Phys. Fluids*, **3**, 418 (1960).
26. H. Alfvén, *Arkiv Mat. Astron. Fysik*, **29B**, No. 2 (1942).
27. E. Åström, *Arkiv Fysik*, **2**, 443 (1951).
28. P. L. Auer, H. Hurwitz, Jr., and R. D. Miller, *Phys. Fluids*, **1**, 501 (1958).
29. T. H. Stix and R. W. Palladino, *Phys. Fluids*, **3**, 641 (1960).

Appendix A

Glossary

A.1. Frequently Used Symbols

a	Stefan-Boltzmann radiation constant
$\mathbf{a}$	Acceleration
A	Avogadro's number
A^2	10^{-16} cm^2 (square Angstrom)
A	Nuclear mass number
$\mathbf{A}$	Vector potential
b	Impact parameter;
	B_θ/rB_z
$\mathbf{B}$	Magnetic induction
B_m	Maximum induction in a magnetic mirror
c	Speed of light;
	Speed of a wave
c_p	Phase speed of a wave
C	A general constant;
	Compression ratio
$\mathscr{C}$	Specific heat
d	A distance
$\mathbf{D}$	Electric displacement
$\mathsf{D}_{\alpha\beta}$	Diffusion coefficient (a tensor)
D_a	Ambipolar diffusion coefficient
e	Magnitude of electronic charge
$\mathbf{E}$	Electric intensity
f	Distribution function;
	A generalized function
f_b	Fractional burnup
$\mathbf{F}$	Force
g	A generalized function;
	A summing index

$\mathbf{g}$	Gravitational acceleration
$\mathbf{G}$	$\nabla_0 \times (\boldsymbol{\xi} \times \mathbf{B}_0)$
h	Planck's constant;
	A summing index
$\mathbf{H}$	Magnetic intensity
i	$\sqrt{-1}$
I	Electric current
$\mathbf{j}$	Electric current density
$\mathbf{J}$	Mass current density;
	Action integral
k	Boltzmann's constant
$K_{\mathscr{L}}$	Cyclotron radiation absorption coefficient
$\mathbf{K}$	Surface current density
l	Mean free path;
$\bar{l}$	Mean chord length
L	Distance;
	Loss rate density
$\mathscr{L}$	$\omega_p{}^2 L/\omega_b c$
m	Mass
m_r	Reduced mass
M	Magnetic moment;
	Mass in amu
	Molecular weight
$\mathbf{M}$	Magnetization
n	Number density
n_λ	Number of electrons in a Debye sphere
$\mathbf{n}$	Unit vector
N	Density of scattering centers;
	Number of particles per unit length
p	Pressure
P	Probability of collision;
	Packing fraction
$\mathbf{P}$	Polarization
$\mathsf{P}_{\alpha\beta}$	Kinetic stress tensor
q	Charge
Q	Potential energy;
	10^{18} Btu
$\mathsf{Q}_{\alpha\beta\gamma}$	Heat flux tensor
r	Radius
r_b	Radius of gyration (Larmor radius)
$\mathbf{r}$	Position vector
$\mathbf{r}_g$	Position vector of guiding center

R	Radius; Mirror ratio
R_m	Maximum value of mirror ratio
$\mathbf{R}$	Radius of curvature (directed from center of curvature)
$\mathsf{R}_{\alpha\beta}$	Electrokinetic stress tensor
$\mathscr{R}$	Resistance
s	Area
S	Source rate density; Sputtering ratio particles/incident ion
$\mathbf{S}$	Poynting vector; Radiation intensity
$\mathscr{S}$	Entropy
t	Time
t_C	Compression time
T	Temperature degrees Kelvin or equivalent kilovolts, as shown in context
T	Electromagnetic stress tensor
T^e	Electric stress tensor
T^m	Magnetic stress tensor
u	Kinetic energy in electron volts
U	Kinetic energy
v	Speed
$\mathbf{v}$	Velocity
$\mathbf{v}_b$	Velocity of a particle about its guiding center
$\mathbf{v}_g$	Velocity of guiding center
$\mathbf{v}_r$	Random velocity
V	Potential in volts
$\mathbf{V}$	Macroscopic velocity
$\mathscr{V}$	Volume
w	Power density
w_c	Power density of cyclotron radiation
$w_{c\mathscr{L}}$	Effective value of w_c including self absorption
w_L	Power loss density caused by escaping particles
w_x	Power density of bremsstrahlung
W	Power
x, y, z	Cartesian coordinates
Z	Ionic charge number
α	Townsend ionization coefficient; Energy absorption coefficient; Summation index

β	Ratio of kinetic pressure to magnetic pressure; Summation index
γ	Ratio of specific heat at constant pressure to that at constant volume
γ_i	Secondary electrons/incident positive ion
$\mathbf{\Gamma}$	Particle current
δ	Secondary emission coefficient by electrons
ϵ	Permittivity
ϵ_0	Permittivity of free space
ζ	A summation index
η	Resistivity
θ	A general angle; Pump speed
θ_m	Vertex half-angle of escape cone
ι	Rotational transform angle
$\boldsymbol{\kappa}$	Propagation vector
λ	Debye shielding length
Λ	Argument of logarithm in Coulomb scattering (Eq. 8.18)
μ	Mobility; Permeability
μ_0	Permeability of free space
ν	Frequency of an event/particle
ν_a	Attachment frequency/particle
ν_c	Collision frequency/particle
ν_i	Ionization frequency/particle
ν_m	Momentum transfer frequency/particle
ξ	Summation index
$\boldsymbol{\xi}$	Plasma displacement
π	$3.14159\cdots$
ρ	Charge density
ρ_m	Mass density
σ	Cross section
σ_a	Cross section for attachment
σ_i	Cross section for ionization
σ_m	Cross section for momentum transfer
σ_r	Cross section for recombination
σ_t	Cross section for charge transfer
σ_x	Cross section for excitation
$s_{\alpha\beta}$	Conductivity (a tensor)
τ	Characteristic time
τ_u	Energy relaxation time
τ_θ	Angular relaxation time

Υ	Dummy variable
ϕ	Potential
φ	Azimuthal angle
Φ	Magnetic flux
χ	Deflection angle; Nonreturn probability of an ion as a neutral
ψ	Probability of particle escape; Arbitrary phase angle
ω	Angular frequency
ω_p	Plasma frequency
$\boldsymbol{\omega}_b$	Cyclotron frequency vector
Ω	Solid angle
$\boldsymbol{\Omega}$	Direction in space
$\mathfrak{d}$	A small distance
$\mathfrak{h}$	Efficiency
$\mathfrak{p}_\theta$	Angular momentum
$\boldsymbol{\mathfrak{p}}$	Momentum
$\mathfrak{U}$	Total energy

A.2. Frequently Used Subscripts

Υ_e	Υ for electrons
Υ_i	Υ for ions
$\Upsilon_\parallel$	Component of $\mathbf{\Upsilon}$ parallel to a unique axis (usually to **B**)
$\Upsilon_\perp$	Components of $\mathbf{\Upsilon}$ perpendicular to a unique axis (usually to **B**)
Υ_T	Component of $\mathbf{\Upsilon}$ transverse to a unique axis (usually to **B**), and usually parallel to a second orthogonal direction (usually parallel to **E** or ∇p)
Υ_H	Component of $\mathbf{\Upsilon}$ perpendicular to two orthogonal axes (usually perpendicular to **B** and **E**). The Hall component

A.3. Mathematical Notation

$d\ell$	Element of length
$d\mathbf{r}$	Element of volume $dx\,dy\,dz$ in configuration space
$d\mathbf{s}$	Directed element of area (a vector)
ds	Magnitude of $d\mathbf{s}$
$d\mathbf{v}$	Element of volume $dv_x\,dv_y\,dv_z$ in velocity space
$\mathcal{O}(\Upsilon)$	Order of Υ
$\overline{\Upsilon}$	Average of Υ (usually over velocity)
Υ^*	Complex conjugate of Υ
cc	Complex conjugate of the expression preceding cc
$\hat{\Upsilon}$	Υ in the vacuum, as distinct from in the plasma
$[\![\Upsilon]\!]$	Change in Υ across a boundary

Appendix B

Conversion of units

In this book the rationalized mks system of units is used exclusively except for incidental quotations and cross sections. In Table B.1 conversion factors are presented to convert from the unrationalized electrostatic and electromagnetic system of units and the unrationalized mks system of units to the rationalized system.

Any quantity may be regarded as the product of the number of units of that quantity times the size of the unit. Thus, in changing from one system to another, the number of units of a quantity is proportional to the *reciprocal* of the size of the unit. For example, the charge of an electron is

$$-4.8 \times 10^{-10} \text{ statcoulomb} = \left(-\frac{10}{c}\right) 4.8 \times 10^{-10} \text{ coulomb}$$
$$= -1.6 \times 10^{-19} \text{ coulomb.}$$

The factors listed in the table are those to be used in multiplying the number of units of a quantity in the electrostatic or electromagnetic system to get the number of units of the quantity in the mks rationalized system. The *reciprocal* of the factor listed must be used to find the relative size of the mks rationalized unit itself. For instance, a formula in esu can be rewritten in mks units by replacing each quantity in esu by the symbol representing the quantity times the *reciprocal* of the factor listed in the table. Equations in the mks unrationalized system may be found from the esu and emu unrationalized systems by omitting the 4π factor where it appears in the esu and emu columns of Table B.1, and by replacing polarization **P** and magnetization **M** by $4\pi\mathbf{P}$ and $4\pi\mathbf{M}$, respectively. Finally, the Gaussian system uses esu for electric quantities and emu for current and magnetic quantities. Thus, in the Gaussian system, quantities related to resistance and conductance are peculiar. To get mks rationalized numbers, multiply the Gaussian quantities by the conversion factors: resistance (statvolt/abamp) $c/10^9$, resistivity (statvolt-cm/abamp) $c/10^{11}$, conductance (abamp/statvolt) $10^9/c$, and conductivity (abamp/statvolt-cm) $10^{11}/c$.

Table B.1. Conversion to mks rationalized system of units from other systems†

Quantity and Symbol		Dimensions in Mks System	Mks Unit	Mks Unrationalized: Multiply NUMBER of unrationalized units by factor below to get mks rationalized number	Unrationalized Esu: Unit	Unrationalized Esu: Multiply NUMBER of electrostatic units by factor below to get mks rationalized number	Unrationalized Emu: Unit	Unrationalized Emu: Multiply NUMBER of electromagnetic units by factor below to get mks rationalized number
Mass	m	m	kg	1	g	$\frac{1}{10^3}$	g	$\frac{1}{10^3}$
Length	$\mathbf{r}$	l	m	1	cm	$\frac{1}{10^2}$	cm	$\frac{1}{10^2}$
Time	t	t	sec	1	sec	1	sec	1
Density	ρ_m	$\frac{m}{l^3}$	$\frac{\text{kg}}{\text{m}^3}$	1	$\frac{\text{g}}{\text{cm}^3}$	10^3	$\frac{\text{g}}{\text{cm}^3}$	10^3
Speed	v	$\frac{l}{t}$	$\frac{\text{m}}{\text{sec}}$	1	$\frac{\text{cm}}{\text{sec}}$	$\frac{1}{10^2}$	$\frac{\text{cm}}{\text{sec}}$	$\frac{1}{10^2}$
Momentum	$\mathfrak{p}$	$\frac{ml}{t}$	$\frac{\text{kg-m}}{\text{sec}}$	1	$\frac{\text{g-cm}}{\text{sec}}$	$\frac{1}{10^5}$	$\frac{\text{g-cm}}{\text{sec}}$	$\frac{1}{10^5}$
Force	F	$\frac{ml}{t^2}$	newton	1	dyne	$\frac{1}{10^5}$	dyne	$\frac{1}{10^5}$
Pressure	p	$\frac{m}{lt^2}$	$\frac{\text{newton}}{\text{m}^2}$	1	$\frac{\text{dyne}}{\text{cm}^2}$	$\frac{1}{10}$	$\frac{\text{dyne}}{\text{cm}^2}$	$\frac{1}{10}$
Energy	U	$\frac{ml^2}{t^2}$	joule	1	erg	$\frac{1}{10^7}$	erg	$\frac{1}{10^7}$
Power	W	$\frac{ml^2}{t^3}$	watt	1	$\frac{\text{erg}}{\text{sec}}$	$\frac{1}{10^7}$	$\frac{\text{erg}}{\text{sec}}$	$\frac{1}{10^7}$
Specific heat	$\mathscr{C}$	$\frac{l^2}{t^2T}$	$\frac{\text{joule}}{\text{kg-° K}}$	1	$\frac{\text{erg}}{\text{g-° K}}$	$\frac{1}{10^4}$	$\frac{\text{erg}}{\text{g-° K}}$	$\frac{1}{10^4}$
Charge	q	q	coulomb	1	statcoulomb	$\frac{10}{c}$	abcoulomb	10
Charge density	ρ	$\frac{q}{l^3}$	$\frac{\text{coulomb}}{\text{m}^3}$	1	$\frac{\text{statcoulomb}}{\text{cm}^3}$	$\frac{10^7}{c}$	$\frac{\text{abcoulomb}}{\text{cm}^3}$	10^7
Displacement	$\mathbf{D}$	$\frac{q}{l^2}$	$\frac{\text{coulomb}}{\text{m}^2}$	$\frac{1}{4\pi}$	$\frac{\text{statcoulomb}}{\text{cm}^2}$	$\frac{10^5}{4\pi c}$	$\frac{\text{abcoulomb}}{\text{cm}^2}$	$\frac{10^5}{4\pi}$
Electric intensity	$\mathbf{E}$	$\frac{ml}{t^2q}$	$\frac{\text{volt}}{\text{m}}$	1	$\frac{\text{statvolt}}{\text{cm}}$	$\frac{c}{10^6}$	$\frac{\text{abvolt}}{\text{cm}}$	$\frac{1}{10^6}$
Potential	ϕ	$\frac{ml^2}{t^2q}$	volt	1	statvolt	$\frac{c}{10^8}$	abvolt	$\frac{1}{10^8}$

Capacitance		$\frac{t^2q^2}{ml^2}$	farad	1	statfarad	$\frac{10^9}{c^2}$	abfarad	10^9
Electric flux		q	coulomb	$\frac{1}{4\pi}$	statcoulomb	$\frac{10}{4\pi c}$	abcoulomb	$\frac{10}{4\pi}$
Polarization	**P**	$\frac{q}{l^2}$	$\frac{\text{coulomb}}{\text{m}^2}$	1	$\frac{\text{statcoulomb}}{\text{cm}^2}$	$\frac{10^5}{c}$	$\frac{\text{abcoulomb}}{\text{cm}^2}$	10^5
Permittivity	ϵ	$\frac{t^2q^2}{ml^3}$	$\frac{\text{farad}}{\text{m}}$	$\frac{1}{4\pi}$	—	$\frac{10^{11}}{4\pi c^2}$	—	$\frac{10^{11}}{4\pi}$
Current	I	$\frac{q}{t}$	ampere	1	statampere	$\frac{10}{c}$	abampere	10
Current density	**j**	$\frac{q}{l^2t}$	$\frac{\text{ampere}}{\text{m}^2}$	1	$\frac{\text{statampere}}{\text{cm}^2}$	$\frac{10^5}{c}$	$\frac{\text{abampere}}{\text{cm}^2}$	10^5
Resistance	$\mathscr{R}$	$\frac{ml^2}{tq^2}$	ohm	1	statohm	$\frac{c^2}{10^9}$	abohm	$\frac{1}{10^9}$
Resistivity	η	$\frac{ml^3}{tq^2}$	ohm-m	1	statohm-cm	$\frac{c^2}{10^{11}}$	abohm-cm	$\frac{1}{10^{11}}$
Conductance		$\frac{tq^2}{ml^2}$	mho	1	statmho	$\frac{10^9}{c^2}$	abmho	10^9
Conductivity	s	$\frac{tq^2}{ml^3}$	$\frac{\text{mho}}{\text{m}}$	1	$\frac{\text{statmho}}{\text{cm}}$	$\frac{10^{11}}{c^2}$	$\frac{\text{abmho}}{\text{cm}}$	10^{11}
Magnetic intensity	**H**	$\frac{q}{lt}$	$\frac{\text{ampere-turn}}{\text{m}}$ (praoersted)	$\frac{1}{4\pi}$	—	$\frac{10^3}{4\pi c}$	oersted	$\frac{10^3}{4\pi}$
Magnetic induction	**B**	$\frac{m}{tq}$	$\frac{\text{weber}}{\text{m}^2}$	1	—	$\frac{c}{10^4}$	gauss	$\frac{1}{10^4}$
Vector potential	**A**	$\frac{ml}{tq}$	$\frac{\text{weber}}{\text{m}}$	1	—	$\frac{c}{10^6}$	$\frac{\text{gauss}}{\text{cm}}$	$\frac{1}{10^6}$
Inductance	L	$\frac{ml^2}{q^2}$	henry	1	stathenry	$\frac{c^2}{10^9}$	abhenry	$\frac{1}{10^9}$
Magnetic flux	Φ	$\frac{ml^2}{tq}$	weber (volt-sec)	1	—	$\frac{c}{10^8}$	maxwell	$\frac{1}{10^8}$
Magnetization (dipole)		$\frac{m}{tq}$	$\frac{\text{weber}}{\text{m}^2}$	4π	—	$\frac{4\pi c}{10^4}$	gauss	$\frac{4\pi}{10^4}$
(loop)	**M**	$\frac{q}{lt}$	$\frac{\text{amp}}{\text{m}}$	1	—	$\frac{10^3}{c}$	$\frac{\text{abamp}}{\text{cm}}$	10^3
Permeability	μ	$\frac{ml}{q^2}$	$\frac{\text{henry}}{\text{m}}$	4π	—	$\frac{4\pi c^2}{10^7}$	$\frac{\text{gauss}}{\text{oersted}}$	$\frac{4\pi}{10^7}$
Magnetic moment (dipole)		$\frac{ml^3}{tq}$	weber-m	4π	—	$\frac{4\pi c}{10^{10}}$	maxwell-cm	$\frac{4\pi}{10^{10}}$
(loop)	**M**	$\frac{l^2q}{t}$	ampere-m^2	1	—	$\frac{1}{10^3c}$	abamp-cm^2	$\frac{1}{10^3}$
Magnetomotive force		$\frac{q}{t}$	ampere-turn (pragilbert)	$\frac{1}{4\pi}$	—	$\frac{10}{4\pi c}$	gilbert	$\frac{10}{4\pi}$

† In this table, c is the numeric 2.99793×10^{10}.

Appendix C

Frequently used vector relations

$$\mathbf{A} \times \mathbf{B} = -\mathbf{B} \times \mathbf{A} \tag{C.1}$$

$$\mathbf{A} \cdot \mathbf{B} \times \mathbf{C} = \mathbf{B} \cdot \mathbf{C} \times \mathbf{A} = \mathbf{C} \cdot \mathbf{A} \times \mathbf{B} = (\mathbf{ABC}) \tag{C.2}$$

$$\mathbf{A} \times (\mathbf{B} \times \mathbf{C}) = (\mathbf{A} \cdot \mathbf{C})\mathbf{B} - (\mathbf{A} \cdot \mathbf{B})\mathbf{C} \tag{C.3}$$

$$(\mathbf{A} \times \mathbf{B}) \times \mathbf{C} = (\mathbf{A} \cdot \mathbf{C})\mathbf{B} - (\mathbf{B} \cdot \mathbf{C})\mathbf{A} \tag{C.4}$$

$$\begin{aligned}(\mathbf{A} \times \mathbf{B}) \times (\mathbf{C} \times \mathbf{D}) &= (\mathbf{ABD})\mathbf{C} - (\mathbf{ABC})\mathbf{D} \\ &= (\mathbf{ACD})\mathbf{B} - (\mathbf{BCD})\mathbf{A}\end{aligned} \tag{C.5}$$

$$(\mathbf{A} \times \mathbf{B}) \cdot (\mathbf{C} \times \mathbf{D}) = (\mathbf{A} \cdot \mathbf{C})(\mathbf{B} \cdot \mathbf{D}) - (\mathbf{A} \cdot \mathbf{D})(\mathbf{B} \cdot \mathbf{C}) \tag{C.6}$$

$$(\mathbf{A} \cdot \mathbf{B})\mathbf{C} = \mathbf{A} \cdot (\mathbf{BC}) \tag{C.7}$$

$$(\mathbf{A} \cdot \nabla)\mathbf{B} = \mathbf{A} \cdot (\nabla \mathbf{B}) \tag{C.8}$$

$$\nabla(\phi + \psi) = \nabla\phi + \nabla\psi \tag{C.9}$$

$$\nabla(\phi\psi) = \phi\nabla\psi + \psi\nabla\phi \tag{C.10}$$

$$\nabla \cdot (\phi\mathbf{A}) = \phi\nabla \cdot \mathbf{A} + \mathbf{A} \cdot \nabla\phi \tag{C.11}$$

$$\nabla \times (\phi\mathbf{A}) = \phi\nabla \times \mathbf{A} + \nabla\phi \times \mathbf{A} \tag{C.12}$$

$$\nabla \cdot (\mathbf{A} \times \mathbf{B}) = \mathbf{B} \cdot \nabla \times \mathbf{A} - \mathbf{A} \cdot \nabla \times \mathbf{B} \tag{C.13}$$

$$\nabla \times (\mathbf{A} \times \mathbf{B}) = (\mathbf{B} \cdot \nabla)\mathbf{A} - (\mathbf{A} \cdot \nabla)\mathbf{B} + \mathbf{A}(\nabla \cdot \mathbf{B}) - \mathbf{B}(\nabla \cdot \mathbf{A}) \tag{C.14}$$

$$\nabla(\mathbf{A} \cdot \mathbf{B}) = (\mathbf{A} \cdot \nabla)\mathbf{B} + (\mathbf{B} \cdot \nabla)\mathbf{A} + \mathbf{A} \times (\nabla \times \mathbf{B}) + \mathbf{B} \times (\nabla \times \mathbf{A}) \tag{C.15}$$

$$\nabla \times \nabla\phi = 0 \tag{C.16}$$

$$\nabla \cdot \nabla \times \mathbf{A} = 0 \tag{C.17}$$

$$\nabla \times (\nabla \times \mathbf{A}) = \nabla(\nabla \cdot \mathbf{A}) - \nabla \cdot \nabla\mathbf{A} \tag{C.18}$$

$$\mathbf{r} = \mathbf{i}x + \mathbf{j}y + \mathbf{k}z \tag{C.19}$$

$$\nabla \cdot \mathbf{r} = 3 \tag{C.20}$$

$$\nabla \times \mathbf{r} = 0 \tag{C.21}$$

$$(\mathbf{A} \cdot \nabla)\mathbf{r} = \mathbf{A} \tag{C.22}$$

$$\int_{\substack{\text{closed}\\\text{volume}}} dr\, \nabla\phi = \int_{\substack{\text{bounding}\\\text{surface}}} d\mathbf{s}\, \phi \tag{C.23}$$

$$\int_{\substack{\text{closed}\\\text{volume}}} dr\, \nabla \cdot \mathbf{A} = \int_{\substack{\text{bounding}\\\text{surface}}} d\mathbf{s} \cdot \mathbf{A} \qquad \text{(divergence, Gauss's theorem)} \tag{C.24}$$

$$\int_{\substack{\text{closed}\\\text{volume}}} dr\, \nabla \times \mathbf{A} = \int_{\substack{\text{bounding}\\\text{surface}}} d\mathbf{s} \times \mathbf{A} \tag{C.25}$$

$$\int_{\substack{\text{open}\\\text{surface}}} d\mathbf{s} \times \nabla\phi = \oint_{\substack{\text{bounding}\\\text{circumference}}} d\boldsymbol{\ell}\, \phi \tag{C.26}$$

$$\int_{\substack{\text{open}\\\text{surface}}} d\mathbf{s} \cdot \nabla \times \mathbf{A} = \oint_{\substack{\text{bounding}\\\text{circumference}}} d\boldsymbol{\ell} \cdot \mathbf{A} \qquad \text{(Stokes's theorem)} \tag{C.27}$$

Appendix D

Frequently used physical constants

c	Speed of light in vacuum	2.99793×10^{8} m/sec
ϵ_0	Permittivity of vacuum	8.854×10^{-12} farad/m
μ_0	Permeability of vacuum	1.257×10^{-6} henry/m
a	Stefan-Boltzmann constant	5.67×10^{-8} joule/m^2-sec-(° K)^4
h	Planck constant	6.625×10^{-34} joule-sec
k	Boltzmann constant	1.3804×10^{-23} joule/° K
A	Avogadro constant	6.025×10^{23}/mole (physical)
e	Charge of proton	1.6021×10^{-19} coulomb
1 ev	Electron volt	1.6021×10^{-19} joule
e/k		11,600° K/volt

Particle	Mass (kg $\times 10^{-27}$)	Rest Energy (Mev)
e	9.108×10^{-4}	0.511
n	1.674	939.512
H^1 atom	1.673	938.730
H^2 atom	3.343	1876.017
H^3 atom	5.006	2809.272
He^3 atom	5.006	2809.250
He^4 atom	6.643	3728.189
Li^6 atom	9.984	5602.735
Li^7 atom	11.64	6534.995

Index